Textbook of Pathologic Anatomy

AF147486

Kuisheng Chen • Li Liang • Mincai Li
Yun Pan

Editors

Textbook of Pathologic Anatomy

For Medical Students

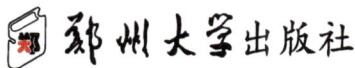

Editors
Kuisheng Chen
Department of Pathology
The First Affiliated Hospital of Zhengzhou
University
Zhengzhou, China

Li Liang
Department of Pathology
Nanfang Hospital, Southern Medical
University
Guangzhou, China

Mincai Li
Departments of Pathology
Hubei University of Science and
Technology
Xianning, China

Yun Pan
Department of Pathology
First Affiliated Hospital of Dali University
Dali, China

ISBN 978-981-99-8444-2   ISBN 978-981-99-8445-9   (eBook)
https://doi.org/10.1007/978-981-99-8445-9

© Zhengzhou University Press 2024

Jointly published with Zhengzhou University Press
The print edition is not for sale in China (Mainland). Customers from China (Mainland) please order the
print book from: Zhengzhou University Press.
This work is subject to copyright. All rights are solely and exclusively licensed by the Publisher, whether
the whole or part of the material is concerned, specifically the rights of reprinting, reuse of illustrations,
recitation, broadcasting, reproduction on microfilms or in any other physical way, and transmission or
information storage and retrieval, electronic adaptation, computer software, or by similar or dissimilar
methodology now known or hereafter developed.
The use of general descriptive names, registered names, trademarks, service marks, etc. in this publication
does not imply, even in the absence of a specific statement, that such names are exempt from the relevant
protective laws and regulations and therefore free for general use.
The publishers, the authors, and the editors are safe to assume that the advice and information in this
book are believed to be true and accurate at the date of publication. Neither the publishers nor the authors
or the editors give a warranty, express or implied, with respect to the material contained herein or for any
errors or omissions that may have been made. The publishers remain neutral with regard to jurisdictional
claims in published maps and institutional affiliations.

This Springer imprint is published by the registered company Springer Nature Singapore Pte Ltd.
The registered company address is: 152 Beach Road, #21-01/04 Gateway East, Singapore 189721,
Singapore

If disposing of this product, please recycle the paper.

Preface

Pathology is a medical discipline that focuses on studying the etiology, pathogenesis, pathological changes, and outcomes of diseases, revealing the laws of disease occurrence and development, elucidating the essence of diseases, and providing a basis for precise diagnosis and treatment of diseases. Although each organ is different in function and structure, different organs can present the same basic response and structural changes under the influence of various pathogenic factors, which is the research object and content of pathology general theory. Each disease has its own etiology, pathogenesis, prone site, morphological changes, and corresponding clinical manifestations.

Anatomical pathology and clinical pathology are two major branches of pathology. Clinical pathology involves collecting blood, urine, feces, secretions, and cytological samples and using certain equipment and techniques to make clinical diagnoses of diseases. Anatomical pathology combines tissue sample processing and autopsy techniques to make pathological judgments on the etiology and development of diseases.

This book consists of 18 chapters, including 6 chapters on general pathology and 12 chapters on systematic pathology, focusing on the basic cellular and tissue responses to pathological stimuli and particular responses of specific organs, respectively.

Thirty-two pathologists from 16 Medical Universities participated in the writing of this book, and their meticulous and professional work ensured the smooth publication of the book. Here, I sincerely thank all the pathologists for their hard work.

Errors and omissions are inevitable in the compilation of textbooks. We sincerely hope that readers can provide valuable opinions and suggestions for revision and improvement during the reprint.

Zhengzhou, China

Guangzhou, China

Xianning, China

Dali, China

Kuisheng Chen

Li Liang

Mincai Li

Yun Pan

Contents

Chapter 1
Adaptation and Injury of Cells and Tissues

Kuisheng Chen, Miaomiao Sun, and Na Wei

Contents

K. Chen (✉) · M. Sun · N. Wei
School of Basic Medical Sciences, Zhengzhou University, Zhengzhou, China

© Zhengzhou University Press 2024
K. Chen et al. (eds.), *Textbook of Pathologic Anatomy*,
https://doi.org/10.1007/978-981-99-8445-9_1

Objectives
1. To establish the significant types and characteristics of adaptation.
2. To establish the significant types and characteristics of degeneration.
3. To determine the characteristics of mucous degeneration, amyloidosis, pigmentation, and pathologic calcification.
4. To describe the types, pathological changes, and characters of cell death.
5. To determine the difference between necrosis and apoptosis.
6. To establish the concept of cellular aging.

Key Concepts
1. For adaptations: its concept, types, and pathological changes.
2. For degenerations: its mechanism and pathological changes.
3. For cell death: its types, end, and effects.

Introduction
Cells and tissues can react and adjust due to different forms of stimuli, either functional or metabolism-related in vitro and in vivo. The cells, tissues, and organs are capable of adapting when experiencing an excessive or insufficient physiological load, or when a mild and persistent pathological stimulus is present. Damage to cells, tissues, and organs will result in modifications to their morphology, function, and metabolism if the stimulation exceeds their capacity for tolerance and adaptation. Most of the mild damage to cells is reversible, but an irreversible injury to the cell can occur in severe cases. Morphologically, normal cells, adaptive cells, reversible damaged cells, and irreversibly damaged cells are ongoing processes. Under certain conditions, they can transform each other, and their boundaries are sometimes unclear. A specific stimulus can induce adaptation, reversible injury, or irreversible injury. However, which one it induces depends on a number of factors, including the type and degree of stimulation as well as the cells' susceptibility, differentiation, blood supply, nutrition, and previous state. Adaptation and damage change are fundamental pathological changes in the development of most diseases.

1.1 Adaptation

Adaptation is a noninvasive response of cells, tissues, and organs to a variety of adverse stimuli in both the internal and external environment. The goal of adaptation, which encompasses both functional metabolism and morphological structure, is to prevent damage to cells and tissue, which reflects adjustment and responsiveness to a degree. Adaptation can generally manifest as atrophy, hypertrophy, hyperplasia, and metaplasia, including changes in the cell's number, volume, and differentiation. Adaptive response mechanism includes upregulation or downregulation of cell-specific receptor function, synthesis of new proteins, the transformation of one protein to another, or the overexpression of some original protein. Therefore, the adaptive response of cells and tissues can occur in any of the following aspect: gene expression and regulation, signal transduction combined with the receptor, protein transcription, transport, and output.

Adaptation is essentially the result of the adjustment of cell growth and differentiation. It can be considered as a state between normal and damaged. A new balance of metabolism, function, and morphology in the internal and external environment can form through a series of adaptive changes. In general, most of the adaptive cells can be gradually restored to normal once the source of the insult is removed.

1.1.1 Atrophy

Atrophy is the reduction in the volume of normal cells, tissues, or organs. In this condition, cell synthesis and metabolism are decreased, as well as the energy demand for normal cellular functions. Tissue and organ atrophy can also be accompanied by a decrease in the number of parenchymal cells, as well as a reduction in the volume of substance in the parenchymal cells as a result of substance loss. The non-development or development of the organs is not categorized as atrophy.

1.1.1.1 Types of Atrophy

Atrophy can be divided into two types: physiological and pathological atrophy

1. **Physiological Atrophy**: Examples include thymus atrophy at puberty and postmenopausal atrophy of the ovary, and uterus in women, and atrophy of the testis in men due to aging. In most cases of atrophy, the decrease in the number of cells is achieved through cell apoptosis.
2. **Pathological Atrophy**: It can be divided into different forms depending on the cause; atrophy caused by inadequate nutrition, which includes inadequate nutrition, excessive consumption, and lack of blood supply. It can further be divided into (1) systemic malnutrition atrophy, for example, in cases of diabetes, tuberculosis, cancer, and other chronic wasting diseases. Long-term malnutrition can cause muscle atrophy, known as cachexia. (2) Next is local malnutrition atrophy; this can be caused by insufficient blood supply to the organs, leading to inadequate nutrients. For example, after cerebral arteriosclerosis, vascular walls thicken, and the lumen narrows, which leads to the lack of sufficient blood supply to the brain causing brain atrophy. The cells and organs of the cell adjust the cell volume, quantity, and function to adapt to the reduced supply of blood and nourishment.

   Atrophy due to pressure: This type is caused by long-term compression of tissues and organs. The mechanism is hypoxia and ischemia. This type of atrophy can be caused by large tumors pressing onto the adjacent tissue(s) or organ(s) and causing an obstruction. When this occurs in the urinary tract, a condition known as hydronephrosis can occur leading to impaired renal flow and atrophy

Fig. 1.1 Atrophy of renal due to pressure. Hydronephrosis and dilation cause atrophy of renal due to pressure

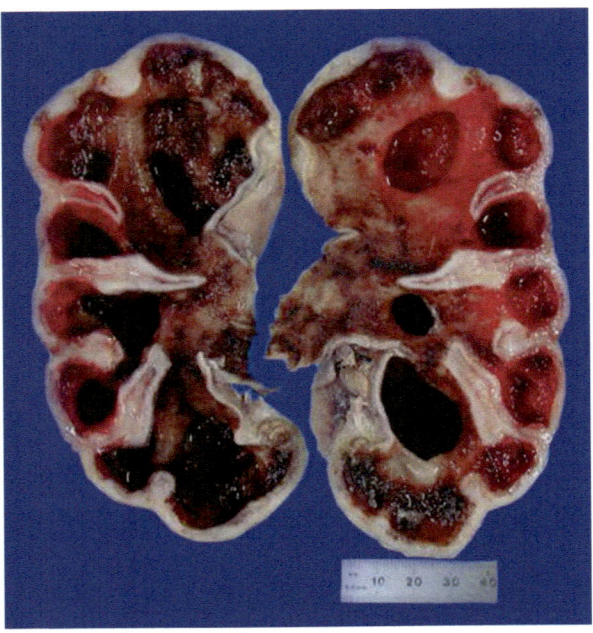

of the renal cortex and medulla (Fig. 1.1). When the right ventricular function is not complete, the central vein of the hepatic lobule and blood sinus congestion also cause the adjacent hepatocytes to atrophy due to compression.

Atrophy due to decreased workload: This may result from low function and a decrease in the organ tissues' long-term workload. For example, prolonged resting of a limb after fracture can lead to muscular atrophy and osteoporosis in that limb. As the limb resumes regular activity, the corresponding skeletal muscle cells will return to normal size and function.

Atrophy due to loss of innervation: This type of atrophy occurs when motoneuron or axon atrophy causing effectors like muscles to lose their normal functions. For example, muscle atrophy can occur due to brain or spinal cord injury. The mechanism involves the loss of neural regulation of muscle movement, reduced activity, and increased catabolism of skeletal muscle cell.

Atrophy due to loss of endocrine stimulation: The decline in endocrine gland function causes the target organ cells to atrophy. For example, pituitary ischemic necrosis in the thalamus can cause a decrease in the adrenocortical hormone leading to theatrophy of the adrenal cortex. The hypofunction of the anterior pituitary can cause atrophy in the thyroid gland, the adrenal gland, the gonadal gland, etc. In addition, tumor cells can also undergo atrophy. For example, prostate cancer cells can undergo atrophy when treated with estrogen treatment.

Atrophy due to aging and injury: The atrophy of brain and muscle cells are the leading causes of aging in the brain and heart. In addition, chronic inflammation caused by viruses and bacteria is also a common cause of cell, tissue, or organ atrophy. For example, the gastric mucosa may atrophy in chronic gastritis, while the mucous villi of the small intestine may atrophy in chronic enteritis. Apoptosis also induces atrophy of tissues and organs. For example, brain atrophy in Alzheimer's disease (AD) is caused by the apoptosis of many nerve cells.

Atrophy can be caused by pathological conditions or it can be physiological. For example, muscle atrophy after fracture may result from various factors like nerve, nutrition, decreased workload, or even pressure factors. However, aging atrophy in the heart or brain can involve both physiological and pathological processes.

1.1.1.2 Pathological Change

In the atrophied cells, tissues, and organs, the volume and overall weight become decreased, and the color and luster deepen and intensify. Lipofuscin granules can accumulate in the cytoplasm of myocardial cells and liver cells. Lipofuscin is a membrane-covered organelle rich in phospholipids that are not completely digested in the cell. Atrophied cells exhibit a decrease in protein synthesis, an increase in decomposition, and significant degradation of the organelle. The functions of atrophied cells, tissues, and organs primarily decline. This is due to reductions in cell volume and quantity as well as reductions in functional metabolism. This leads to the establishment of a new equilibrium between nutrition, hormones, growth factors, and neurotransmitters. After removing the cause, the cells with mild pathological atrophy may recover to the normal state. However, cells with persistent atrophy may eventually die.

1.1.2 Hypertrophy

Hypertrophy is to the volume expansion of cells, tissues, or organs as a result of elevated function and metabolism. Hypertrophy in tissues or organs is not only often caused by volume enlargement of parenchymal cells but also may be accompanied by an increase in the number of parenchymal cells.

1.1.2.1 Types of Hypertrophy

Natural hypertrophy can be classified into two types: pathological hypertrophy and physiological hypertrophy. Based on the causes of hypertrophy, it can be further classified into two categories: compensatory hypertrophy and endocrine hypertrophy. The latter is also known as hormonal hypertrophy because it results from the overuse of endocrine hormones on the effector.

1. **Physiological Hypertrophy**

   Compensatory hypertrophy: This type can be observed in the skeletal muscle of the upper limb in the weightlifter is thickened due to hypertrophy. Overwhelming demand and increased load are the most frequent causes.

   Endocrine hypertrophy: This type is observed during pregnancy causes the uterine smooth muscle cells to become hypertrophic and increase because of the effects of progesterone, estrogen, and its receptors. The uterus can grow from 0.4 to 5 cm thick and from 100 to 1000 g in weight.

2. **Pathological Hypertrophy**

   Compensatory hypertrophy: This may occur during increased cardiac overload during hypertension or functional compensation of normal myocardial function after left ventricular partial necrosis can cause left ventricular hypertrophy (Fig. 1.2). Organ hypertrophy may also occur as a response to the loss of function or absence of organs. For example, the contralateral kidney may experience compensatory hypertrophy following a unilateral nephrectomy or unilateral renal artery occlusion, which causes the loss of renal function.

   Endocrine hypertrophy: Hyperthyroidism increases thyroxine secretion, causing thyroid follicle epithelial cells to hypertrophy. Hypophysis eosinophil adenoma increases the adrenocorticotropic hormone's secretion, resulting in adrenocortical cells' hypertrophy.

Fig. 1.2 Myocardial hypertrophy (HE, low power). Part of the myocardial cell thicken, the nuclei are irregular and hyperchromatic, presenting a compensatory hypertrophy state

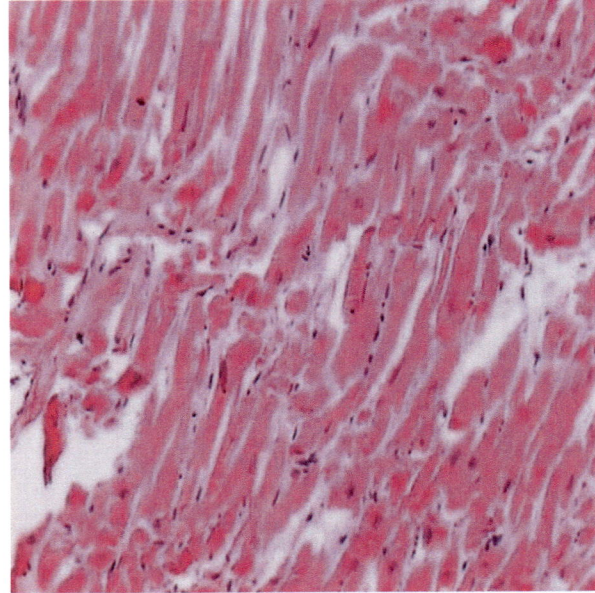

1.1.2.2 Pathological Changes of Hypertrophy

The volume of hypertrophy cells increased, its nucleus hypertrophy and anachronisms, uniform enlargement in hypertrophy tissues and organs. The activation of many pro-oncogenes in cells with hypertrophy results in increased DNA content and organelles, active structural protein synthesis, and enhanced cell function. However, the function of compensatory hypertrophy cells is limited. For example, in myocardial hypertrophy, the blood supply of myocardial cells is relatively lacking; the normal contraction protein in myocardial cells, also due to activation of embryonic genes, transforms into naive contractile protein whose contraction efficiency is poor; the myocardial fiber contraction component even dissolves or disappears, leading to reversing injury. Those changes eventually lead to myocardial overload, inducing dysfunction (decompensation).

In conditions of certain pathological factors, during the atrophy of parenchymal cells, interstitial adipocytes can proliferate to maintain the original volume of organs and even increase the volume of organs and tissues. This is called pseudohypertrophy.

1.1.3 Hyperplasia

Hyperplasia is defined as the phenomenon in which the number of tissues and organs increases due to active mitosis. It often an increase in volume and enhanced function in the affected tissues or organs. It is often caused by excessive cell stimulation and overexpression of growth factor and its receptor and may also be related to inhibition of cell apoptosis. It is usually finely regulated by proliferating genes, apoptotic genes, hormones, and various peptide growth factors and their receptors.

1.1.3.1 Types of Hyperplasia

According to its nature, hyperplasia can be divided into physiological hyperplasia and pathological hypertrophy. According to the causes, it can be divided into compensatory hyperplasia and endocrine hypertrophy, also called hormonal hyperplasia.

1. **Physiological Hyperplasia**
    Compensatory hyperplasia: For example, it includes the hyperplasia of residual hepatocytes after partial liver resection. Because the oxygen content in the air at high altitudes is low, the bone marrow erythrocyte precursor cells and peripheral blood erythrocyte increase in compensation.

Endocrine hypertrophy: Example includes hyperplasia of lobular gland epithelium in the typical female during puberty and endometrium glands in the menstrual cycle.

2. **Pathological Hypertrophy**

Compensatory hyperplasia: In wound healing after tissue injury, fibroblasts and capillary endothelial cells are often proliferated because of increased growth factor stimulation. Chronic or prolonged exposure to physical and chemical factors often causes the proliferation of tissue cells, especially the skin and some organ-coated cells.

Endocrine hyperplasia: The most common cause of pathological hyperplasia is excessive hormone or growth factor. The absolute or relative hyperplasia of the endometrium glands. An increase in estrogen may cause the endometrium glands' hyperplasia to grow too long, resulting in functional uterine bleeding.

Hyperplasia is an important adaptive response of interstitial tissue. For example, fibroblasts and capillary endothelial cells can achieve the purpose of repair through hyperplasia. Inflammation and the hyperplasia of tumor interstitial fibroblasts are important histology and cytological manifestations of anti-inflammatory and antitumor mechanisms of the body. The hyperplasia of parenchymal cells and interstitial cells is not uncommon. For example, the estrogen metabolite dihydrotestosterone can cause hyperplasia of the prostate gland and interstitial fibrous tissue in men; excessive estrogen secretion can cause hyperplasia in women's mammary terminal ducts, acinus epithelium, and interstitial fibrous tissue.

1.1.3.2   Pathological Changes of Hyperplasia

The number of cells increase, and the shape of the cells and cell nucleus is either normal or slightly enlarged. Hyperplasia of cells can be classified as diffuse or limited hyperplasia, or localized. A proliferation of tissue, a homogeneous and diffuse enlargement of the organ, or a single or multiple proliferative nodules in the tissues and organs characterizes it. Most pathological (such as inflammation) cell hyperplasia can stop due to the removal of the trigger factor. If cell hyperplasia is uncontrolled, it may become a neoplastic hyperplasia.

1.1.3.3   The Relationship Between Hypertrophy and Hyperplasia

Although hypertrophy and hyperplasia are two pathological processes, the causes of hypertrophy and hyperplasia of cells, tissues, and organs are similar. So they often go hand in hand. For example, when cell mitosis is blocked in the G2 phase, it will have hypertrophic polyploid cells that are not divided. If the cells progress through successive phases, the processes of, division, and multiplication are completed. In general, the proliferation of cells (permanent, stable, unstable cells) determines whether it is simple hypertrophy or hyperplasia. For organs that are active in cell division and proliferation, like the uterus and the mammary gland, its hypertrophy can be the result of both cell volume increase and cell number increase. However, in the myocardium and skeletal muscle, which have low cell division and proliferation ability, the hypertrophy of tissue organs is only caused by cell hypertrophy.

1.1.4 Metaplasia

Metaplasia is defined as a process in which one type of mature differentiated cell is replaced by another type of mature differentiated cell type, which is only in the cell with more active splitting and proliferation ability. It is not caused by directly transform from mature cells in the original. Still, it is the result of transdifferentiation of the stem cells with the and capacity of proliferation and multidirectional differentiation such as immature differentiated cells and reserve cells, which are (transdifferentiation) the product of some gene activation or inhibition and cell reprogramming expression caused by environmental factors. That is the morphological expression of histomorphological cell change in differentiation and growth regulation. This process may be reached by methylation and demethylation of particular gene DNA.

1.1.4.1 Types of Metaplasia

Metaplasia has many types, usually occurring in homologous cells, namely in epithelial or mesenchymal cells. It is generally preformed as cell types with higher specificity replacing cell types with lower specificity. The metaplasia of the epithelial tissue may recover after the causes are eliminated, but the mesenchymal tissue's metaplasia is mostly irreversible.

1. **The Metaplasia of the Epithelial Tissue**
   Squamous metaplasia: Squamous metaplasia is the most common metaplasia of covering epithelia. For example, the bronchial pseudostratified ciliated columnar epithelium is prone to squamous metaplasia (Fig. 1.3). When salivary glands, pancreas, pelvis, bladder, liver, and gallbladder produce stones or insufficient vitamin A, columnar epithelium, cuboidal epithelium, or transitional epithelium can be turned into the squamous epithelium.

Fig. 1.3 Squamous metaplasia of columnar epithelium. Reserve cells in columnar epithelial cells proliferate. Differentiating and forming stratified squamous epithelium

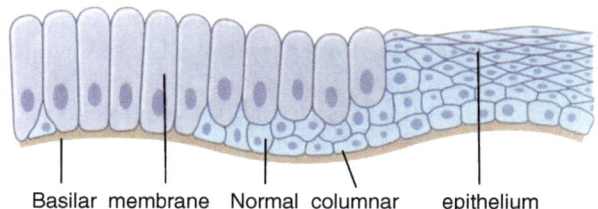

Basilar membrane   Normal columnar   epithelium

Metaplasia of columnar epithelium: The metaplasia of the epithelial tissue of the gland is also more common. When chronic gastritis occurs, metaplasia of gastric mucosal epithelial to the small intestine or colon epithelial tissue with Paneth cells or goblet cells may occur, which is called intestinal epithelial metaplasia; if gastric gland of antrum and body are replaced by pyloric glands, it is called pseudopyloric gland metaplasia. In the case of chronic reflux esophagitis, the squamous epithelium of the lower esophageal segment can also be transformed into a gastric or intestinal columnar epithelium. In the case of chronic cervicitis, the squamous epithelium of the cervix is replaced by the columnar epithelium of the cervix mucosa, forming cervical erosion in the naked eye.

2. **Metaplasia of Mesenchymal Tissue**: The infantile fibroblasts in the mesenchymal tissue can be transformed into osteoblasts or chondrocytes after injury. It is called bone or cartilaginous metaplasia. This type of metaplasia is seen mostly in damaged soft tissues such as ossifying myositis and in the interstitial area of some tumors.

1.1.4.2 The Significance of the Metathesis

The biological meaning of metathesis had both pros and cons. For example, the epidermal epithelium of respiratory mucosa is formed by epidermal metaplasia on the scale; local resistance to external stimulation can be enhanced due to the increase and thickening of cell layers. However because the surface of the squamous epithelium has no ciliary structure like columnar epithelium, the self-purification ability of mucosa may be weakened. In addition, malignancy may occur. For example, bronchial squamous metaplasia and intestinal metaplasia of the gastric mucosa are related to the occurrence of squamous cell carcinoma of the lung and gastric adenocarcinoma. Columnar metaplasia is the histologic source of certain esophageal adenocarcinoma when chronic reflux esophagitis. In this sense, some metaplasia is a precancerous lesion associated with the evolution of multistep tumor cells.

1.1.4.3 Epithelial-Mesenchymal Transition

Epithelial-mesenchymal transition (EMT) mainly refers to the biological process of epithelial cells transforming into specific mesenchymal cells. It is essential in embryonic development, tissue reconstruction, chronic inflammation, tumor growth and metastasis, and various fibrotic diseases.

The transformation of epithelial cells into interstitial cells is characterized by gradual loss of epithelial phenotype, such as the decrease of E-cadherin and cytoskeletal keratin expression, and the expression of interstitial cell phenotypes, such as vimentin, fibronectin, and N-cadherin. When EMT in epithelial malignant tumor occurs, epithelial cell polarity and basement membrane connection are lost, migration and invasion ability increase, making tumor cells to infiltrate growth to peripheral tissues easily and making it easier to move to distant sites by blood flow to form metastases sites. The tyrosine kinase receptor signaling pathway, the integrin signaling pathway, the Wnt signaling pathway, the NF signaling pathway, and the TGF β signaling pathway may be involved in the regulation of EMT.

1.2 The Cause and Mechanism of Cell Tissue Damage

When changes in the outside environment of the body exceed the tissue and cell adaptability, metabolism, histochemistry, ultrastructure, even microscope, and abnormal changes visibly may occur in damaged cells and stromal cells, which is called the injury. The way and result of injury depend not only on the nature, duration, and intensity of the injury factors but also on the type, status, adaptability, and heredity of the damaged cells.

1.2.1 The Causes of Injury in Cell and Tissue

The disease's causes are also the leading causes of cellular tissue damage. It can be divided into the biophysical chemistry nutrition immune-neuroendocrine genetic variation, congenital internal factors such as age, gender, and social psychological, psychological behavior, and iatrogenic external pathogenic factors such as social psychological factors.

1.2.1.1 Hypoxia

Hypoxia and ischemia are the most common reasons leading to injury of cells and tissue. Cardiorespiratory failure causes arterial oxygenation insufficiency, the blood oxygen carrying capacity decreased caused by anemia and carbon monoxide poisoning, or blocked blood decreases the supply. All those changes above can lead to a reduced supply of oxygen and nutrition to cells and tissues, causing damage to structure and loss of function in cells and tissue.

1.2.1.2 Biological Factors

Biological factors are the most common reason for cell injury, including pathogenic organisms, like bacteria, viruses, Rickettsia, mycoplasma, spirals, fungi, protozoa, and worms. Pathogenic organisms invade the body, cause mechanical injury, induce allergy, release internal and external toxins, or secrete certain enzymes, which may damage the structure and function of cells and tissues.

1.2.1.3   Physical Factors

When various physical factors in the environment go beyond the physiological tolerance of the body, cell damage can be caused. For example, high temperature and high radiation can cause heatstroke, scald, or radiation damage, and cold causes frostbite, and the current solid impact causes electrical injury; mechanical damage can cause trauma, fracture, and so on.

1.2.1.4   Chemical Factors

Chemical factors including exogenous substances, such as acid, alkali, lead, mercury, and other inorganic toxicants; organic phosphorus, cyanide, and other organic toxicants; venom, muscarine, and other biological toxins; endogenous substances some metabolites, such as decomposition products of cell necrosis, urea, free radicals, can cause injury to cells. Drugs, health agents, etc. can but also treat and prevent some cell damage, which can have side effects on cells.

1.2.1.5   Nutritional Imbalance

Insufficient or excessive intake of nutrients can cause corresponding lesions in the body. For example, lack of vitamin D, protein, and iodine leads to rickets, malnutrition, and endemic goiter, respectively; lack of iron selenium and other trace elements causes developmental disorders of red blood cells and brain cells; long-term intake of high calorie and high fat is an essential cause of the obesity, hepatic steatosis, and atherosclerosis.

1.2.1.6   Neuroendocrine Factors

Primary hypertension and peptic ulcer have a relationship with excessive excitability of vagus; when hyperthyroidism occurs, the body cells and tissues become more sensitive to infection and poisoning. The secretion of insulin complicates the body, especially the subcutaneous tissues, with bacterial infection.

1.2.1.7   Immune Factors

The body cells react to some antigens, and allergic or hypersensitivity reactions such as asthma and allergic shock may occur; self-antigen can cause tissue injury, such as systemic lupus erythematosus, rheumatoid arthritis; immunodeficiency diseases such as AIDS can cause damage of immune function and damage of lymphocyte.

1.2.1.8 Hereditary Defect

The role of genes in damage is mainly reflected in two aspects: one is the gene mutation or chromosomal aberrations that directly cause progeny genetic diseases such as Down syndrome, hemophilia, and acute hemolytic anemia; the other is a genetic defect that makes the offspring tend to induce certain diseases (genetic susceptibility).

1.2.1.9 Social Psychological Factors

Coronary heart disease, primary hypertension, peptic ulcer, and even some tumors are closely related to social and psychological factors. They are called psychosomatic diseases. For medical workers, it is necessary to prevent iatrogenic injury caused by improper health services, such as hospital-acquired infection or drug-induced injury.

1.2.2 The Injury Mechanism of Cells and Tissues

The mechanism of cell damage is mainly in the cell membrane damage, increased reactive oxygen species and cytosolic free calcium, hypoxia-ischemia, chemical toxicity, and genetic mutation. They interact with each other or are reciprocal causation, causing the occurrence and development of cell injury.

1.2.2.1 Injury of Cell Membrane

The direct effect of mechanical stress, enzyme dissolution, hypoxia-ischemia, reactive oxygen species, bacterial toxins, complement components, ion pump, and ion channel chemical damage can destroy the permeability and integrity of cell membrane structure, effect functions such as membrane information and material exchange, immune response, cell division, and differentiation. The early change is a selective loss of membrane permeability and, eventually, apparent cell membrane damage. The crucial mechanism of cell membrane damage is the formation of free radicals and secondary lipid peroxidation, resulting in the progressive decrease of membrane phospholipids and accumulation of phospholipid degradation production of cytotoxicity. The cell membrane is separated from the cytoskeleton, and the cell membrane is vulnerable to tension damage.

Morphologically, the damage to the cell membrane structure causes the cells and mitochondria, endoplasmic reticulum, and other organelles to be swollen, the microvilli on the surface of the cells disappear, and the vesicles form. The lipid of the cell membrane and organelle membrane degenerated, curling like spiral or concentric round figures, forming myelin figures. The lysosome membrane is broken, releasing a large amount of acid hydrolase, causing cell dissolving. Most of the cell necrosis starts with the dysfunction of cell membrane permeability and ends with the loss of cell membrane integrity. Therefore, cell membrane destruction is often the critical link of cell injury, especially the irreversible damage at the early stage of the cell.

1.2.2.2 Mitochondrial Damage

Mitochondria are the main sites of intracellular oxidative phosphorylation and ATP production, and they are also involved in some processes such as cell growth and differentiation, information transmission, and cell apoptosis. After the mitochondrial damaged, the mitochondria is swollen and vacuolated, the mitochondrial crista becomes shorter, sparse, or even disappeared, and the calcium amorphous dense body appears in the matrix. Mitochondrial ATP production decreases and consumption increases, resulting in the cell membrane sodium pump and calcium pump dysfunction. Transmembrane transport protein and lipid synthesis decreased, and phospholipid deacylation and reacylation stopped. Mitochondrial damage is often accompanied by the infiltration of mitochondrial pigments into the cytoplasm, which can initiate cell apoptosis. When the ATP energy supply is reduced by 5–10%, the cell will have a significant damage effect. Cell necrosis may occur after the suspension of mitochondrial oxidative phosphorylation, and eventually, cell necrosis occurs. Mitochondrial damage is an important early marker of irreversible cell injury.

1.2.2.3 Damage of Activated Oxygen Species

Activated oxygen species (AOS), also called reactive oxygen species, include oxygen in the free radical state (such as O^{2-}), OCl_3, NO, and H_2O_2, which is hydrogen peroxide that does not belong to the free radicals. A free radical is a group formed by the loss of an electron of an atom in the outer layer of the atom; it has intense oxidation activity. A copper-containing enzyme can activate it. AOS can be the product of normal cell metabolism and can also be produced by exogenous factors. It is elementary for it to react with the surrounding molecules, release energy, and cause cell damage. It also causes the surrounding molecules to produce toxic free radicals, forming a chain amplification reaction and further cause cell damage.

There is a system that produces AOS in the cell and an antioxidant system that is resistant to its formation. Average small amounts of AOS generated will be cleared by intracellular antioxidants such as superoxide dismutase, glutathione peroxidase, catalase, and vitamin E. During the process of hypoxia-ischemia, cell phagocytosis, chemical radiation injury, inflammation, and aging, AOS production increases. Lipid, protein, and DNA peroxide. By several target points such as the membrane lipid peroxidation, non-oxidative mitochondrial damage, DNA damage and protein crosslinking, changes in lipid and carbohydrate, protein, and nucleic acid molecular structure may occur, the stability of membrane lipid bilayer structure decrease, DNA single-strand damage, and breaks, which promotes mutual crosslinking of sulfur-containing protein and can directly cause the polypeptide broke into pieces. The strong oxidation of AOS is the primary link to cell damage.

1.2.2.4 Damage of Intracellular Free Calcium in the Cytoplasm

Phospholipids, proteins, ATP, and DNA will be degraded by phospholipase, protease, ATP enzyme, and nuclease in the cytoplasm. This process requires the activation of free calcium. In normal cases, intracellular calcium is combined with intracellular calcium transporters and stored in calcium pools, such as calcium pools in the endoplasmic reticulum and mitochondria. The cell membrane ATP calcium pump and calcium channel are involved in the regulation of low free calcium concentration in the cytoplasm. When cells are anoxic and poisoned, ATP decreases, and Ca^{2+} exchange proteins are activated directly or indirectly. The cell membrane's permeability to calcium increases, the calcium pumping from cells decreases, the calcium flows intracellular increases, and mitochondria and endoplasmic reticulum release calcium repeat, resulting in the rise in intracellular free calcium (intracellular calcium overload), which promotes the activation of these enzymes and damaged the cells. The intracellular calcium concentration is often related to the degree of functional damage to the cell structure, especially the mitochondria. A large amount of intracellular free calcium is a terminal link between many factors that damage cells, and it is a potential mediating agent for cell death and biochemical and morphological changes.

1.2.2.5 The Injury of Ischemic Anoxia

Ischemia is defined as insufficiency of the arterial blood supply in local cell tissue. Ischemia can cause nutrient and oxygen supply disorders; the former is known as malnutrition, and the latter is known as hypoxia. Anoxia means that cells cannot get enough oxygen or oxygen utilization barriers. According to the reasons, it can be divided into (1) hypotonic hypoxia, oxygen partial pressure or airway breathing disorders; (2) hematologic anoxia, the quality, and quantity of hemoglobin is abnormal; (3) circulatory hypoxia, cardiopulmonary failure or local ischemia; (4) histogenous hypoxia, mitochondrial biological oxidation, mainly oxidative phosphorylation, and other internal respiratory dysfunction. In this significance, ischemia is one reason for the anoxia.

Cell hypoxia-ischemia leads to inhibition of mitochondrial oxidative phosphorylation, decrease of ATP formation, and activation of phosphofructokinase and phosphorylase. The sodium and potassium pump function in the cell membrane is low, and the sodium-calcium ion accumulates in the cell and increases with the water molecules. After that, protein synthesis and fat transport disorders in the cytoplasm, anaerobic glycolysis increases, cell acidosis may occur, lysosomal membrane ruptures, DNA chain may be damaged, and nuclear chromosome may be agglutinated. Ischemic anoxia also causes the increase of reactive oxygen species, causing lipid disintegration and destruction of the cytoskeleton. Blood flow blocking is the most common cause of ischemia and hypoxia. Usually, ischemia is more rapid and severe for tissue damage. Because anaerobic glycolysis can continue when anoxia, but it stops during ischemia. Mild anoxia can cause cell edema and fatty change; mild continuous hypoxia can lead to cell apoptosis; severe continuous anoxia can lead to cell necrosis. In some cases, the recovery of blood flow after ischemia can cause the peroxidation of living tissue, aggravating the tissue damage, which is called ischemia-reperfusion injury, and is commonly seen after myocardial infarction and cerebral infarction. Ischemic and anoxia are the most common and critical central links in cell injury.

1.2.2.6 Chemical Damage

Many chemicals, including drugs, can cause cell damage. Chemical damage can be systemic, like chloride poisoning, or local damage, like damage to the skin and mucous membrane due to exposure to strong acids and strong alkalis. The effects of some chemicals may also be organ-specific, such as the liver damage caused by CCl4. The ways of chemical damage are: (1) the chemical substance itself has the direct cytotoxic effect. For example, cyanide can quickly seal the cytochrome oxidase system of the mitochondria and cause sudden death. When mercury is poisoned, mercury binds with sulfur-containing protein in the cell membrane, damaging ATP enzyme-dependent membrane transport function. Chemical antitumor drugs and antibiotics can also hurt cells by similar direct effects; (2) the cytotoxic effects of metabolites on target cells. The liver, kidney, bone marrow, and myocardium are the target organs of toxic metabolites, and cytochrome P450 complex function enzyme plays a vital role in the metabolic process. If CCl4 itself is not active, it will be transformed into toxic-free radicals CCl3 in liver cells, which will cause swelling of the smooth endoplasmic reticulum and lipid metabolism disorder. (3) It will induce immune injury such as anaphylaxis. For example, penicillin leads to type I allergy; (4) it will cause DNA damage. The degree, speed, and location of chemical injury are affected by the dosage of chemical substances and drugs, the time of action, the area of absorption, accumulation, and the individual difference of metabolic rate.

1.2.2.7 Genetic Variation

The damage of genetic variation may occur by congenital heredity or during embryogenesis and can also occur after birth. Chemical substances and drugs, viruses, rays, and so on can damage the DNA in the nucleus, induce gene mutation and chromosome aberration, and make the cell genetic variation. Those changes are caused by: (1) the low synthesis of protein structure and lack of the essential protein in cells; (2)

prevention of nuclear division in important functional cells; (3) the synthesis of abnormal growth regulator; (4) induction of enzyme synthesis disorders of congenital or acquired cell dies due to lack of necessary metabolism mechanism for life.

1.3 Reversible Injury

Reversible injury, whose morphological change is called denaturation, refers to the phenomenon of the accumulation of normal substances or the appearance of abnormal substances in cells or intercellular interstitial due to metabolic disorders after being damaged, usually accompanied by low cell function. The accumulation is caused by excessive or too fast production of these normal or abnormal substances. The cell itself lacks the corresponding mechanism of metabolism, clearance, or transport, which makes them accumulate in organelles, cytoplasm, and nuclear or intercellular substances. After removing the cause, most of these injuries can be restored to normal. Therefore, it is a nonfatal and reversible damage.

All the harmful factors play their role at the molecular level first. It can identify cellular adaptation, reversible injury or irreversible damage, and other morphological changes, which depend on the nature of cell lesions and the sensitivity of observation methods. But in general, the affected cells show biochemical metabolic changes and histochemical and ultrastructural organization (such as a few minutes to dozens of minutes after ischemia), and then appeared morphological changes under a light microscope and visible (like a few hours to a few days after ischemia). Most of the mild damage can be restored to normal after the cause is eliminated, usually called reversible damage. Severe cell damage is irreversible, directly or ultimately leading to cell death.

1.3.1 Cellular Swelling

Cellular swelling, also called hydropic degeneration, is often the initial change in cell damage. It is caused by the function decline of cell volume and cytoplasmic ion concentration regulation mechanism.

1. **Mechanism**: Due to the damage of mitochondria, the decrease of ATP production, and the dysfunction of the potassium pump in the cell membrane, sodium ions accumulate excessively, and a large amount of water is attracted into the cell to make intracellular and extracellular ion isosmotic. After that, the accumulation of the metabolites of inorganic phosphate, lactic acid, and purine nucleotides can increase the osmotic pressure and further aggravate the edema of the cells. Any damage that can cause changes in cell fluid and ionic homeostasis can lead to cell edema, which is commonly seen in the parenchymal cells of the liver, kidney, heart, and other organs during hypoxia, infection, and poisoning.

Fig. 1.4 Swelling of
hepatocyte (HE, low
power). The liver cells are
significantly swollen, the
cytoplasm is light dying,
and some of the
hepatocytes are swollen
like balloon samples

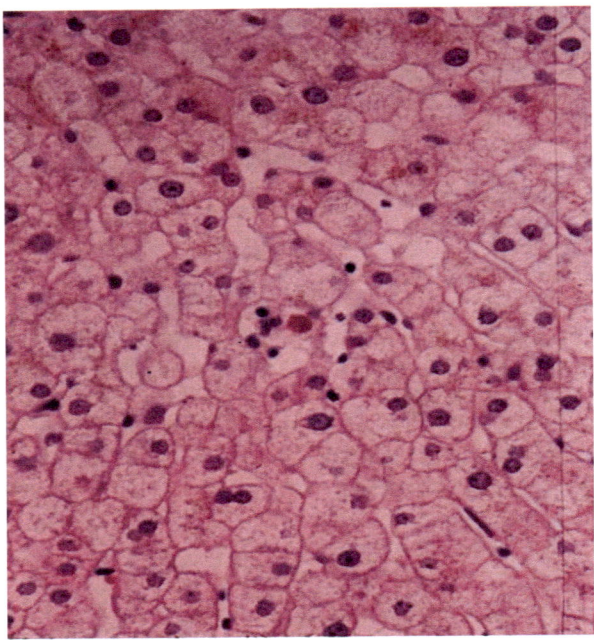

2. **Pathological Changes of Cellular Swelling**: At the beginning of the lesion, the
   mitochondria and the endoplasmic reticulum become swollen and form a red-
   dyed granular substance that appeared in the cytoplasm under a microscope. If
   water and sodium accumulate further, the cells will obviously be swollen. The
   cytoplasm is highly porous and vacuolar, and the nucleus can also be swollen.
   There are vesicles on the surface of the plasma membrane, and the microvilli
   disappear. It is called the ballooning variant in fastigium (Fig. 1.4). Sometimes
   the changes in cell edema are not easily identified by microscopy, but the changes
   in the whole organ may be more prominent. The volume of the affected organs
   increases by the naked eye, the envelope is tense, the cut surface is valgus, and
   the color becomes pale.

1.3.2 Fatty Change

Steatosis is defined as the phenomenon which triglycerides accumulate in the cytoplasm of cells that are not adipocytes. It occurs mainly in hepatocytes, cardiomyocytes, renal tubular epithelial cells, skeletal muscle cells, and so on, which are
related to infection, alcoholism, poisoning, hypoxia, malnutrition, diabetes, and
obesity.

Fig. 1.5 Fatty changes of hepatocyte (HE, low power). In the cytoplasm of the liver, the vacuoles of different sizes are seen as lipid droplets. Some of the nuclei are partial to one side

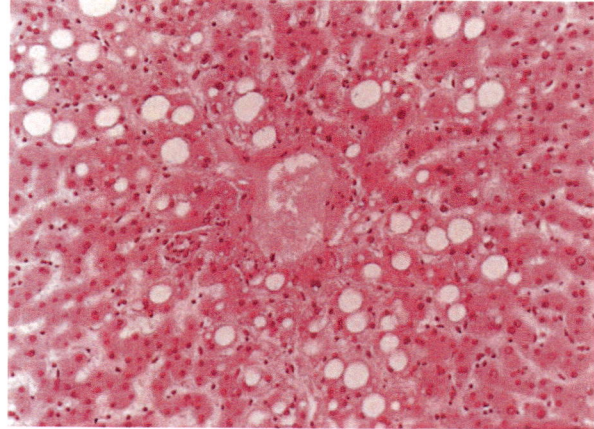

1. **The Pathology of Fatty Change**: There is no obvious change in the organs of the naked eye with a slight fatty change. With the aggravation of the lesion, the volume of the fat-changing organs increases; its color is yellow, the edge is blunt, and the section has a greasy sense. Under the electron microscope, the fat in the cytoplasm aggregates into the fat corpuscle and then fuses into the lipid droplets. In the microscope, there are different sizes of spherical lipid droplets in the cytoplasm of fatty change. The large one can fill the whole cell and squeeze the nucleus to one side. In the paraffin section, the fat is vacuolated as the fat is dissolved in organic solvents (Fig. 1.5). In the frozen section, the special dyeing of Sultan III and Sultan IV can distinguish fat from other substances.

   Hepatocyte is an essential place for fat metabolism, in which fatty changes most often occur, but mild hepatic steatosis usually does not cause morphological changes and dysfunction of the liver. The distribution of fat in the lobule of the liver has a particular relationship with the cause of the disease. For example, when chronic liver congestion, centrilobular hypoxia is severe, so the fat first occurs in centrilobular sites; when phosphorus poisoning occurs, peripheral lobular liver cells are more sensitive to phosphorus poisoning, so the peripheral lobular liver cell is involved more obviously; when severe poisoning and infectious disease, fatty change often involves the entire liver cell. Significant diffuse hepatic steatosis is called fatty liver, and severe hepatic steatosis may develop into liver necrosis and cirrhosis.

   Chronic alcoholism or anoxia can cause myocardial steatosis, often involving the left ventricular endocardium and papillary muscles. Myocardial steatosis is yellow. It is between the dark red of the normal myocardium, forming the yellow-red markings, which is called the tigroid heart. Sometimes, the adipose tissue of the epicardial hyperplasia can extend between the human cardiac myocytes. It is called myocardial fatty infiltration, but not the degeneration of the myocardium.

Severe myocardial fatty infiltration can cause a rupture of the heart, causing sudden death.

The renal tubular epithelial cells can also undergo steatosis. Under the light microscope, lipid droplets are mainly located in the proximal part of renal tubular cells, which are excessive reabsorption of lipoproteins in the original urine, and those of severe cases can involve the distal renal tubule cells.

2. **The Mechanism of Hepatic Steatosis**: The mechanism of hepatic steatosis is as follows: (a) Liver cytoplasmic fatty acid increases: as high-fat diet or malnutrition, fat tissue in vivo is decomposed. Excess free fatty acids enter the liver by blood flow; a large number of lactic acid turn into fatty acid oxidation in liver cells due to hypoxia; obstacles of oxidation decrease fatty acid utilization, fatty acid increases relatively; (b) triglyceride synthesis increases; for example, excessive drinking can change function of mitochondria and smooth endoplasmic reticulum, promote synthesis of a new triglyceride; (c) the lipoprotein and apolipoprotein decrease: when ischemic, anoxia, poisoning or malnutrition occur, lipoprotein and apolipoprotein synthesis in liver cells decrease, fat output is blocked and accumulates in the cell.

In addition, when atherosclerosis or hyperlipidemia syndrome occurs, excess cholesterol and cholesterol can exist in some non-adipocytes, such as macrophages and smooth muscle cells, which can be regarded as a particular type of intracellular lipid accumulation. When the macrophages are significantly increased and gathered in subcutaneous tissue, it is called yellowish tumors.

1.3.3   Hyaline Change

The accumulation of translucent protein in cells or interstitial cells is called hyaline change or hyaline degeneration; HE staining presents eosinophilic homogenization. Hyaline change is a group of lesions that have morphologically similar physical properties, but their chemical composition and pathogenesis are different.

1. **Mechanism of Hyaline Change**: The mechanism of it may be the congenital genetic disorder of protein synthesis or the acquired defect of protein folding, which makes the third-level structure and amino acid sequence of some proteins mutate, resulting in the accumulation of denatured collagen, plasma protein, and immunoglobulin.

2. **Pathological Changes of Hyaline Change**: According to the lesion sites, it can be divided to

   The hyaline changes in the cells: they are usually rounded corpuscles of homogeneous red dye located in the cytoplasm. For example, there are wicking vesicles in the renal tubular epithelial cells, which re-absorp of the urine protein and fuse with lysosomes, forming hyaline droplets; immunoglobulin in the rough endoplasmic reticulum of plasmocyte cytoplasm accumulates, forming Russell bodies; alcoholic liver disease. Degeneration of the intermediate fila-

Fig. 1.6 Hyaline change in central arteries of spleen (HE, low power). In the primary hypertension, the central artery wall was thickened, the lumen was relatively small, and the wall of the artery was stained with red dye, and the homogenous glass-like substance

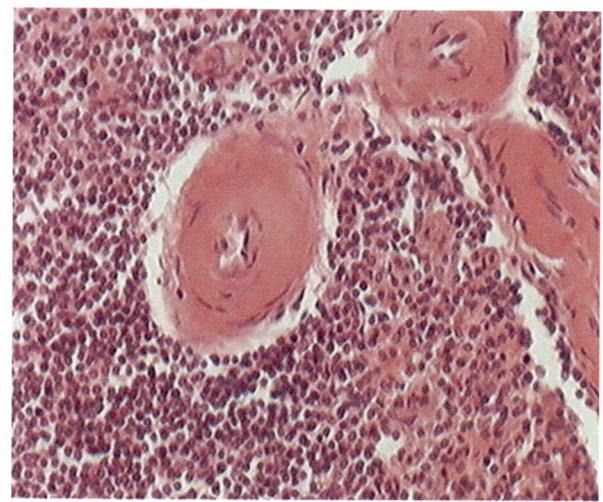

ments prekeratin of the cells in the cytoplasm of the hepatocyte occurs, forming the Mallory body.

Hyaline changes in the fibrous connective tissue occur in the physiological and pathological connective tissue proliferation, which is a manifestation of the aging fibrous tissues. It is characterized by collagen crosslinking, denaturation, and fusion. The collagen fibers proliferated and thickened, and there were few vascular and fibroblast cells. Grossly, it is gray and white, and its quality is rigid and translucent. It can be seen in the atrophic uterus and breast stromal, scar tissue, atherosclerotic fibrous plaque, and the organization of various necrotic tissues.

Arterioles hyalinization, also known as arteriolosclerosis, is common in the fine arterial wall of the kidney, brain, spleen, and other organs with chronic hypertension and diabetes (Fig. 1.6). Due to the infiltration of plasma protein and deposition of metabolism of the basement membrane, the fine arterial wall thickens, the tube becomes narrow, blood pressure rises, and organs become ischemic. The elasticity of the arteriole wall with hyaline change is weakened, the brittleness increases, and the dilatation, rupture, and bleeding are easily secondary.

1.3.4 Amyloid Change

Amyloidosis is an accumulation of amyloid protein and sticky polysaccharide in the mesenchymal cell, named with starch dyeing characteristics. Amyloidosis is also a class of morphological and unique coloring, but the changes in chemical structure and production mechanism are different.

1. **Mechanism of Amyloid Change**: The amyloid protein is derived from the light chain of an immunoglobulin, peptide hormone, calcitonin precursor protein, and serum amyloid A protein. The new polypeptide chain of amyloid is composed of ribosomes and can be arranged in chain α and chain β. Because the body does not contain the enzyme that can digest large molecules with a β-folding structure, β-folding protein, and its precursors are easy to accumulate in the tissue in the body.

2. **Pathological Changes in Amyloid Change**: Amyloidosis is mainly deposited in the cell interstitium, under the small vascular basement membrane, or along the reticular fibrous scaffold. Under the microscope, HE staining is characterized by a light red homogeneous substance and shows a coloration reaction of amyloid. It is orange-red when Congo red, is brown when it meets iodine, and turns blue when diluted sulfuric acid is added.

Amyloidosis can be divided into localized amyloidosis and systemic amyloidosis. Localized amyloidosis can be found in the skin, conjunctiva, tongue, larynx, and lung. It can also be found in the interstitial tissue of Alzheimer's disease brain and Hodgkin's disease, multiple myeloma, medullary thyroid carcinoma, and other tumors. Systemic amyloidosis can be divided into primary and secondary amyloidosis and the former mainly originates from the serum α-immunoglobulin light chain, involving the liver, kidney, spleen, heart, and other organs; the latter from unknown origin. Main components are non-immunoglobulin synthesis in the liver (amyloid protein), which can be found in the elderly, and chronic inflammation such as tuberculosis and tumor stroma.

1.3.5 Mucoid Change

Mucoid change or mucoid degeneration is defined as the accumulation of protein (glucosaminoglycans, hyaluronic acid, etc.) in the cytoplasm, which are commonly seen in mesenchymal tissue tumors, atherosclerotic plaques, rheumatic foci, and malnourished bone marrow and adipose tissue. The characteristics under a microscope are that in the loose interstitial, there are many protruding star-shaped fibrous cells scattered in the grayish-blue mucous matrix. Hyaluronidase activity is inhibited, and hyaluronan mucoid substance and moisture accumulate in the skin and subcutaneous tissue, forming characteristic mucinous edema when hypothyroidism.

1.3.6 Pathological Pigmentation

The average human body contains hemosiderin, lipofuscin, melanin, bilirubin, and other endogenous pigments; carbon dust, coal dust, tattoo pigment, and other exogenous pigments sometimes enter the body. In pathology, some of the above

Fig. 1.7 Pigmentation of hemoflavin (HE, low power). In chronic pulmonary blood stasis, a large number of macrophages in the alveolar cavity phagocyte and decompose red blood cells, and many gold or brown hemosiderin granules are formed in the cytoplasm

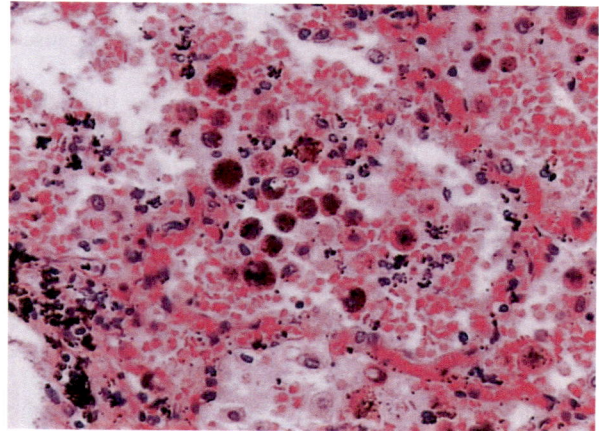

pigments will increase and accumulate inside and outside the cell, known as pathological pigmentation.

1. **Hemoflavin**: It is an aggregation of ferritin particles produced by macrophage's phagocytosis and degradation of erythrocyte hemoglobin. It is a combination of Fe^{3+} and protein. It is golden or brown under a microscope (Fig. 1.7) and can be dyed blue by Prussian blue. The presence of hemoflavin reflects the destruction of the erythrocyte and the residual iron-containing substances, systemic or local. After the rupture of macrophages, this pigment can also be seen outside the cell. In physiological conditions, a small amount of hemoflavin can be formed in the liver, spleen, lymph nodes, and bone marrow. In pathological conditions, such as old bleeding and hemolytic disease, the accumulation of hemoflavin in cell tissue is found in the cell tissue.
2. **Lipofuscin** is the undigested organelle debris residues in autophagy lysosomes, and it is brown, fine granular under microscopic. Its composition is a mixture of phospholipids and proteins. It comes from peroxidation, which is when free radicals catalyze unsaturated fatty acid peroxide with cell membrane phase structure. Typically, a small amount of lipofuscin may be seen in the epididymal epithelial cells, leydig cells, and cytoplasm of ganglion cells. In elderly patients and patients with nutrient depletion, a large number of lipofuscin can be seen in atrophic myocardial cells and around the hepatic nucleus. It is a sign that the cell has been damaged by free radical peroxidation. Therefore, it is known as consumptive pigment. When most cells contain lipofuscin, prominent organ atrophy can usually be seen there.
3. **Melanin**: It is a black-brown fine particle in the cytoplasm with melanin, which is produced from tyrosine by the polymerization of levodopa. Its formation is promoted by the pituitary ACTH (adrenocorticotropic hormone) and MSH (melanocytic stimulating hormone). In addition to melanocytes, melanin can also be clustered in the keratinocytes of the skin basal cells and in the macrophages of the dermis. In some chronic inflammation and pigmented nevus,

melanoma, and basal cell carcinoma, melanin can be locally increased. Addison's disease with low adrenocortical function can cause melanosis of the whole body and mucous membrane.

4. **Bilirubin**: It is the main pigment in the bile duct, which is the product of the aging and destruction of red blood cells in the blood. It is also derived from hemoglobin, but it does not contain iron. The pigment in the cytoplasm is rough, golden, and granular. When the blood bilirubin is increased, the patient has jaundice in the skin and mucous membrane.

1.3.7 Pathological Calcification

Pathological calcification is defined as solid calcium salt deposition in tissues besides the bone and teeth. It can be located within or outside the cell. The main components of calcium salts are calcium phosphate, calcium carbonate, a small amount of iron, magnesium, or other minerals.

1. **Types of Pathological Calcification** has two types:

Dystrophic calcification: It is called malnutrition calcification when calcium salts are deposited in necrotic, soon necrotic tissues or foreign bodies. The metabolism of calcium and phosphorus is normal in the body. It can be seen in tuberculosis, thrombosis, atherosclerotic plaques, heart valve disease, and scar tissue (Fig. 1.8), which may be related to increased local alkaline phosphatase.

Metastatic calcification: It is called metastatic calcification when calcium salts deposit in normal tissue due to disorders of calcium and phosphorus metabolism (hypercalcemia), mainly in hyperparathyroidism, excessive intake of vitamin D, renal failure and some bone tumors, often occurs in interstitial tissue of blood vessel, kidney, lung, and stomach.

Fig. 1.8 Dystrophic calcification of the arterial wall (HE, low power). Under low power lens, atherosclerosis occurs in the arterial wall, then secondary dystrophic calcification. Blue granulated calcium salt deposits

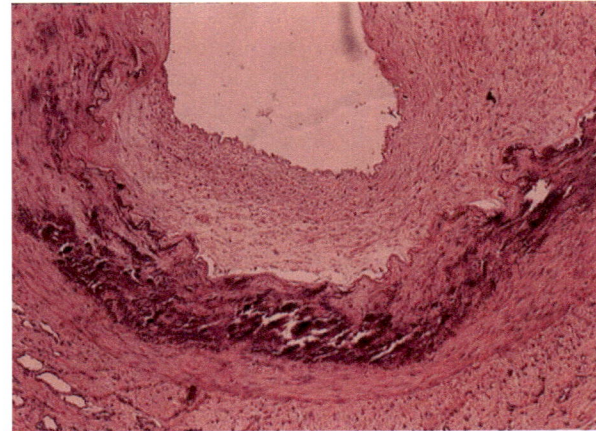

2. **Pathological Changes of Calcification**: The pathological calcification is blue, granular to flaky under the microscope and is fine particles or lumps grossly feel like sand and gravel. A large area of pathological calcification can cause deformation, sclerosis, and dysfunction of tissues and organs. Another form of pathological calcification is the formation of stones made up of calcium carbonate and cholesterol in the gallbladder, bladder, renal pelvis, ureter, and pancreas.

To sum up, the accumulation of different normal or abnormal substances in the cell or in the cytoplasm can cause various types of reversible damage.

1.4 Cell Death

When the cell has fatal metabolism, structure, and dysfunction, irreversible cell damage can occur, which is cell death. Cell death is the most important physiological and pathological change involving all cells. There are two main types of cell death. One is apoptosis and the other is necrosis. Apoptosis is mainly seen in the physiological death of cells, but it is also seen in some pathological processes. Necrosis is the main form of cell pathological death. Both of them have relatively different mechanisms, pathological significance, and morphological and biochemical characteristics. Which way a cell dies depends not only on types, intensity, duration, and the degree of ATP deletion in affected cells but also on the state of programmed expression of genes in cells.

1.4.1 Necrosis

Necrosis is local cell death in vivo characterized by changes in enzyme solubility. Necrosis can be directly induced by strong pathogenic factors, but most of them develop from reversible injury. Its elemental manifestations are cell swelling, the disintegration of cell organelles, and protein denaturation. Lysosomal enzymes are released from necrotic cells and the exudative neutrophils around them. It can promote the further development of necrosis and the dissolution of local parenchymal cells. So, necrosis often includes many cells at the same time.

1.4.1.1 Basic Pathological Changes of Necrosis

1. **The Changes in the Nucleus**: The changes in the nucleus are the primary morphological markers of cell necrosis; they mainly have three forms.
   Pyknosis: Nuclear chromatin DNA is thickening and crinkling, making the nuclear volume reduce, basophilic increase, suggesting that DNA transcriptional synthesis stops.

Karyorrhexis: With nucleus chromatin disintegrating and nuclear membrane rupture, nuclear fragmentation occurs, making nuclear matter dispersed in the cytoplasm, which can also be formed from nuclear condensation to fragmentation.

Karyolysis: Activation of non-specific DNA enzymes and nucleoprotein decompose nuclear DNA and nucleocapsid, decrease acidophilia of chromatin, and make killing nuclei disappear within 1 ~ 2 days.

The three processes are not always progressive. Its nucleus changes are not the same in different diseases and cells.

2. **Cytoplasmic Changes**: Besides the change of nucleus, the acidophilus of cytoplasmic in necrotic cells is enhanced due to the loss of ribosome, the increase of cytoplasmic denatured protein, and the decrease of glycogen granules. The primary ultrastructural morphology of cell irreversible injury is the formation of mitochondrial vacuoles, accumulation of amorphous calcium-dense deposits in mitochondria matrix, and acid hydrolase release by lysosome to dissolve cell components.

3. **Interstitial Changes**: The tolerance of interstitial cells is better than parenchymal cells, so it is later for interstitial cells to show lesions. After necrosis of interstitial cells, the extracellular stroma is also gradually disintegrating and liquefied, finally fused into a flaky, unstructured substance.

As necrosis cell membrane permeability increases. Intracellular lactate dehydrogenase, succinate dehydrogenase, creatine kinase and aspartate aminotransferase, alanine aminotransferase, amylase, and its isoenzyme are released into the blood, resulting in the corresponding reduction in enzyme activity in the cells and increase corresponding enzyme levels. Those can be used as a reference index for clinical diagnosis of some cell necrosis (such as liver, myocardium, and pancreas the reference index of necrosis). Changes in the activity of enzymes in the cell and in the plasma can be detected at the beginning of the necrosis, which is earlier than the changes in the ultrastructure. So, they are helpful to the early diagnosis of cell damage.

1.4.1.2 Types of Necrosis

Due to different positions of enzyme degradation or protein denaturation. The necrotic tissue may appear as distinct morphological changes, usually divided into three basic types: coagulation necrosis, liquefaction necrosis, and fibrinoid necrosis. In addition, caseous necrosis, fat necrosis, gangrene, and other particular types of necrosis can also occur. Generally speaking, after tissue necrosis, the color is pale, elasticity disappears, normal sensory and motor function is lost, blood vessels pulsation disappears, and no fresh blood outflow when cut, clinically called inactivating tissue. It should be excised in time.

1. **Coagulative Necrosis**: When the protein denaturation occurs, lysosomal enzyme hydrolysis is weak, and the necrotic region is gray-yellow, dry, and qualitative,

which is called coagulant necrosis. Coagulation necrosis is the most common, often seen in the heart, liver, kidney, spleen, and other essential organs, often caused by ischemic anoxia, bacterial toxins, and chemical corrosion agents The limit between the necrosis and the healthy tissue is often apparent. The feature under microscopic is the disappearance of the cell microstructure, and the outline of the tissue structure can still be preserved. There is congestion, bleeding, and inflammatory reactions around the necrotic area. The basic outline of tissue structure can be maintained for several days. It may be a continuous acidosis caused by necrosis, which denature the structural protein and enzyme protein of necrotic cells and retards the decomposition process of protein.

2. **Liquefaction Necrosis**: Due to the lack of coagulable protein in the necrotic tissue, the release of a large number of hydrolytic enzymes from the necrotic cells and infiltrated neutrophils, tissue is rich in water and phospholipid, the liquefaction is prone to occur after the necrosis of the cell and tissue, which is called liquefaction necrosis. It can be found in abscesses caused by bacteria or certain fungal infections, softening of the brain caused by ischemia and anoxia, and dissolved necrosis developed from cell edema. The characteristics under a microscope are that the dead cells are completely digested, and the local tissues are rapidly dissolved.

3. **Fibrinoid Necrosis**: It is known as fibrinoid degeneration and is a common form of necrosis in connective tissue and small vascular walls. Filamentous, granular, or small lump-like unstructured material forms in the site of the lesion. Because of its similar dyeing properties with cellulose, it is known as cellulosic necrosis. It can be seen in some allergic diseases such as rheumatism, polyarteritis nodosa, crescentic glomerulonephritis, and accelerated hypertension, the small blood vessels at the bottom of gastric ulcer. The mechanism is related to collagen fiber swelling and disintegration caused by antigen-antibody complex, immunoglobulin deposition of connective tissue, or plasma fibrin exudation and degeneration.

4. **Caseous Necrosis**: In the case of tuberculosis, the necrosis is yellow, like a cheese, due to too much lipid there, which is known as caseous necrosis. It is an unstructured granular red mass under a microscope. No necrotic part of the original organizational structure blurs, even nuclear debris. It is a particular type of complete necrosis. Caseous necrosis is not easy to dissolve and is not easily absorbed because the substance inhibits the activity of hydrolase in the necrotic foci. Caseous necrosis is also found in some infarcts, tumors, tuberculous leprosy, and so on.

5. **Fat Necrosis**: Cells release trypsin decomposing fat when acute pancreatitis and fat cells can be decomposed when breast injury, both of which can cause enzymolysis and traumatic fat necrosis, also included in liquefaction necrosis. After fat necrosis, the released fatty acids are combined with calcium ions to form a gray, white calcium soap that is visible grossly.

6. **Gangrene**: Gangrene is defined as extensive tissue necrosis and infection secondary. It can be divided into dry gangrene, moist gangrene, and gas gangrene, in which the two formers develop from ischemic necrosis caused by disturbance of blood circulation.

Fig. 1.9 Dry gangrene of food. When dry gangrene involves toes, the toes are black, dry, and have clear boundary with the surrounding tissue, is the ischemic necrosis caused by thromboangiitis obliterans. The toe has come off

Dry gangrene is often seen in the terminal limb, whose arteries are obstructed but the veins are not. It is dry, crinkle, and black due to the loss of water (because Fe^{2+} is combined with H_2S in the corrupt tissue, forming the color of iron sulfide). It forms an apparent limit with normal tissue, and the corrupt changes are slight (Fig. 1.9).

Moist gangrene: It often occurs in the viscera interlinked with the outside, such as the lung, intestine, uterus, appendix, and gallbladder, the same as a limb. There is a lot of water in the necrosis tissue, and bacteria are easy to reproduce, so the swelling is blue-green, and the boundaries of the surrounding organization are not clear.

Gas gangrene: It also belongs to moist gangrene and is the open injury to deep muscle, combined with infection of anaerobes such as bacilli perfringens. In addition to necrosis, gas can be produced, which makes necrosis tissue feel like it is twisting hair.

Moist and gas gangrene are often accompanied by system poisoning symptoms. In the type of necrosis, most dry gangrene belongs to coagulative necrosis, and moist gangrene can be a mix of coagulative necrosis or liquefactive necrosis.

1.4.1.3 Outcomes of Necrosis

1. **Dissolution and Absorption**: Necrotic cells and neutrophils around release hydrolases, making necrotic tissue dissolved, liquefied, and absorbed by lymphatic vessels or blood vessels. The fragments that cannot be absorbed are swallowed up by macrophages. When the necrotic liquefaction is extensive, the capsule can be formed. After the necrotic cells dissolve, an acute inflammatory response can be caused in the tissue around them.

2. **Separation and Discharge**: When necrotic foci are more extensive and not completely absorbed, the necrotic substance of the epidermis can be separated and form tissue defects. The shallow is called erosion, and the deep is called ulcer. A deep blind tube only opens on the surface of the skin and mucosa after tissue necrosis, which is called a sinus. A channel-like defect that connects two visceral organs or the organ to the body surface is known as a fistula. After liquefaction, necrosis in the lung, kidney, and other organs can be discharged through bronchia, ureter, and other natural pipelines, the residual cavity called the hole (cavity).

3. **Organization and Encapsulation**: The process in which new granulation tissue takes the place of necrotic tissue, thrombus, pus, and foreign body is called organization. If the necrotic tissue is too large, granulation tissue can hardly grow into or absorb into the central part. It will be wrapped by the granulation tissue around it, which is called encapsulation. The encapsulation of granulation tissue formed by organization and encapsulation can eventually form a fibrous scar.

Calcification: If necrotic cells and cell debris are not promptly removed, the calcium salt and other minerals are easily attracted and deposited, which can cause dystrophic calcification.

1.4.1.4 The Effect of Necrosis

The effect of necrosis on the body is related to the following factors:

1. The physiological importance of necrotic cells, such as the severe consequences of necrosis in the heart and brain tissue.
2. The number of necrotic cells, such as extensive necrosis of the liver cells, can cause the body to die.
3. The regeneration of similar cells around necrotic cells. For example, the liver and epidermis are easy to regenerate. The structure and function of necrotic tissue are easy to recover. However, neurons and cardiomyocytes cannot regenerate after necrosis.
4. The reserve and compensatory capacity of necrotic organs. For example, the kidney, lung, and other paired organs have a substantial reserve and metabolic capacity.

1.4.2 Apoptosis

Apoptosis is a manifestation of programmed cell death in a single cell of local tissue in vivo. It is a way of cell death induced by internal and external factors triggering the cell death process. It is different from necrosis in morphological and biochemical characteristics. Apoptosis plays an irreplaceable role in the development of biological embryogenesis, mature cells, new and old alternation, hormone-dependent physiological degradation, atrophy, and aging, as well as autoimmune diseases and tumor progression. It is not only a product of cell damage.

1. **Morphological and Biochemical Characteristics of Apoptosis**: Morphological of apoptosis (Fig. 1.10) can be: (a) Cell shrunken: The cytoplasm is dense, the water is reduced, the cytoplasm is highly eosinophilic, and the single apoptotic cells are separated from the surrounding cells. (b) Chromatins condensation, nuclear chromatin concentrate to dense clumps (pyknosis), or arrange on the inner surface of the karyotheca assembly (nuclear chromatin edge accumulation), then the nucleus split into pieces (karyorrhexis). (c) The formation of the apoptotic body: the cell membrane invaginates, cytoplasmic buds, and then falls down, forming nuclear debris and apoptotic body wrapped in the membrane of organelles. The apoptotic body is an important morphological marker of apoptosis, and it can be phagocytized and degraded by the macrophage. (d) Integrity of plasma membrane: because of the complete plasma membrane in the apoptotic cells, the recognition with other cells is prevented. Therefore, they neither cause inflammatory reactions nor induce proliferation and repair of surrounding cells. The eosinophilic body in the hepatocytes of viral hepatitis is the expression of hepatocyte apoptosis.

Biochemical characteristics of apoptosis are the activation of aspartic protease containing cysteine, Ca^{2+}/Mg^{2+}-dependent endonuclease, and calpain. Caspases in normal cells are zymogen; when activated, they can crack many important cell proteins, disrupt the cytoskeleton and nuclear skeleton, and then activate restriction endonuclease. DNA degradation fragments of 180–200 dp can be seen in an early stage, and characteristic DNA the ladder can be seen in

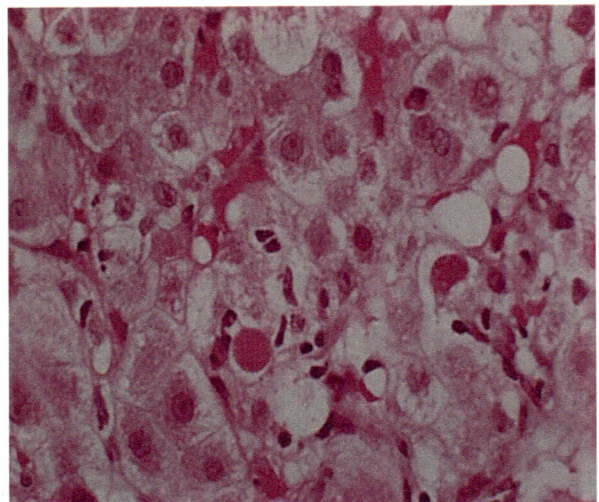

Fig. 1.10 Apoptosis of liver cells (HE, low power). Under high power lens, the apoptosis of single hepatocyte can be seen in the center vision, separated from the adjacent cells, the cytoplasmic eosinophilic is obviously enhanced, the cell is solid, and the apoptosis body is formed

agar gel electrophoresis. The apoptotic protease and endonuclease are the main executors of the program.

2. **Mechanism of Apoptosis**: Apoptosis can be divided into three stages: sign transduction, central control, and structural change. The two former stages are the starting stages, and the last is the execution stage. The signal transduction combines with the related protein Fas (CD95) and Fas ligand (Fas-L) by the exogenous (death receptor initiation) pathway, TNF-α receptor on the cell surface, introducing the apoptotic signals into the cell. Central regulation is activated by mitochondrial permeability and apoptotic molecules such as cytochrome C cytoplasm release; it is through endogenous pathways (mitochondrial pathways). On the basis of the first two, the apoptotic protease activates the cascade reaction, and the apoptotic body appears to be morphologically changed.

The factors regulating apoptosis include inhibitory factors and inducing factors. The former includes growth factors, cell-matrix, sex steroid hormone, some viral proteins, and so on. The latter is growth factor deficiency, glucocorticoid, free radicals, and ionizing radiation. There are dozens of genes involved in the apoptosis process. Among them, Bad\Bax\Bak\p53 and other genes have the effect of promoting apoptosis. Bcl-2\Bcl-xL\Bcl-AL and other genes have the effect of inhibiting apoptosis. C-myc and other genes may have two-way regulation; it induces proliferation when the growth factor is enough and induces apoptosis when the growth factor is lacking.

It is necessary to point out that cell death is similar to apoptosis in that it induces cell death, morphological features, and biochemical characteristics. For example, pyknosis, karyorrhexis, and nuclear chromatin edge accumulation can also be seen in the process of apoptosis in addition to cell death; the ladder feature of agar gel electrophoresis when apoptosis can sometimes be seen in necrosis cells.

In addition, cell death can also be caused by cell autophagy. Cell rough endoplasmic reticulum membrane without ribosomal or lysosomal membrane protrudes, engulfing, encapsulates cells, forming autophagy bodies (autophagic vesicles), and then merges with lysosomes to form autophagic lysosomes, so as to degrade the contents wrapped; this process is called autophagy. Under physiological conditions, cells can eliminate digestive, damaged, denatured, senescent, and dysfunctional cells, organelles, and various biological macromolecules through autophagy, achieve the recycling and utilization of cell materials, and provide raw materials for cell reconstruction and regeneration. In pathological conditions, autophagy can resist the invasion of the pathogen and protect the cells from the damage of poison. Too much or too less autophagy can cause cell death and play an important role in the development of immunity, infection, cardiovascular diseases, neurodegenerative diseases, and tumors. Autophagy has similar stimulant factors and regulates proteins with apoptosis, but a different evoked threshold. Autophagy can cause cell death by inducing apoptosis.

1.5 Cellular Aging

Cell aging is a degenerative change that cell occurs with the growth of the body age. It is the basis of the aging in individual organisms. Biological individuals and their cells must undergo growth, development, aging, and death, and aging is a necessity of life development. It should be said that the aging process begins at birth in any cell.

1.5.1 Characters of Cellular Aging

Cellular aging has some characteristics as follows:

1. Universality: Aging can appear to varying degrees in all cells, tissues, organs, and bodies.
2. Progressivity and irreversibility: As time goes on, aging continues to develop.
3. Endogenous: It is not the direct effect of external causes such as trauma but the decline in the cell's inherent genetic decision.
4. Harmful: When aging, cellular metabolism, adaptation and compensation, and other functions are low, and there is a lack of recovery ability, which leads to the emergence of geriatric diseases. The morbidity and mortality of other diseases are also increasing.

1.5.2 Morphology of Cellular Aging

Synthesis of structural proteins, enzyme proteins, and receptor proteins when cellular aging decreases and the ability to absorb nutrition and repair chromosomal damage decreases. The morphology is characterized by cell volume reduction, water loss, cell and nucleus deformations, mitochondria and Golgi bodies decrease, distortions, or vacuole-shape and cytoplasmic pigmentation (lipofuscin) can be seen. Thus, the weight of organs is reduced, interstitial hyperplasia and hardening can be seen, functional metabolism decreases, and reserve function is insufficient.

1.5.3 Mechanism of Cellular Aging

It is not clear, mainly has two types: genetic programming theory and error accumulation theory.

1.5.3.1 Genetic Programming Theory

According to the genetic programming theory, the aging of cells is determined by the genetic factors of the body; that is, the growth, development, maturation, and aging of cells are completed by a particular gene in the cell gene bank expressed successively by pre-arranged procedures. The final death is the result of exhaustion of genetic information. For example, human fibroblasts in vitro stop splitting after about 50 divisions. The phenomenon that monozygotic twins live and die together supports the genetic programming theory. Some studies have shown that the mechanism that controls a number of cell divisions is closely related to the telomere structure at the end of the chromosomes.

Telomere is a unique structure of eukaryotic cell chromosome ends. It consists of a repeated sequence of non-transcribed short segments of DNA and some binding proteins. Telomeres have the function of avoiding the fusion and degeneration of the chromosome ends. It plays a vital role in chromosome stability, replication, protection, and control of cell growth and longevity and is closely related to apoptosis and cell immortalization.

The telomere at the ends of the chromosomes will gradually shorten with each cell division. This is because the DNA polymerase that copies DNA cannot replicate the DNA at the end of the linear chromosome. Usually, the telomere shortens about 50 ~ 200 nucleotides once the cells divide until the cells are senescent and stop division, so the obviously shortened telomere is the signal of cell aging.

Telomerase is an inverse transcriptase that prolongs the shortened telomere and is a ribonucleoprotein complex (RNP) composed of RNA and protein. It uses self-RNA as a template to synthesize the telomere fragments and connect them to the telomere ends of the chromosomes, which restores and stabilizes the telomere length at the end of the chromosome. Most of the mature somatic cells do not exhibit telomerase activity. In germ cells and some stem cells that need long-term replication, the telomere shortened after cell division can be recovered by the activity of telomerase in the cell and maintained at a certain length. A more significant finding is in immortalized cancer cells, telomerase also shows apparent activity, which brings new hope for tumor therapy research targeting telomerase activity.

Telomere and telomerase theory can explain the aging process of most differentiated mature cells. However, there may be other aging mechanisms for those neurons and cardiomyocytes with a low ability of division and proliferation. In addition, in the lower organisms, the degradation gene clk-1 and the mechanical sensing gene DAF-2 can change the growth rate and time of the cell development process and also play the role of genetic control of aging, but their role in mammalian animals needs to be confirmed.

1.5.3.2 Error Accumulation Theory

In addition to the procedural mechanism of cell heredity, the length of cell life is also dependent on the injury caused by metabolism and the balance between the molecular responses after injury. During cell division, damage from free radicals and other harmful substances can induce lipid peroxidation and damage the mobility, permeability, and integrity of mitochondrion and so on. The DNA breakage mutation causes the mistake in its repair and replication process. When DNA is duplicated, the p53 gene with the function of cell cycle G1 detection and correction is activated, and its protein products induce the transcription enhancement of cyclin-dependent kinase inhibitor (CDKI), p21, and p16. The binding of p21 and p16 to cyclin-dependent kinase (CDK) and cyclin complexes can inhibit CDK activity. The increase of p16 also activates the dephosphorylation of the retinoblastoma gene (Rb gene), which further impedes cell division from multiple links. There is evidence that the expression of P16 and other genes in the stem cells increases with age, and the stem cells themselves gradually lose their self-renewal capacity. At the same time, with the accumulation of errors and abnormal proteins, the function of the original protein peptides and enzymes disappears, leading to the aging of the cells eventually.

In addition, as the age of individuals increases, lymphocytes T and B decrease, NK cell activity decreases, cytokine activity decreases, and immune recognition ability is disordered. On the one hand, foreign bodies, such as pathogens and tumor cells, cannot be eliminated, and on the other hand, it leads to autoimmune disease. Neuroendocrine disorders are also one of the important characteristics of aging. The hypothalamic-pituitary-adrenal system plays an important role in aging. When aging, neurons also lose to varying degrees. The release of catecholamine and other neurotransmitters is reduced, the production of sex hormones is reduced, and the function of the hormone receptors decreases.

To sum up, the mechanism of cell aging includes both the role of genetic programming factors and the effect of harmful factors accumulation in the intracellular and external environment. Cell aging can be carried out at the speed of genetic regulation. The natural lifespan (natural aging) can be achieved. If the harmful factors impede the metabolic function of the cell, the aging process will be accelerated. Therefore, it can be said that in the decisive background of genetic arrangement, cell metabolic disorders are the factors contributing to the aging of cells.

Chapter 2
Tissue Repair

Liqin Ma

Contents

Objectives

1. To master the major types and characters of repair
2. To master the conception and characters of regeneration
3. To master the characters of complete regeneration and fibrous repair
4. To master the conception and characters of granulation tissue
5. To master the major types and characters of healing
6. To comprehend the characters of scar structure
7. To comprehend the characters of fracture healing

Key Concepts

1. For regeneration or repair: its conception, types, pathological changes

L. Ma (✉)
Medical College, Zhejiang University, Hangzhou, China
e-mail: maliqin198@zju.edu.cn

© Zhengzhou University Press 2024
K. Chen et al. (eds.), *Textbook of Pathologic Anatomy*,
https://doi.org/10.1007/978-981-99-8445-9_2

2. For healing: its conception, types, pathological changes, course
3. For granulation tissue: its conception, pathological changes

Introduction
Repair refers to the restoration of tissue/cell[2] architecture and function after any injury or disease.

It contains two types of reactions: complete regeneration (means the injured tissues are able to replace the damaged components and essentially return to a normal state) and fibrous repair or scar formation (means the injured tissue is incapable of complete restitution, or the supporting structure of the tissue is severely damaged, and repair occurs by laying down of connective tissue). After many common types of injury, both types contribute in varying degrees to the ultimate repair. Repair involves the proliferation of various cells and close interactions between cells and extracellular matrix (ECM). The mechanism of repairing depends on the type of inflammation, the extent of tissue necrosis, the types of cells involved, and the regenerative ability of damaged parenchymal cells.

2.1 Regeneration

Regeneration can be physiological regeneration (e.g., in gut epithelium), and it can also be pathological regeneration (e.g., cell and tissue injury, the replacement of lost parenchymal cells by division of adjacent surviving parenchymal cells to restore injured tissue).

2.1.1 Cell Cycle and the Regeneration Capacity of Different Types of Cell

The key processes in the proliferation of cells are DNA replication and mitosis. The cell cycle consists of the presynthetic growth phase 1 (G1), the DNA synthesis phase (S), the premitotic growth phase 2 (G2), and the mitotic phase (M). Any stimulus that initiates cell proliferation, such as exposure to growth factors. Checkpoint controls prevent DNA replication or mitosis of damaged cells and either transiently stop the cell cycle to allow for DNA repair or eliminate irreversibly damaged cells by apoptosis. Once cells enter the S phase, the DNA is replicated, and the cell progresses through G2 and mitosis. The ability of cells to repair themselves is critically influenced by their intrinsic proliferative capacity. Based on this, the cells are divided into three groups.

2.1.1.1   Labile Cell (Continuously Dividing Cell)

These cells are continuously being lost and replaced by maturation from stem cells and by proliferation of mature cells. Labile cells contain hematopoietic cells in the bone marrow and majority of surface epithelia, such as the stratified squamous cells of skin, oral cavity, vagina, and cervix; the cuboidal epithelia of ducts draining exocrine organs; the columnar epithelium of the gastrointestinal tract, uterus, and fallopian tubes; and the transitional epithelium of the urinary tract. These tissues can readily regenerate after injury as long as the stem cells is preserved.

2.1.1.2   Stable Cell (Quiescent Cell)

These cells are quiescent and have only minimal replicative activity in their physiological state. However, they are capable of proliferating in response to injury or/and loss of tissue mass. Stable cells constitute the parenchyma of most solid tissues, such as liver, kidney, pancreas, and so on. They also include endothelial cells and fibroblasts, and the proliferation of these cells is particularly important in wound healing.

2.1.1.3   Permanent Cell (Nondividing Cell)

These cells are considered to be terminally differentiated and non-proliferative in postnatal life. The majority of neurons and cardiac muscle cells belong to this category. Thus, injury to brain or heart is irreversible and results in a scar, because neurons and cardiac myocytes do not divide. Skeletal muscle is usually classified as a permanent tissue, but satellite cells attached to their endomysial sheath provide some regenerative capacity for this tissue. In permanent tissues, repair is typically dominated by scar formation.

2.1.1.4   Stem Cells

In most continuously dividing tissues, the mature cells are terminally differentiated and short-lived. As mature cells die, the tissue is replenished by differentiation of cells generated from stem cells. Stem cells are characterized by two important properties: self-renewal capacity and asymmetric replication. Asymmetric replication of stem cell means that after each cell division, some progeny cells enter the differentiation pathway, while others remain undifferentiated, retaining their self-renewal capacity. Stem cell with the capacity to generate multiple cell lineages (pluripotent stem cells) can be isolated from embryos and is called embryonic stem (ES) cell. Stem cells are normally present in proliferative tissue and generate cell lineages specific for the tissue. However, it is now recognized that stem cell with the capacity to generate multiple lineage are present in the bone marrow and several other tissues of adult individual. These cells are called tissue stem cells or adult stem cells.

2.1.2   Tissue Regeneration

2.1.2.1   Epithelial Regeneration

Cell renewal occurs continuously in labile cells, such as gut epithelium and skin. Damage to epithelia can be corrected by the proliferation and differentiation of stem cells. Tissue regeneration can occur in some parenchymal organs with stable cell populations.

Pancreas, adrenal, thyroid, and lung tissues have some regenerative capacity. Much more dramatic, however, is the regenerative response of liver that occurs after surgical removal of hepatic tissue. As much as 40–60% of liver may be removed in a procedure called living-donor transplantation, in which a portion of the liver is resected from a normal individual or is transplanted into a recipient with end-stage liver's disease, or after partial hepatectomies performed for tumor removal. In all of these situations, the tissue resection triggers proliferative response of the remaining hepatocytes (which are normally quiescent) and subsequent replication of hepatic nonparenchymal cells.

2.1.2.2   Regeneration of Bone and Cartilage

Injury of bone tissue is followed by rapid regeneration. In mature cartilage, the injury defects are likely to be filled with fibrous tissues, or the lesion may precipitate a degeneration of adjacent cartilage. Fibroblasts from the sheath of the injured tendon and other sources proliferate, become active, and lay down orderly collagen fibers, which can restore most of the original strength of the tendon. Injury to permanent cells is always followed by connective tissues/scar formation.

2.1.2.3   Angiogenesis

Angiogenesis refers to the healing process of blood vessel at the injury sites. It begins with the degradation of the basement membrane by proteases secreted by activated endothelial cells that migrate and proliferate. This leads to the formation of endothelial cell sprouts and vascular loops, and the capillary tubes develop with formation of tight junctions and deposition of a new basement membrane (Fig. 2.1). But permeability of the new capillary wall is increasing undergoing the incomplete basement membrane.

Fig. 2.1 Model of angiogenesis. ① degradation of the basement membrane. ② migration and proliferation of endothelial cells; ③ formation of endothelial cell sprouts and vascular loops. ④ increasing permeability

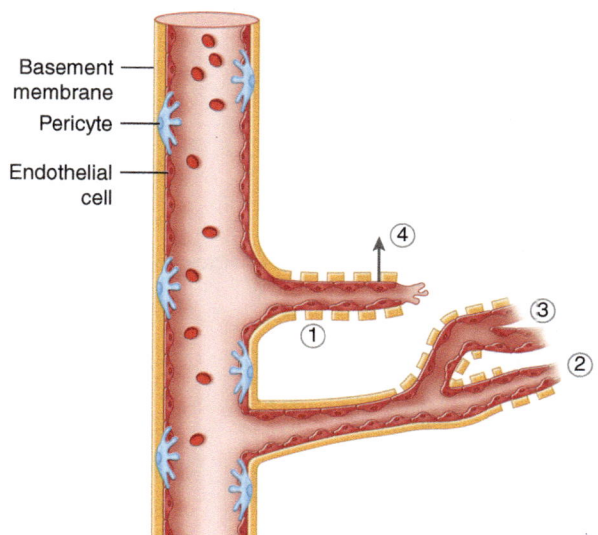

2.1.2.4 Regeneration of Peripheral Nerves

If the peripheral nerve has been transected, the axon distal to the injury site rapidly degenerates and eventually disappears. The myelin sheath and axon of the remaining intact nerve degenerates back to the next node of Ranvier. Then macrophages enter this area to remove the myelin and axonal debris. Schwan cells line up in the basement membrane tube and synthesize growth factors, which induce axonal sprouts formed at the terminal end of the proximal segment of severed axon. The basement membrane tubes provide pathways for the regeneration axons to follow to muscles and skin (Fig. 2.2). Axonal regeneration may be accompanied by recovery of function in the denervated area. If there has been severe trauma to nerve and disruption of its fascicular architecture or long distance between two ends of nerve disconnection, a fibrous scar can form and obstruct regenerated axons.

2.1.3 Influence Factors of Regeneration

2.1.3.1 Growth Factors

Most growth factors have pleiotropic effects; they stimulate the cellular proliferation, migration, differentiation, and contractility and enhance the synthesis of specialized proteins (such as collagen in fibroblast). A growth factor may act on a specific cell type or multiple cell types. They induce cell proliferation by binding to specific receptors and affecting the expression of genes, in which products typically

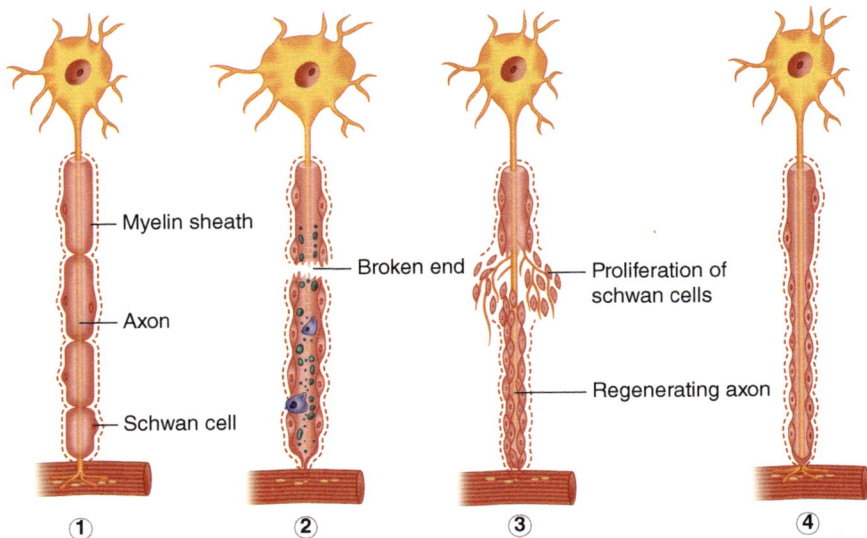

Fig. 2.2 Model of regeneration of peripheral nerves. ① Normal nerve fiber. ② Broken nerve fiber. ③ Regeneration of nerve fiber. ④ After regeneration

have several functions; they prevent apoptosis and enhance the synthesis of cellular proteins in preparation for mitosis. A major activity of growth factor can stimulate the function of growth control genes, many of which are called proto-oncogenes because mutations in them lead to unrestrained cell proliferation, characteristic of cancer (oncogenesis). Some growth factors stimulate the proliferation of some cells and inhibit cycling of other cells. In fact, growth factor can have opposite effects on the same cells depending on their concentration. There is a list of mainly growth factors and their functions (Table 2.1).

2.1.3.2 Cell Surface Receptors

Receptor proteins are generally located on the cell surface, but the ligands must be sufficiently hydrophobic to enter the cells (e.g., vitamin D, or steroid and thyroid hormone). The binding of ligand to cell surface receptor leads to a cascade of secondary intracellular events that culminate in transcription factor activation or repression, leading to cellular responses. These are usually transmembrane molecules with an extracellular ligand-binding domain; ligand binding causes the stable dimerization with subsequent phosphorylation of the receptor subunits.

Table 2.1 Growth factors involved in regeneration and wound healing

Cytokine	Source	Functions
Epidermal growth factor (EGF)	Activated macrophages, salivary glands, keratinocytes	Mitogenic for keratinocytes and fibroblasts; stimulates keratinocyte migration and granulation tissue formation
Transforming growth factor a (TGF-α	Activated macrophages, T lymphocytes keratinocytes	Similar to EGF; stimulates replication of hepatocytes and many epithelial cells
Transforming growth factor β (TGF-β)	Macrophages, endothelial cells, keratinocytes, SMCs, fibroblasts platelets, T lymphocytes	Chemotactic for PMNs, macrophages, lymphocytes, fibroblasts, and SMCs; stimulates angiogenesis, and fibroplasia; inhibits production of MMPs and keratinocyte proliferation; regulates integrin expression and other cytokines
Vascular endothelial cell growth factor (VEGF)	Mesenchymal cells	Increases vascular permeability; mitogenic for endothelial cells
Platelet-derived growth factor (PDGF)	Platelets, macrophages, endothelial cells, keratinocytes, fibroblasts, endothelial cells, SMCs	Similar to TGF-β; mitogenic, for and SMCs; stimulates production of MMPs, fibronectin, and HA; stimulates angiogenesis and wound remodeling; regulates integrin expression
Keratinocyte growth factor (KGF)	Fibroblasts	Stimulates keratinocyte migration, proliferation, and differentiation

SMCs smooth muscle cells; *PMNs* polymorph nuclear neutrophils; *MMPs* matrix Metalloproteinases; *HA* hyaluronic acid

2.1.3.3 Extracellular Matrix (ECM)

Tissue repair depends not only on growth factor activity but also on interactions between cells and ECM components. By supplying a substratum for cell adhesion and serving as a reservoir for growth factors, ECM regulates the proliferation, movement, and differentiation of the cells living within it. Synthesis and degradation of ECM accompanies morphogenesis, wound healing, chronic fibrotic processes, tumor invasion, and metastasis. Its various functions include: (1) mechanical support for cells anchorage, cell migration, and maintenance of cells polarity; (2) control of cell growth; (3) maintenance of cells differentiation; (4) scaffolding for tissue renewal; (5) establishment of tissue microenvironments, in which basement membrane acts as boundary between epithelium and underlying connective tissue and also forms part of the filtration apparatus in the kidney; and (6) storage and presentation of regulatory molecules.

Collagen: The collagens are composed of three separate polypeptide chains braided into a ropelike triple helix. The collagen proteins are rich in hydroxyproline and hydroxylysine. About 30 collagen types have been identified, some of which are unique to specific cells and tissues. The fibrillar collagens form major proportion of the connective tissue in healing wounds and particularly in scars. The tensile strength of fibrillar collagens derive from their cross-linking, which is the result of covalent bonds catalyzed by the enzyme lysyl-oxidase. Genetic defects in these collagens cause diseases such as osteogenesis imperfecta and Ehlers-Danlos syndrome. Other collagens are nonfibrillar and may form basement membrane (type 4), or be components of other structures such as intervertebral discs (type 9) or dermal-epidermal junctions (type 7).

Elastin: Although tensile strength is derived from fibrillar collagens, the ability of tissues to recoil and return to baseline structure after physical stress is conferred by elastic tissue. This is especially important in the walls of large vessels, as well as in the uterus, skin, and ligaments.

Elastins require a glycine in every third position, but they differ from collagen by having fewer cross-links. The fibrillin meshwork serves as a scaffold for the deposition of elastin and assembly of elastic fibers; defect in fibrillin synthesis leads to skeletal abnormalities and weakened aortic walls (such as Marfan syndrome).

Proteoglycans and Hyaluronan: Proteoglycans consist of long polysaccharides called glycosaminoglycans linked to the protein backbone. Hyaluronan, a huge molecule composed of many disaccharide repeats without protein core, is also an important constituent of ECM. Because of its ability to bind water, it forms viscous, gelatin-like matrix. Besides providing compressibility to the tissues, proteoglycans also serve as reservoirs for growth factors secreted into ECM. Proteoglycans can also be integral cell membrane proteins and have the roles in cell proliferation, migration, and adhesion.

Adhesive Glycoproteins and Adhesion Receptors: They are structurally diverse molecules involved in cell-to-cell adhesion, the linkage between cells and ECM, and binding between ECM components. The adhesive glycoproteins include fibronectin (major component of interstitial ECM) and laminin (major constituent of basement membrane). The adhesion receptors are also known as cell adhesion molecules (CAMs), grouped into four families: immunoglobulins, cadherins, selectins, and integrins.

Fibronectin is a large (450 kD) disulfide-linked heterodimer synthesized by a variety of cells.

Fibronectin messenger RNA (mRNA) has two splice forms, which generate tissue and plasma fibronectin. Fibronectins have specific domains that bind to a wide spectrum of ECM components and can also attach to cell integrins via a tripeptide arginine-glycine-aspartic acid motif. Tissue fibronectin forms fibrillar aggregates at wound healing sites; plasma fibronectin binds to fibrin to form the provisional blood clot of a wound.

Laminin is the most abundant glycoprotein in basement membrane. It is a 820 kD cross-shaped heterotrimer that connects cells to underlying ECM components such as type 4 collagen and heparin sulfate. It can also modulate the cell proliferation, differentiation, and motility.

Integrins are a family of transmembrane heterodimeric glycoproteins composed of chains that are the main cellular receptors for ECM components. They bind to many ECM components through RGD motifs, initiating signaling cascades that can affect cell locomotion, proliferation, and differentiation. Their intracellular domains link to actin filaments at focal adhesion complexes, through adaptor protein such as talin and vinculin.

2.2   Fibrous Repair

If tissue injury is severe or chronic and results in damage to parenchymal cells and epithelia as well as the stromal framework, or if nondividing cells are injured, repair cannot be accomplished by regeneration alone. Under these conditions, repair occurs by replacement of the nonregenerated cells with connective tissue, or by a combination from regeneration of some cells or scar formation.

2.2.1   Granulation Tissue

Repair begins within 24 h of injury by emigration of fibroblasts and the induction of fibroblast and endothelial cells proliferation. By 3–5 days, a specialized type of tissue that is characteristic healing, called granulation tissue, is apparent. The granulation tissue derives from the pink, soft, granular gross appearance, such as that seen beneath the scab of skin wound. Its histological appearance is characterized by proliferation of fibroblasts and new thin-walled, delicate capillaries and inflammatory cells such as the macrophages, neutrophils, and lymphocytes (Fig. 2.3). Macrophages can secrete PDGF, FGF, TGF, TNF, and IL-1, with the PDGF released by platelets when blood clotting on the surface of wound can stimulate hyperplasia of fibroblasts and capillary. Macrophages and neutrophils can engulf bacteria and tissue pieces. Granulation tissue has some important roles in the process of tissue repair:

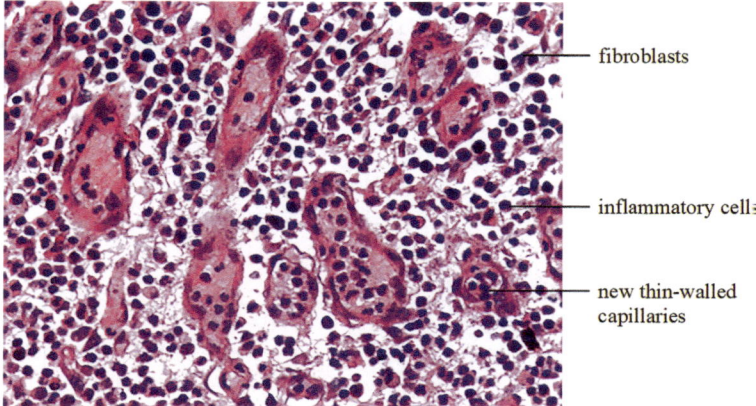

fibroblasts

inflammatory cells

new thin-walled
capillaries

Fig. 2.3 Granulation tissue (400×, HE stain)

(1) anti-infection and wound protection; (2) filling the wound and other defective tissues; and (3) organization or wrapping the necrosis, thrombosis, inflammatory exudation, and other foreign tissues.

2.2.2 Scar Tissue

Granulation tissue then progressively accumulates connective tissue matrix, eventually resulting in the formation of scar tissue (Fig. 2.4). So scar tissue refers to the mature fibrous connective tissue redeveloped by granulation tissue. Repair by connective tissue deposition consists of four sequential processes: formation of new blood vessels (angiogenesis), migration and proliferation of fibroblasts, deposition of ECM (scar formation), and maturation and reorganization of the fibrous tissue (remodeling). So the histological appearance of scar tissue is characterized by proliferation of a mass of collagen fiber and a little of fibrocytes and blood vessels. The gross appearance of scar tissue is characterized by local contraction, white or gray, translucent, strong, tough, and inelastic.

Scar tissue has some beneficial roles in the process of tissue repair: (1) filling the wound and other defective tissues, keeping the tissues integrity, and (2) keeping the tissues and organs strong. But the scar tissue has also some harmful effects: (1) Scar contraction, because of tough, inelastic, deformation caused by contraction, scar tissue can cause the organs dysfunction, especially for the joints and some important organs. (2) Cicatricial adhesion, it often occurs between organs or organ and coelomic cavity wall and then affects their function, such as extensive fibrosis, hyaline, and sclerosis. (3) Excessive hyperplasia of scar tissue, also called "hypertrophic scar." If this kind of hypertrophic scar protrude even sprawl from the skin, known as "keloid." Generally, it is considered related to skin tension and physique.

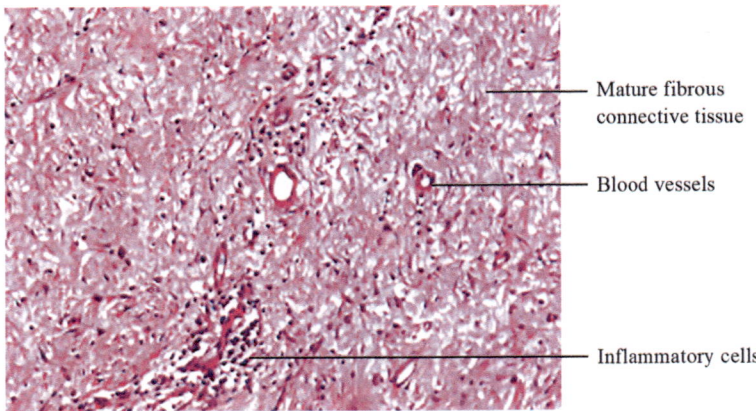

Mature fibrous connective tissue

Blood vessels

Inflammatory cells

Fig. 2.4 Scar tissue (100×, HE stain)

2.2.3 Formation Course of Granulation Tissue and Scar

2.2.3.1 Angiogenesis

Blood vessels are assembled by two processes: vasculogenesis, in which the primitive vascular network is assembled from angioblasts (endothelial cell precursors) during embryonic development and angiogenesis, or neovascularization, in which preexisting vessels send out capillary, sprouts to produce new vessels. Angiogenesis is a critical process in healing tissue at sites of injury, in the development of collateral circulations at sites of ischemia, in allowing tumors to increase in size beyond the constraints of their original blood supply. It has recently been found that endothelial precursor cells may migrate from bone marrow to areas of injury and participate in angiogenesis at these sites.

The main steps that occur in angiogenesis from preexisting vessels are listed below.

1. Vasodilation in response to nitric oxide and increased permeability of the preexisting vessel induced by vascular endothelial growth factor (VEGF)
2. Migration of endothelial cells toward the area of tissue injury
3. Proliferation of endothelial cells just behind the leading front of migrating cells
4. Inhibition of endothelial cells proliferation and remodeling into capillary tubes
5. Recruitment of periendothelial cells (pericytes for small capillaries and smooth muscle cells for larger vessels) to form the mature vessel

New vessels formed during angiogenesis are leaky because of incompletely formed interendothelial junctions and because VEGF increases vessel permeability. This leakiness explains why granulation tissue is often edematous and accounts in part for the edema that may persist in healing wounds long after acute inflammatory response has resolved. Structural ECM proteins participate in the process of vessel sprouting in the angiogenesis, largely through interactions with integrin receptors in endothelial cells. Nonstructural ECM proteins contribute to angiogenesis by destabilizing cell-ECM interactions to facilitate continued cell migration or degrade the ECM to permit remodeling and in growth of vessels.

Growth Factors and Receptors Involved in Angiogenesis:
Several factors induce angiogenesis, but the most important are VEGF and basic fibroblast growth factor (FGF). In angiogenesis, originating from preexisting local vessels, VEGF stimulates both proliferation and motility of endothelial cells, thus initiating the process of capillary sprouting. VEGFs are dimeric glycoproteins with many isoforms and with different properties, such as VEGFs bind to a family of receptors (VEGFR-1, VEGFR-2, and VEGFR-3) with tyrosine kinase activity. The most important of these receptors for angiogenesis is VEGFR-2, which is restricted to endothelial cells. VEGF acts through VEGFR-2 to mobilize these cells from the bone marrow and to induce proliferation and motility of these cells at the sites of angiogenesis.

Targeted mutations in this receptor result in lack of vasculogenesis. Several agents can induce VEGFs, the most important being hypoxia. Other inducers are platelet-derived growth factor (PDGF) and TGF. Regardless of the process that leads to capillary formation, new vessels need to be stabilized by the recruitment of pericytes and smooth muscle cells and by the deposition of connective tissue. Angiopoietins-1, Angiopoietins-2, PDGF, and TGF-participate in the stabilization process. In particular, Angiopoietins-1 interacts with a receptor on endothelial cells called Tie2 to recruit periendothelial cells. PDGF participates in recruitment of smooth muscle cells; TGF-enhances the production of ECM proteins. FGFs constitute a family of factors with more than 20 members. The best characterized are FGF-1 (acidic FGF)and FGF-2 (basic FGF). Released FGF can bind to heparin sulfate and be stored in ECM. FGF-2 participates in angiogenesis mostly by stimulating proliferation of endothelial cells. It also promotes the migration of macrophages and fibroblasts to the damaged area and stimulates epithelial cell migration to cover epidermal wounds.

2.2.3.2 Scar Formation

Scar formation builds on the granulation tissue framework of new vessels and loose ECM that develop early at the repair site. It occurs in two steps: (1) migration and proliferation of fibroblasts into the site of injury and (2) deposition of ECM by these cells.

As healing progresses, the number of proliferating fibroblasts and new vessels decreases; however, the fibroblasts progressively assume a more synthetic phenotype, and hence, there is increased deposition of ECM. Collagen synthesis, in particular, is critical change to the development of strength in a healing wound tissue. Collagen synthesis by fibroblasts begins early in wound healing (days 3–5) and continues for several weeks, depending on the size of the wound. Ultimately, the granulation tissue scaffolding evolves into a scar composed of largely inactive, spindle-shaped fibroblasts, dense collagen, and fragments of elastic tissue. As the scar matures, there is progressive vascular regression, which eventually transforms a highly vascularized granulation tissue into a pale, largely avascular scar.

Growth Factors Involved in ECM Deposition and Scar Formation:
Many growth factors are involved in these processes, including TGF-, PDGF, and FGF. TGF-belongs to a family of homologous polypeptides (TGF-1, TGF-2, and TGF-3) that includes other members such as bone morphogenetic proteins, activins, and inhibins. In the context of inflammation and repair, TGF-has two main functions: (1) TGF-is a potent fibrogenic agent. (2) TGF-inhibits lymphocyte proliferation and can have a strong anti-inflammatory effect. PDGF causes migration and proliferation of fibroblasts, smooth muscle 7 cells, and macrophages. Cytokines may also function as growth factors and participate in ECM deposition and scar formation. IL-1 and TNF, for example, induce fibroblast proliferation and can have

a fibrogenic effect. They are also chemotactic for fibroblasts and stimulate the synthesis of collagen and collagenase by these cells.

ECM and Tissue Remodeling:
The transition from granulation tissue to scar involves shifts in the composition of ECM; even after synthesis and deposition, the scar ECM continues to be modified and remodeled. The outcome of the repair process is, in part, a balance between ECM synthesis and degradation. The degradation of collagens and other ECM components is accomplished by a family of matrix metalloproteinases (MMPs), which are dependent on zinc ions for their activity. MMPs include interstitial collagenases, which cleave fibrillar collagen (MMP-1, MMP-2 and MMP-3); gelatinases (MMP-2 and MMP-9), which degrade amorphous collagen and fibronectin; and stromelysins (MMP-3, MMP-10, and MMP-11), which degrade a variety of ECM constituents, including proteoglycans, laminin, fibronectin, and amorphous collagen. The synthesis of MMPs is inhibited by TGF-and may be suppressed pharmacologically with steroids. In addition, activated collagenases can be rapidly inhibited by some specific tissue inhibitors of metalloproteinases (TIMPs), produced by most mesenchymal cells.

2.3   Wound Healing

2.3.1   Cutaneous Wound Healing

Cutaneous wound healing has four main phases: (1) inflammation, (2) wound contract, (3) formation of granulation tissue and scar, and (4) ECM deposition and remodeling.

Re-epithelialization of the wound surface. Based on the nature of the wound, the healing of cutaneous wounds can occur by first or second intention.

2.3.1.1   Healing by First Intention

One of simplest examples of wound repair is the healing of a clean, uninfected surgical incision. This is referred to as primary union, or healing by first intention. The incision causes only focal disruption of epithelial basement membrane continuity and death of a relatively few epithelial and connective tissue cells. As a result, the epithelial regeneration predominates over fibrosis. A small scar is formed, but there is minimal wound contraction. The narrow incisional space first fills with fibrin-clotted blood, which is rapidly invaded by granulation tissue and covered by new epithelium. Within 24 h, neutrophils are seen at the incision margin, migrating toward the fibrin clot. Basal cells at the cut edge of epidermis begin to show increased mitotic activity. Within 24–48 h, epithelial cells from both edges have begun to migrate and proliferate along the dermis, depositing basement membrane components as they progress. By day 3, neutrophils have been largely replaced by macrophages, and granulation tissue progressively invades the incision area. Collagen fibers are now evident at the incision margin, but these are vertically oriented and not bridge the incision. Epithelial cell proliferation continues, yielding a thickened epidermal covering layer. By day 5, neovascularization reaches its peak as granulation tissue fills the incisional area. Collagen fibrils become more abundant and begin to bridge the incision. And the epidermis recovers its normal thickness as differentiation of surface cells yields a mature epidermal architecture with surface keratinization (Fig. 2.5).

2.3.1.2 Healing by Second Intention

When cell or tissue loss is more extensive, such as in large wounds, abscess formation, and ulceration, the repair process is more complex. In second-intention healing, also known as healing by secondary union, the inflammatory reaction is more intense; there is abundant development of granulation tissue, and the wound contracts by the action of myofibroblasts. This is followed by accumulation of ECM and formation of a large scar. In larger wound, inflammation is more intense because large tissue defects have a greater volume of necrotic debris, exudates, and fibrin that must be removed. Consequently, large defects have a greater potential for secondary, inflammation-mediated injury. Larger defects require a greater volume of granulation tissue to fill in the gaps and then provide the underlying framework for regrowth of tissue epithelium. A greater volume of granulation tissue generally results in a greater mass of scar tissue (Fig. 2.5).

Secondary healing involves wound contraction. Within 6 weeks, for example, large skin defects maybe reduced to 5–10% of their original size, largely by contraction. This process has been ascribed to a presence of myofibroblasts, which are modified fibroblasts exhibiting many of ultrastructural and functional features of contractile smooth muscle cells.

2.3.2 Fracture Healing

Fracture, the most common bone lesion, is defined as a break in the continuity of a bone. If the break occurs at the site of previous disease (e.g., a bone cyst, a malignant tumor), the result is a pathologic fracture. Traumatic fracture may be the result of an excessive impact, rotation, bending or other mechanical force action on previously normal bone. The repair of a fracture is a highly regulated process that can be artificially separated into overlapping histological, biochemical, and biomechanical stages (Fig. 2.6):

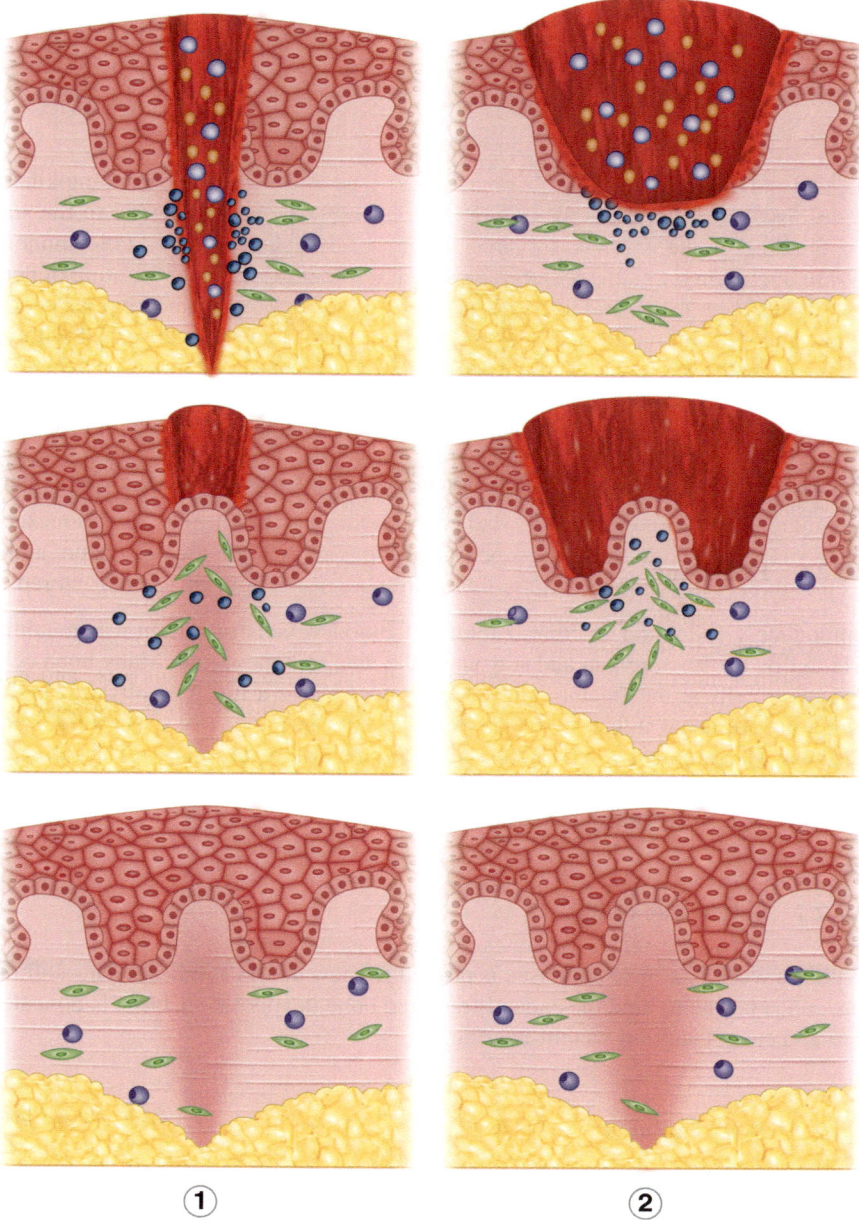

Fig. 2.5 Model of Wound Healing. ① Healing by First Intention. ② Healing by Second Intention

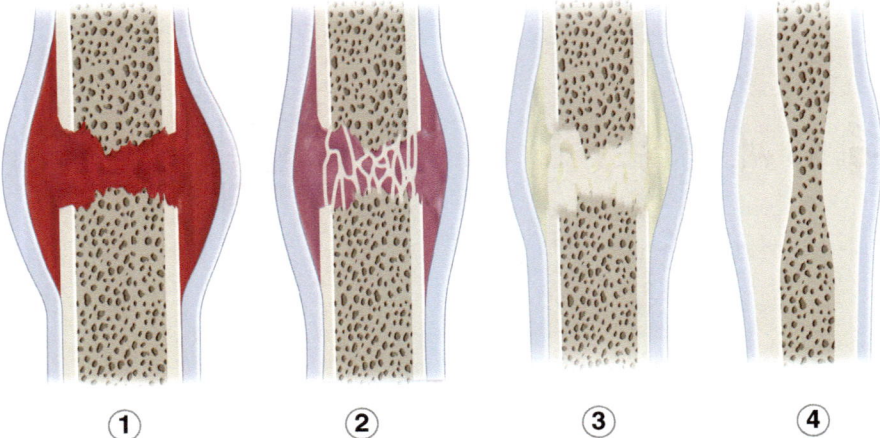

Fig. 2.6 Model of fracture healing. ① Hematoma. ② Soft tissue callus. ③ Bony callus. ④Bone remodeling

2.3.2.1 Hematoma

Immediately following fracture, rupture of blood vessels results in a hematoma, which fills the fracture gap and surrounds the area of bone injury. Many bone cells and other cells at the fracture site undergo necrosis as a result of physical injury ischemia. An acute inflammatory response occurs in regions of tissue injury and necrosis. Hematoma also provides a fibrin mesh, which helps seal off the fracture site and at the same time serves as a framework for the influx of inflammatory cells and ingrowth of fibroblast and capillary buds.

2.3.2.2 Soft Tissue Callus

Within a week, the involved tissue is primed for new matrix synthesis. This soft tissue callus is able to hold the ends of fractured bone in apposition, but it is noncalcified and cannot support weight-bearing.

2.3.2.3 Bony Callus

Bone progenitors in the medullary cavity deposit new foci of woven bone, and activated mesenchymal cells at the fracture site differentiate into cartilage synthesizing chondroblasts. In uncomplicated fracture, this early repair process peaks in 2–3 weeks. The newly formed cartilage acts as a nidus for endochondral ossification, recapitulating the process of bone formation in epiphyseal growth plates. This bony callus bridges the fractured ends.

2.3.2.4   Bone Remodeling

Subsequent weight-bearing leads to resorption of the callus and at the same time there are fortification of regions that support greater loads. This callus remodeling restores the original size and shape of bone, including the spongy cancellous architecture of the medullary cavity.

The healing of a fracture healing process can be disrupted by many factors. such as displaced fracture, inadequate immobilization, too much motion, infection, and calcium insufficiency. Generally, healing process varies tremendously among patients. Also, it is closely related to the age, health, fracture types, and the bone involved.

2.3.3   Influence Factors of Repair

Wound healing can be altered by a variety of influences, frequently reducing the quality or adequacy of the reparative process. Particularly important are infections and diabetes. Variables that modify wound healing may be extrinsic (e.g., infection) or intrinsic to the injured tissue.

Infection is the single most important cause of delay in healing; it prolongs the inflammation phase of the process and potentially increases the local tissue injury. Nutrition has profound effects on wound healing; protein deficiency, for example, and particularly vitamin C deficiency, inhibits collagen synthesis and retards healing. Glucocorticoids (steroids) have well documented anti-inflammatory effects, and their administration may result in poor wound strength due to diminished fibrosis. In some situations, however, the anti-inflammatory effects of glucocorticoids are desirable. For example, in corneal infections, glucocorticoids are sometimes prescribed (along with antibiotics) to reduce the likelihood of opacity that may result from collagen deposition. Mechanical variables such as increased local pressure or torsion may cause wounds to pull apart, or dehisce. Poor perfusion, due either to arteriosclerosis and diabetes or to obstructed venous drainage, also impairs healing. Finally, foreign bodies such as fragments of steel, glass, or even bone impede healing.

The location of the injury and the character of the tissue in which the injury occurs are also important. For example, inflammation arising in tissue spaces (e.g., pleural, peritoneal, and synovial cavities) can develop extensive exudates. Subsequent repair may occur by digestion of the exudates, initiated by proteolytic enzymes of leukocytes and resorption of the liquefied exudate. This is called resolution, if, in the absence of cellular necrosis, normal tissue architecture is generally restored. However, in the setting of larger accumulations, exudates undergo organization: granulation tissue grows into the exudates, and a fibrous scar ultimately forms.

Chapter 3
Regional Hemodynamic Disorders

Mincai Li and Yaping Gan

Contents

M. Li (✉) · Y. Gan
School of Basic Medical Sciences, Hubei University of Science and Technology,
Xianning, China

© Zhengzhou University Press 2024
K. Chen et al. (eds.), *Textbook of Pathologic Anatomy*,
https://doi.org/10.1007/978-981-99-8445-9_3

Objectives
1. To master the morphology and outcomes of congestion
2. To master the primary influences predispose to thrombosis
3. To master the types of thrombus and its effects on the host
4. To master the morphology and y types of embolism and infarction
5. To comprehend the morphology and developments of chronic congestion of liver and lung
6. To comprehend the relationship between thrombosis, embolism, and infarction

Key Concepts
The congestion consequences, the conditions and the process of thrombosis, the types of embolization, the route and the outcome of thromboembolism, the characteristics, and the types of infarction

Introduction
The function and structure of cells and tissues require normal fluid homeostasis, which delivers oxygen and nutrients and removes wastes. Normal fluid homeostasis depends on the integrity of the vessel wall, the maintenance of intravascular pressure, and the osmolarity of regional tissue. Abnormal fluid homeostasis is associated with the vascular volume or pressure, the plasma protein content, and the endothelial function, which changed by the tissue injury or disease. For example, the increased vascular volume is called hyperemia or congestion. The vessel wall integrity depends on the endothelial function and structure, which is involved in the thrombosis formation. In general, the inappropriate clotting (thrombosis) or movement of clots (embolism) can block blood supplies and cause cell or tissue infarction.

3.1   Hyperemia and Congestion

The terms hyperemia and congestion imply an increased volume of blood in a local tissue. Hyperemia usually happens in arteriolar dilation, which is an active process resulting from augmented blood flow. Congestion usually happens venous, which is a passive hyperemia resulting from impaired venous return out of a tissue obstruction with the distal veins, venules, and capillaries. The affected tissue has a red-blue color owing to the accumulation of deoxygenated hemoglobin.

Congestion might be caused by three factors: the pressure from the vessel wall, the obstruction in the vessel, and the heart failure. The left ventricular affected the lung congestion and the right-side failure affected the systemic organs, such as the liver and the limb. The congestive heart failure was the systemic phenomenon in the left and right ventricular failure.

Congestion of capillary beds is closely associated with the edema development, so edema and congestion usually happen together. In a long-standing congestion, called chronic passive congestion, the condition of poorly oxygenated blood results in chronic hypoxia, which could cause parenchymal cells degeneration or death, subsequently developed tissue fibrosis. A capillary rupture in chronic congestion may result in small foci of hemorrhage; phagocytosis and digestion of the erythrocyte debris can cause accumulations of hemosiderin-laden macrophages.

3.1.1 Morphology

Edema is also called congestion edema. In acute congestion, the cut surfaces of tissues and organs are excessively wet and hemorrhagic. In chronic passive congestion, the hypoxia results in the atrophy, degeneration, or even death of the parenchymal cells. The deposition of microhemorrhages with hemosiderin and the formation of fibrous scarring commonly appear. These important organs, namely, the lungs, liver, and spleen, develop the most obvious manifestation of chronic passive congestion.

3.1.2 Important Organ Congestion

3.1.2.1 Lungs Congestion

Lungs congestion is mainly associated with left ventricular failure, for example, myocardial infarction, myocarditis, and cardiomyopathy. There is elevated left atrial pressure and consequent elevated pulmonary venous pressure. Microscopically, alveolar capillaries engorged with blood characterize acute pulmonary congestion; there may also be associated with alveolar septal edema or focal minute intraalveolar hemorrhage. In chronic pulmonary congestion, the septa become thickened and fibrotic, and numerous hemosiderin-laden macrophages ("heart failure cells") (Fig. 3.1) appear in the alveolar spaces. In time, the fibrotic septa and together with the hemosiderin pigmentation constitute the basis for the designation brown indurations of lungs. The longtime congestion and consequent pulmonary may cause progressive thickening of the walls in the pulmonary arteries and arterioles.

3.1.2.2 Liver Congestion

Acute and chronic hepatic congestion is associated with right-sided heart failure. In acute hepatic congestion, the liver is dark-red in grossly. Microscopically, the central vein and sinusoids are distended with blood. Therefore, there may even be central hepatocyte degeneration or necrosis; the periportal hepatocytes endure less severe hypoxia and may develop only fatty change. In chronic passive congestion, the hepatic lobule regions are grossly red-brown and slightly depressed (because of cell death) and are accentuated against the surrounding zones of uncongested tan, sometimes fatty, liver ("nutmeg liver"; Fig. 3.2). Microscopically, there is centrilobular necrosis with the hepatocyte dropout, hemorrhage, and hemosiderin-laden

Fig. 3.1 Lung with
chronic passive congestion;
the alveolar spaces are
filled with edematous
fluids and heart failure
cells (hemosiderin-laden
macrophages) (400×, H.E.)

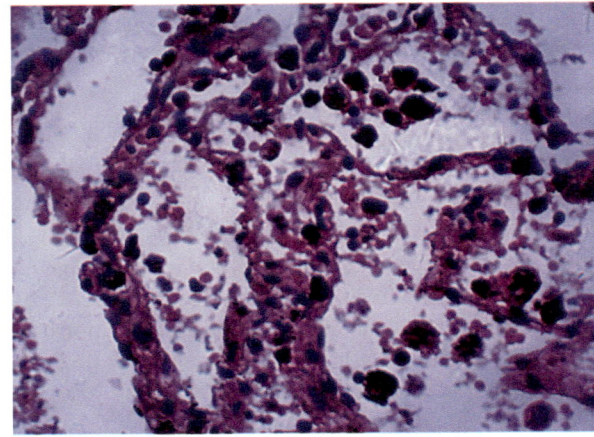

Fig. 3.2 Liver with
chronic passive congestion;
central areas of the hepatic
lobules are red-brown and
slightly depressed
compared with the
surrounding zones viable
parenchyma, forming a
"nutmeg liver" pattern

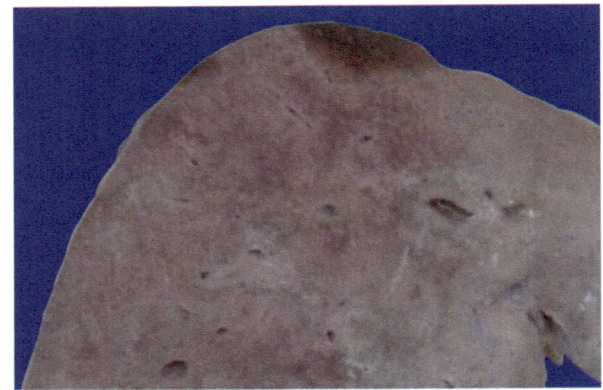

macrophages (Fig. 3.3). In long-standing, severe hepatic congestion, the hepatic hemorrhagic fibrosis may develop. And the central portion of the hepatic lobule is the last to receive blood. The centrilobular necrosis occurs whenever there is reduced hepatic blood flow; there need not be previous hepatic congestion.

3.2   Hemorrhage

Hemorrhage means the rupture of blood vessels or the heart wall and the blood overflow from vessels or the heart wall into the extravascular space. Capillary bleeding can occur under conditions of chronic congestion. An increased tendency to hemorrhage with insignificant injury happens in a wide variety of clinical disorders collectively called hemorrhagic diathesis. However, the rupture of a large artery or

Fig. 3.3 Liver with chronic passive congestion, centrilobular necrosis with degenerating hepatocytes and sinusoidal congestion (100×, H.E)

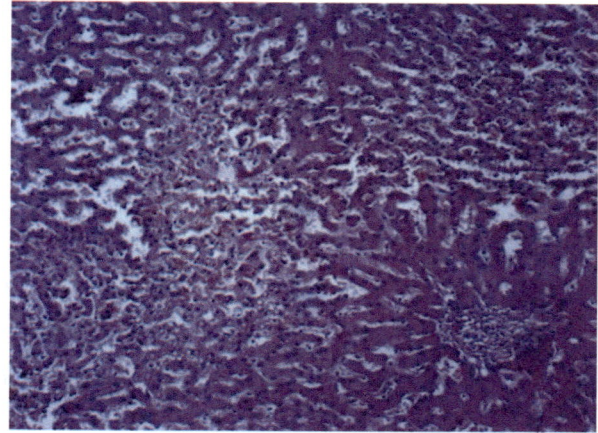

vein is always due to vascular injury, including trauma, atherosclerosis, inflammatory, or neoplastic erosion of the vessel wall, and results in severe hemorrhage.

Hemorrhage can be confined within a tissue or can be external; any accumulation within the tissue is referred to the hematoma. Hematomas might be relatively insignificant (as in the bruise), or might involve so much bleeding as to result in death. Minute (1–2 mm) hemorrhages into the skin, mucous membranes, or serosal surfaces are called petechiae, and the slightly larger (3–5 mm) hemorrhages are called purpura. Larger (1–2 cm in diameter) subcutaneous hematomas are called ecchymoses.

The red blood cells in local hemorrhages are phagocytosed and are degraded by macrophages; the hemoglobin (red-blue color) is enzymatically transformed to bilirubin (blue-green color) and eventually to hemosiderin (golden-brown), constituting of the characteristic color changes in the hematoma. Large accumulations of blood in the body cavities are called hemothorax, hemopericardium, hemoperitoneum, or hemarthrosis (in joints). Sometimes, patients with extensive hemorrhages develop jaundice from the breakdown of massive erythrocytes and the systemic increased bilirubin.

The clinical significance of hemorrhage depends on the rate and the volume of blood loss, the site of hemorrhage. The rapid loss of as much as 20% of the blood volume, or the slow losses of larger amounts, may have little impact on healthy adults; nevertheless, the greater losses can cause hemorrhagic shock. The hemorrhage site is also important; bleeding that would be trivial in the subcutaneous tissues may cause death if located in the brain or heart. The net iron loss if hemorrhage can be external. Patients with chronic or recurrent external blood loss result in an iron deficiency anemia. In contrast, when red blood cells are retained in tissues or body cavities, the iron can be reutilized for hemoglobin synthesis.

3.3   Hemostasis and Thrombosis

3.3.1   Normal Hemostasis

Normal hemostasis is a consequence of regulated processes that maintain blood in a fluid, clot-free state in normal vessels while inducing the rapid formation of a localized hemostatic plug at the site of an injured vessel. The pathologic form of hemostasis is thrombosis, which involves the formation of thrombus in uninjured vessels or the thrombotic occlusion after relatively injury.

After the initial injury, the arteriolar vasoconstriction occurs because of reflex neurogenic mechanisms and is augmented by the local secretion of factors such as endothelium. The effect is transient, and bleeding would resume if not for activation of the platelet and coagulation factor. Three key contributors to hemostasis are the vascular wall, particularly endothelium and underlying connective tissue; platelets; and the coagulation cascade.

3.3.1.1   Endothelium

Endothelial cells modulate several aspects of normal hemostasis. The balance between endothelial anti- and prothrombotic activities determines whether thrombus formation, propagation, or dissolution occur.

1. **Antithrombotic Properties**

   (a) Antiplatelet effects: An intact endothelium prevents platelets from interacting with the highly thrombogenic subendothelial ECM. Endothelial prostacyclin (PGI_2) and nitric oxides are potent vasodilators and inhibitors of platelet aggregation.

   (b) Anticoagulant effects: Anticoagulant effects are mediated by membrane-associated, heparin-like molecules and thrombomodulin. Thrombomodulin binds to thrombin, converting it from a procoagulant to an anticoagulant capable of activating the anticoagulant protein C.

   (c) Fibrinolytic properties: Endothelium synthesizes tissue plasminogen activator (t-PA), promoting fibrinolytic activity to clear fibrin deposits from endothelial surfaces.

2. **Prothrombotic Properties**

   Endothelium can become prothrombotic, with activities that affect platelets, coagulation protein, and the fibrinolytic system. The endothelial injury causes platelet adhesion to subendothelial collagen through won Willebrand factor (VWF). Loss of endothelium allows circulating VWF to bind to the basement membrane and induce platelets to adhere.

3.3.1.2 Platelets

Platelets play a critical role in normal hemostasis. After vascular injury, platelets encounter ECM constituents and undergo three reactions:

1. **Platelet Adhesion**

    Adhesion to ECM is mediated by interactions with VWF. Failure of the normal proteolytic processing of VWF leads to aberrant platelet aggregation in the circulation: this defect in VWF processing can cause thrombotic microangiopathies.

2. **Secretion**

    Secretion of both granule types occurs soon after adhesion. The release of dense body contents is especially important to platelet aggregation. Finally, platelet activation increases surface expression of phospholipid complexes and provide coagulation factors.

3. **Platelet Aggregation**

    Aggregation follows platelet adhesion and granule release. ADP and TXA promote the formation of the primary hemostatic plug. Thrombin converts fibrinogen to fibrin within the platelet plug to contribute to the stability of the clot. Thrombin contributes by directly stimulating neutrophil and monocyte adhesion from the cleavage of fibrinogen.

3.3.2   The Pathogenesis of Thrombosis

We discuss the dysregulation that underlies pathologic thrombus formation. Three primary influences predispose to thrombus formation, the so-called Virchow's triad: (1) endothelial injury, (2) turbulence of blood flow, and (3) hypercoagulability.

3.3.2.1 Endothelial Injury

The endothelial injury might lead to thrombosis. It is particularly important for thrombus formation occurring in the heart or arterial circulation, where the normal high flow rates might hamper clotting by preventing platelet adhesion or diluting coagulation factors. Thus, thrombus formation within the cardiac chambers, over ulcerated plaques in atherosclerotic arteries, or at sites of traumatic or inflammatory vascular injury (vasculitis) is largely due to endothelial injury. Regardless of the loss cause of endothelium, the end results will lead to exposure of subendothelial ECM, adhesion of platelets, the release of tissue factor, and local depletion of prostacyclin (PGI2) and plasminogen activator (PAs).

3.3.2.2   Turbulence of Blood Flow

Turbulence contributes to arterial and cardiac thrombosis by causing endothelial injury or dysfunction as well as by forming countercurrents and local pockets of stasis. Stasis is a major factor in the development of venous thrombi.

Normal blood flow is that the platelets flow centrally in the vessel lumen, separated from the endothelium by a slower-moving clear zone of plasma. Stasis and turbulence therefore (1) disrupt laminar flow and bring platelets into contact with the endothelium, (2) prevent the dilution of activated clotting factors, (3) retard the inflow of clotting factor inhibitors and permit the buildup of thrombi, and (4) promote endothelial cell activation, predisposing to local thrombosis and leukocyte adhesion.

Turbulence and stasis clearly contribute to thrombosis in a number of clinical settings. Ulcerated atherosclerotic plaques expose subendothelial ECM and are sources of turbulence. Abnormal aortic and arterial dilations called aneurysms cause local stasis and are favored sites of thrombosis. Myocardial infarctions have associated with endothelial injury and have regions of noncontractile myocardium, adding an element of stasis in the formation of mural thrombi. Mitral valve stenosis results in left atrial dilation.

3.3.2.3   Hypercoagulability

Hypercoagulability is an important component and contributes less to thrombotic states. The causes of hypercoagulability may be primary (genetic) and secondary (acquired) disorders.

1. **Primary Hypercoagulability**

   Of the inherited causes of hypercoagulability, mutations in the factor V gene and prothrombin gene are the most common. The factor V mutation substitutes a glutamine for the normal arginine residue and renders the protein resistant to cleavage by protein C. Patients with an inherited deficiency of anticoagulants present with venous thrombosis and recurrent thromboembolism in adolescence or early adult life. The mutations underlying these inherited thrombophilias may be co-inherited and have a much higher risk than normal individuals of developing venous thrombosis.

2. **Secondary Hypercoagulability**

   The pathogenesis of acquired thrombotic diatheses is more complicated. Hypercoagulability is associated with the hyperestrogenic state of pregnancy, probably related to increased hepatic synthesis of coagulation factors and reduced synthesis of antithrombin Ill, the release of procoagulant tumor products predisposes to thrombosis in disseminated cancers. The hypercoagulability attributes to increase platelet aggregation and reduce endothelial PGL_2 release. Antiphospholipid antibody syndrome has protean manifestations, including recurrent thrombosis repeated miscarriages, cardiac valve vegetations, and thrombocytopenia; it is associated with autoantibodies directly against anionic phospholipids.

3.3.3 Morphology of Thrombi

Thrombi may develop anywhere in the cardiovascular system. Their size and shape depend on the site of origin and the circumstances leading to their development. Arterial or cardiac thrombi usually begin at a site of endothelial injury or turbulence; venous thrombi characteristically occur in sites of stasis. Arterial thrombi tend to grow in a retrograde direction from the point of attachment, whereas venous thrombi extend in the direction of blood flow. The propagating tail may not be well attached and is prone to fragment to create an embolus.

3.3.3.1 Pale Thrombi

Thrombi may have grossly apparent pale or brown white. The surface of pale thrombi is rough and attachment closely to vessel wall. Microscopically, there are produced platelets admixed with some fibrin and called platelets thrombi.

3.3.3.2 Mixed Thrombus

Cardiac or aortic thrombi may have grossly apparent laminations called lines of Zahn. The thrombi is firmly adherent to the injured arterial wall and are produced by alternating pale layers of platelets admixed with some fibrin and darker layers containing more erythrocytes (Fig. 3.4). Lines of Zahn are significant only in that they imply thrombosis at a site of blood flow. Thrombi formed in the sluggish flow of venous blood usually resemble statically coagulated blood. Nevertheless, careful evaluation generally reveals irregular, somewhat ill-defined laminations.

Thrombi arising in heart chambers or in the aortic lumen are termed mural thrombi. Abnormal myocardial contraction (arrhythmias, dilated cardiomyopathy, or myocardial infarction) leads to cardiac mural thrombi, while ulcerated atherosclerotic plaque and aneurysmal dilation are the precursors of aortic thrombus formation. Arterial thrombi are usually occlusive, and the most common sites are coronary, cerebral, and femoral arteries. Although the thrombus is usually superimposed on an atherosclerotic plaque, the other vascular injury may be involved.

Fig. 3.4 Mixed thrombus, the thrombi alternating lamination with fibrin and erythrocytes (100×, H.E)

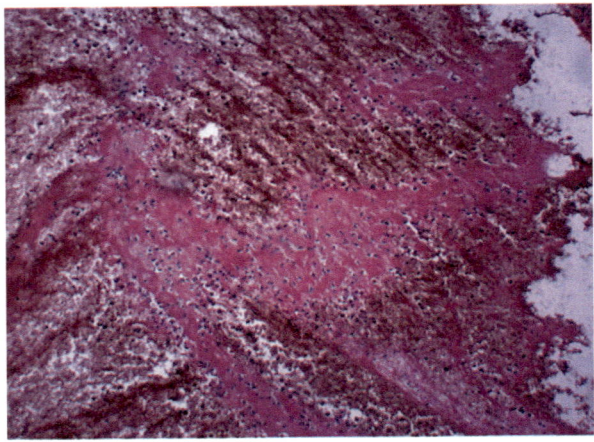

3.3.3.3 Red Thrombi

Because these thrombi forms in a relatively static environment, they tend to contain more enmeshed erythrocytes and are called red, or stasis, thrombi. Venous thrombosis (phlebothrombosis) is almost invariably occlusive, and the thrombus often creates a long cast of the vein lumen. Phlebothrombosis most commonly (90% of cases) affects the veins of the lower extremities. Less commonly, venous thrombi may occur in the upper extremities, periprostatic plexus, or the ovarian and periuterine veins; under special circumstances, they might be found in the dural sinuses, portal vein, or hepatic vein.

Postmortem clots may be mistaken at autopsy for venous thrombi. Postmortem clots are gelatinous with a dark red dependent portion where erythrocytes have settled by gravity and a yellow chicken fat supernatant, and they are commonly not attached to the underlying wall. In contrast, red thrombi are firmer and are focally attached to vessel walls, and sectioning reveals strands of gray fibrin.

Thrombi on heart valves are termed vegetations. Bacterial or fungal blood-borne infections may lead to valve damage and the development of large thrombotic masses. Sterile vegetations can also develop on noninfected valves in patients with hypercoagulable states, so-called nonbacterial thrombotic endocarditis. Less commonly, noninfective, verrucous endocarditis may occur in patients with systemic lupus erythematosus.

3.3.4 *Fate of the Thrombus*

If a patient survives an initial thrombotic event, thrombi undergo some combination of the following four events in the following days to weeks:

1. **Propagation**

The thrombus may accumulate more platelets and fibrin, increasing vascular occlusion.

2. **Embolization**

   Thrombi dislodge and transport elsewhere in the vasculature.

3. **Dissolution**

   Activation of fibrinolytic pathways can cause rapid shrinkage and total lysis of recent thrombi. With older thrombi, extensive fibrin polymerization renders the thrombus substantially more resistant to proteolysis, and lysis is ineffectual. It is important that therapeutic infusions of fibrinolytic agents such as t-PA are effective for a short time after thrombi form.

4. **Organization and Recanalization**

   Thrombi can induce inflammation and fibrosis (organization) and may eventually become recanalized. Older thrombi commonly become organized. The recanalization means to re-establish vascular flow or be incorporated into a thickened vascular wall. Although the channels may not successfully restore significant flow to many obstructed vessels, recanalization can convert the thrombus into a vascularized mass of connective tissue. With time and contraction of the mesenchymal cells, the connective tissue may be incorporated into the vessel wall.

3.3.5   Clinical Correlations

Thrombi can cause obstruction of arteries and veins and are possible sources of embolus. The significance of thrombi depends on where the thrombus occurs. Although venous thrombi may cause congestion and edema in vascular beds distal, a graver consequence is that the thrombi may embolize to the lungs and cause death. Conversely, arterial thrombi can embolize and cause tissue infarction.

3.3.5.1   Venous Thrombosis (Phlebothrombosis)

Most venous thrombi occur in the superficial or the deep veins of the leg. Superficial venous thrombi usually occur in the saphenous system. Such thrombi may cause local congestion, edema, and pain but rarely embolize. Deep thrombi in the larger leg veins at or above the knee joint are more serious because they may embolize. Although these venous thrombi may cause local pain and distal edema, the venous obstruction may be rapidly offset by collateral channels. Therefore, deep vein thromboses are entirely asymptomatic in approximately 50% of affected patients and are recognized only in retrospect after they have embolized.

3.3.5.2 Arterial and Cardiac Thrombosis

Atherosclerosis is a major initiator of thromboses, related to the abnormal vascular flow and the loss of endothelial integrity. Cardiac mural thrombi can arise in the setting of myocardial infarction related to dyskinetic contraction of the myocardium and damage to the adjacent endocardium. Rheumatic heart disease may cause atrial mural thrombi due to mitral valve stenosis. Both cardiac and aortic mural thrombi can also embolize peripherally. Virtually, any tissue may be affected, but the brain, kidneys, and spleen are prime targets because of their large flow volume.

3.3.5.3 Disseminated Intravascular Coagulation (DIC)

DIC, the insidious or sudden onset of widespread fibrin thrombi in the microcirculation, complicate a variety of disorders ranging from obstetric complications to advanced malignancy. Although these thrombi are not visible on gross inspection, they are readily apparent microscopically and may cause diffuse circulatory insufficiency, particularly in the brain, lungs, heart, and kidneys. With the development of the multiple thrombi, there is a rapid consumption of platelets and coagulation proteins. On the other hand, fibrinolytic mechanisms are activated, and an initially thrombotic disorder can evolve into a serious bleeding disorder. It is important that DIC is a potential complication of any condition associated with widespread activation of thrombin.

3.4 Embolism

The embolism is known as the process that an abnormal substance appears intravascular and obstructs vascular by following blood flow, so the abnormal substance is called the embolus.

3.4.1 The Kinds of Embolus

An embolus is a detached intravascular solid, liquid, or gaseous mass that is carried by the blood to a site distant from its origin point. The embolus consequences usually cause the downstream tissues dysfunction or infarction. Thromboembolism means the vast majority of emboli derived some part of a dislodged thrombus. Rare forms of emboli are composed of fat droplets, air or nitrogen bubbles, atherosclerotic debris, tumor fragments, amniotic fluid, or foreign bodies such as bullets. Usually, an embolism should be thought to be thrombotic in origin. Depending on the original site, emboli can lodge anywhere in the vascular tree. So the clinical outcomes are best understood from the standpoint of whether emboli lodge in the pulmonary or systemic circulation.

3.4.2 Pulmonary Thromboembolism

Pulmonary emboli are the most common form of the thromboembolic disease. The incidence of pulmonary embolism was 2–4 per 1000 hospitalized patients. Almost all venous emboli originate from thrombi with deep leg vein above the knee level. Fragmented thrombi are carried through larger veins and pass through the right side of the heart. Depending on the size, the embolus may occlude the main pulmonary artery, impact across the bifurcation (saddle embolus), or pass out into the smaller branching arterioles. Frequently, there are multiple emboli as a shower of smaller emboli from a single large thrombus. A patient with having one pulmonary embolus is at high risk of having more.

The pathophysiologic outcomes of pulmonary embolism depend on the size of embolus and on the cardiopulmonary status of patient. There is two important consequences of pulmonary arterial occlusion: (1) an increase in pulmonary artery pressure from flow blockage and vasospasm caused by neurogenic mechanisms and/or release of mediators (e.g., TXA2) and (2) ischemia of the downstream pulmonary parenchyma. Therefore, occlusion of a major vessel causes an abrupt increase in pulmonary artery pressure, diminished cardiac output, right-sided heart failure, and sometimes sudden death.

Most pulmonary emboli are small and silent in clinical findings. They eventually undergo organized and become incorporated into the vascular wall. Sudden death, right ventricular failure, or cardiovascular collapse occurs when a major pulmonary circulation is blocked by emboli. Embolic obstruction of medium-sized arteries may result in pulmonary hemorrhage. However, a similar embolus in the setting of left-side cardiac failure may result in a large infarct. Embolic obstruction of small end-arteriolar pulmonary branches usually results in infarction. Multiple emboli occurring over a period may cause pulmonary hypertension with right ventricular failure.

3.4.3 Systemic Thromboembolism

Systemic thromboembolism refers to emboli the arteries of the systemic circulation. Most systemic emboli originate from intracardiac mural thrombi, which are associated with left ventricular infarcts and left atrial dilation. The remainder originates from aortic aneurysms, ulcerated atherosclerotic plaques, thrombi, or fragmented valvular vegetations. About 10% systemic emboli are of unknown origin.

In contrast to venous emboli, arterial emboli can travel to a wide variety of sites; the arrest point depends on the embolus origin and the relative blood flow through the downstream tissues. The major sites for arteriolar embolization include the lower extremities and the brain. The outcomes of embolization depend on the vulnerability to anoxia, the caliber of the occluded vessel, and the collateral blood supply. In general, arterial emboli cause infarction.

3.4.4   Fat Embolism

Fat enters the circulation by rupture of the marrow vascular sinusoids or small venules in injured tissues. Microscopic fat globules can be found in the pulmonary vasculature after fractures of long bones or after soft-tissue crush injury. Although fat and marrow embolism occur in most individuals with severe skeletal injuries, fewer than patients show any clinical findings. Fat embolism syndrome is characterized by pulmonary insufficiency, neurologic symptoms, anemia, and thrombocytopenia and is fatal in 10% of cases. The typical symptoms appear 1–3 days after injury as the sudden onset of tachypnea, dyspnea, and tachycardia.

The pathogenesis of fat emboli syndrome involves mechanical obstruction and biochemical injury. Fat microemboli occlude pulmonary and cerebral microvasculature, which is aggravated by platelet aggregation. This effect is further exacerbated by fatty acid release from lipid globules, causing local toxic endothelial injury. The microscopic demonstration of fat microglobules typically requires specialized techniques (frozen sections and fat stains).

3.4.5   Air Embolism

Gas bubbles within the circulation can coalesce and obstruct vascular flow and cause distal ischemic injury. Air may enter the pulmonary circulation during obstetric or laparoscopic procedures, or as a consequence of chest wall injury. Usually, more than 100 mL of air produce a clinical effect; bubbles can coalesce to form frothy masses sufficiently large to occlude major vessels.

Decompression sickness, a particular form of gas embolism, is caused by sudden changes in atmospheric pressure. Scuba divers, underwater construction workers, and individuals in unpressurized aircraft are all at risk. When air is breathed at high pressure, increased amounts of gas become dissolved in the blood and tissues. If the diver ascends too rapidly, the nitrogen comes out of the solution in the tissues and the blood to form gas emboli, which cause tissue ischemia. Rapid formation of gas bubbles within skeletal muscles and supporting tissues in and about joints is responsible for the painful condition called the bends. Gas bubbles in the lungs vasculature cause edema, hemorrhages, and focal atelectasis or emphysema, resulting in respiratory distress called the chokes. A more chronic form of decompression sickness is called caisson disease, where the persistence of gas emboli in the bones causes multiple-focal ischemic necrosis. The heads of femurs, tibiae, and humeri are most commonly affected.

Treating acute decompression sickness requires placing affected individuals in a high-pressure chamber to force the gas back into solution. Subsequent slow decompression permits gradual resorption and exhalation so that the obstructive bubbles do not reform.

3.4.6 Amniotic Fluid Embolism

An amniotic fluid embolism is a grave, uncommon complication of labor and the immediate postpartum period. The onset is characterized by sudden severe dyspnea, hypertensive shock, and cyanosis, followed by seizures and coma. If the patient survives the initial crisis, pulmonary edema typically develops, along with disseminated intravascular coagulation, due to the release of thrombogenic substances.

The underlying cause is the infusion of amniotic fluid into the maternal circulation via tears in the placental membranes and rupture of uterine veins. In the pulmonary microcirculation, histology shows squamous cells shed from fetal skin, lanugo hair, fat from vernix caseosa, and mucin derived from the fetal respiratory or gastrointestinal tracts.

3.5 Infarction

An infarct is an area of ischemic necrosis caused by occlusion of the vascular supply to the affected tissue. Infarction affecting the heart and the brain is the common and important cause of clinical illness. More than half of all deaths are an outcome of cardiovascular disease and originated from myocardial or cerebral infarction. Pulmonary infarction is a common clinical complication, bowel infarction is frequently fatal, and ischemic necrosis of the extremities is a serious problem in the diabetic population.

The vast majority of infarctions result from arterial thrombotic or arterial embolism. Uncommon causes include vessel torsion, vascular compression, or traumatic vessel rupture. Venous thrombosis can cause obstruction and congestion. Infarcts caused by venous thrombosis occur likely in organs with a single venous outflow channel (e.g., ovary).

3.5.1 *Morphology*

Infarctions are classified on the basis of the color (reflecting the hemorrhage amount) and the presence or absence of microbial infection. Thus, infarcts may be either anemic (white) or hemorrhagic (red) and may be either bland or septic.

3.5.1.1   Anemic Infarcts

Anemic infarcts occur with arterial occlusions in solid organs with end-arterial circulation (e.g., heart, spleen, and kidney), where the solidity limits the hemorrhage amount to seep into the ischemic area from adjoining capillary beds (Fig. 3.5). All infarcts tend to be wedge-shaped, with the occluded vessel at the apex and the periphery of the organ forming the base. When the base is a serosal surface, there can be an overlying fibrinous exudate. The margins of infarctions are defined and slightly hemorrhagic by the narrow rim of congestion due to inflammation.

Fig. 3.5 Anemic infarct; the pale infract appears at the apex of the spleen, and the margin of infract shows the hemorrhage

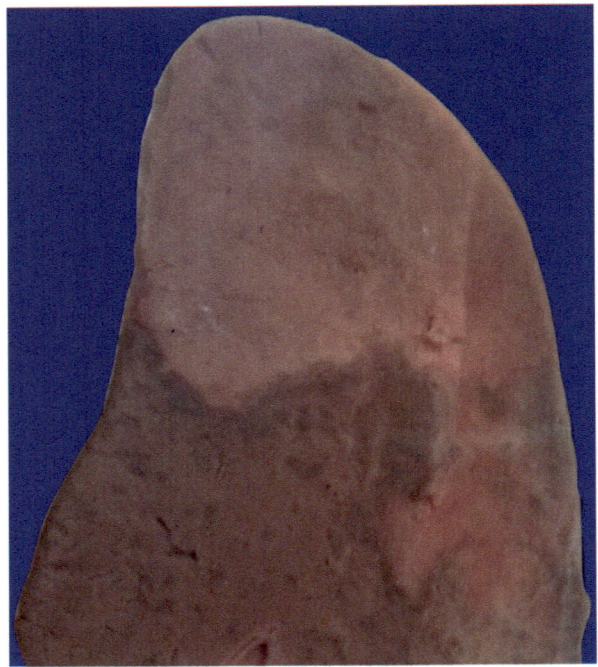

3.5.1.2 Hemorrhagic Infarcts

Hemorrhagic infarcts (Fig. 3.6) occur (1) with venous occlusions (such as in ovarian torsion); (2) in loose tissues (e.g., lung) where blood can collect in infarcted zones; (3) in dual circulations tissues (e.g., small intestine), permitting flow of blood from a parallel supply into a necrotic area; (4) in previously congested tissues; and (5) when flow is reestablished to a site of previous arterial occlusion.

In solid organs, the few extravasated red cells are lysed, and the released hemoglobin remains in the form of hemosiderin. Thus, infarcts resulting from arterial occlusions typically become progressively paler and more sharply defined with time. In comparison, the hemorrhagic infarcts are too extensive to permit the lesion ever to become pale. After a few days, extensive hemorrhages become firmer and browner, reflecting the accumulation of hemosiderin pigment.

The main histologic characteristic of infarction is ischemic coagulative necrosis. Acute inflammatory begins to develop along the margins of infarcts within a few hours and is well defined within 1–2 days. Eventually, the inflammatory response is followed by a reparative response in the preserved margins. In some tissues, parenchymal regeneration can occur at the periphery. However, most infarcts are ultimately replaced by scar. In the central nervous system, ischemic injury results in liquefactive necrosis.

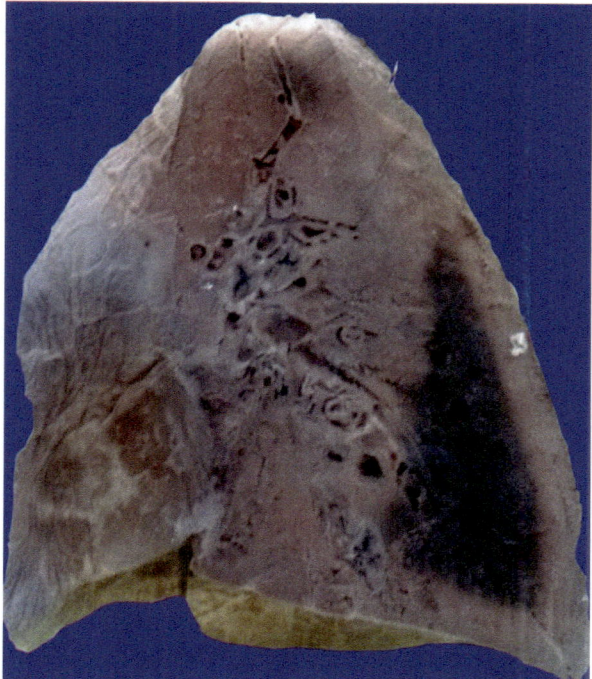

Fig. 3.6 Hemorrhagic infarct; the infracted pulmonary appears congested and purple-red

3.5.1.3 Septic Infarcts

Septic infarcts occur when infected heart valve embolize or when microbes seed an area of necrotic tissue. In these cases, the infarct is converted into an abscess, with a correspondingly inflammatory response and organization.

3.5.2 Factors that Influence the Development of an Infarct

Vascular occlusion can range from minimal effect to tissue necrosis, even result in organ dysfunction and sometimes death. The outcome is influenced by the following key determinants.

3.5.2.1 Anatomy of the Vascular Supply

The availability of an alternative blood supply is the most important factor of whether occlusion of an individual vessel causes damage. For example, the dual blood supply of lung by the pulmonary and bronchial artery indicates that the obstruction of pulmonary arterioles does not cause infarction until the bronchial circulation is compromised. Similarly, the liver with its hepatic artery and portal vein and the hand and forearm with their parallel radial and ulnar arterial supply are resistant to infarction. By contrast, the kidney and the spleen are end-arterial circulations, and vascular obstruction generally causes infarction.

3.5.2.2 Development Rate of Occlusion

Slowly developing occlusions is less to cause infarction because they allow time for the development of alternative blood supplies. For example, small inter-arteriolar anastomoses—with minimal functional flow—interconnect the three major coronary arteries in the heart. If one of the coronary arteries is slowly occluded, flow within this collateral circulation may increase sufficiently to prevent infarction, even though the major artery is completely occluded.

3.5.2.3 Tissue Vulnerability to Hypoxia

The susceptibility of a tissue to hypoxia influences the likelihood of infarction. Neurons undergo irreversible damage when deprived of their blood supply for only 3–4 min. Myocardial cells die after only 20–30 min of ischemia. By contrast, fibroblasts within myocardium remain viable after many hours of ischemia.

3.5.2.4 Oxygen Content of Blood

The partial pressure of oxygen in blood determines the consequence of vascular occlusion. Partial flow obstruction of a small vessel in an anemic patient could cause tissue infarction, whereas it would be without effect under conditions of normal oxygen tension. With compromised flow and ventilation, congestive heart failure could lead to infarction in the setting of an otherwise inconsequential obstruction.

3.6 Edema

Edema is characterized by the accumulation of interstitial fluid within tissues. Collections of fluid are variously designated hydrothorax, hydropericardium, or hydroperitoneum in different body cavities. Anasarca is a severe and generalized edema marked by profound swelling of subcutaneous tissues and accumulation of fluid in body cavities.

3.6.1 Pathogenesis

Normally, the fluid outflow at the arteriolar end of the microcirculation is balanced by inflow at the venular end; a small net outflow of fluid is drained by the lymphatics. Either increased hydrostatic pressure or diminished colloid osmotic pressure result in the increased interstitial fluid. The increased hydrostatic and plasma osmotic pressures achieve a new equilibrium. Excess edema fluid is removed by lymphatic drainage. Clearly, lymphatic obstruction attenuates fluid drainage and cause edema. Finally, sodium retention in renal disease can also cause edema.

The edema fluid is a protein-poor transudate, with a specific gravity less than 1.012. Conversely, inflammatory edema is usually a protein-rich exudate with a specific gravity s greater than 1.020.

3.6.1.1 Increased Hydrostatic Pressure

Localized increases in hydrostatic pressure can result from an impaired venous return. For example, lower extremity deep venous thrombosis might cause edema restricted to the affected leg. With resultant systemic edema, generalized increases in venous pressure occur commonly in congestive heart failure. In congestive heart failure, the reduced cardiac output increases the capillary hydrostatic pressure and causes reduced renal perfusion.

3.6.1.2 Reduced Plasma Osmotic Pressure

Albumin is responsible for maintaining intravascular colloid osmotic pressure. When albumin is either lost or inadequately synthesized, the decreased osmotic pressure occurs. An important cause of albumin loss is the nephrotic syndrome, in which glomerular capillary walls become leaky to lose albumin in the urine. Reduced albumin synthesis occurs in the setting of diffuses liver diseases and protein malnutrition. Regardless of the case, reduced plasma osmotic pressure leads to a net movement of fluid into the interstitial tissues.

3.6.1.3 Lymphatic Obstruction

Lymphedema of impaired lymphatic drainage is usually localized and can result from neoplastic or inflammatory obstruction. For example, the parasitic infection filariasis can cause lymphatic and lymph node fibrosis. Women with breast cancer might be treated by resection and/or irradiation of the associated axillary lymph nodes, and the loss of lymphatic drainage can cause upper extremity edema. In breast carcinoma infiltration and obstruction of lymphatics can cause edema of the overlying skin, the so-called peaud'orange (orange peel) appearance.

3.6.1.4 Sodium and Water Retention

Increased salt retention causes the increased hydrostatic pressure and the reduced vascular osmotic pressure. Excessive salt and water retention can occur with any compromise of renal function, including post-streptococcal glomerulonephritis and acute renal failure.

3.6.2 Morphology

Edema is easily recognized grossly; microscopically, edema fluid is reflected as a clearing and separation of the extracellular matrix elements with subtle cell swelling. Although any tissue in the body may be involved, edema is commonly encountered in subcutaneous tissues, lungs, and brain.

3.6.2.1 Subcutaneous Edema

Subcutaneous edema can be diffuse in regions with high hydrostatic pressures. Diffuse edema is more prominent in certain body areas because of the gravity effects, termed dependent edema. Dependent edema is a prominent feature of cardiac failure, particularly of the right ventricle. Edema due to renal dysfunction or nephrotic syndrome often manifests first in loose connective tissues. Finger pressure over edematous subcutaneous tissue displaces the interstitial fluid and leaves a finger-shaped depression, so-called pitting edema.

3.6.2.2 Pulmonary Edema

Pulmonary edema is a clinical problem frequently seen in the left ventricular failure. The lungs typically are two to three times their normal weight, and sectioning reveals frothy, blood-tinged fluid consisting of a mixture of air, edema fluid, and extravasated red cells.

3.6.2.3 Edema of the Brain

Brain edema may be localized (e.g., abscesses or neoplasms) or generalized, depending on the nature and extent of the pathologic process or injury. With generalized edema, the sulci are narrowed as the gyri showing signs of flattening against the skull.

3.6.3 Clinical Correlation

The edema effects may range from merely annoying to rapidly fatal. Subcutaneous edema is important to recognize primarily because it indicates underlying cardiac or renal disease; however, when significant, it can also impair wound healing or the clearance of infection. Pulmonary edema can cause death by interfering with the normal ventilatory function and creates a favorable environment for bacterial infection. Brain edema is serious to be rapidly fatal or is severe to cause the herniation as the increased intracranial pressure.

Further Reading

1. Kumar V, Abbas A, Aster J. Robbins basic pathology, 10th Ed: Elsevier 2017: 97–115.
2. 翟启辉, 周庚寅. 病理学:英文. 北京: 北京大学医学出版社. 2009:47–64.
3. 陈莉. 病理学(双语版, 修订版). 北京: 科学出版社. 2005:82–101.

Chapter 4
Inflammation

Zhang Xu and Zhang Chenli

Contents

Objectives

1. To master the concept of inflammation and granulomatous inflammation
2. To master basic lesions of inflammation and the process of exudation
3. To master pathological types and changes of acute inflammation
4. To master composition and morphological characteristics of granuloma
5. To master local manifestations and systemic reactions of inflammation
6. To master leukocyte response during acute inflammation
7. To master the outcome of acute inflammation

Z. Xu · Z. Chenli (✉)
School of Basic Medicine, Lanzhou University, Lanzhou, China
e-mail: zhangchl@lzu.edu.cn

© Zhengzhou University Press 2024
K. Chen et al. (eds.), *Textbook of Pathologic Anatomy*,
https://doi.org/10.1007/978-981-99-8445-9_4

8. To master pathological changes of general chronic inflammation
9. To master the concept of inflammatory polyp and pseudotumor
10. To understand the classification of inflammation
11. To understand the role of the inflammatory mediators
12. To understand common types and formation condition of granulomatous inflammation

Introduction

Inflammation is a very common and important basic pathological process. It occurs to the body surface of the trauma infection and the most common diseases as well as frequently occurring disease of internal organs, such as furuncle, carbuncle, pneumonia, gastritis, hepatitis, and nephritis. The modern pathology and immunology research have shown that inflammation is not only an adaptive response to the body but also a protective process of the body including the immune system. It has the effect of reducing the damage to the body, preventing the damage factors from spreading in the body and repairing the damaged tissue. But under some circumstances, the effect of inflammatory response to the body also leads to a different degree of harm, such as severe allergic reaction caused by drugs and toxicant and fibrinous pericarditis caused by cardiac fibrous adhesion. Therefore, dialectical analysis of the two sides of inflammation is of great importance in understanding the nature of inflammation and guiding clinical practice.

4.1 Overview

4.1.1 Conception of Inflammation

Inflammation is a defensive response to injury by living tissue with a vascular system. Although invertebrates (including single-celled animals and other nonvascular multicellular animals) can also respond to injury factors, for example, phagocytosis or scavenging of harmful factors, none of these are inflammation. Only the species with blood vessels can be characterized by vascular response, preserving complex and perfect inflammatory reaction of phagocytic scavenging process. Therefore, the vascular response is the central part of the inflammatory process. The vascular reaction causes plasma and leukocyte exudation, and infiltrating of leucocytes was activated, which plays the role of the dilution and limitations and kill damage factor in injury, and eliminate; absorb the necrotic tissue, at the same time, the parenchymal cells and stromal cells regeneration; and repair damaged tissue. Inflammation is essentially a complex pathological process that begins at the injury and ends at the healing or recurrence. The damage and anti-damage are consistent throughout the inflammatory response.

4.1.2   Causes of Inflammation

Inflammatory factors are the causes of tissue damage caused by inflammation. There are many types of inflammatory factors, which can be summarized as the following categories.

4.1.2.1   Biological Pathogens

Biological factors are the most common and important sources of inflammation, including bacteria, viruses, rickettsia, fungi, spirochetes, and parasites. Inflammation caused by biological pathogens is also called infection. Bacteria and their endotoxins or exotoxins can directly damage cells and tissues; viral replication in infected cells leads to cell necrosis; some antigens are infected with the body and induce tissue damage through immune responses such as parasite infections and tuberculosis.

4.1.2.2   Physical Agents

Physical agents include high temperature (burns), low temperature (frost), electric shock, ultraviolet radiation, radiation damage, and mechanical trauma (cutting and crushing injury).

4.1.2.3   Chemical Agents

Chemical factors include exogenous and endogenous chemicals. Exogenous chemicals include acids, bases, oxidants, mustard gas, and certain heavy metals (mercury). Endogenous chemicals have the breakdown products of tissue necrosis that accumulate in the body's metabolites, such as urea and uric acid.

4.1.2.4   Allergic Reaction

When the body's immune response is abnormal, it can cause inappropriate or excessive immune reactions, causing tissue and cell damage to induce inflammation, which include allergic rhinitis, urticaria, glomerulonephritis, and diseases caused by abnormal autoimmune reaction, such as rheumatoid arthritis and systemic lupus erythematosus.

4.1.2.5 Tissue Necrosis

Ischemia, hypoxia, or other reasons can cause tissue necrosis. Necrotic tissue is a potentially inflammatory factor.

4.1.2.6 Foreign Bodies

The foreign bodies that enter the body through various channels, such as various kinds of metallic wood debris, particles, and surgical sutures, can cause inflammatory response due to their different antigenicity.

4.1.3 The Basic Pathological Changes of Inflammation

The basic lesion of inflammation is alteration, exudation, and proliferation. In general, alteration is a damage process. Exudation and proliferation are the processes of anti-injury and repair, and exudation is the most characteristic lesion of inflammation.

4.1.3.1 Alteration

Alteration is referred to degeneration or necrosis of local tissues or cells. Alteration can occur in both parenchymal cells and interstitial cells. Parenchymal cell alteration is characterized by cell edema, fatty degeneration, or necrosis. Mucous degeneration or fibrinous necrosis can occur in interstitial connective tissue.

Alteration can be caused by direct effect of inflammatory agents and can be mediated by blood circulation disorders or immune mechanisms. It can also be caused by indirect effects of inflammatory reaction products. Therefore, the degree of deterioration depends on the inflammatory agents and the reaction state of the organism.

4.1.3.2 Exudation

Exudation is the process inflammation in which the fluid or cell components of the local tissue blood vessels are emitted through the vessel wall into the tissue interstitial, body surface, mucosal surface, or serous cavity. The fluid and cells are known collectively as diffusate or exudate. The exudate accumulates in the tissue gap, known as inflammatory edema. When the exudate concentrates in serous cavity, it is called inflammatory effusion (such as peritoneal effusion and pleural effusion). The inflammatory exudate and non-inflammatory transudate are different from mechanism and composition (Table 4.1), and it is of great significance to distinguish the exudate and transudate in the differential diagnosis of disease.

Table 4.1 The comparison of exudate and transudate

Difference exudate leakage		
Cause	Inflammation	Non-inflammation
Protein amount	>30 g/L	<30 g/L
The proportion	>1.018	<1.018
Nucleocyte number	>1000 × 10⁶/L	<300 × 10⁶/L
Rivalta test *	Positive	Negative
Coagulability	Self-coagulation	Non self-coagulation
Appearance	Cloudy	Clarify

Note: The Rivalta test* was a qualitative experiment of mucin

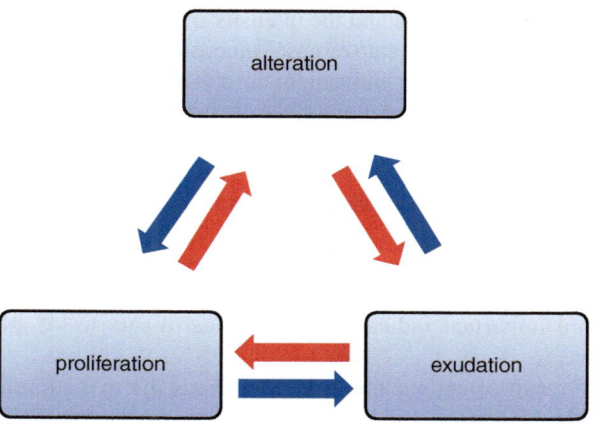

Fig. 4.1 Schematic diagram of basic lesions and interrelationship in inflammation

4.1.3.3 Proliferation

Proliferation is a regeneration of local tissue cells under the stimulation of inflammatory cytokines, disintegrating products or some physicochemical factors. In general, the proliferation of mesenchymal cells, such as endothelial cells, macrophages, and fibroblasts, can produce a large number of collagen fibers, forming inflammatory fibrosis. In some cases, the epithelial or parenchymal cells may also proliferate, such as the proliferation of epithelial cells and glands in chronic inflammation of the nasal mucosa.

The above basic lesions are consistent and interrelated throughout the inflammation (Fig. 4.1. In different types or periods of inflammation, the levels of the three lesions were different. In general, acute or early inflammation is dominated by alteration and exudation; chronic or later inflammation appears mainly proliferation. Although pathological changes of inflammatory response are diverse and complex, the basic lesion is the response of vascular nerve and body fluid.

4.1.4 The Local Manifestations and Systemic Responses of Inflammation

4.1.4.1 The Local Manifestations of Inflammation

Local manifestations of inflammation include redness, swelling, heat, pain, and dysfunction. Redness and fever in the inflammatory region are caused by increased blood flow in local vessels, increased metabolism, and enhanced heat production. A local swelling of inflammation is associated with local congestion and exudation. The effect of the exudation and the accumulation of hydrogen ions in local lesions can cause pain by stimulating nerve endings. The severity of the inflammation depends on the nature of the site and the intensity of the response. For example, the infected upper respiratory tract causes nasal mucosal swelling to nasal congestion; joint swelling or pain restricts movement in acute arthritis. Liver inflammation causes liver cell metabolic disorder and liver dysfunction.

4.1.4.2 The Systemic Responses of Inflammation

The inflammatory agents can cause systemic responses of the body. The systemic acute reaction of inflammation mainly includes the number change of leukocyte and fever, which are important indications for clinical diagnosis of inflammatory or infectious diseases.

Fever is the result of endogenous and exogenous thermal stimulation of hypothalamus. Interleukin-1 (IL-1, IL-6, and tumor necrosis factor (TNF)) is the most important cytokines mediating acute-phase inflammation. IL-1 and TNF act on the thermoregulatory center of the hypothalamus, causing fever by locally producing prostaglandin E (PGE). Fever can promote the formation of antibodies, the proliferation, and phagocytosis of mononuclear phagocyte system, which enhanced the body's defensive function. However, excessive or long-term fever can affect the body's normal metabolic process, leading to dysfunction of various systems, especially nervous system dysfunction.

Peripheral blood leukocytosis is of defensive significance, mainly related to the accelerated release of leukocytes from bone marrow depots due to IL-1 and TNF, and may often reflect the body resistance and the infection severity. Most bacterial infections lead to neutrophil increase and parasitic infections, or certain allergies lead to eosinophil increase, and some viral infections (such as infectious mononucleosis, mumps, and rubella) selectively result in lymphocyte increase. However, the peripheral blood leukocytes decrease under infection with certain viruses, rickettsia, parasites, and bacteria (such as *Salmonella typhi*).

Some inflammation (such as typhoid) can stimulate the proliferation of mononuclear macrophage because bacteria or toxins enter the blood and induce mononuclear macrophage to disseminate to liver, spleen, and lymph nodes.

4.1.5 The Clinical Type of Inflammation

The inflammation is generally divided into four types: peracute, acute, subacute, and chronic inflammation.

4.1.5.1 Peracute Inflammation

The course of the disease is only a few hours to a few days. The inflammatory response is intense and can cause severe damage of tissue/organs, even leading to death. This type of inflammation is usually caused by allergic reactions, such as penicillin.

4.1.5.2 Acute Inflammation

The acute inflammation is a rapid response to injury, and the symptoms are obvious. In addition, the course of acute inflammation is short and usually lasts several days, generally not more than 1 month. Local lesions are usually dominated by exudative changes, and the most cells of exudation are neutrophils. Sometimes, acute inflammation can be characterized by alteration (such as acute viral hepatitis) or proliferation (such as acute glomerulonephritis).

4.1.5.3 Subacute Inflammation

The course of subacute inflammation is between acute and chronic inflammation, and it lasts about a month to months. Most subacute inflammations are developed from acute inflammation, such as subacute severe hepatitis.

4.1.5.4 Chronic Inflammation

Chronic inflammation is characterized by slow onset, mild symptoms, and long duration. It can last for months to years. Chronic inflammation can be developed from acute inflammation, or it can be chronic at the beginning. Proliferation is the main pathological morphology. But alteration and exudation are less. The infiltrated inflammatory cells are mainly lymphocytes, plasma cells, and macrophages.

4.2   Acute Inflammation

Acute inflammation is an early, rapid response to the body's stimulation of proinflammatory cytokines. In acute inflammatory reactions, there are three characteristic responses: hemodynamic changes, increased vascular permeability, and leukocyte exudation.

4.2.1   Changes in Vascular Reactivity and Fluid Exudation

Inflammatory vascular response is the earliest changes in inflammatory, including changes in hemodynamics and increased vascular permeability. Increased blood flow caused hemodynamics. Increased vascular permeability leads to leakage of plasma proteins and leukocytes to extravascular tissue.

4.2.1.1   Hemodynamic Changes

Hemodynamic changes are the basis for exudation. After tissue injury occurs during acute inflammation, hemodynamic changes occur rapidly, which include changes in vascular caliber and flow.

1. **Arteriole Spasms Shortly**

   Immediate vascular response is of transient vasoconstriction of arterioles. Inflammatory cytokines act on the local blood vessels of the body through the nerve reflex or inflammatory mediators, which cause transient arteriole spasm and transient ischemic tissue. This phase lasts only a few seconds in generally.
2. **Vasodilation and Blood Flow Acceleration**

   Vasodilatation results in increased blood volume in microvascular bed of the area, which is responsible for local redness and fever in the inflammatory tissue. The mechanism of inflammatory hyperemia is related to neurological and humoral factors. Duration of vasodilation depends on the length of time that proinflammatory agent is damaged as well as the type and severity of the lesion.
3. **Slower Blood Flow**

   Slower blood flow results in increased vascular permeability. Extravasation of protein-rich fluids out of the blood vessels leads to an increased concentration of intravascular red blood cells, blood viscosity, and blocked or even stasis of blood flow, which creates conditions for leukocyte adhesion.

4.2.1.2 Increased Vascular Permeability

Increased vascular permeability is the most important cause of exudative fluid. The maintenance of normal microcirculation vascular permeability mainly depends on the integrity of vascular endothelial cells, and increased vascular permeability is mainly related to the following changes of vascular endothelial cells during inflammation (Fig. 4.2).

1. **Endothelial Contraction**

    This is the most common mechanism of increased vascular permeability and is mainly associated with immediate transient response and structural reorganization of cytoskeleton.

    Endothelial cells contract rapidly induced by histamine, bradykinin, and other inflammatory mediators, and lead to endothelial cells appear between 0.5 and 1.0 µm micropores. The phase can also be called epicardial transient response because of the short half-life of these inflammatory mediators, which lasts only for 15–30 min. Endothelial cytoskeletal remodeling is mainly induced by cytokines (such as IL-1, TNF, and IFN-γ), hypoxia, and other factors.

2. **Increased Endothelial Cell Penetration**

    There are vesicles composed of interconnected vesicles that form the transcytoplasmic channel in the cytoplasm near the junction of endothelial cells. The process by which protein-rich liquid traverses endothelial cells through the cell channel is called transcytosis. Vascular endothelial growth factor (VEGF) and histamine, bradykinin, and other inflammatory mediators can increase vascular permeability by increasing the number of pericytes and expanding the caliber.

3. **Endothelial cell damage**

Fig. 4.2 The main mechanism of vascular permeability increases. (**a**) The endothelial cells contract. (**b**) Leukocyte mediate endothelial cells injury. (**c**) Endothelial cell penetration enhanced. (**d**) Neonatal capillary permeability increased

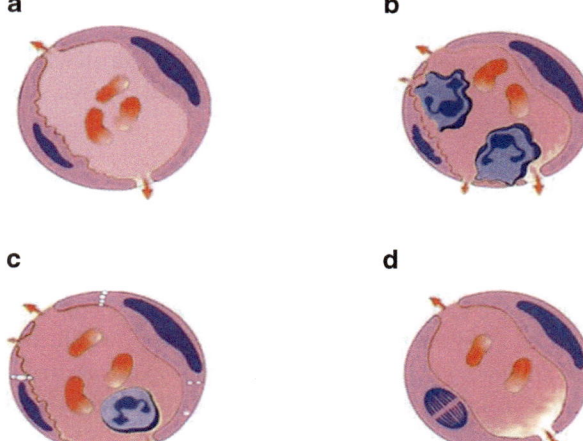

(a) Direct endothelial injury: It is usually caused by severe injuries, such as severe burns and pyogenic bacteria infection, which lead to endothelial necrosis and rapid increase of vascular permeability.

(b) Leukocyte-mediated endothelial cell injury: Toxic oxygen metabolites and proteolytic enzymes are released by leukocytes that adhere to the vascular endothelium, which lead to endothelial cell injury and detachment.

4. **Leakage of Newly Formed Capillaries**

In the inflammatory repairing phase, endothelial cell connection of newly formed capillaries is not perfect. The vessel structure makes vascular permeability increase.

4.2.1.3 Liquid Leakage

It is also known as liquid exudation. Because of increased vascular wall permeability, blood components exude from blood vessel wall. Fluid exudation in acute inflammation is induced by vasodilatation and accelerated blood flow. Thus, a large quantity of proteins exosmosis in the blood reduced plasma colloid osmotic pressure and increased osmotic pressure of the tissue.

Exudation is an important feature of acute inflammation and plays an important defensive role in the following ways: (1) Protein-rich exudates dilute the harmful substances locally. (2) The antibodies, complement, and lysozyme contained in the exudate are beneficial to the killing of pathogens. (3) The fibrin in the exudates get into a network to facilitate the confining of inflammatory lesions and can be a scaffold for repair. In addition, fibrin in exudates is beneficial to fibroblasts to produce collagen fibers in the later stage of inflammation. (4) The pathogens and toxins in the exudates are brought to the lymph nodes with lymph flow, which helps to activate the body's cellular and humoral immune responses.

However, if there is too much exudates, it would increase the pressure and block to the organ, for example, pericardial effusion can result in pericardial packing, and acute inflammatory edema of the larynx can cause suffocation. If the fibrin in the exudates cannot be dissolved and absorbed, it will form abnormal structure and result in dysfunction of local tissues and organs, such as pericardial adhesion or pericardial occlusion.

4.2.2 Leukocyte Extravasation and Phagocytosis

Leukocyte extravasation is the movement of leukocytes out of the circulatory system and toward the site of tissue damage or infection, which are also known as inflammatory cells. Inflammatory cells extravasating into the tissue space are also called as inflammatory cellular infiltration, which is the most important inflammatory morphological features.

Leukocyte extravasation is a very complex and continuously active process and plays an important role in the local defense. It is divided into four stages: (1) margination and rolling, (2) adhesion and transmigration, (3) chemotaxis and activation, and (4) phagocytosis and degradation (Fig. 4.3). The extravasated leukocyte plays an important role in local phagocytosis and immunity.

4.2.2.1 Margination and Rolling of Leukocytes

Normally leukocytes are predominantly located in the bloodstream. Because of the increased vascular dilatation and permeability as well as slower blood flow, leukocytes in capillaries go to the side flow from the axial flow in the early stage of inflammation, which is known as leukocytic margination. Then leukocytes roll on the surface of endothelial cells and do not attach to endothelial cells, known as the leukocytic pavement (Fig. 4.4).

4.2.2.2 Leukocytes Adhere and Transmigration

Peripherally marginated and pavemented neutrophils slowly roll over the endothelial cells lining the vessel wall (rolling phase). It is followed by transient bond between the leucocytes and endothelial cells becoming firmer (adhesion phase). Adhesion molecules play an extremely important role in leukocyte adhesion. Adhesion molecules include selectins, immunoglobulins, integrins, and mucin-like glycoproteins. Leukocyte adhesion molecules are mostly integrins or selectins.

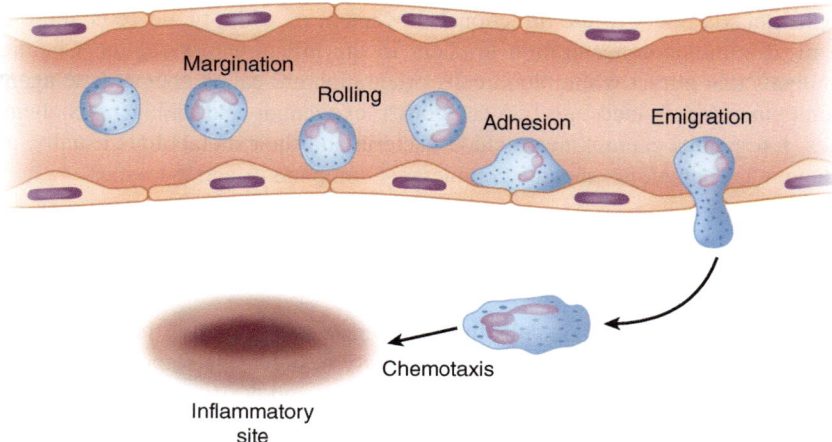

Fig. 4.3 A schematic diagram of neutrophilic infiltration process

Fig. 4.4 Neutrophilic
margination and pavement

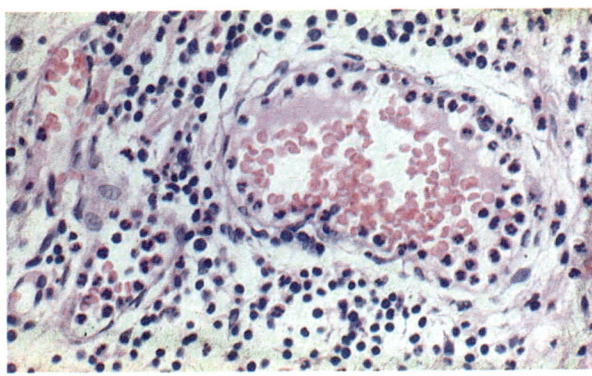

Leukocytes travel mainly in the lesion site of the small veins. Part cytoplasmic protrusions of the adherent leukocytes formed pseudopods and insert into the endothelial cell gap. The whole leukocyte moves through the endothelium in the way of the amoeba, then stops for a moment, secretes the collagen enzyme that breaks down the basement membrane, and finally enters the surrounding tissue. The process that leukocytes penetrate the vascular wall and get into the extravascular space is called transmigration.

4.2.2.3 Chemotaxis and Activation

A key function of leukocytes in accumulating inflammatory responses is the tropism of leukocytes to the site of injury. The aggregation of white blood cells into the inflammatory lesion is affected by their chemotactic action, and chemokines play a role in the whole process of leukocyte aggregation, which is of special significance in inflammatory response.

Chemotaxis refers to the directional movement of white blood cells along the chemical stimulus concentration gradient in the inflammatory region. Chemical stimulants that attract white blood cells orientation are called chemotactic agents. Chemokines are divided into two categories: exogenous and endogenous chemokines. Exogenous chemokines include bacteria and their metabolites usually, and endogenous chemokines include complement components and cytokines.

4.2.2.4 Phagocytosis and Degradation

In one hand, the white blood cells that congregate in the area of the inflammation are largely consumed and have immune activity in defensive response, and in the other hand, the white blood cells can cause damage and destruction to local tissue.

1. **Phagocytosis**

Phagocytosis refers to the process that white blood cells engulf and degrade pathogens, tissue fragments, and foreign bodies, and it is extremely important in the inflammatory defense response. Phagocytosis is another way for white blood cells to kill pathogens other than lysosomal enzymes.

There are mainly two kinds of phagocytes in human body: (1) Neutrophils, also known as "small phagocytes," usually appear in the early stages of inflammation, acute inflammation, and suppurative inflammation. Its cytoplasm contains rich neutral particles (the equivalent of lysosomal granules under the electron microscope), which contains a variety of enzymes (such as lysozyme and acid hydrolase), and it can dissolve the bacteria on the surface of the glycoprotein. (2) Macrophages, also known as "large phagocyte," usually participates in specific immune response in the late inflammation, chronic inflammation, and nonsuppurative inflammation (tuberculosis, typhoid, etc.); the macrophages in the inflammatory region are mostly monocytes, which extradavated from the blood, and differentiated into macrophages (histiocytes) in local tissues; macrophages contain many vacuoles and lysosomes, which are rich in lysozyme, acid phosphatase, and peroxidase.

Phagocytosis of the microbe by macrophages involves the following three steps (Fig. 4.5). (1) Recognition and attachment: Phagocytes first identify and adhere to a pathogen by the opsonin, which is a type of protein found in serum that enhances the phagocytic activity of phagocytes, mainly Fc fragment of immunoglobulin IgG and complement C3b. Phagocytes recognize bacteria that are coated with antibodies or complement and then bind to the corresponding receptors via antibodies or compliment by virtue of their surface Fc receptors (FcRs) and C3b receptors. (2) Engulfment: The opsonized particle or microbe bound to the surface of phagocyte is ready to be engulfed. This is accomplished by formation of cytoplasmic pseudopods around the particle due to activation of actin filaments beneath cell wall, enveloping it in a phagocytic vacuole, which

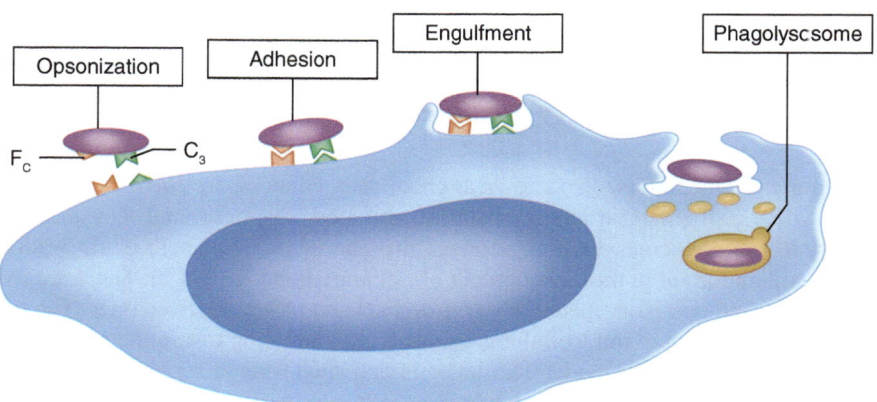

Fig. 4.5 The process of neutrophil phagocytosis

are called phagosome. Phagosomes gradually break away from the cell membrane and fuse with the primary lysosomes to form a phagolysosome. Lysosomal contents play the role of killing and degradation of pathogens by the process of degranulation. (3) Killing and degradation: bacteria that enter the phagocytes are mainly killed by the active oxidative metabolites. The phagocytic process causes white blood cells to produce the intermediate metabolites of oxygen during aerobic metabolism, namely, superoxide anions (O_2^-). Most of the superoxide anions are converted to H_2O_2 by spontaneous disproportionation. MPO exists in the eosinophilic granules of neutrophils, and in the presence of chlorides, the enzyme can reduce H_2O_2 to form hypochloric acid (HOCl-); HOCl- is also a strong oxidizer and sterilization factor that can damage the normal physiological condition of bacterial cell membrane by halide or protein/lipid oxidation, or make bacteria finally killed by enzymes inactivated.

2. **Immunity**

There are two types of immunity: specific and nonspecific immunity. Cells involved in the immune process are mainly lymphocytes, plasma cells, and macrophages. Lymphocytes are often seen in chronic inflammation or viral infections, and they are mainly from the blood or local lymphatic tissue. Macrophages uptake antigens and deliver antigenic information to T or B lymphocytes, and then sensitized T lymphocytes can produce and release lymphokines, which lead to cellular immunity, and B lymphocyte can transform into plasma cell and produce antibodies that causes the humoral immune response.

Natural killer (NK) cell is an important immune cell, it does not have T-cell receptor or produce antibodies, but it has natural killing activity. NK cells are not only associated with antitumor, antiviral infection, and immune regulation but also play roles in hypersensitivity and autoimmune diseases.

3. **The Role of Tissue Damage**

In the process of phagocytosis, chemokines, lysosomal enzymes, reactive oxygen radicals, prostatin, and leukotrienes can be released, which can strongly induce endothelial cells and tissue damage.

4.2.3   The Inflammatory Mediators

Inflammatory mediators, also known as chemical mediators, are chemical activated factors that participate in and mediate inflammatory response, and they have the ability to cause vascular dilation, permeability, and white blood cells extravasation, which are important in the development of inflammation (Table 4.2). There are two sources for inflammatory mediators: cell-derived and plasma-derived ones. The former usually present in intracellular granules in inflammation to stimulate the secretion or synthesis of the body. The latter is activated after a series of hydrolysis catalyzed by proteolytic enzymes, which usually present in the body. Most inflammatory mediators conduct their biological activity by binding to specific receptors on the surface of target cells, and one mediator can act on a variety of target cells

Table 4.2 The main medium and its role in inflammation

For use	Major inflammatory mediators
Vasodilation	Histamine, bradykinin, prostaglandin (PGI2, PGE2, PGD2, PGF2), NO
Increase the permeability of blood vessel walls	Histamine, slow excitation peptide, C3a and C5a, LTC4, LTD4, LTE4, PAF, substance P
Chemotaxis	LTB4, C5a, bacterial products, cationic proteins, chemical factors
Fever	IL-1, IL-2, TNFα, PGE$_2$
The pain	PGE$_2$, bradykinin
Tissue damage	Oxygen free radicals, lysosomal enzymes, NO

and produce different biological effects on different cells and tissues. Most inflammatory mediators have a short half-life and will degrade rapidly. They will be inactivated or eliminated once are activated or released. So the existence of mediators is in a dynamic equilibrium through the regulatory system or self-stability mechanism in our bodies. Most inflammatory mediators have the potentially damaging effects.

4.2.3.1 Inflammatory Mediators Released by Cells

1. **Vasoactive Amines**
   Vasoactive amines include histamine and serotonin, and they are present prostatin and are released rapidly and exert their effect once stimulated. They are often the "protagonists" of the early inflammatory response, so they are also called Fast Media. (1) Histamine: It's mainly found in the mast cells in the connective tissue around the blood vessels and in the granules of basophils and platelets in the blood. When stimulated, it is released in the form of degranulation. Factors that cause histamine release include complement fragments (such as C3a and C5a), cytokines (such as IL-1 and IL-8), and neuropeptides. Histamine receptors (H1, H2, and H3) are present in the cell membrane, and the receptor plays a biological role in combination with histamine. The H1 receptor excitation results in the contraction of bronchial and vascular smooth muscle and leads to vascular endothelial contraction and vascular permeability. Histamine also has a chemotactic effect on eosinophils. (2) 5-hydroxytryptamine (5-HT): It is also known as serotonin, mainly in platelets and enterochromaffin cells. Collagen, thrombin, ADP, platelet-activating factor (PAF), and immune complexes stimulate platelet aggregation to release 5-HT. 5-HT and histamine play a similar role, and they mainly function in increasing vascular permeability.

2. **Arachidonic Acid (AA) and Its Metabolites**
   AA is an unsaturated fatty acid and mainly in the phospholipids of various organs, for example, the prostate, brain, lung, and intestine. Phospholipase A$_2$ (PLA$_2$) was activated and released from the membrane phospholipid under stimulation of inflammatory factors. AA itself is not inflammatory mediators, but when released, it produces prostaglandins and leukotrienes via cyclooxygenase

and lipoxygenase pathways, respectively. It can also generate metabolites such as lipoxin through other pathways to play an inflammatory mediator role. (1) PG: PG are the metabolites of AA by cyclooxygenase (COX) pathway. The important products associated with the inflammatory process are PGE_2, PGD_2, PGF_2, and PGI_2, which are synthesized by specific enzymes. Platelets, for example, contain thrombin synthase, so TXA_2 is mainly produced by platelets, which can cause platelet aggregation and vasoconstriction to initiate the clotting process. PGE_2 is a hypersensitive substance that causes pain in the inflammatory process by increasing the sensitivity of the skin. PGE2 is also a strong heating agent, which is used in medicine. (2) Leukotriene (LT): LT is a metabolic product of AA produced by the lipoxygenase (LOX) pathway, of which the main inflammatory-related products are LTA_4, LTB_4, LTC_4, LTD_4, and LTE_4. LTB_4 is the activation factor of neutrophils and white blood cell function, which is attached to the endothelial cells, producing oxygen free radicals and releasing lysosomes; LTC_4, LTD_4, and LTE_4 can cause strong vasoconstriction, bronchospasm, and vascular permeability increased. (3) Lipoxin (LX): It's also the metabolic product produced by AA through the pathway of fat and oxygenase, which has the dual function of promoting and inhibiting the inflammatory response, and its inflammatory-related products mainly include LXA_4 and LXB_4. LX may be the negative regulator of LT activity in vivo as it inhibits chemotactic reaction of neutrophils.

3. **Leukocyte Products**

It mainly includes oxygen free radicals and lysosomal enzymes released by neutrophils and monocytes. (1) Oxygen free radicals (OFR): This mediator mainly includes superoxide anions (O_2^-), hydrogen peroxide (H_2O_2), and hydroxyl radicals (-OH), which can combine with NO in cells to form active nitrogen intermediate. When these media are released in small amounts, they can increase the expression of IL-8, certain cytokines, endothelial cell, and leukocyte adhesion molecules, causing inflammation cascade and amplifying the effect. The release of these active substances will cause serious damage to the tissue. Of course, human serum, tissue fluid, and host cells themselves have antioxidant mechanisms that protect the body against potential oxygen free radicals. Therefore, whether the OFR cause damage in the inflammatory response also depends on the balance between OFR and antioxidants. (2) Lysosomal enzymes: Both the death of phagocytes and the exhalation of enzymes in the phagocytic process can lead to enzyme release in the lysosome. Lysosomal enzymes, such as neutral protease (elastase, collagenase, cathepsin, etc.), play an important role in destroying purulent inflammation tissue, including collagen fibers, fibrin, basement membrane, elastin, and cartilage; the neutral protease also cleans C3 and C5 directly, releasing anaphylaxis and kinin. On the other hand, the antiprotease system is present in human serum and tissue fluid, such as alpha 1-antitrypsin (AAT), which is the key factor in inhibiting elastic protease in the neutrophils. If AAT is absent in lung tissues, neutral protease cannot be inhibited, which will lead to emphysema throughout the whole lung.

4. **Cytokine**

Cytokines are small molecules of polypeptide or glycoprotein, which is produced by the immune cells (lymphocytes and monocytes), and certain nonimmune cells (endothelial cells, epithelial cells, and fibroblasts). Cytokines include colony-stimulating factor (CSF), IL, IFN, TNF, TGF-β, chemokine family, and other cytokines (PDGF, FGF, EGF, VEGF, IGF, NGF, HGF, TGF-α, etc.), which play important roles in cellular physiology various biological effects like the immune response and inflammation.

5. **PAF**

PAF is a potent bioactive phospholipid derived from platelets, basophils, mast cells, neutrophils, monocytes, and endothelial cells. By binding to the PAF receptor on the target cell membrane, PAF cause platelet aggregation and neutrophilic aggregation, adhesion and release. It also can directly act on target cells or stimulate white blood cells to synthesize other inflammatory mediators (such as reactive oxygen and LT). Clinically, PAF receptor blockers are used to prevent the binding of PAF to receptors and are therefore of therapeutic interest in diseases associated with PAF overproduction such as asthma and septic shock.

6. **NO**

NO is formed by the action of l-arginine, molecular oxygen, NADPH, and other auxiliary factors in different types of nitric oxide synthase (NOS), and NO is derived from endothelial cells, macrophages, and specific nerve cells in the brain. There are three types of NOS isozymes in the body, including the inducible nitric oxide synthase (iNOS), endothelial cell nitric oxide synthase (eNOS), and nerve cell type nitric oxide synthase (nNOS). iNOS are mainly found in macrophages and lung endothelial cells, which catalyze the production of excessive NO and participate in the inflammatory reaction. However, eNOS and nNOS participate in the production of NO in physiological state and maintain normal physiological function of the body. The main function of NO as an inflammatory medium is to relax the vascular smooth muscle and expand the blood vessels. In addition, it can reduce the aggregation and adhesion of platelets and inhibit the inflammatory response induced by mast cells.

7. **Neuropeptide**

Neuropeptide is a kind of special information substance that is broadly referred to as an endogenous active substance in nerve tissue. It is characterized by low level, high activity, and extensive and sophisticated function. It regulates a variety of physiological functions in the body, like pain, sleep, emotion, learning, memory, and the development of the nervous system itself. For example, the presence of substance P in the lung and the nerve fiber of the gastrointestinal tract not only participates in regulating pain signal, blood pressure, and the activation of immune cells as well as endocrine cells but also has the effect on increasing vessel wall permeability at the initial stage of inflammation. The G protein is a p-specific receptor, and a mouse without the receptor cannot respond to the stimulation, which is sufficient to increase the permeability of the pulmonary capillaries.

4.2.3.2   The Inflammatory Mediators of Humoral Origin

1. **Kinin System**

   There is a wide range of biological activity for the kininogen-kinin system. Kinin system is closely related to coagulation system, complement system, renin-angiotensin system, and other cytokine system as well as various vascular activity factors, to jointly maintain normal physiological functions. The final product of the system, bradykinin, is an important inflammatory medium under kininogenase activation. Its main function is to expand the arterial artery, increase blood vessels permeability, and reduce smooth muscle contraction (such as bronchial smooth muscle). Kininases are divided into plasma and tissue types, and the molecular weight, physiological function, physical and chemical properties, and immunological properties are all different for these two types of kininase. Plasma type of kininase usually activates peptide enzyme into the form of the activated peptide enzyme (prekininogenase), which existed in circulating bloodstream, and the activation center link is XII factor (Hageman factor) activation. First, XII factor produces a fragment (prekininogenase) under the activation by collagen and basement membrane, which changes the prekininogenase into activated kininase XII factor, and kininogen will eventually be cracked into bioactive bradykinin under the action of kininogen. Tissue kinase is present in various secretions (saliva, pancreatic juice, and tear), urine, and feces. It can hydrolyze the peptides, which are converted to bradykinin by aminopeptidase.

2. **Complement System**

   The complement system is composed of a series of enzymes active in the serum and tissue fluid and has the effect of increasing the permeability of blood vessels, chemotaxis, and modularization. Complement is mainly synthesized by hepatocyte in plasma while it is mainly released from macrophages in the inflammatory tissue. The complement in plasma exists in an inactivated form and can be activated by classical pathway (antigen-antibody complex); alternative pathway (pathogenic microorganism surface molecule, such as endotoxin or lipopolysaccharide); and agglutinin pathway. The activation of C3 and C5 in the complement system is the most important process. The cleavage fragments C3a, C5a, and C3b are important mediators in the inflammatory process. They mainly function in three aspects: (1) Anaphylaxis: The membranes of mast cells and basophils can release histamine, LT, and PG and act as mediators, causing vasodilation and increased vascular permeability; C3a and C5a cause similar pathological changes in allergic reactions, which is the so-called C5a and C3a allergy toxin (anaphylatoxin). (2) Chemokines and adhesion: C5a, C3a, and C4a are the potent chemokines of neutrophils, eosinophils, basophils, and monocytes and can activate white blood cells and increase the affinity of protein molecules on the surface of white blood cells to promote their adhesion to endothelial cells. (3) Phagocytosis: C3b has the effect of modulating the cell wall of bacteria, which can enhance the phagocytosis of neutrophils and monocytes because there are C3b receptors on the surface of these phagocytes.

3. **Clotting System and Fibrinolytic System**

Factor XII not only can activate the kinase system but also activate blood coagulation and fibrinolysis. The blood coagulation system is activated to produce thrombin, fibrin polypeptide, and factor Xa with activity of inflammatory mediators. Thrombin can promote leukocyte adhesion and fibroblast proliferation by combining to the protease-activated receptors (PAR) of platelets, vascular endothelial cells, and smooth muscle cells. Thrombin also functions in the release of cytokines and inflammatory mediators, microvascular exudation, neutrophil chemotaxis, and other pathological processes. Fibrin polypeptide can increase vascular permeability, and it is also a chemotactic factor of leukocyte. Thrombin Xa binds to effector cell protease receptor-1 (ECPR-1), an agent that mediates acute inflammation, resulting in increased vascular permeability and enhanced leukocyte exocytosis. Activated fibrinolytic system degrades C3 and produce C3a, increasing vasodilatation and vascular permeability; fibrin degradation product (FDP) produced by fibrinolytic process has the effect of increasing vascular permeability, and fibrinolytic enzyme activates blood coagulation system by activating factors XII.

4.2.4 Morphologic Patterns of Acute Inflammation

Inflammations include three basic pathological changes: alteration, exudation, and proliferation. According to the characteristics of inflammation lesions, acute inflammation may be generally divided into three types: alterative inflammation, exudative inflammation, and proliferative inflammation.

4.2.4.1 Alterative Inflammation

Alterative inflammation is characterized by degeneration or necrosis of cells or tissues at the main lesion. Slight exudation and proliferation are common in severe infections, poisonings occurred in the liver, kidney, heart, brain, and other parenchymal organs. The organ often exhibits the obvious dysfunction for the cellular degeneration or necrosis. For example, the epidemic encephalitis cause serious dysfunction of the central nervous system; in the case of toxic myocarditis caused by diphtheria exotoxin, cardiomyocyte degeneration, and necrosis cause severe heart dysfunction; extensive necrosis of liver cells leads to serious liver dysfunction in acute severe viral hepatitis.

4.2.4.2 Exudative Inflammation

Exudative inflammation is the most common inflammation. A large amount of exudates are the main feature in inflammatory lesions. The exudation is different from the variance of the inflammatory factors and the reaction of the body. According to the main components of exudation, the exudative inflammation is generally divided into serious inflammation, fibrinous inflammation, suppurative inflammation, hemorrhagic inflammation, and other types of inflammation.

1. **Serous Inflammation**

   Serous inflammation is a major inflammation with the exudation of serous fluid. Exudation is mainly composed of serum, a high concentration of albumin (3–5%), and a low globulin content. The electrolyte amount is the same with blood, mixed with a small number of neutrophils, fibrin, and epithelial cells. Physical factors (such as high temperature), chemical factors (such as acid and alkali), biological factors (such as bacterial toxins), and snake venom can cause serious inflammation, and it can also be found in the early stages of acute inflammation. Serous inflammation often occurs in the serosa (pleura, peritoneum, and pericardium), mucosa, synovium, and loose connective tissue and skin. For example, after poisonous snake bites or bee stings, plasma leakage gathered in the connective tissue space to form local inflammatory edema; for tuberculosis, rheumatic disease involving the serosa or synovium, a large number of serous exudation can cause effusion of the chest, peritoneal, pericardial, or joint; the serous exudation accumulates in the epidermis in second-degree skin blisters. The serous inflammation of mucosa, also known as serous catarrhal inflammation, is characterized by an amount of serous discharge. For example, a large amount of serous discharge from the nasal mucosa during early stage of a cold. The term catarrh originates from Greek and means the downward drip of the serous, which is used to describe exudation that exudes outward along the mucosal surface. So catarrhal inflammation is an exudative inflammation of mucosal tissue.

   Serous inflammation prognosis is well. The effusion of the serous lymphatic vessels and blood vessels can be absorbed, and the local minor damage of epithelial tissue is easy to repair. Excessive oozing of the plasma can have adverse effects and may cause serious consequences. For example, laryngeal seroma can cause asphyxia in severe cases; pleural or pericardial effusion can result in dysfunction of lung and heart.

2. **Fibrinous Inflammation**

   Fibrinous inflammation is characterized by accumulation of fibrin in the exudation. Fibrinous inflammation is caused by bacterial toxins such as diphtheria, dysentery, and *Streptococcus* pneumonia, or various endogenous, exogenous toxins, such as urea and mercury poisoning in uremia. The toxins lead to a large amount of fibrinogen leakage from the blood vessels, which is converted into fibrin in the necrotic tissue. With HE staining, the fibrin is red and stained with a granular, sort-like, or interwoven network, accompanied with neutrophils or necrotic tissue.

   Fibrinous inflammation often occurs in the mucosa (pharynx, larynx, trachea, and intestine), serous membrane (pleura, peritoneum, and peritoneum), or lung (Fig. 4.6). When the mucosa occurs fibrinous inflammation (such as diphtheria and bacterial dysentery), the exudation of fibrin, white blood cells, necrotic

Fig. 4.6 Lobar pneumonia

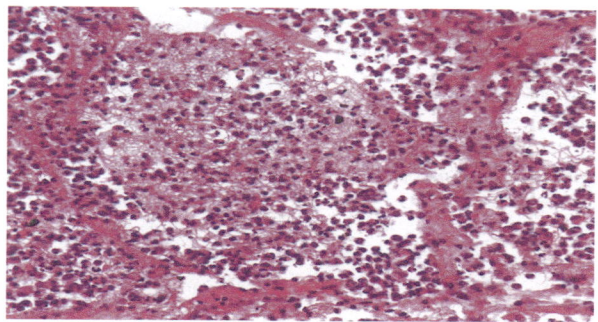

mucosal tissue, and pathogenic bacteria can form a layer of gray and white film (pseudomembrane) covering the mucous surface. Therefore, it is also called pseudomembranous inflammation. In diphtheria, the membrane of the pharynx and trachea is white. Because of the firm attachment of pharyngeal and diphtheria pseudomembrane with deep tissue, so the pseudomembrane is not easy to fall off (solid membranous inflammation) (Fig. 4.7a, b). However, there is a loose connection between diphtheria and mucosal exudations of the trachea, and the membranous is easy to fall off (floating membranous inflammation), which may obstruct the bronchial tube and cause suffocation. Serofibrinous inflammation is common in pleural and pericardial membranes, such as tuberculous fibrous pleurisy and rheumatic pericarditis. A large amount of fibrin exudates from the heart, forms numerous villi, and covers the pericardium surface, which is called a shaggy heart. Fibrous inflammation occurs in red and gray hepatic changes stage of lobular pneumonia with a large amount of fibrin exudate in the alveolar cavity. And the small amount of fibrin can be dissolved by proteolytic enzyme released by neutrophils. With fewer neutrophils or increased antitrypsin activity, the excessive fibrin can cause malabsorption and result in adhesion or transformation of lobular pneumonia.

3. **Purulent Inflammation**

   Suppurative inflammation is characterized by a large number of neutrophil exudation, varying degrees of tissue necrosis and pus formation. It is usually caused by the infection of the septic bacteria, such as *Staphylococcus*, *Streptococcus*, *Meningococcus*, and *Escherichia coli*. It also can be induced by certain chemicals, such as turpentine, ba-bean, and the necrotic tissue, which is called aseptic inflammation. Pyogenesis is the process of dissolving the necrotic tissue by a lysosomal enzyme released by neutrophils. Pus mainly contains a large number of neutrophils, pus cells (denaturation and necrotic neutrophils), a small amount of serous, liquefactive necrotic tissue, and bacteria. There are three types of suppurative inflammation:

1. **Abscess**

   An abscess is a localized suppurative inflammation of the organ or tissue. An abscess is characterized by the formation of liquefactive necrosis and the cavity

Fig. 4.7 (**a**) Diphtheria pseudomembrane in tonsillitis (Gross appearance). (**b**) Diphtheria pseudomembrane in tonsillitis (Microscopic appearance)

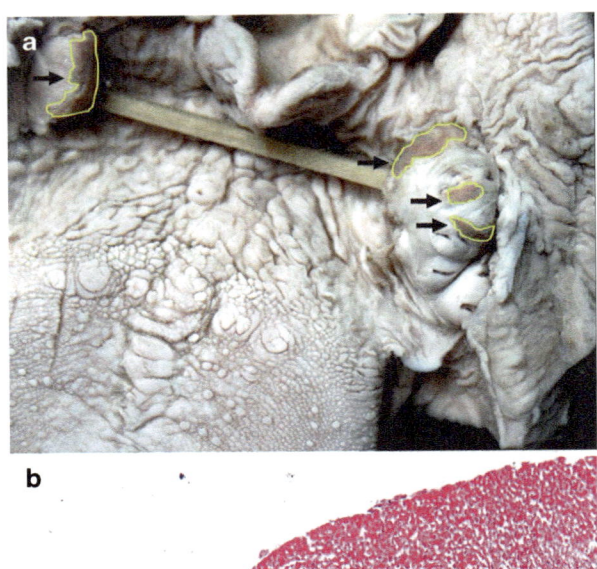

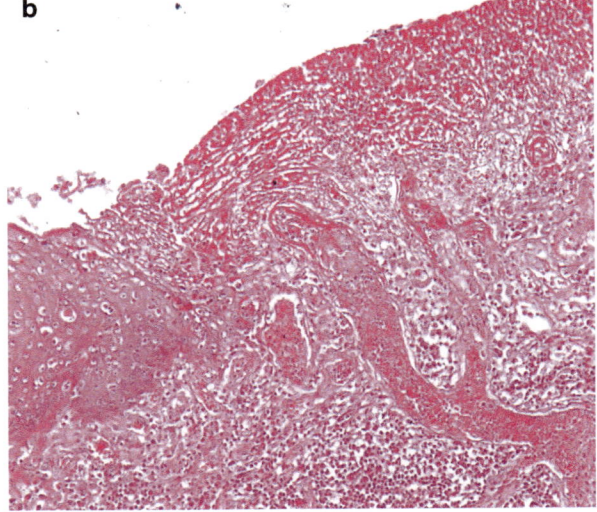

(pus cavity). The abscess occurs mainly in the skin and internal organs (such as the lung, liver, kidney, and brain) (Fig. 4.8a, b). *Staphylococcus aureus* is the common pyogenic bacteria that can produce plasma coagulase and convert the fibrinogen into fibrin to prevent bacteria spread. *Staphylococcus aureus* also produces the adhesion protein receptors, which may migrate abscesses into the distance through the blood vessels. At the early stage of abscess, it is a focal collection of neutrophils and necrotic tissues. Then it becomes surrounded by vascular and fibroblastic proliferation, to form a so-called abscess membrane, which may absorb pus and limit the spread of inflammation. If the pathogen is exterminated, the pus will be absorbed, and the abscess will be replaced by the granulation tissue. If the abscess does not heal, a large amount of fibrous tissue forms a thick wall of chronic abscess, which often requires an incision of pus.

Fig. 4.8 (**a**) Hepatic abscess (Gross appearance). (**b**) Hepatic abscess (Microscopic appearance)

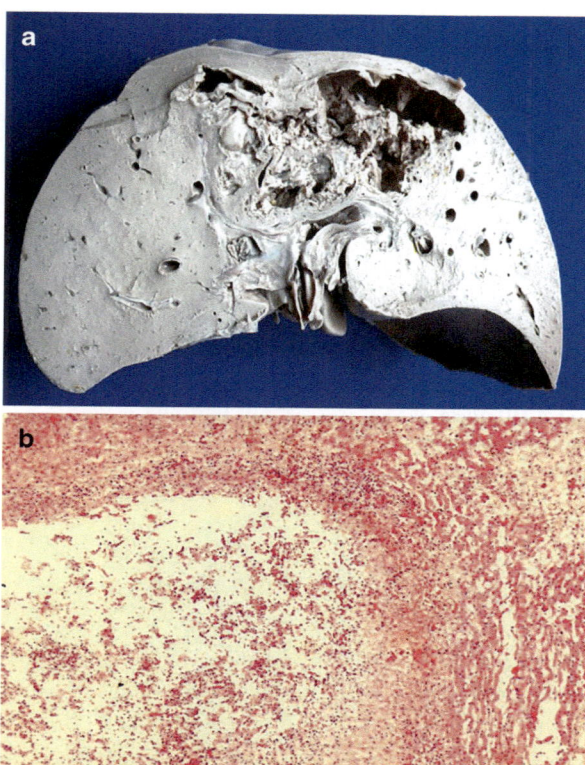

It may lead to complications: (1) Ulcers: Skin, mucous membrane, or synovial joints of suppurative inflammation may form deep ulcers, because of the local tissue necrosis or the limitation of collapse loss. (2) Sinusitis: A deep tissue abscess would break into a form or a natural tube to form a sinus. (3) Fistulas: The perianal abscess puncture to the skin to form the anal purulent sinus. When the abscess penetrates the inner wall of the intestine, the bowel can communicate with the skin of the body surface and form a purulent fistula (Fig. 4.9).

Some of the common examples of abscess formation are as follows: (1) Furuncle is an abscess occurring in a single hair follicle and its associated sebaceous glands. Furuncle appears in the areas of the rich hair follicles and sebaceous glands (such as the neck, head, face, and back). (2) Carbuncle is the fusion of multiple abscesses in the subcutaneous fat, fascia formation of multiple communication abscesses. Carbuncle common appears on the neck, back, waist, hip, and other thick and tough skin. Carbuncle often requires multiple drainages.

2. **Phlegmonous Inflammation**

Cellulitis is the diffuse inflammation of soft tissues (such as subcutaneous, mucous membrane, muscle, and appendix) resulting from spreading effects of hemolytic *Streptococcus*. It can secrete hyaluronidase to decompose hyaluronic

Fig. 4.9 The abscess around the anus is a pattern of sinus and fistula

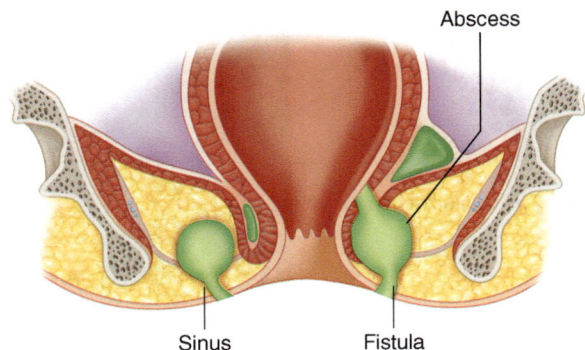

Fig. 4.10 Acute phlegmonous appendicitis

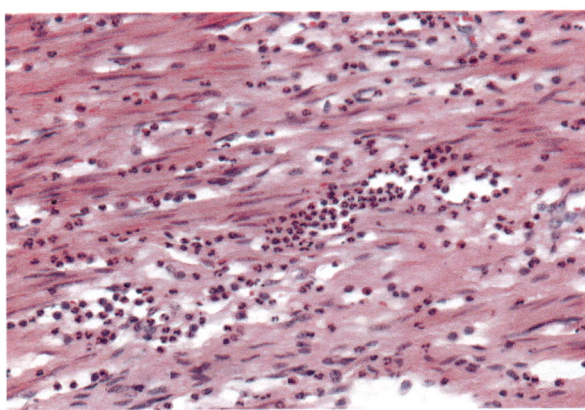

acid and disintegrate matrix in connective tissue. It also can secrete streptokinase to dissolve fibrin. Thus, hemolytic *Streptococcus* is easy to spread to the surrounding tissue through the interstitial and lymphatic tissue. Phlegmonous inflammation is characterized by severe tissue edema, diffuse infiltration of neutrophils (Fig. 4.10), and unclear surrounding tissues. There is no obvious necrosis and dissolution in the local tissue. Light cellulitis can be absorbed without traces, but severe cellulitis can spread rapidly and cause systemic toxicity.

3. **The Surface of the Pus and Empyema**

   Superficial suppuration refers to the purulent inflammation, which is characterized by infiltration of neutrophils on the surface of mucosa or serous membrane, but inflammatory cells in deep tissue are not obvious. The suppurative inflammation of mucosa is also called purulent catarrhs, the pus exudation, such as suppurative urethritis and suppurative bronchitis. When the pus occurs in the membrane, the gallbladder, the fallopian tube, or the appendix, pus is stored in the cavity, called empyema.

4. **Hemorrhagic Inflammation**

   Some infections (such as anthrax, bubonic plague, leptospirosis, and epidemic hemorrhagic fever) induce vascular damage followed by hemorrhage.

The above inflammation can occur alone or can be combined, such as serous fibrosis, hemorrhagic hemorrhage of serous, and fibrinolysis. During the development of inflammation, one type of inflammation may turn into another kind of inflammation. For example, severe inflammation may become fibrinous or suppurative inflammation.

4.2.4.3 Proliferative Inflammation

Although most of the acute inflammation is alteration and exudation, some acute inflammation is also characterized of cell proliferation, known as proliferative inflammation. Lesions were mainly manifested as proliferation of vascular endothelial cells, tissue cells, and fibroblasts. For example, *Streptococcal* acute glomerulonephritis shows glomerular vascular endothelial cells and mesangial cells proliferation; mononuclear macrophage hyperplasia occurs in typhoid fever.

4.2.5 Outcome of Acute Inflammation

The variation in inflammatory response depends upon factors pertaining to the organisms (type, virulence, dose, and route of entry) or host factors (systemic diseases, immune status, defect in neutrophil function, and type of tissue). Acute inflammation may have variety of outcomes.

4.2.5.1 Resolution and Healing

Resolution means complete return to normal tissue following acute inflammation. This occurs when tissue changes are slight and the cellular changes are reversible; the structure and function of the original tissue are fully restored through the regenerative repair. However, when the tissue destruction is extensive, then healing occurs by fibrosis.

4.2.5.2 Chronic Inflammation

Under certain instances, persisting or recurrent acute inflammation may progress to chronic inflammation. The reasons are the followings: (1) Inflammatory factor continuously exists or repeatedly acts on the body, (2) the body's resistance is low, and (3) the treatment is not complete.

4.2.5.3 Extension and Spread

When pathogenic microorganisms are highly toxic and the body has low resistance, pathogenic microorganisms can propagate continuously, from tissue interspace or vascular system to surrounding or whole body tissues or organs.

1. **Local Extension**

   The pathogenic microorganism of local inflammation can spread to surrounding tissues and organs through the natural channels of tissue interspace or organs. For example, *Mycobacterium tuberculosis* can spread to the surrounding tissue along the interstitial space. It can also spread along the bronchus to form new tuberculosis in other parts of the lung.
2. **Lymphatic Spread**

   Pathogenic microorganisms can invade lymphatic vessels through the interstitial space and spread into local lymph nodes, causing local lymphadenitis. For example, foot suppurative inflammation can cause inguinal lymphadenitis, tuberculosis, and disseminated hilar lymph node.
3. **Hematogenous Spread**

   The pathogenic microorganisms may invade the blood circulation, or toxins are absorbed into the blood, causing bacteremia, toxemia, septicemia, or sepsis. (1) Bacteremia is defined as presence of small number of bacteria in the blood, which do not multiply significantly. The patient may have no symptoms of systemic toxication. Some inflammatory diseases in the early stages are bacteremias, such as typhoid, epidemic cerebrospinal meningitis, and lobular pneumonia. (2) Toxemia means the presence of bacterial toxins and metabolites in the blood. It can cause high fever and other systemic toxic symptoms, but the blood culture cannot find bacteria. The toxemia is often accompanied by the degeneration or necrosis of the heart, liver, kidney, and other parenchymal cells. (3) Septicemia means the presence of rapidly multiplying, highly pathogenic bacteria in the blood, resulting in systemic toxic symptoms. The pathogenic bacteria can be found in blood culture. Clinically, patients often have chills, high fever, skin and mucous bleeding spots, swollen spleen, swollen lymph nodes, or even shock. (4) Pyemia is the dissemination of small septic thrombi in the blood, which cause their effects at the site where they are lodged. This can result in pyemic abscesses or septic infarcts, which is often found in the lungs, liver, kidney, brain, skin, etc. These abscesses are small-sized, distributed more evenly. Bacterial colonies can be observed in the central and small blood vessels. This abscess is also known as an embolic abscess.

4.3 Chronic Inflammation

Chronic inflammation is defined as prolonged process (months or even years) in which tissue destruction and inflammation occur at the same time. Most chronic inflammations are originated from acute inflammation, partly due to the mild and persistent stimulation of the pro-inflammatory cytokines, such as certain virulent pathogenic microorganisms (M. tuberculosis, *Treponema pallidum*, and fungi) or autoimmune diseases (such as rheumatoid arthritis and systemic lupus erythematosus). Some chronic inflammations are due to long-term exposure to potentially toxic substances, for example, silicosis is due to long-term inhalation of silica (SiO_2). Recurrent and continuous progress is an important clinical feature of chronic inflammation, and its acute phase is similar to acute inflammation. Chronic inflammation is characterized by cellular proliferation (Fig. 4.11). Based on the morphological characteristics, chronic inflammation can be divided into two categories: nonspecific chronic inflammation and granulomatous inflammation.

4.3.1 Chronic Nonspecific Inflammation

It is also called general chronic inflammation and is common in clinical. The proliferative cells are fibroblasts, vascular endothelial cells, and parenchymal cells, accompanied by infiltrative cells, such as lymphocytes, plasma cells, and macrophages (Fig. 4.12). Local epithelial cells and glandular epithelium can also proliferate. Some repairing changes may occur. For example, granulation tissue hyperplasia, fibrosis development, and scar formation play an important role in absorbing and healing chronic abscess, sinus, fistula, and chronic mucosal ulcers. However, this repairing process might cause tissue or organ adhesions or sclerosis. For example, chronic Crohn's disease may cause intestinal stenosis and intestinal obstruction.

There are two special types of chronic nonspecific inflammation: (1) Inflammatory polyp, a pedicled mass formed by the hyperplasia of local mucosal epithelium, glandular, and granulation tissue under the long-term effect of inflammatory factors. It occurs in the cavity organs, such as the cervix and gastrointestinal mucosa. Inflammatory polyps are generally small with a diameter of no more than 2 cm; under the microscope, the hyperplasia of mucosal epithelium, glandular, and granulation tissue is obvious, and interstitial edema accompanied by chronic inflammatory cell infiltration might be observed. (2) Inflammatory pseudotumor, a tumorlike

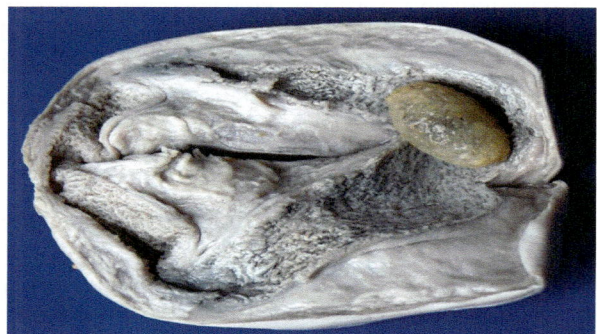

Fig. 4.11 Chronic cholecystitis, with gallstone

Fig. 4.12 Chronic
nonspecific inflammation
in the mucosa

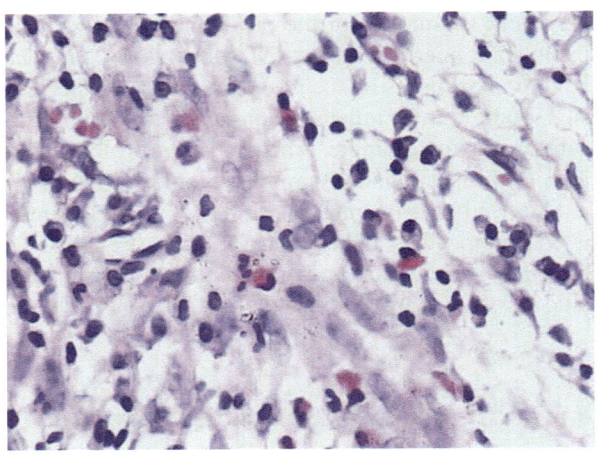

mass formed by the chronic inflammatory hyperplasia of tissue, often in the lung
and eye socket. Under imaging examination, the morphological characteristics are
similar to the tumor, known as an inflammatory pseudotumor. Under microscopy,
inflammatory pseudotumor is composed of granulation tissue, fibrous tissue, inflammatory cells, and hyperplasia parenchyma. The inflammatory pseudotumor of the
lung is relatively complex with the chronic inflammatory cell infiltration, the significant hyperplasia of alveolar epithelial and fibrous tissue, as well as the different
level fibrosis. It is not easy to distinguish inflammatory pseudotumor from lung
tumor. Sometimes, it can only be diagnosed by pathological examination.

4.3.2 Granulomatous Inflammation

It is a chronic proliferative inflammation characterized by formation of granuloma,
which is a nodular lesion characterized by the limited infiltration and hyperplasia of
macrophages. The essence of granuloma is inflammation caused by delayed hypersensitivity, and the involved cells are macrophages and epithelioid cells by the
immune response. Therefore, granuloma can be defined as the aggregation of macrophages and their derived cells (such as epithelioid cells and multinucleated giant
cells), with or without the presence of other inflammatory cells. The special morphological manifestations of various granulomas have important pathological diagnostic value.

4.3.2.1   The Causes of Granulomatous Inflammation

1. Bacterial infections include tuberculosis and leprosy.
2. Spirochetes infections include *Treponema pallidum* and syphilis.

3. Fungal infections include candidiasis, hairy mycosis, cryptococcosis, actinomycosis, and histoplasmosis.
4. Parasitic infections include schistosomiasis, filariasis, and ascariasis.
5. Endogenous and exogenous foreign bodies: The former are the endogenous foreign bodies such as urate in the nodules of gout. The latter includes various metals or nonmetallic substances that enter the body from the outside, such as beryllium, zirconium, surgical sutures, talcum powder, wood spines, iron chips, dust, asbestos, silica gel, and mineral oil.
6. The cause is unknown for sarcoidosis.

4.3.2.2   The Types of Granuloma

1. **Infective Granuloma**
   It is caused by pathogen infection such as treponema pallidum, fungi, and parasites Granuloma is of diagnostic significance. For example, tuberculosis is the granulomatous inflammation caused by *Mycobacterium tuberculosis*, which is characterized by the formation of a typical tuberculous granuloma (tubercle). Tubercle consists of central caseous necrosis, surrounded by proliferative epithelioid cells, Langhans multinucleate giant cell, lymphocytes, and fibroblasts (Fig. 4.13).
2. **Foreign Body Granuloma**
   It is caused by foreign bodies, for example, surgical suture, dust, talcum powder, and wood spines. Foreign bodies are in the lesion center, surrounded by numerous macrophages, foreign body giant cells, fibroblasts, and lymphocytes (Fig. 4.14).
3. **Sarcoidosis Granuloma**
   It is a non-necrotic epithelioid granuloma that occurs during sarcoidosis. Sarcoidosis is a kind of systemic diseases with unknown etiology. Sarcoidosis can be observed in multiple systems and organs such as lymph nodes, skin, upper respiratory tract, lungs, liver, eyes, heart, nervous system, salivary glands, muscles, and bones. The granuloma is mainly composed of epithelioid cells, multinucleated giant cells, and lymphocytes, without caseous necrosis.

Fig. 4.13 Tubercle

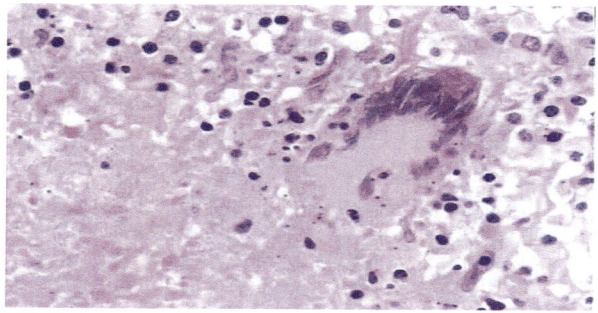

Fig. 4.14 Tophus

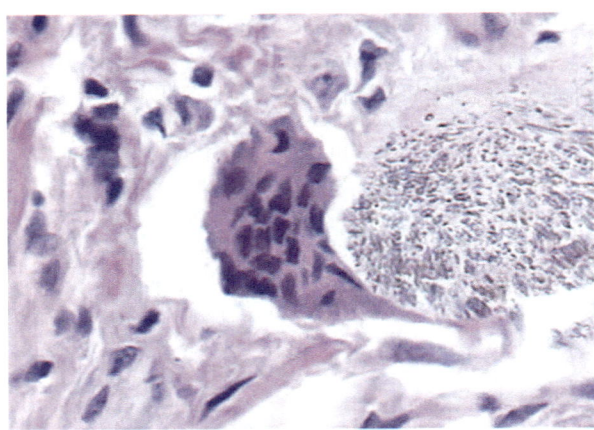

4.3.2.3 The Histogeny of Granuloma (Pathogenesis of Granuloma)

Formation of granuloma is a type IV granulomatous hypersensitivity reaction. It is a protective defense reaction by the host but eventually causes tissue destruction because of persistence of the poorly digestible antigen, for example, *Mycobacterium tuberculosis*, leprosy bacilli, etc.), or difficult to be degraded, for example, suture, dust, and other foreign bodies. Macrophages play an important role in secrete chemokines and aggregate in inflammatory lesions. Macrophages, being antigen-presenting cells, can present antigen to CD4 + T lymphocytes. These lymphocytes get activated and elaborate lymphokines, which also can activate macrophages to improve their phagocytosis and bactericidal activity. Macrophage morphology is modified to be epithelial cell-like appearance (epithelioid cells). Epithelioid cells are weakly phagocytic. Fusion of adjacent epithelioid cells can form multinucleated giant cell. Like epithelioid cells, these giant cells are weakly phagocytic but produce secretory products, which help in removing the invading agents.

4.3.2.4 The Components of Granuloma

1. **The Epithelioid Cells**

   They are modified macrophages/histiocytes, which are somewhat elongated cells having slipper-shaped nucleus with 1–2 small nucleoli. The nuclear chromatin of these cells is vesicular and lightly staining, while the cytoplasm is abundant, pale-staining with hazy outlines so that the cell membrane of adjacent epithelioid cells is closely apposed. Mitochondria, endoplasmic reticulum, ribosomes, Golgi complex, and lysosomes are abundant in the cytoplasm. Although epithelioid cells are weakly phagocytic because of the absence of FC and C3b receptors on cell membrane, they also can secrete degradation enzyme and cytokines (TNF, IL-1, etc.).

2. **Multiple Nuclear Giant Cells**

Multiple nuclear giant cells are fused by epithelioid cells, and their size is large (40–50 mm), with abundant cytoplasm and eosinophilic acid, with nuclei ranging from tens to hundreds. These nuclei may be arranged at the periphery like the horseshoe or as a flower ring, or may be clustered at the two poles (Langhans' giant cells), or they may be present centrally (foreign body giant cells). The former are commonly seen in tuberculosis, while the latter are common in foreign body tissue reactions.

Chapter 5
Neoplasm

Rui Zhou and Li Liang

Contents

R. Zhou · L. Liang (✉)
School of Basic Medical Sciences, Southern Medical University, Guangzhou, China
e-mail: lli@fimmu.com

© Zhengzhou University Press 2024
K. Chen et al. (eds.), *Textbook of Pathologic Anatomy*,
https://doi.org/10.1007/978-981-99-8445-9_5

Objectives

1. Definition, nomenclature and classification of neoplasm
2. Pleomorphism, differentiation and anaplasia of neoplasm
3. General gross features of neoplasm and effects of neoplasm on host
4. Growth patterns and metastatic pathways of neoplasm
5. Differential characteristics of benign and malignant neoplasms
6. Definition of precancerous lesions, dysplasia, and carcinoma in situ
7. Pathological features of common neoplasms
8. Differential characteristics of carcinoma and sarcoma
9. Common moleculartargets of neoplasm
10. The key carcinogenesis of neoplasm: Four classes of gene disorders
11. Common carcinogens and oncogenic tumor virus
12. The well-identified inherited cancer syndromes
13. Common tumor antigens

Key Concepts

neoplasm, cancer, tumor, benign neoplasm, malignant neoplasm, carcinoma, sarcoma, pleomorphism, atypia, differentiation, anaplasia, pathological mitotic figures, invasion, metastasis, angiogenesis, progression, heterogeneity, cancer stem cell, precancerous lesion, atypical hyperplasia, dysplasia, carcinoma in situ, carcinogen, carcinogenesis, oncogenesis, oncogenes, proto-oncogenes, oncoprotein, tumor suppressor gene, inherited cancer syndrome, tumor antigen, cancer cachexia, paraneoplastic syndrome.

Introduction

A neoplasm or tumor is a group of cells that have undergone unregulated growth and will often form a mass or lump and may be distributed diffusely. They have different biological behaviors and clinical manifestations. The so-called cancer usually refers to these malignant tumors, which are seriously harmful to human health.

Recent evidence shows that neoplasm is the first leading cause of death in urban residents in China. In 2015, about 90.5 million people suffer from cancer. About 14.1 million new cases occur per year (not including skin cancer other than melanoma). It causes about 8.8 million deaths (15.7% of deaths) worldwide every year. The most common types of cancer are lung cancer, prostate cancer, colorectal cancer, and stomach cancer in males and breast cancer, colorectal cancer, lung cancer, and cervical cancer in females. In children, acute lymphoblastic leukemia and brain tumors are most common, except in Africa where non-Hodgkin lymphoma occurs more often.

This chapter describes the basic pathologic properties of neoplasia, including the nature of benign and malignant neoplasms and the molecular basis of neoplastic transformation. We also discuss the host response to neoplasms and the clinical features of neoplasia.

5.1 Definition and General Morphology of Neoplasm

Neoplasm is an abnormal growth of tissue, which usually (but not always) forms a local mass. When it forms a mass, neoplasm is commonly referred to as a tumor. *Tumor*, Latin for *swelling*, originally meant any form of swelling, neoplastic or not. The word *"neoplasm"* is from ancient Greek *neo-*"new" and *plasm-*"formation." Some neoplasms do not form a tumor, such as leukemia and most forms of carcinoma in situ.

The process of the neoplasm formation is called neoplasia, which is the result of severe disorders of cell growth regulation with kinds of tumorigenic factors. Neoplasia literally means "new growth," which is caused by an abnormal proliferation of cells, usually named as neoplastic proliferation. The characteristics of neoplastic proliferation are (1) uncoordinated, uncontrolled and harmful growth of cells, (2) clonal proliferation of cells, (3) varying degrees of dedifferentiation, and (4) a certain degree of autonomy.

The concept contrary to the neoplastic proliferation is nonneoplastic proliferation. It is usually a controlled and limited process that is coordinated with the needs of the body and occurs in the conditions of normal cell renewing, defense response caused by injury, and reparation. For example, the proliferation of endothelium and fibroblast in the inflammatory granulation tissues is nonneoplastic. It generally will not continue after the cessation of the stimuli, which evokes the changes. Cells or tissues of nonneoplastic proliferation are polyclonal and capable of differentiation and maturation.

Various factors that cause tumor formation are called tumorigenic agents. Substances that cause the formation of a malignant tumor are called carcinogens. Research over the past few decades has shown that neoplasia is a complicated process resulting from the accumulation of multiple genetic mutations.

The nature of a neoplasm is mainly determined by pathological observations (including gross and microscopic morphological examinations) of specimens from biopsy and surgical excision.

5.1.1 Gross Appearance of Neoplasm

Attention should be paid to the number, size, shape, color, and texture of tumor in gross observation because the above information is helpful for determining the type and benign and malignant tumor.

5.1.1.1 Number

The tumor can be single or multiple. Some tumors, such as carcinoma of digestive tract, are usually single; other types of tumor often show as multiple masses, for example, the patients with neurofibromatosis can have dozens or even hundreds of masses. When physical examination is performed for tumor patients or surgical specimens are examined, it should be avoided noticing obvious tumor and neglecting the possibility of multiple tumors.

5.1.1.2 Volume

The volume of the tumor can be very different. Tiny tumors, such as thyroid microcarcinoma, are difficult to observe by naked eye. They need to be observed under microscope. A large tumor can weigh up to several kilograms or even tens of kilograms, such as cystadenoma occurring in the ovary. The volume of tumor is related to many factors such as the nature (benign or malignant), location, and growth rate. Tumor that occurs on the body surface or large body cavity has plenty of space and large volume. On the contrary, tumor that occurs in a narrow and closed cavity like cranial cavity and spinal canal is usually small because of growth restriction.

Generally speaking, the larger the malignant tumor, the easier it is to metastasize. Therefore, the volume of malignant tumor is an important index of tumor stage. In some tumor types (gastrointestinal stromal tumor), volume is also an important index to predict tumor biological behavior.

5.1.1.3 Shape

The shape of tumor can be different because of its different histological type, location, growth mode, and good and evil nature. In medicine, we use some typical terms to describe the shape of tumor, such as papillary, villous, polypoid, nodular, lobular, infiltrative, ulcerative, and cystic. Common shapes of tumors are shown in Fig. 5.1.

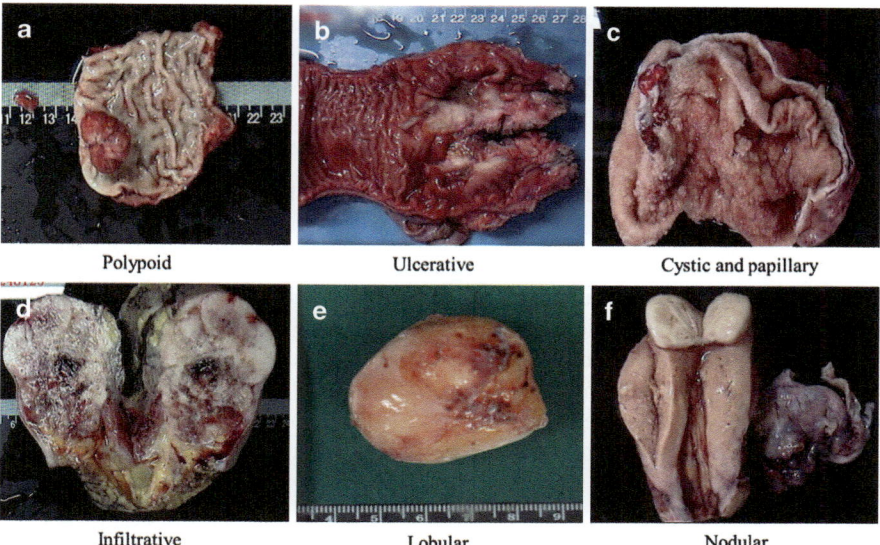

Fig. 5.1 Common shapes of neoplasms

5.1.1.4 Color

The color of the tumor is determined by the color of tumor tissue, cells, and products. The cut surface of fibrous tumor is mostly gray-white. The lipoma is yellow, and the hemangioma is often red. Some secondary changes occur in tumor, such as degeneration, necrosis, and bleeding, which can change the original color of the tumor. Melanoma cells produce melanin, which makes the tumors dark brown.

5.1.1.5 Texture

Types and the proportion of tumor cells to interstitial tissue determine the texture of the tumor. Lipoma is soft. Less interstitial tumors, such as colorectal adenoma, are soft. Invasive carcinoma with fibrous tissue reaction is hard.

5.1.2 Microscopic Morphology of Neoplasm

The tissue morphology of the tumor is ever-changing and the basis of histopathological diagnosis of tumor. All tumors have two basic components: the parenchyma and the host-derived stroma. Neoplastic cells constitute the parenchyma, and their morphology, composition, and products are the main basis for judging the differentiation and histological classification of tumors. Tumor stroma is usually made up of connective tissue and blood vessels, which supports and nourishes the parenchyma of the tumor. Neoplastic cells stimulate angiogenesis, which is the key factor for tumor to continue to grow. Infiltration of lymphocytes is also seen in tumor stroma, which may be related to the body's immune response to tumor tissue.

5.2   Differentiation and Atypia of Neoplasm

Tumor differentiation refers to the similarity between tumor tissue and some normal tissue in morphology and function. The degree of similarity is called tumor differentiation. For example, tumors similar to adipose tissue suggest that they differentiate into adipose tissue. The more the neoplastic cells resemble their normal forebears in morphology and function, the better the differentiation of the neoplasm, which is termed as well-differentiated. On the contrary, the smaller the similarity with normal tissue, the less differentiated or poorly differentiated. A poorly differentiated tumor, which is unable to determine the direction of its differentiation, is called undifferentiated tumor.

Tumor atypia refers to the difference between the neoplastic parenchyma and the corresponding normal tissues in tissue structure and cell morphology (Fig. 5.2). Architectural atypia of the neoplasms is the difference in tissue structure and spatial arrangement between neoplastic cells and the corresponding normal tissues. For example, in squamous cell carcinoma or carcinoma in situ of esophagus, there is an obvious disorder of squamous epithelium. In gastric adenocarcinoma, glandular epithelium is formed with irregular glands or glandular structure. In endometrial adenocarcinoma, normal endometrial stroma disappears between glands.

Cellular atypia of the neoplasms has many manifestations, including (1) cell volume abnormality. Some of neoplastic cells are characterized by increased cell volume, some of which are primitive small cells. (2) Pleomorphism. The size and shape of tumor cells are very different. Sometimes, tumor giant cells appear in tumor tissue. (3) The increase of nuclear-to-cytoplasm ratio. It is the prominent hallmark of neoplasm, especially of malignant neoplasm, that the nuclei are large

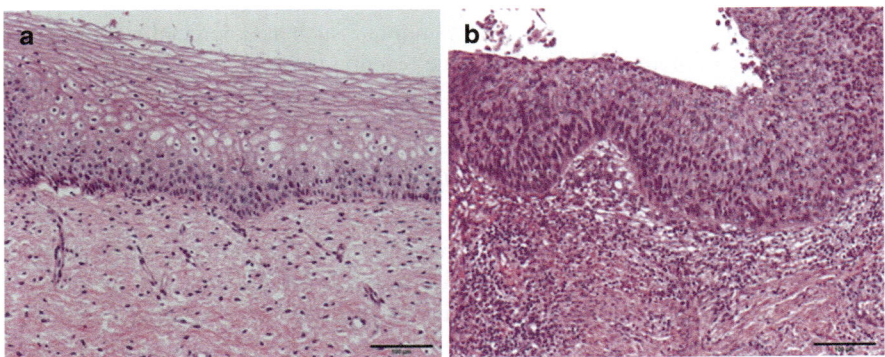

Fig. 5.2 Normal stratified squamous epithelium (**a**). Architectural atypia (**b**) (H&E, 200×)

Fig. 5.3 Nuclear
pleomorphism. The tumor
cells present marked
pleomorphism in size and
shape, with multiple or
bizarre nuclei and
multipolar mitoses (arrow)
(H&E, 400×)

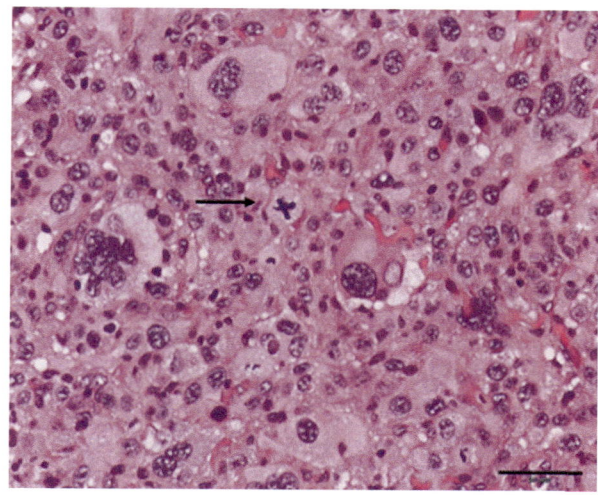

disproportionately for the cells with the nuclear-to-cytoplasm ratio that increases from the normal 1:4 or 1:6 to approach 1:1. (4) Nuclear pleomorphism. The nuclear shape is variable and often irregular with the appearance of giant nuclei, double nuclei, multiple nuclei, and weird or bizarre nuclei. The chromatin is often coarsely clumped and unevenly distributed under the nuclear membrane (hyperchromatic), which results from the increase of DNA in nuclei. (5) Distinctive and large nucleolus. The number of nucleolus is increased. (6) Increase in nuclear mitosis. Abnormal mitoses (pathological mitotic figures) can be observed sometimes, such as asymmetrical or multipolar mitoses (Fig. 5.3).

Atypia is a manifestation of disturbance of maturation and differentiation in neoplastic tissues and cells and also an important indicator to distinguish between benign and malignant neoplasms. Benign neoplasms usually show light inconspicuous cellular atypia but still have different degrees of architectural atypia. For malignant neoplasms, both cellular atypia and architectural atypia are obvious. The greater the atypia, the lower the maturity and differentiation, and the greater the difference between the neoplasms and the corresponding normal tissues. The obvious atypia is called anaplasia. Most of the anaplastic neoplasms are highly malignant, which show the characteristics of anaplasia.

5.3 Nomenclature and Classification of Neoplasm

The nomenclature and classification of neoplasm are important parts of pathological diagnosis of neoplasm and also important for clinical practice. Medical staff should understand the meaning of these names in tumor pathological diagnosis and use them properly.

5.3.1   The Principle of Nomenclature

There are a wide variety of human neoplasm and complex naming. Neoplasm is usually named according to its tissue or cell type and biological behavior.

5.3.1.1   General Principles of Tumor Nomenclature

The Nomenclature of Benign Neoplasm
In general, benign neoplasms are designated by attaching the suffix –oma to the tissue or cell type from which the neoplasm arises. For example, a benign tumor arising from glandular epithelium is called an adenoma; a benign tumor of smooth muscle is leiomyoma.

The Nomenclature of Malignant Neoplasm
Malignant neoplasm of the epithelial tissue is called carcinoma. Carcinomas show some features of epithelial differentiation, designated by adding the "carcinoma" after the name of the epithelium. For example, the malignant neoplasm of the squamous epithelium is termed as squamous cell carcinoma; the malignant neoplasm of the glandular epithelium is named adenocarcinoma. Some carcinomas have more than one kind of epithelial differentiation, for example, adenosquamous carcinoma of lung has both squamous cell carcinoma and adenocarcinoma component. Undifferentiated carcinoma refers to a cancer, which can be identified as cancer with morphology or immunophenotype but lacking specific epithelial differentiation characteristics.

Malignant neoplasm of the mesenchymal tissues is called sarcoma. Mesenchymal tissues include fibrous tissue, adipose tissue, muscle, blood vessel, lymph vessel, bone, and cartilage. Sarcomas show the characteristics of differentiation to some kinds of mesenchymal tissues, designated by adding the "sarcoma" after the name of the mesenchymal tissue, such as fibrosarcoma, liposarcoma, leiomyosarcoma, and osteogenic sarcoma. Undifferentiated sarcoma refers to a sarcoma that can be identified with morphology or immunophenotype but lacking specific mesenchymal differentiation characteristics. Carcinosarcoma is a malignant neoplasm containing two components of carcinoma and sarcoma. It should be emphasized that, in pathology, carcinoma is a malignant neoplasm of epithelial tissue. Cancer refers to all malignant neoplasms, including carcinoma and sarcoma.

5.3.1.2   Exceptional Conditions of Tumor Nomenclature

In addition to the general rules mentioned above, neoplasm is sometimes named in combination with the morphological characteristics. For example, adenoma that forms papillary and cystic structures is called papillary cystadenoma; adenocarcinoma with papillary and cystic structures is named papillary cystadenocarcinoma.

Due to historical reasons, the names of a few tumors have been established by usage, not completely following the above principles. (1) The morphology of some tumors is similar to that of some immature cells or tissues during development. They are called blastoma. Benign blastoma includes osteoblastoma and malignant blastoma, such as neuroblastoma, medulloblastoma, and nephroblastoma. (2) Leukemia, lymphoma, and seminoma are malignant neoplasms actually, despite the end of –oma and –mia in name. (3) Some malignant tumors are neither carcinoma nor sarcoma but are directly called malignant –oma, such as malignant melanoma, anaplastic (malignant) meningioma, malignant teratoma, and malignant schwannoma. (4) Some tumors are named after the scholars who initially described or studied the tumor, such as Ewing sarcoma and Hodgkin lymphoma. (5) Several neoplasms are named in the form of neoplastic cells, for example, clear cell sarcoma. (6) "-omatosis" in the names of neurofibromatosis, lipomatosis, angiomatosis, etc. mainly refers to the state of multiple tumors. (7) Teratoma is an omnipotent tumor occurring in the gonadal or embryonic remaining parts, often occurring in the gonadal gland, usually containing a variety of components of more than two germ layers. It is divided into two types of benign and malignant teratoma.

5.3.2 Classification

The classification of neoplasm is mainly based on the tissue type, cellular type, and biological behavior, including clinicopathological characteristics and prognosis of various neoplasms. The classification of common tumors is shown in Table 5.1. There are more detailed classifications of tumors in each organ system.

Different types of neoplasms have different clinicopathological features, therapeutic response, and prognosis. The correct classification of neoplasms is an important basis for the formulation of treatment plans and the prognosis of patients. Classification is also the basis of the diagnosis and research of neoplasms. Proper classification helps to define diagnostic criteria and unify diagnostic terminology, which is the premise of clinical pathological diagnosis. Unified diagnostic criteria and terms are also the basic requirements for disease statistics, epidemiological investigation, etiological and pathogenetic study, and comparative analysis of the results of different institutions.

The World Health Organization (WHO) invites experts worldwide to classify neoplasms in different systems, constantly revising it according to the progress of clinic and basic research. Now, the WHO tumor classification is widely used in the world.

Pathological diagnosis of neoplasm relies not only on the clinical manifestation of patients, medical imaging, and morphologic features but also on the expressions of specific molecules in tumor cells. For example, the expressions of desmin in myogenic tumor, cluster of differentiation (CD) antigen on the surface of lymphatic cells, various cytokeratin (CK) in epithelium, and human melanoma black 45

Table 5.1 Nomenclature of neoplasms

Tissue of origin	Benign	Malignant
Tumors of epithelial origin		
Stratified squamous cell	Squamous cell papilloma	Squamous cell carcinoma
Basal cells		Basal cell carcinoma
Glandular epithelial cell	Adenoma	Adenocarcinoma
Urinary tract epithelium	Urothelial papilloma	Urothelial carcinoma
Tumors of mesenchymal origin		
Fibrous tissue	Fibroma	Fibrosarcoma
Adipose tissue	Lipoma	Liposarcoma
Smooth muscle	Leiomyoma	Leiomyosarcoma
Striated muscle	Rhabdomyoma	Rhabdomyosarcoma
Blood vessels	Hemangioma	Angiosarcoma
Lymph vessels	Lymphangioma	Lymphangiosarcoma
Bone, cartilage	Osteoma, chondroma	Osteogenic sarcoma, chondrosarcoma
Lymphoid hematopoietic issue		
Lymphatic cells		Lymphomas
Hematopoietic cells		Leukemias
Nervous tissue, meninges		
Gliocyte		Diffuse astrocytomas
Neuron	Gangliocytoma	Neuroblastoma, medulloblastoma
Meninges	Meningioma	Anaplastic meningioma
Schwann cells	Schwannoma	Malignant schwannoma
Other neoplasms		
Melanocytes	Nevus	Malignant melanoma
Placental trophoblast cells	Hydatidiform mole	Choriocarcinoma
Germ cells		Seminoma, embryonal carcinoma
Totipotential cells in gonads or in embryonic rests	Mature teratoma	Immature teratoma, teratocarcinoma

(HMB45) in malignant melanoma cells can be detected by immunohistochemical (IHC) method (Fig. 5.4). Ki67 and other markers can be used to detect the proliferative activity of tumor cells, which helps to assess their biological behavior and prognosis (Fig. 5.5). These markers are important tools for modern pathological diagnosis. Immunological markers play an increasingly important role in the diagnosis of tumor tissue.

Table 5.2 lists common immunological markers and cell or tissue type that usually expresses these markers in tumor diagnosis. It must be noted that most of the immunological markers do not have absolute specificity. It is usually necessary to use a set of markers, good positive and negative controls at the same time,

Fig. 5.4 Expression of
HMB45 in malignant
melanoma (IHC, 200×)

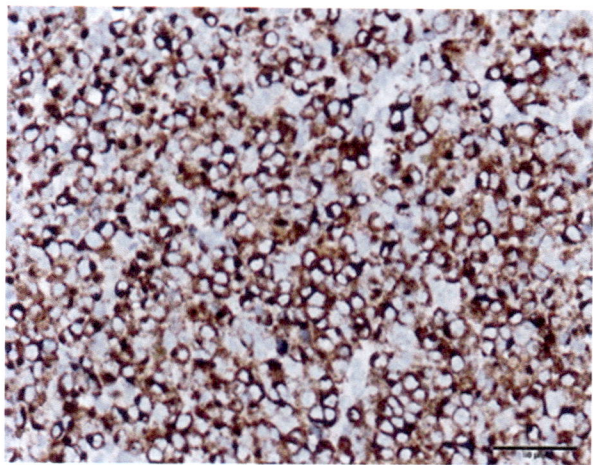

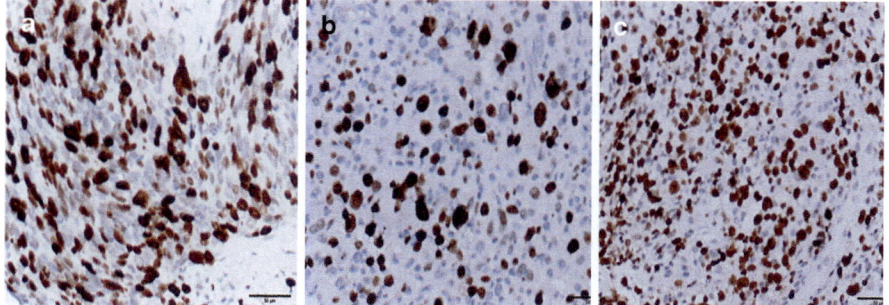

Fig. 5.5 Strong nuclear positivity of Ki67 in poorly differentiated carcinoma (**a**), sarcoma (**b**), and
lymphoma (**c**) (IHC, 200×)

to help the diagnosis of histology. Otherwise, it can easily lead to inappropriate
conclusions.

An increasingly in-depth study of the molecular mechanism of tumorigenesis
provides a new direction for the classification, diagnosis, and treatment of neoplasms. In addition to the morphological and biological behavior of various tumors,
the latest version of WHO's organ system tumor classification also takes into consideration the characteristic cytogenetics and molecular genetic changes. In recent
years, the DNA chip technology has been used to detect the gene expression profiles of tumor cells in a large scale, and the characteristic expression profiles associated with biological behavior or treatment response and prognosis are also shown
in some tumors. Molecular diagnosis had been one of the important means of
tumor pathological diagnosis. Table 5.3 lists cytogenetic changes of some
neoplasms.

Table 5.2 Common immunohistochemical markers of neoplasm

Marker	Common positive cells or tumor types
Alpha fetal protein, AFP	Fetal liver, yolk sac, hepatic carcinoma, yolk sac tumor
CD3	T-lymphatic cell, T-cell lymphoma
CD15	Granulocyte, R-S cell (Hodgkin lymphoma), some adenocarcinoma
CD20	B-lymphatic cell, B-cell lymphoma
CD30	R-S cell (Hodgkin lymphoma), large cell anaplastic lymphoma, embryonal carcinoma
CD31	Endothelium, neoplasm of blood vessel
CD34	Endothelium, neoplasm of blood vessel, gastrointestinal stromal tumors, solitary fibrous tumor
CD45	Leukocyte, neoplasm of lymphoid hematopoietic tissue
CD45RO	T-lymphatic cell, T-cell lymphoma
CD68	Macrophage
CD79a	B-lymphatic cell, B-cell lymphoma
Calcitonin	Thyroid parafollicular cells, thyroid medullary carcinoma
Chromogranin A, CgA	Neuroendocrine cell, neuroendocrine neoplasm, pituitary adenoma
Cytokeratin, CK	Epithelium, mesothelial cell, carcinoma, mesothelioma
Desmin	Muscle cell, leiomyoma, leiomyosarcoma, rhabdomyosarcoma
Epithelial membrane antigen, EMA	Epithelium, carcinoma, meningioma
Glial fibrillary acidic protein, GFAP	Gliocyte, astrocytoma
HMB45	Melanoma, angiomyolipoma, PEComa
Ki67	Proliferative cell
Placental alkaline phosphatas, PLAP	Germ cell tumor
Prostate specific antigen, PSA	Prostate epithelium, prostate carcinoma
S-100	Nervous tissue, adipose tissue, Langerhans histocyte, schwannoma, adipose tissue tumor, melanoma
Smooth muscle actin, SMA	Smooth muscle cell, myofibroblast, leiomyoma, myofibroblast tumor
Synaptophysin, Syn	Neuron, neuroendocrine cell, neuron tumor, neuroendocrine neoplasm

Table 5.3 Cytogenetic changes of some neoplasm

Type of tumor	Cytogenetic changes
Carcinoma of the lung	del(3)(p14–23)
Renal carcinoma	del(3)(p14–23), t(3:5)(p13;q12)
Nephroblastoma(Wilm's tumor)	del(11)(p13)
Dermatofibrosarcoma protuberant	t(17;22)(q22;q13)
Myxoid liposarcoma	t(12;16)(q13;p11),t(12;22)(q13;q11–12)
Synovial sarcoma	t(x;18)(p11;q11)
Rhabdomyosarcoma	t (2;13)(q35–37),t(1;13)(q36;q14)
Myxoid chondrosarcoma	t (9;22)(q22;q12)
Astrocytoma	del(9)(p13–24)
Neuroblastoma	del(19)(p32–26)
Retinoblastoma	del(13)(q14)
Primitive neuroectodermal tumors(PNET)	t(11;12)(q24;q12),t(21;22)(q22;q12), t(7;22) (p22;q12),t(17;22)(q12;q12), t(2;2 2)(q33;q12)

5.4   Growth and Spread of Neoplasm

In addition to continuous growth, malignant tumor also has local invasion and even spreads to other parts through metastasis.

5.4.1   The Growth of Neoplasm

5.4.1.1   The Growth Pattern of Neoplasm

There are three main growth patterns of neoplasm: expansile, exophytic, and invasive growth patterns.

Benign tumors in solid organ are mostly expansile. The growth rate of benign tumor is slower. As the volume increases, the tumor pushes but not invades surrounding tissues. The benign tumor is well circumscribed and can form a complete fibrous capsule around the tumor. Encapsulated tumor can often be pushed and easily removed without recurrence. The effect of expansile growth pattern on local organ and tissue is mainly squeezing.

Exophytic growth refers to the growth of a neoplastic protrusion on the surface of body or organ (such as the digestive tract) and in the cavity (such as thoracic cavity and peritoneal cavity). The neoplastic protrusion can be papillary, polypoid, fungating, or cauliflower. Both benign and malignant neoplasms can grow exogenously, but malignant neoplasms often infiltrate at the base of neoplastic protrusion. Due to the rapid growth of malignant tumor, the blood supply of the central part of the tumor is relatively inadequate. So tumor cells are prone to necrosis, and the necrotic tissues fall off to form an ulcer with an uneven bottom and a marginal bulge.

Invasive growth usually is a hallmark of malignancy. Invasion refers to the phenomenon that neoplastic cells grow into and destroy the surrounding tissue surrounding tissue including tissue space, lymph, and blood vessels. Invasive tumor has no capsule and no clear borderline with the adjacent normal tissue. It isn't easily moved with palpation. During the operation, a larger area of the surrounding tissues should be removed. If the resection is not complete, postoperative recurrence is easy. In surgery, it is helpful for the surgeon to determine if it needs to be expanded through rapid frozen section examination of the marginal tissues by pathologist.

5.4.1.2   The Growth Characteristics of Neoplasm

The growth rate of different tumors is very different. Benign tumors usually grow slowly, for years or even decades. Malignant tumors grow faster, especially those with poor differentiation. They can form an obvious mass in short time. There are many factors that affect the growth rate of tumor, such as the doubling time of tumor cells, growth fraction, the proportion of tumor cell formation, and death.

The doubling time of tumor cells refers to the time required for cell division to propagate into two progeny cells. The doubling time of most malignant tumor cells is not faster than that of normal cells, so the rapid growth of malignant tumor may not be caused by the shortened time of tumor cell doubling. Growth fraction refers to the proportion of cells that proliferate in the tumor cell population. The cells that proliferate are constantly dividing and proliferating. Every time this process of division and proliferation is called a cell cycle, including G1, S, G2, and M stages. In the early stage of malignant tumor, cell division and proliferation are active, and the growth fraction is high. With the growth of tumor, some tumor cells enter quiescent stage (G0 stage) and stop dividing and proliferating. Many antitumor chemotherapeutic agents play a role in interfering with cell proliferation. Therefore, tumors with high growth fraction are sensitive to chemotherapy. If a tumor has large number of non-proliferative cells, it may be less sensitive to chemotherapeutic drugs. Radiation therapy or surgery can be performed to reduce or remove most of the tumor. At this time, the residual G0 tumor cells can enter the proliferation period and increase the sensitivity of the tumor to chemotherapy. To promote tumor cell death and inhibit tumor cell proliferation are two important ways of tumor therapy.

5.4.1.3   The Angiogenesis of Neoplasm

Neoplasm cannot continue to grow beyond the size of 1–2 mm in diameter unless it has the ability to induce neo-angiogenesis. Neoplastic cells themselves and inflammatory cells (mainly macrophages) can produce angiogenesis factors such as vascular endothelial growth factor (VEGF) to induce neovascularization. The combination of angiogenic factors and their receptors, which are located on the surface of vascular endothelial cells and fibroblasts, can stimulate the endothelial cells division and capillary sprouting growth. Recent evidence shows that neoplastic cells themselves can form a vascular-like tubular structure with the basement membrane, which can communicate with the blood vessels as a microenvironment or microenvironmental component that does not depend on angiogenesis. It is called vasculogenic mimicry. Inhibition of tumor angiogenesis or "vasculogenic mimicry" is an important topic of antitumor research.

5.4.1.4   The Progression and Heterogeneity of Neoplasm

The increasing invasion of malignant tumors during growth is called the progression of tumors, showing as the speed of growth, the infiltration of the surrounding tissue and the occurrence of distant metastasis. Tumor progression is related to its increasing heterogeneity. Although neoplasms are derived from monoclonal proliferation of a single malignant transformation cell, progeny cells may have the different changes of genes or other large molecules, variations in the growth rate, invasive ability, the response of growth signal, and the sensitivity of anticancer drugs in the process of growth. At this time, the tumor cell population is no longer made up of exactly the same tumor cells but heterogeneous subclones with their respective characteristics. In the process of tumor progression, which acquires heterogeneity, cells with growth advantages and strong invasiveness overwhelm cells with no growth advantages and weak invasiveness.

In recent years, the study of leukemia, breast cancer, glioma, and other tumors shows that although a tumor is made up of a large number of tumor cells, there are a few cells that have the ability to start and maintain tumor growth and self-renewal, which are called cancer stem cells (CSC), tumor stem cells (TSC), and tumor-initiating cells (TIC). CSCs were first described in tumors of hematopoietic origin and have now been identified in several types of solid tumors, such as cancers arising in the breast, lung, prostate, colon, brain, head and neck, pancreas, and skin. Long-term self-renewal potential, quiescence, and resistance to chemotherapy and radiotherapy are proprieties associated with CSCs. Further research on CSCs will help to understand the tumorigenesis, growth, and response to treatment and explore the new therapeutic treatment.

5.4.2   The Spread of Neoplasm

Malignant tumors not only infiltrate in the primary site, adjacent organs, or tissues but also spread to other sites of body by various ways. Local invasion and metastasis are the most important biological characteristics of malignant tumors.

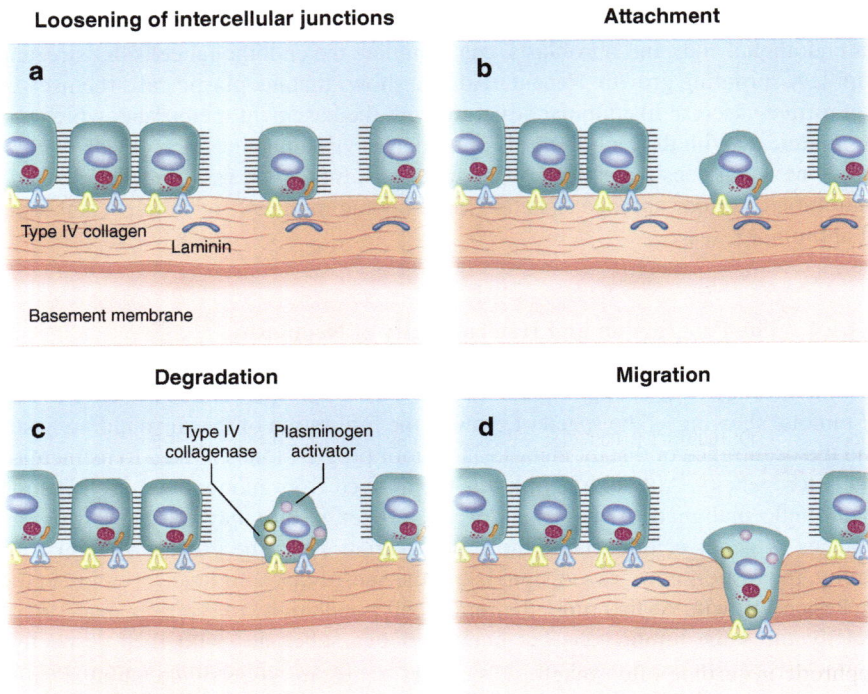

Fig. 5.6 The mechanisms of local invasion and direct spreading of neoplasm

5.4.2.1 Local Invasion and Direct Spreading

Direct spreading denotes the phenomenon that as malignant tumors continue to grow, tumor cells often infiltrate along tissue gaps or nerve bundle to damage adjacent organs or tissues. For example, advanced cervical cancer directly spreads to rectum and bladder.

The mechanisms of local invasion and direct spreading are quite complicated. Taking cancer as an example, it can be roughly summed up in four steps (Fig. 5.6). (1) The reduction of adhesion molecules on the surface of cancer cells. On the surface of normal epithelium, there are a variety of cell adhesion molecules (CAMs). The interaction between them helps to bind cells together and prevent cell migration. The reduction of adhesion molecules on the surface of cancer cells causes cells to separate from each other. (2) The increased attachment of cancer cells to the basement membrane. The attachment of normal epithelium to the basement membrane is mediated by some molecules at the basal surface of the epithelium such as laminin (LN). Cancer cells express more LN receptors, increasing the attachment of cancer cells to the basement membrane. (3) The degradation of extracellular matrix (ECM). Cancer cells produce proteinase such as type IV collagenase to dissolve the extracellular matrix components, form the partial defect in the basement membrane

Fig. 5.7 Lymphatic
metastasis of papillary
thyroid carcinoma (H&E,
100×)

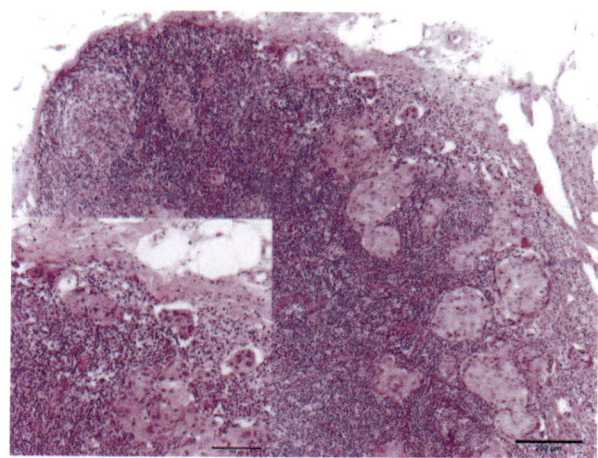

and subsequently help cancer cells to pass through. (4) The migration of cancer
cells. Cancer cells remove through the basement membrane defect by amoebic
movement. After that, cancer cells further dissolve interstitial connective tissue and
move in the interstitium. When cancer cells reach the wall of the vessel, they pass
through the basement membrane of blood vessels in a similar way into the blood
vessels.

Metastasis

The term metastasis denotes the process that malignant tumor cells migrate to other
site and continue to grow to form the same type of tumor by invading lymph vessels,
blood vessels, or body cavity from the original site. Tumor formed by metastasis is
called metastatic tumor or secondary tumor. Metastasis is an important feature of
malignant tumors; however, not all malignant tumors can metastasize. For example,
basal cell carcinoma of the skin can cause damage to local tissues but rarely
metastasize.

Malignant tumors metastasize through the following ways:

1. Lymphatic metastasis. Neoplastic cells invade the lymph vessels (Fig. 5.7) and
   reach the local lymph nodes (regional lymph nodes) by lymphatic flow. For
   example, the carcinomas of the upper quadrant of the breast usually metastasizes
   to the ipsilateral axillary lymph nodes to form metastatic breast cancer of the
   lymph nodes. Tumor cells first gather at marginal sinus and then involve the
   whole lymph node, leading to the swelling of lymph nodes and the hardening of
   the texture. The tumor invades the capsule and allows the neighboring lymph
   nodes to merge into clusters. The metastasis of local lymph nodes can continue
   to transfer to other lymph nodes at the next stop of the lymphatic circulation and
   finally enter the blood flow through the thoracic duct.
2. Hematogenous metastasis (Fig. 5.8). After cancer cells invade the blood vessels,
   they can reach distant organs with blood flow, continue to grow, and form metastases. Because the venous wall is thinner and the pressure in the tube is relatively

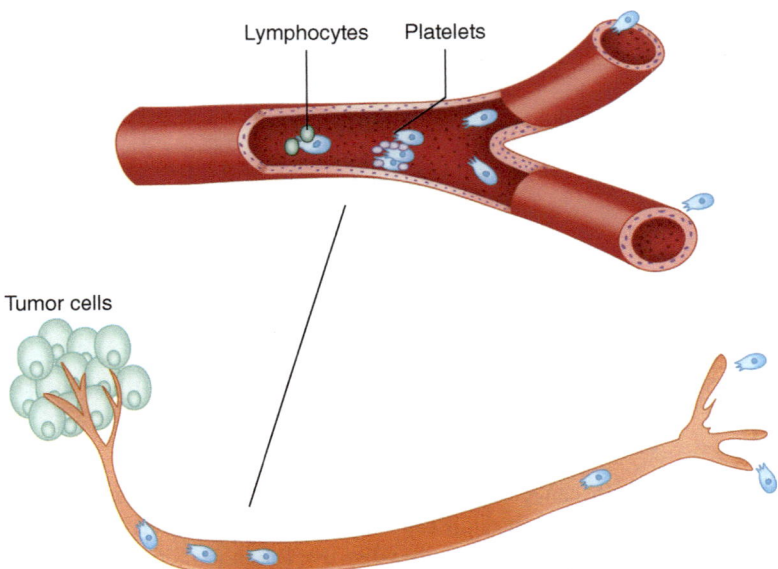

Fig. 5.8 The mechanism of hematogenous metastasis of malignant tumors

low, tumor cells often pass through the vein into the blood, and a few can also be connected to the lymphatic vessels. The tumor cells that invade the systemic circulatory vein reach the lung through the right heart and form a metastatic tumor in the lung, such as lung metastases from osteosarcoma. The tumor cells that invade the portal venous system first form hepatic metastasis, such as hepatic metastasis of gastrointestinal carcinoma. Tumor cells of primary lung tumor or intrapulmonary metastatic tumor can invade the pulmonary vein directly or enter the pulmonary vein through the pulmonary capillary. Then they metastasize to all organs through the left heart with the blood flow of the aorta, such as the brain, bone, kidney, and renal adrenal gland. Thus, the metastatic tumors of these organs usually occur after lung metastasis. Additionally, tumor cells that invade the chest, waist, and pelvic vein can enter the vertebral venous plexus through anastomotic branches, for example, prostate cancer can metastasize to the spine through this pathway and then to the brain without lung metastasis.

Malignant tumors can involve many organs through the blood channel, but the most frequently involved organs are lung and liver. So, clinically, it is necessary to carry out imaging examination of lung and liver for judging whether there is a hematogenous metastasis or not and determining the patient's clinical staging and treatment plan. The morphologic features of metastatic tumor are clearly demarcated, often multiple, scattered, much closer to the surface of the organ. Metastatic tumors located at the surface of the organ form cancer umbilicus due to the central hemorrhage, necrosis, and subsidence of cancer nodules.

Malignant tumor cells that enter the blood vessels do not always migrate to other organs to form metastases. Most of the single cells are destroyed by natural killer cells, but the tumor cells agglutinated with platelets form an uneasily eliminated tumor cell plug, which can be adhered to the vascular endothelial cells and then pass through the vascular endothelium and the basement membrane to form a new metastatic tumor. In the course of tumor progression, there are subclones with different invasiveness. Highly invasive subcellular clones are easy to form widespread hematogenous dissemination.

The location of tumor metastasis is affected by the location of primary tumor and blood circulation. However, some tumors show affinity for some organs. For example, lung cancer easily metastasizes to the adrenal and brain; thyroid carcinoma, renal carcinoma, and prostate cancer easily metastasize to the bone; and breast cancer often metastasizes to the lung, liver, bone, ovary, and so on. These phenomena may be related to the following factors: (1) The ligands on the endothelial cells of these organs can specifically identify and bind the adhesion molecules on the surface of some cancer cells. (2) These organs release the chemotactic factors that attract some cancer cells. (3) This is the result of negative selection, that is, the environment of some tissues or organs is not suitable for tumor growth, for example, enzyme inhibitors in tissues are not conducive to the formation of metastases.

3. Seeding metastasis. When a malignant tumor occurring in the thoracic or abdominal cavity penetrates to the organ surface, the tumor cells can fall off and grow on the surface of the other organs of the body like sowing, forming multiple metastatic tumors. This dissemination is called seeding metastasis. It is common in malignant tumor of abdominal organ. For example, mucinous carcinoma of the gastrointestinal tract invades the serosa and can be implanted into the greater omentum, peritoneum, and pelvic organs, such as the ovary. In the ovary, there is a bilateral ovarian growth and a diffuse infiltration of carcinoma of the signet ring cells with mucus. This special type of ovarian metastatic tumor is called Krukenberg tumor. It should be noted that Krukenberg tumors are not necessarily seeding metastases but also metastases through lymphatic and blood channels.

The implant metastasis of the serous cavity is often accompanied by serous effusion, caused by the block of subserous lymphatics or capillaries by tumor embolus, the increase of capillary permeability, the blood leakage, and the bleeding induced by tumor cells destroying blood vessels. The effusion of the body cavity may contain unequal tumor cells. It is one of the important ways for the diagnosis of malignant tumor to extract the effusion of the body cavity for cytological examination in order to find the malignant tumor cells.

5.5 Grading and Staging of Neoplasm

Li Liang

Grading of malignancies is an indicator of malignancy. The malignancies are graded on the basis of the degree of differentiation, atypia, and the number of mitoses. Three categories of grading are used mostly. I grade is well differentiated with low malignancy. II grade is moderately differentiated. III grade is poorly differentiated. Some tumors use two categories (low grade and high grade). It should be noted that I, II, and III grades are not equivalent to biological behavior codes in the international classification of diseases ICD-O.

Tumor staging refers to the growth and spread of malignant tumors. The larger the tumor size, the wider the growth and dissemination, and the worse the prognosis. The following factors should be considered in tumor staging: the size of the primary tumor, the depth of infiltration, the scope of infiltration, the involvement of adjacent organs, the local and distant lymph node metastasis, the distant metastasis, and so on.

The staging system widely used in the world is TNM classification system. T refers to the primary tumor. N refers to regional lymph node involvement, and M refers to metastasis. TNM staging varies for specific forms of cancer, but there are general principles. The primary lesion is characterized as T1 to T4 with the increase of tumor volume and the extent of adjacent tissue involvement. T0 is used to indicate an in situ lesion. N0 means no nodal involvement, whereas N1 to N3 represent the increase of the degree and range of lymph node involvement. M0 means no distant metastasis, whereas M1 indicates the presence of metastasis.

5.6   Effects of Neoplasms on the Host

Benign tumors are composed of well-differentiated tumor cells. They grow slowly in local sites and do not infiltrate and metastases. Therefore, benign tumors generally have relatively little influence on the body, mainly manifested by local compression and obstructive symptoms. The occurrence or severity of these symptoms is mainly related to the location and secondary changes of the tumor. For example, benign tumors on the body surface do not have significant effect on the body except for a few local symptoms. However, if they occur in the cavity or an important organ, they may cause more serious consequences. For example, a leiomyoma that enters the intestinal cavity can cause severe intestinal obstruction or intussusception. Intracranial benign tumors may lead to increased intracranial pressure and the corresponding neurological symptoms due to oppression of the brain tissue and obstruction of the ventricular system. Sometimes, benign tumors have secondary changes and also affect the body to varying degrees. For example, uterine submucosal leiomyomas are often accompanied by superficial erosion or ulceration of the endometrium, which may lead to bleeding and infection. Benign tumors of the endocrine gland can secrete too many hormones and cause symptoms. For example, pituitary growth adenomas secreting excessive growth hormone can cause gigantism or acromegaly.

Malignant tumors are poorly differentiated, rapidly growing, infiltrating, and destroying the structure and function of organs. They also metastasize. The effects of malignant tumors on the host are serious, and the therapeutic effect is not satisfactory. The patient's mortality is high, and the survival rate is low. In addition to the symptoms of local compression and obstruction, malignant tumors are also prone to ulcers, bleeding, and perforation. Tumors involving local nerves can cause stubborn pain. Tumor products or combined infection can cause fever. Malignant tumors of the endocrine system, including those of the diffuse neuroendocrine system (DNES) such as carcinoids and neuroendocrine carcinomas, produce biogenic amines or polypeptide hormones and cause endocrine disorders. Patients with advanced malignant tumors often develop cancer cachexia, which is characterized by severe body weight loss, anemia, anorexia, and general weakness. The occurrence of cancer cachexia may be mainly the result of tumor tissue itself or cytokines produced by body reactions.

Some non-endocrine tumors may also produce and secrete hormones or hormonelike substances, such as adrenocorticotropic hormone (ACTH), calcitonin, growth hormone (GH), and parathyroid hormone (PTH), resulting in endocrine symptoms, which is called ectopic endocrine syndrome. This type of tumors is mostly malignant, with a majority of carcinomas, such as lung carcinoma, stomach carcinoma, and liver carcinoma. The production of heterotopic hormones may be related to abnormal gene expression in tumor cells.

Ectopic endocrine syndrome belongs to paraneoplastic syndrome. The generalized paraneoplastic syndrome refers to some lesions and clinical manifestations that cannot be explained by the direct spread or distant metastasis of the tumor and are caused by tumor products (such as ectopic hormones) or abnormal immune responses (such as cross-immunity). Patients exhibit abnormalities of many systems, including endocrine, neural, digestive, hematopoietic, bone, joint, kidney, and skin. It should be noted that endocrine tumors (such as pituitary adenomas) produce lesions or clinical manifestations that are caused by endocrine-inherent hormones (such as growth hormone), which do not belong to paraneoplastic syndromes.

Some tumor patients show paraneoplastic syndrome before finding tumors. The tumors will be found in time if the paraneoplastic syndrome is considered carefully and further looked for its causes by medical staff. In addition, when the diagnosed tumor patients have such symptoms, the possibility of paraneoplastic syndrome should be taken into account, and it should not be mistaken as metastasis.

5.7 Identification Between Benign and Malignant Neoplasms

The biological behaviors of tumors and their influences on hosts vary greatly. Most tumors are divided into benign or malignant tumor. Benign tumors are generally easy to treat and with a good therapeutic reaction. Malignant tumors are harmful, treatment measures are complex, and the results are not satisfactory. If a malignancy is misdiagnosed as a benign tumor, the treatment may be delayed or incomplete. Conversely, misdiagnosing of a benign tumor as a malignancy can lead to overtreatment. Therefore, it is of great significance to distinguish benign and malignant tumors. The main differences between benign and malignant tumors are listed in Table 5.4.

Table 5.4 Differences between benign and malignant neoplasms

	Benign neoplasm	Malignant neoplasm
Differentiation	Well differentiated, minor atypia	Poorly differentiated, obvious atypia
Mitosis	No or less, without pathological mitotic figures	Many, with pathological mitotic figures
Growth rate	Slow	Fast
Growth pattern	Expansive or exogenous growth	Invasive or exogenous growth
Secondary changes	Uncommon	Common, such as bleeding, necrosis, ulceration, etc.
Metastasis	No	Yes
Recurrence	No or less	Often
Impact on hosts	Less, local oppression or obstruction	More, destroy the tissue of the original site and metastatic sites, necrosis, hemorrhage, and infection; cachexia

Some tumors cannot be classified as benign or malignant directly. They need to be evaluated for the risks of recurrence and metastasis (low, medium, and high) based on their morphological characteristics. In some types of tumors (such as ovarian serous tumors), in addition to typical benign tumors (such as ovarian serous papillary cystadenoma) and typical malignant tumors (such as ovarian serous papillary cystadenocarcinoma), there are also borderline tumors with the tissue morphology and biological behavior between benign and malignant, such as ovarian borderline serous papillary cystadenoma. Some borderline tumors have a tendency to develop malignant; the malignant potential of some borderline tumors is difficult to determine at present. Further studies are needed to further understand its biological behavior.

Tumorlike lesions or pseudoneoplastic lesions are not true tumors. However their clinical manifestations or histological morphology are similar to those of tumors. Some tumorlike lesions may be mistaken for malignant tumors. Therefore, it is important to recognize this type of lesion and have full consideration in differential diagnosis.

It must be emphasized that benign or malignant tumors refer to their benign or malignant biological behaviors. Pathologically, in most cases, it is feasible to judge benign or malignant tumors by their morphological changes and subsequently estimate their biological behavior and prognosis. At present, pathologic diagnosis is the most important in all tumor detection methods. However, it must be recognized that there are many complex factors that affect the biological behavior of a tumor. Though pathologists observe some of their characteristics (tumor morphology, immunophenotype, etc.), many factors (especially changes at molecular level) are less known to us. In addition, the histological diagnosis will inevitably encounter technical problems such as whether the sample of the tissue is representative. Therefore, this prognosis assessment is not very accurate. When performing pathological diagnosis, pathologists not only need to rely on the diagnostic criteria generally accepted in the field at that time and their own experience and judgment but also pay attention to clinicopathological correlation, which means taking full account of the patient's clinical condition, imaging data, and other test results. In medical practice, the role of the pathologist is a consultant, who provides diagnosis by considering clinical and pathological information comprehensively. When formulating treatment plans, each physician has the responsibility to fully consider the clinical and pathological links and make reasonable judgments and decisions. Patients and their families often lack understanding of the complexity of the diagnosis and treatment of diseases (including tumors) and often expect simple and definitive answers to all questions (benign or malignant, treatment response, survival, etc.). All physicians (including pathologists) have the responsibility to use various opportunities to educate the public and make them fully aware of the complexities of the diagnosis and treatment of these diseases.

5.8 Brief Introduction of Common Neoplasms

This section describes the general clinical and pathological features of some common tumors. In the chapters of systemic diseases in this book, a more detailed introduction to common tumors in various organ systems is provided.

5.8.1 Epithelial Tumors

Epithelial tissues include epithelial and glandular epithelia. Epithelial tumors are very common. Most malignant tumors of the human body are malignant epithelial tumors (carcinomas), which are extremely harmful to human beings.

5.8.1.1 Benign Epithelial Tumors

Papilloma

Papilloma occurs in the areas covered by squamous epithelium, urinary tract epithelium, and so on, known as squamous cell papilloma (Fig. 5.9), urothelial papilloma, etc. It grows exogenously to the surface of the body or cavity, forming finger or papillary protrusion, and may also be cauliflower or villous. The root of the tumor may have pedicles connected to normal tissue. Microscopically, the axis of the papilla consists of interstitial components, such as blood vessels and connective tissue, and the surface is covered with epithelium.

Fig. 5.9 Squamous cell
papilloma of skin (H&E,
100×)

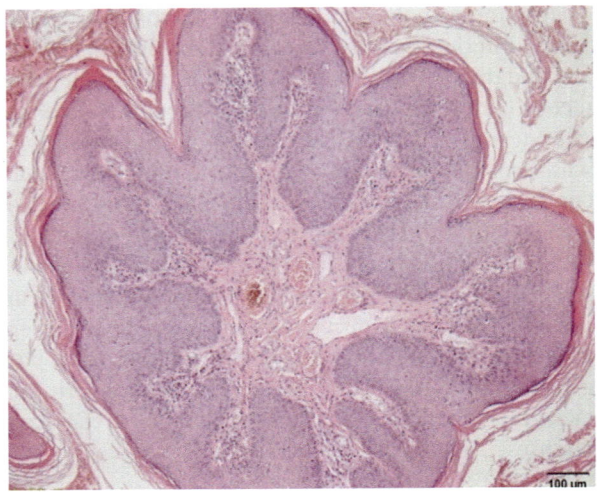

Adenoma

Adenoma is a benign tumor of the glandular epithelium and commonly occurs in
intestine, breast, and thyroid. Mucosal adenomas are mostly polypoid; adenomas in
the glandular organs are mostly nodular with clear boundaries to the surrounding
normal tissue. The glands of adenomas are similar to those of the corresponding
normal tissues and may have secretory functions.

According to the composition or morphological characteristics of adenomas,
they are divided into tubular adenoma, villous adenoma, cystadenoma, fibroadenoma, pleomorphic adenoma, etc.

Tubular Adenoma and Villous Adenoma

Tubular adenoma and villous adenoma are common in colon and rectal mucosa,
often polypoid, and may have pedicle connected to the mucosa. However, some
adenomas are sessile or flat. Microscopically, the glandular epithelium of tumor
forms well-differentiated tubules or villous structure or mixed structures (called
tubulovillous adenomas). Villous adenomas have a higher risk of developing carcinoma, especially those with larger volumes. In familial adenomatous polyposis
(FAP), the risk of adenoma developing carcinoma is extremely high, and patients
with carcinoma are younger.

Cystadenoma

Cystadenoma results from the accumulation of glandular secretions in the adenoma
and enlargement of the lumen. Grossly, the cysts with different sizes can be seen,
often occurring in the ovary and so on. There are two main types of ovarian

cystadenoma: serous papillary and mucinous cystadenomas. In serous papillary cystadenoma, the glandular epithelium shows papillary growth inside the cyst and secretes serous fluid. In mucinous cystadenoma, the glandular epithelium secretes mucus with a multilocular and smooth wall and less papillae.

5.8.1.2 Malignant Epithelial Tumors

Carcinoma is the most common human malignancy. The incidence of carcinoma increases significantly among people over the age of 40.

Carcinomas that occur on the surface of the skin and mucosa are polypoid, umbellate-like or cauliflower-like, often with necrosis and ulcers on the surface. Carcinomas in the organ often show irregular nodule, infiltrating the surrounding tissues like tree roots or crab feet. The texture is hard, and the cut surface is often grayish white. Microscopically, cancerous cells are arranged in nests, acini, ducts, or cords with a clear boundary to the surrounding interstitium. Sometimes, cancerous cells diffusely infiltrate in the interstitium. Metastasis of cancer usually occurs in the early stage of lymphatic metastasis and late stage of hematogenous metastasis.

Squamous Cell Carcinoma

Squamous cell carcinoma often occurs in the areas covered by squamous epithelium, such as skin, mouth, lips, esophagus, larynx, cervix, vagina, and penis. Some regions, such as bronchus and bladder, are not covered by squamous epithelium normally, but they can develop squamous cell carcinoma on the basis of squamous metaplasia. Grossly, squamous cell carcinoma is generally cauliflower-like or ulcerative. Microscopically, in well-differentiated squamous cell carcinomas, lamellar keratins (called keratin pearls or cancer pearls) appear in the center of cancer nests (Fig. 5.10). Intercellular bridges can also be seen between cells. In poorly

Fig. 5.10 Well-differentiated squamous cell carcinoma. Keratin pearls present in the center of cancer nests (H&E, 200×)

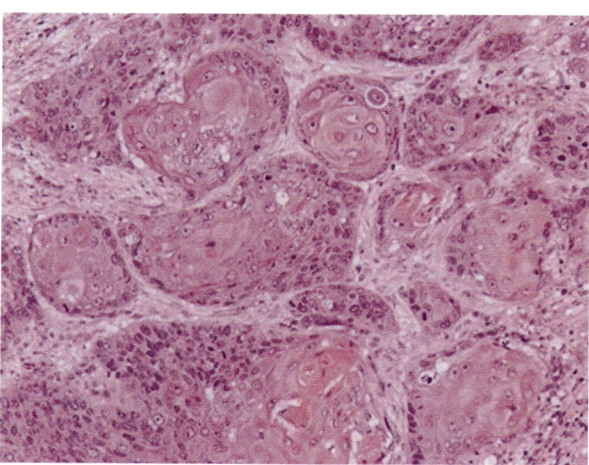

differentiated squamous cell carcinomas, there is no keratinization and less or no intercellular bridges.

Adenocarcinoma

Adenocarcinoma is a malignant tumor of glandular epithelium, which is commonly seen in the gastrointestinal tract, lung, breast, and female reproductive system. The cancerous cells form glands or adenoids with different sizes, irregular shapes, and arrangement. Cells are often arranged in irregular layers, with different sizes of nuclei and many mitotic figures (Fig. 5.11). Papillary structure predominant adenocarcinoma is called papillary adenocarcinoma. Adenocarcinoma with highly dilated cystic cavity is called cystadenocarcinoma. Cystic adenocarcinoma with papillary growth is called papillary cystadenocarcinoma.

Adenocarcinoma that secretes large amounts of mucus is called mucinous carcinomas or colloid carcinomas, commonly seen in the stomach and large intestine. Grossly, the cancerous tissue is grayish white, moist, and translucent like jellies. Microscopically, the dilated glandular cavities contain a large amount of mucus. Mucus pools can be formed due to disintegration of the glands, and cancerous cells float in the mucus. Sometimes, the mucus accumulates in cancerous cells and pushes the nucleus to one side, making the appearances of signet rings, known as signet-ring cells. Carcinomas with predominant signet ring cells are called signet-ring cell carcinomas.

Fig. 5.11 Adenocarcinoma. The tumor cells are arranged in glandular and cribriform structure (H&E, 200×)

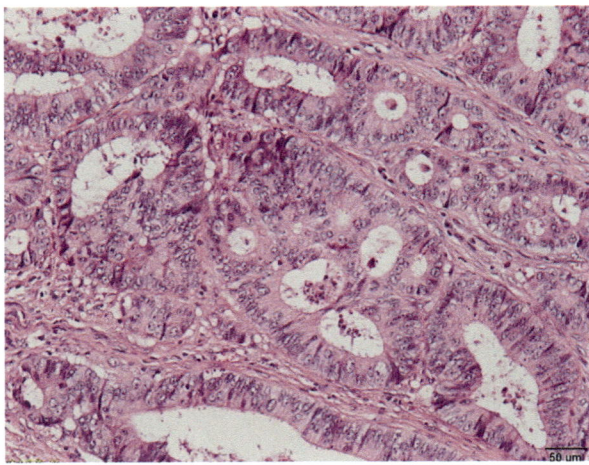

Fig. 5.12 Basal cell
carcinoma of skin (H&E,
200×)

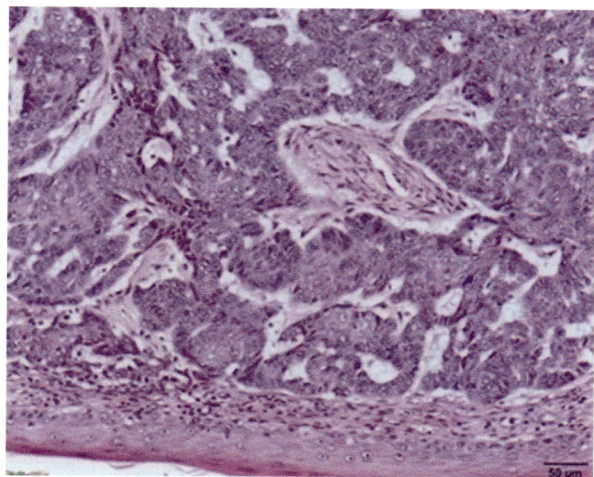

Basal Cell Carcinoma

Basal cell carcinoma often occurs in the head and face of the elderly. Microscopically, cancerous nests consist of deeply stained basal cell-like cancerous cells (Fig. 5.12). There are superficial and nodular subtypes. The carcinoma grows slowly, often forms ulcers on the surface, and infiltrates deep tissues. It rarely metastasizes. Clinically, it shows low-grade malignancy and is sensitive to radiation therapy.

Urothelial Carcinoma

Urothelial carcinoma occurs in the bladder, ureter, or renal pelvis. It can be papillary or non-papillary. It is divided into low-grade and high-grade urothelial carcinoma. The higher tumor grade they have, the more prone to relapse and deep infiltration. Those with lower grade also have a tendency to relapse. In some cases, the tumor grade increases after recurrence.

5.8.2 Mesenchymal Tumors

There are many types of mesenchymal tumors, including those from adipose tissue, blood vessels and lymphatic vessels, smooth muscle, striated muscle, fibrous tissue, and bone tissue. It is customary to classify peripheral nerve tissue tumors into mesenchymal tumors. The tumors of mesenchymal tissue except bone tumors are also known as soft tissue tumors.

Benign tumors are common in mesenchymal tissue tumors, and malignant tumors (sarcomas) are less common. In addition, there are a lot of tumorlike lesions in the mesenchymal tissue, forming a clinically visible "masses," but not true tumors. Some tumorlike lesions mimic sarcomas, making them difficult to diagnosis.

5.8.2.1 Benign Mesenchymal Tumors

Lipoma

Lipoma occurs mainly in adults and is the most common benign soft tissue tumor. It often occurs in the subcutaneous tissues of the back, shoulders, neck, and extremities. Grossly, a lipoma is usually lobulated with capsule. It is soft with yellow cut surface, just like adipose tissue (Fig. 5.13). The diameter varies from a few centimeters to several tens of centimeters. It is single or multiple. Microscopically, it looks like normal adipose tissue with irregular lobules and fibrous septa. It is easy to be removed by surgery.

Hemangioma

Hemangioma is common and occurs in many sites, such as the skin, muscle (muscle hemangioma), and visceral organs. It is divided into capillary hemangioma, cavernous hemangiomas (Fig. 5.14), venous hemangioma, and etc. Hemangioma often has no capsule and is not clearly defined. Skin or mucosal hemangioma usually protrudes as bright red or dark red masses or purple spots above the surface. Visceral hemangioma is mostly nodular. Diffuse cavernous hemangiomas occurring in limb soft tissues may cause limb enlargement. Hemangioma is more common in children and may be congenital. It may grow up with the development of the body, cease to develop in adults, or even regress spontaneously.

Fig. 5.13 Lipoma. Usually lobulated, with intact capsule

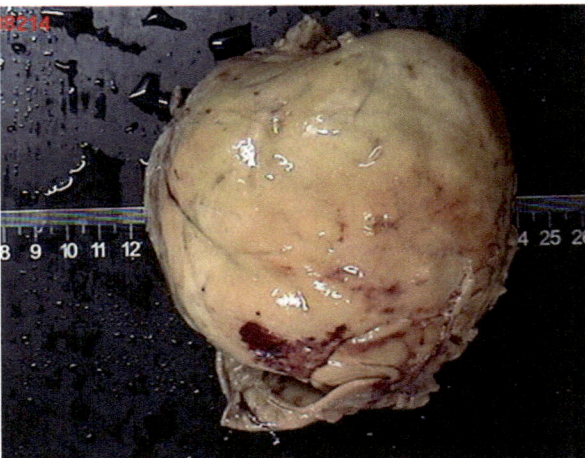

Fig. 5.14 Cavernous hemangiomas. The tumor is composed of thin vessels of different sizes (H&E, 200×)

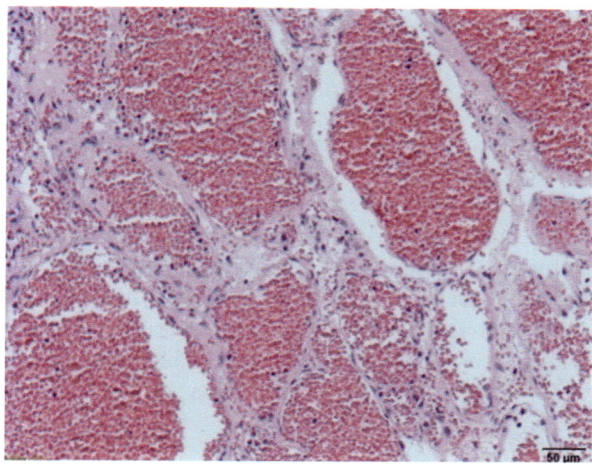

Lymphangioma

Lymphangioma is composed of hyperplastic lymphatic vessels containing lymphatic fluid. Lymphatic vessels exhibit dilated and fused cysts and contain a large amount of lymphatic fluid. It is common in children, also called cystic hygroma.

Leiomyoma

Leiomyoma is common in the uterus and etc. The tumor is composed of relatively uniform spindle cells, resembling normal smooth muscle cells. The nucleus is long-barreled and blunt at both ends. Tumor cells are arranged in bundles and woven. The mitotic figure is rare.

Chondroma

Chondroma originated from periosteum is called periosteal chondroma. Chondroma originated from the bone marrow cavity of the hand, foot, and long bones of the extremities is called enchondroma, which may cause expanding of the bone with thin bone shells outside. The cut surface is light blue or silver white, translucent with calcification or cystic degeneration. Microscopically, the tumor is composed of mature hyaline cartilage and shows irregular lobules, which are surrounded by loose fibrovascular stroma. Chondromas occurring in the pelvis, sternum, ribs, long bones, or vertebrae are prone to malignancy, and those occurring in the finger (toe) bone are rarely malignant. Pathological diagnosis and differential diagnosis of chondroma are dependent on comprehensive analyses of site, imaging, and tissue morphology.

5.8.2.2 Malignant Mesenchymal Tumors

Malignant mesenchymal tumors are collectively referred to sarcomas and less than carcinomas. Some types of sarcomas often occur in children or adolescents. For example, embryonal rhabdomyosarcoma is common in children, and 60% of osteosarcomas are under 25 years of age. Some sarcomas occur mainly in middle-aged and elderly people, such as liposarcoma. Sarcomas are usually large and fishlike in the cut surface, prone to have secondary changes including hemorrhage, necrosis, and cystic degeneration. Microscopically, most sarcoma cells do not form nests. They proliferate diffusely and are poorly demarcated with the interstitium. There are less interstitial connective tissues but more abundant blood vessels in sarcomas. Thus, sarcomas usually metastasize via blood vessels.

The difference between carcinoma and sarcoma is listed in Table 5.5.

Liposarcoma

Liposarcoma is one of the most common sarcomas in adults. It often occurs in deep soft tissues, retroperitoneal region, etc. It is common seen in adults and rare in youngsters. Grossly, most are nodular or lobular, mimicking lipoma. Some may be mucous or fish fleshlike. Tumor cells show a variety of morphology and are characterized by the appearance of lipoblasts. There are lots of lipid vesicles of varying sizes in the cytoplasm, which extrude the nucleus and form a pressure trace (Fig. 5.15). It can be divided into several subtypes: well-differentiated liposarcoma, myxoid/round cell liposarcoma, pleomorphic liposarcoma, and dedifferentiated liposarcoma.

Table 5.5 Difference between cancer and sarcoma

	Carcinoma	Sarcoma
Tissue differentiation	Epithelial tissue	Mesenchymal tissue
Incidence	Higher. It is nine times more than the sarcoma. More common in adults over 40 years old	Lower. Some types mainly occur in young people or children; some types are mainly seen in middle-aged people
General features	Hard texture, grayish-white	Soft, grayish-red, fish-flesh appearance
Microscope features	More cancer nests; a clear boundary between parenchyma and mesenchymal. Hyperplasia of the fibrous tissue in the stroma	Diffuse distribution of tumor cells. Unclear boundary between the parenchyma and the mesenchymal. Abundant blood vessels and less fibrous tissue in the stroma
Reticular fiber	Seen around the cancer nests. There are no reticular fibers around each cancer cells	There are many reticular fibers around each sarcoma cells
Metastasis	Mostly via lymphatic pathway	Mostly via blood pathway

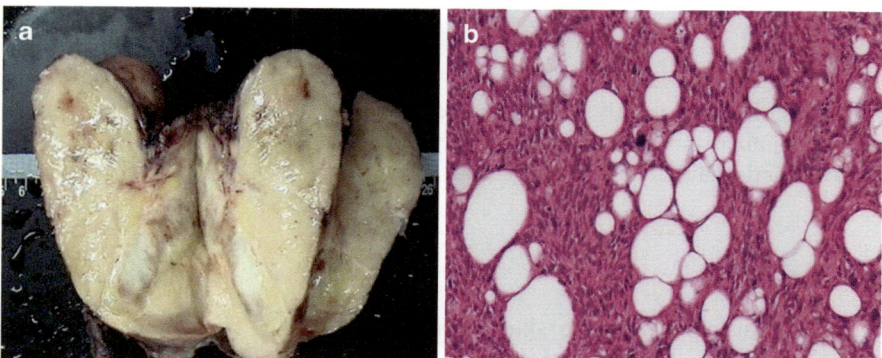

Fig. 5.15 Liposarcoma. Fish fleshlike macroscopically (**a**). Microscopically lipid vesicles of varying sizes in background of sarcoma (**b**, H&E, 200×)

Fig. 5.16 Embryonal rhabdomyosarcoma. Various differentiated rhabdomyoblast with eosinophilic cytoplasm and residual normal skeletal muscle (H&E, 200×)

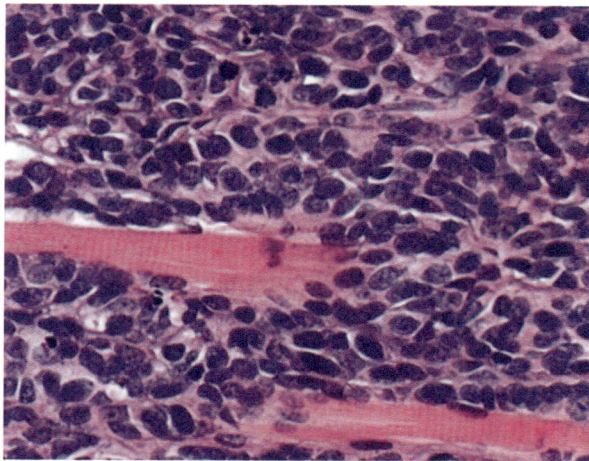

Rhabdomyosarcoma

Rhabdomyosarcoma is common in infants and younger children under 10 years and rare in adults. It occurs frequently in the head and neck, genitourinary tract, etc. and occasionally in the limbs. The tumor is composed of rhabdomyocytes from different stages of differentiation (Fig. 5.16). The cytoplasm of well-differentiated rhabdomyoblast is red. Sometimes, the longitudinal or transverse lines are visible. Histologically, there are subtypes such as embryonal rhabdomyosarcoma (including sarcoma botryoides), alveolar rhabdomyosarcoma, and pleomorphic rhabdomyosarcoma. Rhabdomyosarcoma has a high degree of malignancy, grows rapidly, and is easy to metastasize early with poor prognosis.

Leiomyosarcoma

Leiomyosarcoma occurs in the uterus, soft tissue, retroperitoneum, mesentery, omentum, and skin, mostly seen in middle-aged and elderly. The coagulative necrosis and number of mitotic figures are very important for the diagnosis of leiomyosarcoma and evaluation of its malignancy.

Angiosarcoma

Angiosarcoma occurs in the skin, breast, liver, spleen, bone, and soft tissue. Cutaneous angiosarcoma is more common, especially in the skin of the head and face. The tumors bulge from the surface of the skin, appearing as dark red or grayish white papule or nodule. They are prone to necrosis and hemorrhage. Microscopically, tumor cells show different degrees of atypia, forming vascular cavity-like structures with various sizes and irregular shapes, and vascular cavities are often anastomosed to each other (Fig. 5.17). In poorly differentiated angiosarcoma, tumor cells proliferate in solid. The formation of the vascular cavity was not obvious or only fissured, which contains red blood cells.

Fibrosarcoma

Fibrosarcoma often occurs in the subcutaneous tissue of the extremities. The tumor is infiltrative, grayish-white and fish fleshlike at the cut surface, often accompanied by hemorrhage and necrosis. The typical morphological change is that atypical spindle cells arrange in the "herringbone" pattern. The infantile fibrosarcoma, which occurs in infants and young children, has a better prognosis than adult one.

Fig. 5.17 Angiosarcoma of scalp. The lesions show obvious vasoformative growth, with anastomosing channels (H&E, 200×)

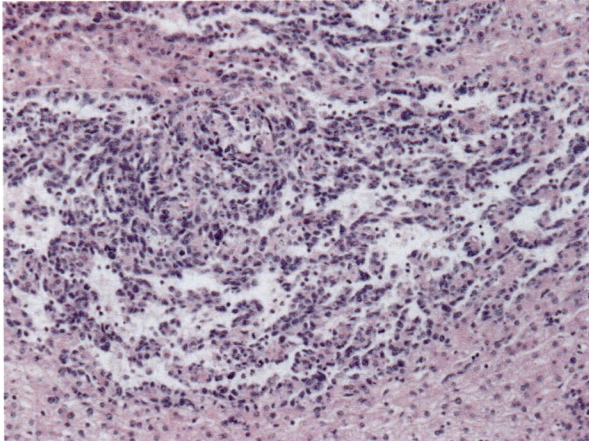

In the past, fibrosarcoma was considered to be a common sarcoma of soft tissue. Later studies showed that many of them were not fibrosarcomas but other sarcomas or sarcoma-like lesions. Most of the "fibrosarcomas" described in the earlier literature have now been classified into other types of tumor. The concept and classification of fibrous tissue and fibroblast neoplasms have undergone great changes and development in recent years.

Osteosarcoma

Osteosarcoma is the most common bone malignancy and often seen in young people. It occurs in the metaphysis of long bones, especially the lower end of the femur and the upper end of the tibia. At the cut surface, it is grayish white and fish flesh-like, commonly accompanied with hemorrhage and necrosis. The tumor destroys the bone cortex and raises the outer membrane of its surface (Fig. 5.18). Triangular bulge is formed between the bone cortex of the upper and lower ends of the tumor and the periosteum, due to the formation of new bone created by the periosteum, thus constituting the Codman triangle observed by X-ray examination. Since the periosteum is lifted, radial reactive new trabecular bones are formed between the periosteum and cortical bone. The trabecular bones are vertical to the bone surface, showing daylight radiate shadow on the X-ray. These imaging features are characteristics of osteosarcoma. Microscopically, the tumor cells are obviously atypical, spindle-shaped or polygonal and directly form neoplastic bone tissue or bone tissue, which is the most pivotal histological evidence for the diagnosis of osteosarcoma (Fig. 5.18). Chondrosarcoma- and fibrosarcoma-like components can also be found in osteosarcoma. Osteosarcoma is highly malignant and grows rapidly. Most of them are found to metastasize via bloodstreams when diagnosis.

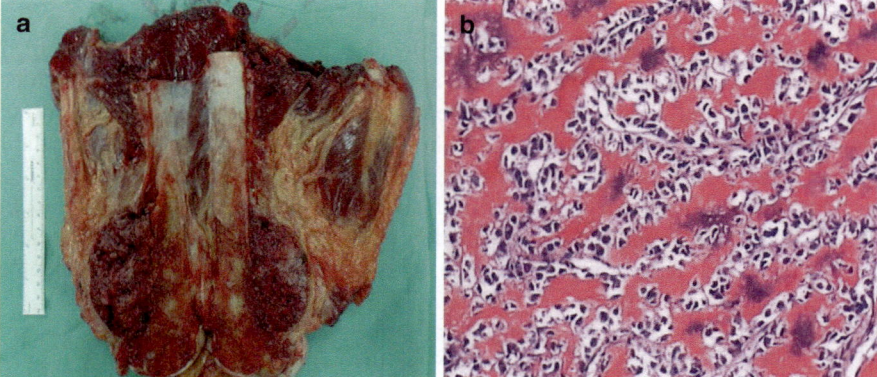

Fig. 5.18 Osteosarcoma. The tumor infiltrates medullary cavity and destroys the cortex (**a**). Microscopically trabeculae of neoplastic woven bone tissue is lined by spindle-shaped or polygonal tumor cells (H&E, 200×)

Fig. 5.19 Chondrosarcoma.
Abnormal chondrocytes
with large and dark nuclei
form cartilage lobule of
different size (H&E, 200×)

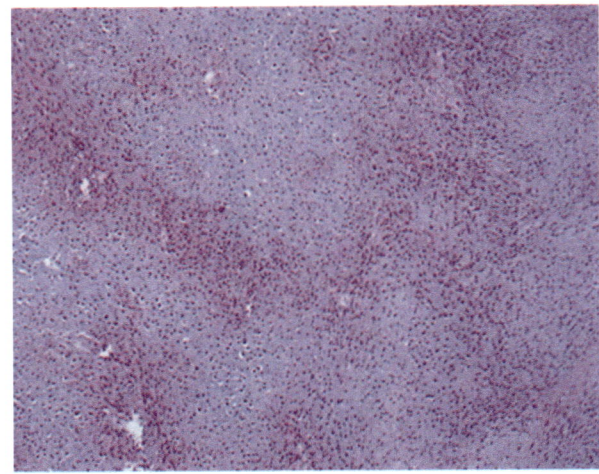

Chondrosarcoma

The age of onset is mostly from 40 to 70 years old. Chondrosarcoma occurs commonly in the pelvis and less in the femur, tibia, and scapula. Grossly, the tumor is located in the marrow cavity, showing a pale, translucent lobular mass. Microscopically, there are abnormal chondrocytes in the cartilage matrix, with large and dark stained nuclei, prominent nucleoli, and much mitosis (Fig. 5.19). Many diploid, megakaryocytic, and multinuclear tumor giant cells appear. Chondrosarcoma usually grows slower than osteosarcoma and metastasizes later.

5.8.3 Neuroectodermal Tumor

Some parts of early embryonic ectoderm develop into the nervous system called neuroectoderm, including the neural tube and neural crest. The neural tube develops into the cerebrum, spinal cord, retinal epithelium, etc. Nerve ganglia, Schwann cells, melanocytes, and adrenal medulla chromaffin cells originated from the neural crest. There are many types of tumors originating from neuroectodermal blasts.

About 40% of the primary tumors of the central nervous system are gliomas. In children with malignant tumors, the incidence of intracranial malignancies is next only to leukemia. The common tumors in the peripheral nervous system are neuromyeloma and neurofibroma.

Retinoblastoma arises from the retinal embryonic basal, and tumor cells are naive small round cells with a morphology similar to that of undifferentiated retinoblasts, showing a characteristic Flexener-Wintersteiner rosette. Most of the tumors are seen in infants under 3 years old and with a poor prognosis.

Fig. 5.20 Malignant melanoma of the skin. The tumor cells showing obvious eosinophilic nucleoli, with pigment in the cytoplasm (H&E, 200×)

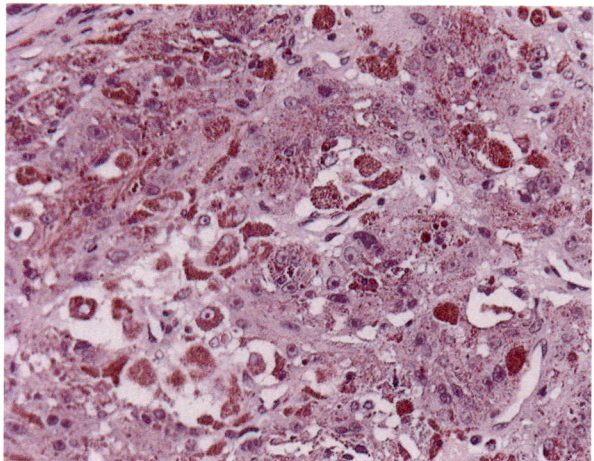

Malignant melanoma (Fig. 5.20) is more common in the skin and mucous membranes, occasionally in visceral organs. Malignant melanoma of the skin can develop from melanocytic nevus. The tumor cells can be pigmented or not. Tumor staging is closely related to prognosis.

5.9   Precancerous Diseases (or Lesions), Atypical Hyperplasia, and Carcinoma In Situ

Although some diseases (or lesions) are not malignant tumors, they have the potential to develop into malignant tumors; thus, the patients have an increased risk of developing the corresponding malignancy. These diseases or lesions are called precancerous diseases or precancerous lesions. It should be noted that precancerous diseases (or lesions) will not certainly develop into malignant tumors.

It will take a long time to develop from precancerous lesions to cancer. In epithelial tissues, sometimes atypical hyperplasia or dysplasia is firstly observed, and then it develops into carcinoma in situ (CIS) and further into invasive carcinoma.

5.9.1   Precancerous Diseases (or Lesions)

Precancerous diseases (or lesions) may be acquired or inherited. People with inherited cancer syndromes have some chromosomal and genetic abnormalities that increase their chances of developing certain tumors. Acquired precancerous diseases (or lesions) may be associated with certain lifestyle habits, infections, or some chronic inflammatory diseases.

5.9.1.1 Adenomas of Large Intestines

It is common. The lesion may be single or multiple. There are two common sub-types: villous adenoma and tubular adenoma. Villous adenomas are more likely to develop into carcinoma. Familial adenomatous polyposis (FAP) almost always develops into carcinoma.

5.9.1.2 Mammary Fibrocystic Disease

There is an increased risk of carcinogenesis in mammary fibrocystic disease with atypical hyperplasia of ductal epithelium. The breast fibrocystic disease is common in women around the age of 40, mainly characterized by cystic dilatation of the breast duct and proliferation of lobular and ductal epithelial cells.

5.9.1.3 Chronic Gastritis and Intestinal Metaplasia

There is a certain relationship between gastric intestinal metaplasia and gastric car-cinogenesis. Chronic gastritis with *Helicobacter pylori* infection is associated with gastric mucosa-associated lymphoid tissue (MALT) lymphoma and gastric adenocarcinoma.

5.9.1.4 Ulcerative Colitis

Ulcerative colitis is a kind of inflammatory bowel disease. Colon adenocarcinoma may occur on the basis of recurrent ulcers and mucosal hyperplasia.

5.9.1.5 Chronic Ulcer

Squamous epithelial hyperplasia and atypical hyperplasia may further develop into cancer due to long-term chronic irritation.

5.9.1.6 Leukoplakia

Leukoplakia often occurs in the oral cavity, vulva, etc. It exhibits hyperplasia, hyperkeratosis, and atypia of squamous epithelium. Grossly, white patches appear. It may develop into squamous cell carcinoma if untreated for a long time.

5.9.2 Dysplasia and Carcinoma In Situ

In the past years, the term of atypical hyperplasia was often used to describe cell proliferation with atypia. It is mostly used in epithelial lesions, including covered epithelium (such as squamous epithelium and urothelium) and glandular epithelium (such as breast duct epithelium and endometrial glandular epithelium). Since atypical hyperplasia is found not only in neoplastic lesions but also in repair, inflammation (so-called reactive atypical hyperplasia), in recent years, the academic community tends to use the term dysplasia to describe atypical hyperplasia associated with tumorigenesis.

Dysplasia is divided into mild, moderate, and severe grades. In the case of covered epithelium, mild dysplasia means less pleomorphic, involving the lower 1/3 layer of the epithelium. Moderate dysplasia involves the lower 2/3 layer of the epithelium; severe dysplasia shows most pleomorphic, involving more than 2/3 layer of the epithelium. Mild dysplasia can return to normal, while moderate to severe dysplasia is difficult to reverse.

The term of carcinoma in situ is usually used for epithelial lesions. It refers to dysplastic cells that are same with cancerous cells in the aspects of morphological and biological characteristics and involves the entire layer of epithelium, but not penetrates the basement membrane. Therefore, sometimes, it is also called intraepithelial carcinoma. In situ carcinoma is commonly seen in squamous epithelium or urinary tract epithelium, such as the cervix, esophagus, skin, and bladder. It is also found in the squamous metaplasia of mucosal surface, such as the squamous metaplasia of the bronchial mucosa. If the ductal epithelium of the breast is cancerous but does not invade the basement membrane and infiltrate into the interstitium, it is called ductal carcinoma in situ. Detection and treatment of a carcinoma in situ in time will prevent it from developing into invasive carcinoma. An important task for cancer prevention and control is to establish a technical method for early detection of carcinoma in situ.

At present, the concept of intraepithelial neoplasia is often used to describe the continuous process of epithelial dysplasia to carcinoma in situ. Mild dysplasia is called intraepithelial neoplasia grade I, moderate dysplasia is called intraepithelial neoplasia grade II, and severe dysplasia and carcinoma in situ is called intraepithelial neoplasia grade III. Severe dysplasia and intraepithelial neoplasia are collectively called grade III, mainly due to the difficulties to distinguish them completely, and the principles of their treatment are basically the same.

5.10   Molecular Detection of Neoplasm

Li Liang

The molecular targeted therapy has achieved great success with the characteristics of specificity and effectiveness. It aims at reversing the malignant biological behavior of tumor cells at molecular levels, such as cell signal transduction pathways, oncogenes and tumor suppressor genes, cytokines and receptors, antiangiogenesis, and suicide genes. As a new mode of biological treatment, the molecular targeted therapy only fights tumor cells and has little effect on normal cells.

The current detection methods include polymerase chain reaction (PCR) and sequencing, immunohistochemistry (IHC) assay, fluorescence in situ hybridization (FISH), or chromogenic in situ hybridization (CISH). The proteomics and genomics approaches are used to detect more precise targets for continuous advancement of molecular targeted therapies.

Currently, there are more than ten kinds of molecular targets. Here, we list some common targets.

5.10.1   HER-2

As a member of the EGFR family and a protein with receptor tyrosine kinase activity, the HER-2 protein plays a pivotal role in controlling the activation of cell conduction pathways, such as epithelial cell growth, differentiation, adhesion, and activation. HER-2 is overexpressed in about 20% of breast cancer patients, which indicates a poor prognosis. In addition, HER-2 overexpression is observed in lung carcinoma, pancreatic carcinoma, gastric carcinoma, and ovarian carcinoma.

The targeted therapeutic drug trastuzumab (Herceptin) antagonizes the growth-promoting effect of HER-2 family by downregulating HER-2 gene expression and mediates antibody-dependent cell-mediated cytotoxicity (ADCC). It has no effect on HER-2 negative tumor cells or normal cells. Therefore, the expression of HER-2 in tumor tissues must be detected before the use of trastuzumab treatment.

In recent years, the study of trastuzumab in the treatment of gastric carcinoma has become a hot topic. HER-2 is overexpressed in 10–55% of gastric carcinoma. About 20% of gastric cancer patients with HER-2 overexpression have poor prognosis. Recently, trastuzumab combined with chemotherapy is used for HER-2 positive gastric and gastroesophageal junctional tumors, which shows survival advantage without increasing the toxicity of chemotherapeutic drugs. Thus, it can be an option for HER-2 positive advanced gastric carcinoma patients.

5.10.2 EGFR Gene Mutation

EGFR is overexpressed and/or mutated in multiple tumors. The mutated EGFR selectively activates Akt signaling proteins and transcriptional activators. Then, the activation of STAT signaling pathway prolongs cell survival, leading to uncontrolled tumor cell growth and increased malignancy. In recent years, molecular targeted drugs targeting EGFR have attracted more and more attention. Among them, EGFR-tyrosine kinase inhibitors (TKI), gefitinib and erlotinib, have been used clinically for the treatment of advanced non-small cell lung cancer (NSCLC). Gefitinib is especially effective in Asian women, nonsmoking, and adenocarcinoma subgroup. It is also effective for the treatment of head and neck tumors, prostate cancer, and breast cancer. Erlotinib is used primarily for the patients with advanced and metastatic NSCLC.

Evidence shows that the EGFR gene status is the most critical predictor of EGFR-TKI efficacy. EGFR gene mutations occur frequently in women, nonsmokers, adenocarcinomas, and Asian populations. The detection of EGFR gene mutation status can be performed by PCR and sequencing.

5.10.3 K-Ras, BRAF Gene Mutations, and EGFR Monoclonal Antibody

Proto-oncogene K-ras plays a key role in the development of multiple tumors. The RAS protein, also known as P21 protein, is a membrane GTP/GDP binding protein. When the K-ras gene is mutated, it is not affected by the status of the upstream EGFR gene and always in an activated state, which continuously stimulates cell growth and leads to tumorigenesis. K-ras mutations are observed in a variety of tumors, including pancreatic and colorectal cancers (30–50%). The point mutations of the K-ras gene are mainly concentrated in codon 12 and codon 13, which account for more than 90% of all mutations.

The BRAF gene is a member of the RAF family and encodes a serine/threonine protein kinase, which is the most critical activating factor for the MEK/ERK signaling pathway. There are different proportions of BRAF mutations in a variety of human malignancies such as malignant melanoma, colorectal cancer, lung cancer, thyroid cancer, liver cancer, and pancreatic cancer. The most common mutations are V600E and V600K, which activate the BRAF protein and subsequently lead to the activation of MEK/ERK signaling pathway to promote cell growth and even tumorigenesis.

Cetuximab and Panitumumab are human murine chimeric IgG monoclonal antibodies directed against the extracellular domain of EGFR that bind to cell surface receptors to produce ADCC activity, regardless of the presence or absence of EGFR mutations in cells. Cetuximab is currently the first line of treatment for advanced metastatic colorectal cancer (mCRC). After adding cetuximab to conventional chemotherapeutics, the patient's objective response rate is significantly improved, and survival is prolonged. BRAF mutations are resistant to both drugs, and the survival rate is significantly lower than other patients. The new version of the NCCN Clinical Guideline for Colorectal Cancer states that all patients with mCRC should detect mutation status of both K-ras and BRAF genes, and only patients with wild-type K-ras and BRAF should be recommended for cetuximab.

5.10.4   C-Kit and PDGFR-α

The KIT protein encoded by the c-kit gene is a transmembrane receptor tyrosine kinase. It is a receptor of stem cell factor (SCF). When combined with SCF, KIT protein activates the corresponding signaling pathway and promotes cell proliferation, division, and survival. C-kit gene mutations occur in more than 90% of patients with gastrointestinal stromal tumors (GIST). Deletions or point mutations of c-kit exons 11, 9, 13, and 17 are common. The activation of KIT protein is no longer regulated by the ligand SCF when c-kit mutated, and the sustained activation of KIT tyrosine kinase results in the inhibition of apoptosis and uncontrolled proliferation and finally results in the formation of tumors. In another 10–15% of GIST patients, there was no c-kit mutation but PDGFR-α gene mutation. The protein PDGFR-α encoded by the PDGFR-α gene shares the same type III tyrosine protein kinase family as c-kit. PDGFR-α is activated by binding to ligand PDGF and activates phosphorylation pathways and other changes of phosphatidylinositol, CAMP, and various proteins, which may promote DNA synthesis, cell division, and proliferation and cause malignant tumor.

Imatinib (also known as Gleevec) is a tyrosine kinase inhibitor that selectively inhibits the proto-oncogenes abl, Abl-Bcr, c-kit, and PDGF receptor tyrosine kinase and inhibits kinase or substrate proteins, and inactivation of tyrosine phosphorylation blocks the signal transduction and achieves therapeutic goals. Most major KIT mutations occur in exon 11, and this mutation is favorable for imatinib therapy; however, patients with a KIT mutation in exon 9, which accounts for 10–20% of GIST cases, were poor responders to imatinib therapy. Detection of mutations in the corresponding exons of c-kit and PDGFR-α gene is an important indicator for predicting the efficacy of imatinib. GIST patients should be tested for c-kit and PDGFR-α prior to treatment with imatinib.

5.11   Molecular Basis of Carcinogenesis

With the development of molecular biology and technical improvement in DNA sequencing methods, the etiology and molecular basis of cancer are coming to light. Carcinogenesis is a multistep process caused by the dysfunction of regulatory genes related to cell growth and proliferation. Accumulation of nonlethal DNA damages caused by environmental and genetic carcinogenic factors synergistically or sequentially is the essence of neoplasm. Four classes of gene disorders including activation of proto-oncogenes, silencing of tumor suppressor genes, mutation of genes that related to programmed cell death (apoptosis), and changes of genes involving in DNA repair lay at the heart of carcinogenesis.

5.11.1 Activation of Proto-oncogene

Proto-oncogenes are generally vital regulatory genes that encode growth factors, growth factor receptors, signal-transducing proteins, and transcription factors in normal cells. The corresponding *oncogenes* are activated or mutated proto-oncogenes, which encode oncoproteins that serve excessive normal function similar to their normal counterparts or impart completely new functions. Oncoproteins can continuously transform normal cells to malignant cells that are self-sufficient in growth. The well-characterized oncogenes and their related human cancers are listed in Table 5.6.

The transformation of proto-oncogenes to oncogenes is called activation. Commonly, this process is accomplished through the following mechanisms: (1) Point mutation. The point mutation of RAS protein is an example. RAS proteins are growth-promoting signal-transducing proteins belonging to membrane-associated small G proteins that bind guanosine nucleotides. They normally bind to guanosine diphosphate (GDP) and stay at the quiescent state. A point mutation (G->T) in coden 12 changes a glycine (Gly) to Valine (Val) and hence lost the activity of GTPase, through which target cells escape the upstream signal control and are persistently activated in proliferation. (2) Gene amplification. Excessive replication of target genes will lead to the overexpression of gene products and constitutive activation of its downstream signal transduction. For example, the abnormal HER2 (*eRBB2*) amplification is observed in certain breast cancers, leading to a persistent downstream tyrosine kinase activity. (3) Chromosomal translocation. The proto-oncogene located chromosomal segment is translocated to another chromosome, resulting in abnormal expression, structure, or functions of the proto-oncogene. In Burkitt lymphoma, cell growth regulator MYC located in chromosome 8 is translocated to chromosome 14 behind a strong promoter, leading to the excessive transcription and overexpression of the transcription factor MYC. Chromosomal translocation might also lead to the formation of fusion proteins, such as the translocation of proto-oncogene *abl* in chronic myelogenous leukemia (CML). The newly formed Bcr/Abl fusion protein is a dysfunctional oncoprotein that promotes the malignant transformation of target cells.

Table 5.6 Selected oncogenes, mode of activation, and associated human tumors

Category	Proto-oncogene	Mode of activation	Associated tumors
Growth factors			
PDF-βchain	*SIS*	Overexpression	Astrocytoma
FGF	*HST-1*	Overexpression	Osteosarcoma, stomach cancer,
	INT-2	Amplification	Bladder cancer, breast cancer, melanoma
Growth factor receptors			
EGF-receptor family	ERBB1(EGFR)	Mutation	Adenocarcinoma of lung
	ERBB2(HER)	Amplification	Breast carcinoma
FMS-like tyrosine kinase 3	FLT3	Point mutation	Leukemia
Proteins involved in signal transduction			
GTP-binding (G) proteins	KRAS	Point mutation	Colon, lung, and pancreatic tumors
	HRAS	Point mutation	Bladder and kidney tumors
	NRAS	Point mutation	Melanomas, hematologic malignancies
	GNAQ	Point mutation	Uveal melanoma
	GNAS	Point mutation	Pituitary adenoma, other endocrine tumors
Non-receptor tyrosine kinase	ABL	Translocation	Chronic myelogenous leukemia
		Point mutation	Acute lymphoblastic leukemia
RAS signal transduction	BRAF	Point mutation, translocation	Melanomas, leukemias, colon carcinoma, others
Nuclear regulatory proteins			
Transcriptional activators	MYC	Translocation	Burkitt lymphoma
	NMYC	Amplification	Neuroblastoma
	LMYC	Amplification	Small cell lung cancer
Cell cycle regulator			
Cyclins	CCND1 (cyclin D1)	Translocation	Mantle cell lymphoma, multiple myeloma
		Amplification	Breast and esophageal cancers
Cyclin-dependent kinase	CDK4	Amplification or point mutation	Glioblastoma, melanoma, sarcoma

5.11.2 Silencing of Tumor Suppressor Gene

Tumor suppressor genes, or anti-oncogenes, are also essential regulatory genes in cellular growth and proliferation. They can be transcription factors, cell cycle inhibitors, signal-transducing molecules, cellular membrane receptors, or DNA damage regulators. Different from oncogenes, tumor suppressor genes exert their opposite function on cell growth and proliferation. Mutated tumor suppressor genes in both alleles will lead to the failure of growth inhibition and induce the occurrence of cell transformation. Many of our concepts of tumor suppressor genes are derived from RB and p53, two well-characterized tumor suppressor genes at present. Their products are both nuclear transcription factors. The well-characterized tumor suppressor genes and their associated human cancers are listed in Table 5.7.

5.11.2.1 Retinoblastoma Gene (*RB Gene*)

RB is the first discovered tumor suppressor gene that was identified by studying a rare disease, familial retinoblastoma. Familial and sporadical retinoblastomas are two different types of retinoblastoma. Interestingly, most familial retinoblastoma patients are predisposed to develop tumor at the early age in both eyes and have high risk of developing other soft-tissue sarcomas. Sporadical retinoblastoma patients, however, have a relative late onset of tumor in one eye and are not at an increased

Table 5.7 Selected tumor suppressor genes and associated human tumors

Gene	Function	Associated somatic cell tumors	Tumors associated with genetic mutations
APC	WNT-signaling inhibitor	Carcinomas of stomach, colon, pancreas; melanoma	Familial colonic polyps and carcinomas
RB	Regulator of cell cycle progression	Retinoblastoma; osteosarcoma carcinomas of breast, colon, lung	Familial retinoblastoma and osteosarcoma
P53	Regulator of cell cycle and transcription, apoptosis induced by DNA damage	Most of the human carcinomas	Li-Fraumeni syndrome, multiple tumor and sarcoma
NF1	RAS/MAPK signaling inhibitor	Neuroblastoma	Neurofibromatosis type 1
WT1	Regulator of transcription	Nephroblastoma	Familial Wilms' tumor
VHL	Hypoxia-induced transcription factors inhibitor	Renal cell carcinoma	Familial renal cell carcinoma, cerebellar hemangioblastoma
P16	Cycle dependent kinase inhibitor, CKI	Carcinomas of pancreas, esophagus, breast, melanoma	Familial malignant melanoma
BRCA1, BRCA2	Repair of double-stranded breaks in DNA		Familial breast and ovarian carcinomas

risk of developing other types of cancer. This could be well explained by the "two-hit hypothesis" proposed by Knudson in 1974. The Knudson's hypothesis suggests that RB gene must be existed and retinoblastoma develops only when two RB gene alleles are both mutated. In the familial case, children are inherited a copy of defective RB allele (first hit) in the germline, and the other copy of RB allele is normal. Retinoblastoma develops when the normal RB allele is mutated (second hit), and it is relatively easy to take place. In the sporadical case, retinoblastoma occurs when two normal RB alleles undergo spontaneous somatic mutation. The possibility of this event is relatively low.

Mutation of both alleles of *RB* gene at chromosome locus 13q14 is required to produce retinoblastoma. In addition, transformation of normal RB gene to retinoblastoma cells can reverse their tumor phenotype. The loss of RB gene is also found in other types of tumors including bladder cancer, lung cancer, breast cancer, and osteosarcoma. The RB protein is a vital governor of proliferation through negatively regulating the G1/S cell cycle transition. In G1 stage, hypophosphorylated RB in complex with E2F transcription factor family inhibits the transcription of proteins that are required in S phase of the cell cycle (Fig. 5.21). The activation of cdk4/6 phosphorylates RB and releases E2F, which can active the transcription of S phase proteins. The mutation of RB can result in an uncontrolled E2F activity and then cells lose their control of G1/S transition.

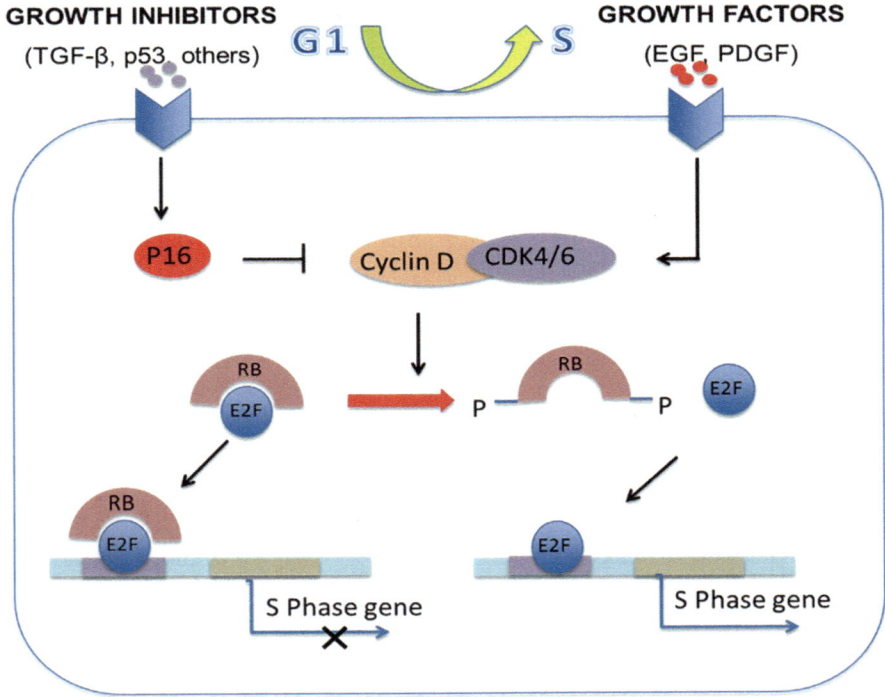

Fig. 5.21 The role of RB in regulating the G1-S checkpoint of the cell cycle

5.11.2.2 P53 Tumor Suppressor Gene

P53 located in chromosome 17 is another extensively studied tumor suppressor gene in human cancers. Loss of p53 function is detected in 50% of human tumors, indicating its critical role in preventing cancer development. Increased p53 protein is detected in cells when there is cellular stress or primarily DNA damage. Once activated, P53 works as a specific transcription factor that stimulates the transcription of p21, which can stop the cell cycle at G1 stage until the damage is repaired or the cells undergo programmed cell death (apoptosis). In healthy cells, p53 is association with MDM2, an E3 ubiquitin ligase, leading to the ubiquitization and degradation of P53. As a result, p53 is virtually undetectable in normal cells. With the loss or mutation of p53 function, DNA damage goes without repair, cells undergo mitotic replication rather than apoptosis, and mutations accumulate in oncogenes and other cancer genes. As a result, cells are predisposed to malignant transformation.

5.11.2.3 Other Tumor Suppressor Genes

Mutations of adenomatous polyposis coli (APC) are associated with familial adenomatous polyposis, in which patients develop thousands of adenomatous polyps in the colon. *APC* is a tumor suppressor that involves in downregulating WNT signaling pathway, which has a major role in controlling cell fate, adhesion, and cell polarity. APC participates in the degradation of β-catenin to stop its nuclear translocation and the following activation of c-myc. Mutation or dysfunction of *NF1* gene located in chromosome 17 develops neurofibromatosis type I. Neurofibromin, the protein product of the *NF1* gene, contains a GTPase-activating domain that keeps RAS stay inactive. On the contrary, loss-of-function mutation in NF1 gene leads to a persistently activated RAS and uncontrolled cell proliferation. A large amount of tumor suppressor genes have been identified during the past decades including E-Cadherin, CDKN2A, PTEN, NF2, WT1, PATCHED, and VHL. There is no doubt that more tumor suppressor genes remain to be discovered.

5.11.3 Dysfunction of Apoptosis-Related Genes

Accumulation of malignant transformed cells may result not only from activated proto-oncogenes or silenced tumor suppressor genes but also from mutated genes that regulate programmed cell death. Dysfunction of apoptosis-regulating genes results in less cell death and enhanced survival of the cells. It's no doubt that apoptosis of cancer cells is frustrated at the key points of apoptotic signaling pathways. Pro-apoptotic and anti-apoptotic members of the BCL2 family of proteins are well-known apoptotic regulatory proteins. The pro-apoptotic proteins BAX and BAK are required for apoptosis and directly promote mitochondrial permeabilization that triggers apoptosis. Overexpression of anti-apoptotic members including BCL2,

BCL-XL, and MCL1 prevents cells from apoptosis. Apoptosis is also initiated when CD95/Fas is activated by its ligand, CD95L/FasL, leading to the recruitment of caspases to cleave DNA and causing cell death. Dysfunction of CD95/Fas or caspase family members prevents cells from apoptosis and leads to their malignant transformation.

5.11.4 Dysfunction of Genes Regulate Repair of Damaged DNA

Both external factors including ionizing radiation, ultraviolet radiation, and internal DNA replication mistakes can cause DNA damages. In normal cells, minor DNA damages can be repaired. Mutations tend to accumulate in cells with mutated DNA repair genes, followed by their malignant transformation. Defective in DNA repair system, xeroderma pigmentosum patients are at an increased risk of developing skin cancers, for the accumulation of UV induced DNA damage.

5.11.5 Activated Telomerase

Telomeres are short repeated DNA sequences located at the end of chromosome that are important for chromosome replication through preventing chromosomes from fusion and degradation. Except germ cells, most somatic cells do not have the activity of telomerase that can recover the length of telomere. Normal somatic cells have limited capacity of replication also because of the telomere attrition. The length of telomere is shortened progressively as the times of DNA replication increase. The telomerase activity is reactivated in most malignant tumor cells, resulting in stable telomere length and indefinite cell proliferation.

5.11.6 Epigenetic Changes

Besides changes in nucleotide sequence, epigenetics changes including histone modification and DNA methylation also regulate gene expression and genetic changes. Numerous mutations involving genes that encode epigenetic regulatory proteins have been identified by cancer genome sequencing. DNA methylation, a modification created by DNA methyltransferases, is considered as a vital regulatory mechanism of gene expression. Abnormal DNA methylation of key promoters is commonly seen in tumor cells, including both hypermethylation of tumor suppressor genes and hypomethylation of oncogenes. DNA hypermethylation leads to the decreased expression of tumor suppressor genes such as RB and VHL. Hypomethylation of CpG-repeated sequences rich in noncoding areas of the genome decreases the stability of DNA molecules and triggers the malignant phenotype. Histone modifications catalyzed by enzymes are associated with chromatin regulatory complexes. Methylation and acetylation of histones are tightly associated with DNA replication, transcription, and damage repair. Dysfunction of histone modification also contributes to carcinogenesis. In recent years, noncoding RNAs have been paid more attention as generalized epigenetic factors that can regulate gene expression. The progress in the regulatory roles of noncoding RNAs contributes to the identification of the complicated mechanisms behind carcinogenesis.

5.11.7 Carcinogenesis Is a Multistep Process

Malignant tumors must develop several fundamental abnormalities resulting from the stepwise accumulation of multiple mutations to produce a fully malignant tumor. In the laboratory, normal epithelial cells bearing a series of mutations including activation of RAS, inactivation of RB, P53 and PP2A, can complete malignant transform. However, these mutations never occur simultaneously but gradually accumulate during the natural development of human cancers. Development of colon carcinoma is a classic example of incremental acquisition of malignant phenotypes. Colon carcinoma involves a series of morphologically identifiable stages. Colon epithelial hyperplasia occurs first followed by the formation of adenomas and then the ultimately malignant transformation. Molecular mechanisms at each of

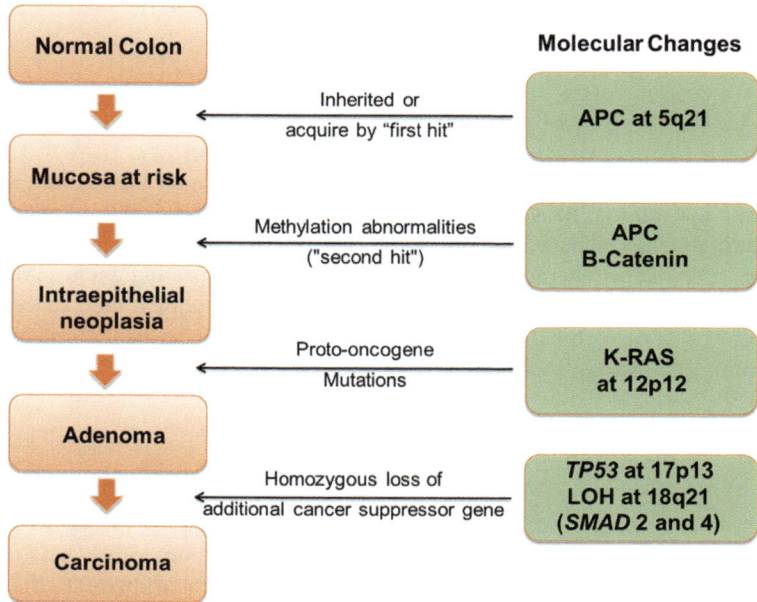

Fig. 5.22 Molecular mechanism for the evolution of colorectal cancer

these stages are well illustrated (Fig. 5.22). Defective of the tumor suppressor gene *APC* and the following activation of RAS occur first to induce epithelial hyperplasia and adenomas. Loss of a tumor suppressor gene on 18q and loss of *P53* contribute to the ultimately malignant transformation. Similar situation also exists in other epithelial cancers although the precise temporal sequence of mutations may be different in each organ and tumor type.

5.12 Etiology of Neoplasms

Environmental tumorigenic factors induce the development of tumors through multiple molecular pathways discussed above. However, the identification of tumorigenic factors is not easy as most tumors develop long times after exposed to those factors. Carcinogens are materials that can trigger carcinogenesis. Carcinogens play a vital role in the initiation and promotion of tumors.

5.12.1 Chemical Carcinogens

Hundreds of chemicals have been identified as carcinogens to both animals and humans. Most of the chemical carcinogens are mutagens that covalently bind to DNAs and induce the following DNA structure changes. They usually involves in the initiation of carcinogenesis.

5.12.1.1 Direct-Acting Carcinogens

Minority of carcinogens are direct-acting carcinogens that are carcinogenic without metabolic conversion. Most of them (mainly alkylating and acylating agents) are weak carcinogens that have little impact on cancer initiation. Some of weak carcinogens are even cancer chemotherapeutic drugs, such as alkylating agents. Although they are carcinogens to acute myeloid leukemia, in some cases, they successfully cure, control, or delay the recurrence of certain types of cancer including leukemia, lymphoma, and ovarian carcinomas.

5.12.1.2 Indirect-Acting Carcinogens

Indirect-acting carcinogens refer to most chemicals that require metabolic conversion to become active carcinogens. Polycyclic hydrocarbons presented in fossil fuels are the most potent indirect-acting carcinogens. Benzo[a]pyrene, the active component of soot, is the product of tobacco combustion in cigarettes and contributes to the high incidence of lung cancer. Polycyclic hydrocarbons produced from smoked meats and fish are associated with gastric cancers. The aromatic amines and azo dyes widely used in the aniline dye and rubber industries are other class of indirect-acting carcinogens. Nitrates and nitrites used in the processing of meat contribute to the development of digestive tract tumors. Aflatoxin from moldy food, especially in moldy peanuts and corns, strongly triggers hepatic cancer through inactivating the tumor suppressor gene p53.

5.12.2 Radiation Carcinogenesis

UV rays of sunlight, as well as electromagnetic and particulate radiation, are all carcinogenic. UV light is one of the clearly identified causation of skin cancer through inducing the formation of pyrimidine dimers in DNA. In normal person, this DNA damage can be repaired by nucleotide excision repair pathway. The hereditary disorder *xeroderma pigmentosum* patients, as discussed previously, tend to develop skin cancer for the loss of DNA damage repair enzymes. Electromagnetic and particulate radiations including x-rays, γ-rays, α-particles, β-particles, protons, and neutrons cause the break, translocation, and mutation of chromosome, which can activate oncogenes and inactivate tumor suppressor genes to stimulate the malignant transformation of damaged cells.

5.12.3 Microbial Carcinogenesis

Most microbial carcinogens are tumor viruses including DNA and RNA tumor viruses. Recent research indicates that some bacteria are important for the development of gastritis and gastric ulcer. For example, *Helicobacter pylori* are tightly associated with some gastric tumors.

5.12.3.1 Oncogenic DNA Tumor Virus

In infected host cells, DNA viruses integrate their genome into host genome and ultimately trigger the malignant transformation of host cells. Various oncogenic DNA viruses that cause tumors in animals and humans have been identified including human papilloma virus (HPV), Epstein-Barr virus (EBV), hepatitis B virus (HBV), Merkel cell polyoma virus, and Kaposi sarcoma herpesvirus (human herpesvirus 8). The oncogenic potential of HPV attributes to the activities of two viral genes encoding E6 and E7, which can bind to RB and P53, respectively, resulting in uncontrolled cellular growth and malignant transformation. EBV is closely associated with the pathogenesis of several human tumors including Burkitt lymphoma and nasopharyngeal carcinoma. EBV-infected B lymphocyte undergo polyclonal proliferation, and the following mutations ultimately cause malignancies. HBV and

HCV infection have been demonstrated closely related with the development of hepatocellular carcinoma, and the chances of HBV infected patients are 200 times easier to develop hepatocellular carcinoma. The oncogenic effects of HBV and HCV are multifactorial, and the dominant one seems to be immunologically mediated chronic inflammation, hepatocyte injury and regeneration, and genomic damage.

5.12.3.2 Oncogenic RNA Tumor Virus

RNA tumor viruses are retroviruses that transform host cells directly through integrating virus oncogene to host genome, or indirectly through overexpression and activation of host oncogenes. Human T-Cell Leukemia Virus Type 1 (HTLV1) is the only human retrovirus that is firmly implicated in the pathogenesis of cancer. HTLV1 causes adult T-cell leukemia/lymphoma (ATLL) that is mainly endemic in certain parts of Japan and the Caribbean basin. The activity of HTLV1 is associated with the products of its Tax gene that actives the transcription of host genes including c-fos, c-sis, GM-CSF, IL-2, and its receptors, which further stimulate the proliferation of T lymphocytes.

5.12.4   Helicobacter pylori

H. pylori are incriminated as the major cause of gastritis and peptic ulcers and the first bacterium that is classified as a carcinogen. *H. pylori* infection has also been demonstrated closely related with the genesis of both gastric adenocarcinomas and gastric lymphomas.

5.13   Neoplasm and Genetics

Besides environmental tumorigenic factors, hereditary factors also account for the development of many caner types. Patients with inherited cancer syndrome are more likely to develop neoplasms because of their defective genes and abnormal chromosomes. The well-identified inherited cancer syndromes are also listed in Table 5.2.

5.13.1   Autosomal Dominant Inherited Cancer Syndrome

The predisposition to autosomal dominant inherited cancers follows an autosomal dominant pattern of inheritance and a defective copy of gene increases the chance of tumorigenesis. Patients with familial retinoblastoma, discussed previously, are inherited a copy of defective RB allele from parental generation. Retinoblastoma develops when the normal RB allele is mutated. Some precancerous lesions including familial adenomatous polyposis and neurofibromatosis are autosomal dominant inherited. In these diseases, tumor suppressor genes, such as RB, APC, and NF1, are mutated or lost.

5.13.2 Autosomal Recessive Inherited Cancer Syndrome

The predisposition to recessive inherited cancers follows an autosomal recessive pattern of inheritance that both gene alleles are mutated. One example is xeroderma pigmentosum mentioned before. Exposed to UV light, xeroderma pigmentosum patients with defective nucleotide excision repair are at the increased risk of developing skin cancer. Besides xeroderma pigmentosum, syndromes involving defects in DNA repair system constitute a group of inherited recessive disorders, such as bloom syndrome, ataxia-telangiectasia, and Fanconi anemia.

5.13.3 Familial Cancer

Some common types of cancers, such as breast, ovary, colon, and brain cancers, have been reported to occur in a familial pattern. Since multiple factors are involved in the inheritance of familial cancers, the transmission pattern of familial cancer is not easy to rule out.

5.14 Host Defense Against Neoplasm: Tumor Immunity

Cells that undergo malignant transformation can be recognized and destroyed by host immune system. Tumor antigen and antitumor effector mechanisms are the main contents of tumor immunology. Immune surveillance is the normal function of the immune system to recognize the merging malignant cells and destroy them. However, immune surveillance is imperfect, and some malignant cells successfully escape the policing through producing a number of factors that promote immune tolerance and immune suppression.

5.14.1   Tumor Antigens

Tumor antigens that elicit host immune response are broadly classified into two categories. They are tumor-specific antigen (TSA) that are present only in tumor cells and tumor-associated antigen (TAA) that are present both in tumor cells and some normal cells. According to their molecular structure and source, tumor antigens can also be classified into the following classes: (1) Products of mutated proto-oncogenes, tumor suppressor genes, and neutral "passenger" genes. These products, such as peptides derived from mutated oncoproteins of RAS, p53, and BCR-ABL, enter the traditional antigen-processing pathway and recognized by CD4+/CD8+ T cells. (2) Aberrantly expressed cellular proteins. One example of such antigen is tyrosinase, which is also expressed in normal melanocytes and melanomas involving in melanin biosynthesis. They fail to be recognized by immune system in normal cells because of their limited amount. (3) Antigens produced by oncogenic viruses. Human immune system plays a surveillance role that cytotoxic T lymphocytes (CTLs) can recognize antigens produced by HPV and EBV. (4) Oncofetal antigens. Oncofetal antigens, such as carcinoembryonic antigen (CEA) and α-fetoprotein (AFP), are highly expressed in cancer cells and fetal tissues. With the development of detection techniques, small amount of them are even found in normal tissues. (5) Altered cell surface glycolipids and glycoproteins. Abnormal expressions or forms of surface glycoproteins and glycolipids are observed in most tumors. Abnormal forms of MUC-1, expressed on both breast and ovarian carcinomas, have been used for diagnostic and therapeutic studies. (6) Cell type-specific differentiation antigens. These antigens are specific for particular lineages or differentiation stages of various cell types.

5.14.2   Antitumor Effector Mechanisms

Although antibodies produced by cancer patients can recognize tumor antigens, cell-mediated immunity is still the dominant antitumor mechanism in vivo. The main effector cells are as follows:

CTLs The antitumor effect of CTLs is mainly accomplished through reacting against virus-associated neoplasms, such as EBV- and HPV-induced tumors.

Natural killer (NK) cells NK cells function as the first line against human tumors. They are lymphocytes that can destroy tumor cells without primary sensitization. With the activation of IL-2 and IL-15, NK cells increase their activities against human tumors.

Macrophages Activated macrophages kill tumors in a manner similar to killing microbes. They can exhibit cytotoxicity even without the appearance of T cells. Interferon-γ, cytokine secreted by T cells and NK cells, is the potent activators of macrophages; as a result, T cells, NK cells, and macrophages may collaborate in the antitumor process.

5.14.3 Immune Surveillance and Escape

The term "immune surveillance" was first proposed as a normal function of the immune system against emerging malignant cells. This has been supported by many observations: One strong argument is that persons with congenital immunodeficiencies are 200 times easier to develop cancers compared with the immunocompetent individuals. Another is the positive response of advanced cancer to the newly developed therapeutic agents that function through stimulating the host T-cell response. Furthermore, transplant recipients and persons with acquired immune deficiency syndrome (AIDS) also have an increased incidence of malignancies.

However, most cancers occur in immunocompetent persons indicating that tumor cells must develop mechanisms to escape extermination by immune system in the hosts. The possible escape mechanisms have been proposed: (1) **Selective outgrowth of antigennegative variants**. Only the weak immunogenic mutants can survive during the tumor progression. (2) **Loss or reduced expression of MHC molecules**. Tumor cells with abnormal levels of HLA class I molecules may escape the attack of CTLs. (3) **Activation of immunoregulatory pathways.** Tumor cells might inhibit tumor immunity by engaging the immune regulatory pathways. (4) **Secretion of immunosuppressive factors by cancer cells**. Host immune responses are suppressed by tumor products, such as TGF-β, a potent immunosuppressant, secreted in large quantities of tumors. (5) **Induction of regulatory T cells (Tregs)**. Tumors products might reach the "immune escape" through favoring the development of immunosuppressive regulatory T cells.

Chapter 6
Environmental and Nutritional Diseases

Ye Wang

Contents

Objectives

1. To comprehend the major environmental diseases, to make sure the environmental pathogens of each disease

Y. Wang (✉)
West China School of Public Health and West China Fourth Hospital, Sichuan University, Chengdu, China
e-mail: yellpath007@scu.edu.cn

© Zhengzhou University Press 2024
K. Chen et al. (eds.), *Textbook of Pathologic Anatomy*,
https://doi.org/10.1007/978-981-99-8445-9_6

2. To master the main clinical manifestations and control measure of the common environmental diseases

Key Concepts
For each disease: pathogen, pathological changes, main clinical manifestations

Introduction
The field of environmental pathology includes all such diseases caused by environment factors, most of which is man-made. In addition, overnutrition and malnutrition are also discussed in this chapter. The World Health Organization (WHO) estimates that 80% cases of cardiovascular diseases and type 2 diabetes mellitus and 40% of all cancers are preventable by avoidance of tobacco, healthy diet, and physical activity. There are many factors that have been considered to affect human health such as: (1) industrial effluents and automobile exhausts, (2) accumulation of wastes, and (3) unsatisfactory disposal of radioactive and electronic waste.

6.1 Environmental Pollution

Environmental pollution refers to natural or man-made destruction, which adds harmful substances to the environment and exceeds the self-purification capacity of the environment. Any agent—chemical, physical, or microbial—that alters the composition of environment is called pollutant. On the other hand, our living environment is affected by smoking, drinking water, and eating food. Therefore, in this part, we mainly discuss briefly the health effects of smoking, environmental compounds, and air pollution.

6.1.1 Air Pollution

People can't live without air. If people are exposed to polluted air for a long time, their health will be damaged. Air pollution is a major cause of high morbidity and mortality worldwide, especially for those who already have lung or cardiac diseases.

Inhaling polluted air can cause various diseases or cause lesions in many organs and systems (Table 6.1). Fine particulate matter (PM 2.5) is a well-known air pollutant threatening public health. Studies have confirmed that long-term exposure to the particles could reduce the pulmonary function, cause exacerbation of asthma and chronic obstructive pulmonary diseases, and increase incidence and mortality of lung cancer. Some of the pollutants can be found in specific locations (such as coal dust, silica, and asbestos), while others are general pollutants present widespread in the ambient atmosphere (e.g., sulfur dioxide, nitrogen dioxide, and carbon monoxide).

The effects of pollutants on human health are related to the following factors:

Table 6.1 Examples of air pollutants

Pollutant	Source(s)	Consequences
Sulfur dioxide (SO$_2$)	Coal smoke Tobacco	Mucosal irritation
Carbon monoxide (CO)	Car exhaust Gas stove	Anoxia (death)
Polychlorinated biphenyls (PCBs)	Air spray Refrigerators	Gastrointestinal dysfunction; jaundice
Formaldehyde	Laboratory fumes House insulation	Mucosal irritation
Carbon	Smog Mining Coal smoke	Anthracosis
Quartz (silica)	Stone cutting	Silicosis
Asbestos	Insulation Shipbuilding	Asbestosis
Fine particulate matter (PM 2.5)	Industrial emission Car exhaust	Interstitial pneumonia, fibrosis, and other diseases

1. Exposure time
2. Total dose of exposure
3. Impaired ability of the host to clear inhaled particles
4. Particle size of 1–5 μm, which can be capable of getting impacted in the distal airways to produce tissue injury

6.1.1.1 Outdoor Air Pollution

Air pollution is a very critical issue worldwide, particularly in developing countries. The six most common pollutants in the air are sulfur dioxide, carbon monoxide, ozone, nitrogen dioxide, lead, and particulate matter. Usually, these agents produce well-known smog, sometimes suffocating large cities such as Beijing, Shanghai, Houston, Cairo, New Delhi, Mexico City, and Hong Kong. Although the respiratory system bears the brunt, all organs in the body would be involved in. Major health effects of outdoor pollutants are summarized in Table 6.2.

Ozone (O3) is produced by the interaction of ultraviolet (UV) radiation and oxygen (O2) in the stratosphere and naturally accumulates in the so-called ozone layer 10–30 miles above the earth's surface. The ozone layer absorbs a lot of ultraviolet radiation from the sun, thus protecting the living things on the earth. Ozone on the earth's surface is a gas formed by the reaction of nitrogen oxides and volatile organic compounds in the presence of sunlight. These chemicals are emitted by industrial emissions and motor vehicle exhaust.

Generally, ozone inhalation only causes upper respiratory tract inflammation and mild symptoms (pulmonary dysfunction and chest discomfort). However, excessive

Table 6.2 Health effects of outdoor air pollutants

Pollutant	Populations at risk	Effects
Sulfur dioxide	Healthy adults Individuals with chronic lung disease Asthmatics	Increased respiratory symptoms Increased mortality Increased hospitalization Decreased lung function
Acid aerosols	Healthy adults Children Asthmatics	Altered mucociliary clearance Increased respiratory infections Decreased lung function Increased hospitalizations
Nitrogen dioxide	Healthy adults Asthmatics Children	Increased airway reactivity Decreased lung function Increased respiratory infections
Ozone	Healthy adults and children	Decreased lung function Increased airway reactivity Lung inflammation Decreased exercise capacity Increased hospitalizations
Particulates	Children Individuals with chronic lung or heart disease Asthmatics	Increased respiratory infections Decreased lung function Excess mortality Increased attacks

exposure or prolonged exposure can pose a risk to patients with asthma or emphysema.

Together with other toxic gases, even low-level exposure to ozone can also damage lung function. Unfortunately, air pollutants often combine to create a veritable "witches' brew" of ozone and other agents such as *sulfur dioxide* and particulates.

1. Sulfur dioxide is produced by power plants burning coal and oil, smelting copper, and paper mills as a by-product. When released into the air, it can be converted to sulfur trioxide, which forms sulfuric acid when exposed to water, which can cause damages of nasopharynx and lead to dyspnea and asthma attacks in susceptible individuals.

2. Particulate matter (PM) is a type of air pollution that comprises a heterogeneous mixture of different particle sizes and chemical compositions. There are various sources of fine PM (PM2.5), and the components may also have different effects on people. Generally speaking, the coarse particles with particle size of 2.5–10 microns mainly come from road dust, and the fine particles with particle size of less than 2.5 microns (PM2.5) mainly come from fossil fuel combustion (such as automobile exhaust and coal combustion), volatile organic compounds, and so on. When the concentration of PM2.5 increased by 10 micrograms per cubic meter, the lung function of residents in air pollution area dropped by about 26 mL. If the concentration of PM2.5 exceeds 75 micrograms per cubic meter, then the risk of chronic obstructive pulmonary disease is 2.53 times than that of 35 micrograms per cubic meter. Although the particles have not been well char-

acterized chemically or physically, fine or ultrafine particles less than 10 μm in diameter are the most harmful. They are readily inhaled into the alveoli, where they are phagocytosed by macrophages and neutrophils, which respond by releasing a number of inflammatory mediators, such as IL-1. In contrast, particles that are greater than 10 μm in diameter are of lesser consequence, because they are usually blocked by nasal hair in the nasal cavity, or are captured by airway mucous epithelium, and are excreted with sputum through cilia swing.

6.1.1.2 Tobacco Smoking

Smoking is a global health problem that endangers health. Cigarette smoke contains more than 60 known carcinogens, as well as toxic metals and formaldehyde. Tobacco contains several harmful constituents including nicotine, many carcinogens, carbon monoxide, and other toxins (Table 6.3). The WHO report shows that smoking is harmful to human beings in many ways, and these include mainly asthma, pneumonia, lung cancer, hypertension, heart diseases, and reproductive dysplasia. The harm caused by smoking is related to many factors, and the most important of which is dose of exposure expressed in terms of pack years (e.g., one pack daily for 20 years equals 20 pack years). It is estimated that the life expectancy of a person who smokes two packs of cigarettes at the age of 30 is 8 years less than a nonsmoker. After quitting smoking, the higher mortality slowly declines, and the beneficial effect reaches the level of nonsmokers after 20 or more of smoke-free years.

1. Mechanism

    Mechanisms of smoking-induced diseases include the following:

    (a) It directly affects the respiratory tract mucosa and causes bronchitis. Cigarette smoke increases the recruitment of leukocytes to the lung, increasing local elastase that would degrade the elastic fibers and lead to emphysema.
    (b) Carcinogenesis. A variety of components in the smoke can cause cancer, particularly polycyclic hydrocarbons and nitrosamines, which are potent carcinogens of lung carcinomas (Tables 6.3 and 6.4). The risk of developing

Table 6.3 Major constituents of tobacco smoke with adverse effects

Constituents	Adverse Effect
Tar Polycyclic aromatic hydrocarbons Nitrosamines Benzopyrene	*Carcinogenesis*
Nicotine	Ganglionic stimulation and depression, tumor promotion
Phenol	Tumor promotion, mucosal irritation
Formaldehyde Nitrogen oxide	Toxicity to cilia, mucosal irritation
Carbon monoxide	*Reduced oxygen transport*

Table 6.4 Organ-specific carcinogens in tobacco smoke

Carcinogen(s)	Organ
Polycyclic aromatic hydrocarbons 4-(Methylnitrosoamino)-1-(3-pyridyl)-1-butanone (NNK) 210Polonium	Lung, larynx
NNK, NNN, 210polonium	Oral cavity: snuff
Polycyclic aromatic hydrocarbons, NNK, NNN	Oral cavity: smoking
N'-Nitrosonornicotine (NNN)	Esophagus
NNK	Pancreas
4-Aminobiphenyl, 2-naphthylamine	Bladder

lung cancer is related to the intensity of exposure, frequently expressed in terms of "pack years" (e.g., one pack daily for 20 years equals 20 pack years) or in cigarettes smoked per day. In addition to lung cancer, smoking can also cause oral, esophageal, pancreatic, and bladder cancer.

Smoking increases the incidence of cancer on the premise of other carcinogens. A typical example is that smoking increases the incidence of lung cancer by ten times among asbestos workers. The combination of tobacco (chewed or smoked) and alcohol consumption has multiplicative effects on the risks of oral, laryngeal, and esophageal cancers.

(c) Atherosclerosis is considered to be associated with smoking. Coronary atherosclerosis can cause myocardial ischemia and even myocardial infarction. The mechanism of atherosclerosis induced by smoking is thought to be related to increased platelet aggregation. Smoking is an important modifiable risk factor for the development of cardiovascular diseases such as coronary artery disease, stable angina, acute coronary syndromes, sudden death, stroke, peripheral vascular diseases, congestive heart failure, erectile dysfunction, and aortic aneurysms via initiation and progression of atherosclerosis, which is responsible for approximately 140,000 premature deaths from cardiovascular diseases each year. Smoking has a multiplicative effect on atherosclerosis risk when combined with hypertension and hypercholesterolemia.

(d) Smoking during pregnancy increases the risk of spontaneous abortion and premature birth, or causes intrauterine growth retardation; however, birth weights of infants whose mothers stopped smoking before pregnancy are normal.

(e) Secondhand smoke, also known as passive smoking and environmental tobacco smoke, is a mixed smog from cigarette or other tobacco products released from the end of the combustion and tobacco smoke exhaled by smokers. It is also the most widespread and serious indoor air pollution. Studies have shown that secondhand smoke contains tar, ammonia, nicotine, suspended particulates, PM2.5, polonium 210, and more than 4000 kinds of harmful chemicals and dozens of carcinogens.

Secondhand smoke is more harmful to passive smokers than active smokers, especially to young children. The survey shows that, in China, the main victims of passive smoking are women and children. Although they do not smoke themselves, they often suffer from secondhand smoke in the family and public places. In addition, workplace, venue, and so on will often become a secondhand smoke flooding place. It is estimated that the relative risk of lung cancer among nonsmokers exposed to environmental smoke is about 1.3 times that of nonsmokers not exposed to smoke. Secondhand smoke greatly increases the risk of atherosclerosis and fatal myocardial infarction. In addition, it also increases the number of respiratory infections and asthmatic attacks in children, the risk of sudden infant death syndrome and middle ear infections in young children, and the number of low birth weight infants born to exposed mothers.

2. Tobacco-Related Diseases

Tobacco contains numerous toxic chemicals having adverse effects varying from mild throat irritation to carcinogenesis. Compared with nonsmokers, smokers have a higher relative risk of major diseases and higher mortality rates (in descending order of frequency):

(a) Lung cancers: Cigarette smoke contains more than 3000 toxic chemicals. According to statistics, smoking ten cigarettes a day would increase the incidence of lung cancer by 13 times. If smoking 20 cigarettes a day, the incidence of lung cancer would increase by 20 times. If smoking 40 cigarettes a day, the incidence of lung cancer would increase 65 times.

(b) Chronic obstructive pulmonary disease (COPD): Smoking is the primary risk factor for the onset of COPD. The longer the smoking time and the greater the amount of smoking, the higher the prevalence rate of the COPD; 25% of the heavy smokers will eventually develop into COPD, while 90% of the COPD patients are smokers. According to statistics, the prevalence rate of smokers who smoke more than 40 cigarettes per day is 75.3%.

(c) Cerebrovascular accidents and cardiovascular diseases: Smoking is a major risk factor for many cardiovascular and cerebrovascular diseases. The incidence of coronary heart diseases, hypertension, cerebrovascular disease, and peripheral vascular disease of smokers increased significantly. Statistics show that 75% of patients with coronary heart diseases and hypertension have smoking history. The incidence and mortality of coronary heart diseases were 3.5 and 6 times of nonsmokers, respectively, while the incidence of myocardial infarction was two to six times of nonsmokers.

(d) Smoking during pregnancy is associated with high risk of lower birth weight of fetus, high perinatal mortality, and intellectual deterioration of newborn.

6.1.1.3 Carbon Monoxide Poisoning

Carbon monoxide (CO) is a nonirritating, colorless, tasteless, and odorless gas. It is produced by the incomplete oxidation of carbonaceous materials. Poisoning due to inhalation of carbon monoxide occurs when the human body is exposed to high concentration of carbon monoxide. The mechanism of poisoning is that the affinity of carbon monoxide and hemoglobin is 200–300 times higher than that of oxygen and hemoglobin, so carbon monoxide is easily combined with hemoglobin to form carboxyhemoglobin, which causes hemoglobin to lose the capacity of oxygen carrying, causing tissue asphyxia. Hypoxia leads to central nervous system (CNS) depression, which develops so insidiously that victims often are unaware of their plight and are unable to help themselves. Systemic hypoxia appears when the hemoglobin is 20–30% saturated with CO, and unconsciousness and death are probable with 60–70% saturation.

1. Acute carbon monoxide poisoning is common in accidents and suicides. The clinical manifestations were mainly anoxia, and its severity was proportional to the saturation of HbCO. In mild poisoning, the patients would suffer from headache, weakness, vertigo, and difficulty in breathing while working, when HbCO saturation is 10–20%. In patients with moderate poisoning, the symptoms were aggravated, and the skin and mucous membranes were cherry red (Fig. 6.1). The patients could have nausea, vomiting, blurred consciousness, deficiency, or coma, and the saturation of HbCO reached 30–40%. In severe poisoning, the patients would fall into deep coma, accompanied by hyperthermia, increased muscle tension, and paroxysmal or tonic spasm, when HbCO saturation was

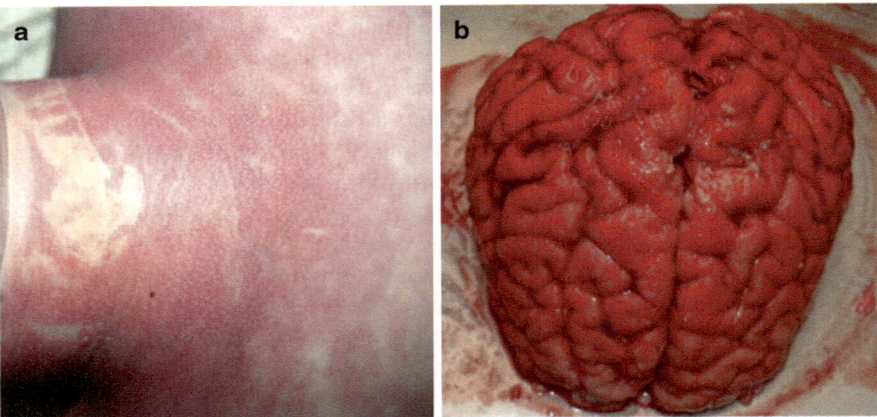

Fig. 6.1 Carbon monoxide (CO) poisoning. A faulty natural gas room heater caused this fatal poisoning. The characteristic cherry-red coloration of skin and brain persists in death. CO binds irreversibly with hemoglobin to produce bright red hemoglobin, which retains its color even though no oxygen is present. (Picture A has been was originated from coroner Luo hui, Police Station of Boluo county, Guangdong Province; Picture B has been was originated from professor Zhang lushun, Chengdu Medical College)

>50%. Some patients have blisters and redness on the chest and extremities, mainly due to autonomic neurotrophic disorders. Some patients with acute CO poisoning were awakened after the coma, and after 2–30 days of false recovery, they would be comatose again, and there were psychosis of dementia with tremor paralysis syndrome, sensorimotor disorder, or peripheral neuropathy, also called acute carbon monoxide poisoning delayed encephalopathy. Patients often have brain edema, pulmonary edema, myocardial damage, arrhythmia, and respiratory depression, which can cause death. If a cadaver is dissected, the visceral organs may appear cherry red (Fig. 6.1).

Of course, if death occurs rapidly, morphologic changes may not be present; with longer survival, the brain may be slightly edematous and exhibit punctate hemorrhages. These changes are not specific; they simply imply systemic hypoxia. In victims who survive CO poisoning, complete recovery is possible; however, impairments of memory, vision, hearing, and speech sometimes remain.

2. When someone has long-term exposure to low concentration of CO, chronic poisoning will happen, and the clinical manifestations are headache, dizziness, memory loss, lack of concentration, and palpitation. As a result, with low-level persistent exposure to CO, carboxyhemoglobin levels will gradually increase, causing chronic poisoning and even endangering life. The slowly developing hypoxia can evoke widespread ischemic changes in the brain, particularly in the basal ganglia and lenticular nuclei. With cessation of exposure to CO, the patient usually recovers, but there may be permanent neurologic damage.

6.1.1.4 Agricultural Exposures

In the process of agricultural activities, pesticide is essential to ensure output. Pesticide refers to chemical agents used in agriculture to control crop diseases and regulate their growth. Many kinds of pesticides can be divided into insecticides, acaricides, rodenticides, nematides, mollusks, fungicides, herbicides, plant growth regulators, and so on. According to the sources of raw materials, they can be divided into mineral source pesticides (inorganic pesticides), biological pesticides (natural organisms, microbes, antibiotics, etc.), and chemical synthesis of pesticides. According to the chemical structure, pesticides can be classified as organic chlorine, organophosphorus, organic nitrogen, organic sulfur, carbamate, pyrethroid, amide, urea, ether, phenols, phenoxy carboxylic acids, amides, three azoles, heterocyclics, benzoic acids, organometallic compounds, and so on.

1. Organophosphorus pesticide (OPS) is the most widely used insecticide in China. It mainly includes dichlorvos, parathion (1605), Phi phoxim (3911), internal phosphorus (1059), dimethoate, dimethoate, and malathion (4049). Every year, millions of people around the world suffer from acute organophosphorus poisoning. About 300,000 of them lose lives, and most of them occurs in developing countries. There are three ways organophosphorus pesticides can enter the body:

entering through the mouth by taking orally (in suicides), by the skin or/and mucous membrane, and by the respiratory tract.

Acute organophosphorus pesticide poisoning (AOPP) refers to a series of injuries caused by a large amount of organophosphorus pesticides that enter the body in a short period of time. It mainly includes cholinergic excitation or crisis in acute poisoning patients, followed by intermediate syndrome (IMS) and delayed peripheral neuropathy (OPIDPN). Before the crisis, the patients often showed obvious adverse reactions of cholinesterase inhibitors, such as nausea, vomiting, abdominal pain, diarrhea, sweaty, tears, wet cold skin, increased oral secretion, muscle fascicular tremor, emotional excitement, anxiety, and other mental symptoms. When the cholinergic crisis occurs, the clinical manifestations include pupil reduction, increased tracheal secretions, heart rate slowing, muscle tremor, spasmodic, and contraction.

2. Organochlorine pesticide poisoning is commonly caused by organochlorine pesticides such as DDT and chloride. Organochlorine pesticides have been widely used. Since then, it is difficult to be degraded and remains in agricultural products for a long time affecting human health. Poisoning is mostly caused by improper handling, food poisoning, suicide, and also for homicide. It can inhibit the metabolism of inositol, stimulate the excitement of central nervous system, and damage organs such as liver and kidney. Clinical manifestations of moderate poisoning include severe vomiting, sweating, salivating, blurred vision, muscle tremors, convulsions, palpitations, and lethargy. When a person is seriously poisoned, the clinical manifestations are epileptic seizures, coma, respiratory failure, heart fibrillation, and potentially liver and kidney damage.

3. Chronic poisoning refers to long-term exposure to small doses of a toxin, causing some changes in the body's physiological, biochemical, and pathological aspects, with a series of clinical signs and symptoms. Chronic human exposure to low level agricultural chemicals is implicated in cancer, chronic degenerative diseases, congenital malformations, and impotence, but the exact cause-and-effect relationship is undefined.

WHO conducted a statistical analysis of 19 major agricultural countries in the world in 2013 and found that there were about three million cases of pesticide poisoning caused by pesticide residues in these countries, of which 500,000 were acute poisoning. More than 75% of the mortality rate occurred in developing countries due to ready availability and indiscriminate use of hazardous pesticides, which are otherwise banned in developed countries. Pesticide residues in food items such as in fruits, vegetables, cereals, and grains are of greatest concern.

6.1.1.5   Industrial Exposure

1. Volatile organic solvents. Volatile organic compounds (VOCs) are those whose melting point is below room temperature and boiling point ranges from 50 to 260 °C. Organic solvents are widely used in huge quantities worldwide. The

harm of VOC is very obvious. When the concentration of VOC exceeds a certain concentration in the living room, people would suffer from headache, nausea, vomiting, and weakness in the extremities for a short time. VOC also can damage human liver, kidneys, brain, and nervous system.

Occupational exposure to benzene and 1,3-butadiene increases the risk of leukemia. Benzene poisoning can be divided into acute and chronic benzene poisoning. Acute benzene poisoning refers to the pathophysiological process that occurs mainly in the central nervous system as an organic solvent containing benzene or inhaling high concentration of benzene vapor. Chronic benzene poisoning refers to benzene and its metabolite phenols, which directly inhibit the nuclear division, cause gene mutation, and affect the hematopoietic function of the bone marrow. The clinical manifestation was a continuous decrease in white blood cell counts, which eventually developed into aplastic anemia or leukemia.

Butadiene is a colorless gas with sweet and aromatic odor. Its molecular weight is 54.1, the freezing point is $-108.9\ ^{\circ}C$, and the boiling point is $-4.4\ ^{\circ}C$. It is soluble in organic solvents and has water solubility of 0.38%. It is low in toxicity and can cause both anesthetic and stimulating effects. Its anesthetic effect is stronger than that of propane, but it is only half of butene.

2. Metals. Heavy metal poisoning refers to poisoning caused by heavy metal elements or compounds whose atomic weight is greater than 65, such as mercury poisoning and lead poisoning. It is because heavy metals can alter the structure of protein irreversibly and affect the function of tissue cells and affect human health.

Mercury is a silver white liquid metal and evaporates at room temperature. Mercury poisoning is often a chronic poisoning. It often occurs in productive process, resulting in long-term inhalation of mercury vapor and mercury compound dust. The main symptoms were mental abnormalities, gingivitis, and tremor. Acute mercury poisoning is caused by inhalation of large amounts of mercury vapor or intake of mercury compounds. For those who are allergic to mercury, poisoning can occur even if they are partially coated with mercury matrix preparations.

Lead is a widespread industrial pollutant that can affect various functions of the human body including the nervous system, the cardiovascular system, the skeletal system, the reproductive system, and the immune system, and it can cause diseases of the gastrointestinal tract, liver, kidney, and brain. As for lead poisoning, there will be special discussion in the following chapter.

3. Bisphenol A(BPA). BPA is used to synthesize polycarbonate (PC), epoxy resin, and other materials in industry. Since the 1960s, BPA has been used to make plastic bottles, infant cups, and the inner coatings of food and beverage cans. BPA is ubiquitous, ranging from mineral water bottles to medical devices to food packaging. Every year, 27 million tons of plastic containing BPA are produced worldwide. BPA can lead to endocrine disorders, threatening the health of the fetus and children. Obesity, cancer, and metabolic disorders are also thought to be related to BPA. It is believed that bottles containing BPA can induce precocious puberty in the European Union countries.

4. Vinyl chloride (VC). VC is an important monomer used in polymer chemical industry. It can be made from ethylene or acetylene. It is a colorless, liquefied gas, with a boiling point of $-13.9\,°C$, critical temperature of $142\,°C$, and critical pressure of 5.22 MPa. It could form an explosive mixture with air. The explosion limit is from 4 to 22% (volume). It is more explosive under pressure. VC is a toxic substance. Exposure to VC for a long time may lead to hepatic angiosarcoma, which is a rare type of liver tumor.
5. Environmental dusts. Dust refers to solid particles suspended in the air. According to the International Organization for Standardization, the solid with particle size less than 75 µm is defined as dust. Too much dust in the atmosphere will have a disastrous effect on the environment. Dust exposure is more common in the workplace. Chronic nonneoplastic lung disease caused by inhalation of mineral dust is called pneumoconiosis. The most common pneumoconioses include coal workers' pneumoconiosis, silicosis, and asbestosis, which are caused by exposures to coal dust (in mining of hard coal), silica (in sandblasting and stonecutting), and asbestos (in mining, fabrication, and insulation work). The increased risk of cancer as a result of asbestos exposure, however, extends to family members of asbestos workers and to other persons exposed outside the workplace. Pneumoconioses and their pathogenesis are discussed in Chap. 8 (Table 6.5).

Table 6.5 Human diseases associated with occupational exposures

Toxicant(s)	Effect(s)
Solvents, acrylamide, methyl chloride, mercury, lead, arsenic, DDT	Peripheral neuropathies
Chlordane, toluene, acrylamide, mercury	Ataxic gait
Alcohols, ketones, aldehydes, solvents	CNS depression
Ultraviolet radiation	Cataracts
CO, lead, solvents, cobalt, cadmium	Heart disease
Isopropyl alcohol, wood dust	Nasal cancer
Radon, asbestos, silica, bis(chloromethyl)ether, nickel, arsenic, chromium, mustard gas	Lung cancer
Grain dust, coal dust, cadmium	Chronic obstructive lung disease
Beryllium, isocyanates	Hypersensitivity
Ammonia, sulfur oxides, formaldehyde	Irritation
Silica, asbestos, cobalt	Fibrosis
Vinyl chloride	Liver angiosarcoma
Mercury, lead, glycol ethers, solvents	Toxicity
Naphthylamines, 4-aminobiphenyl, benzidine, rubber products	Bladder cancer
Lead, phthalate plasticizers	Male infertility
Cadmium, lead	Female infertility
Mercury, polychlorinated biphenyls	Teratogenesis
Benzene, radon, uranium	Leukemia
Polychlorinated biphenyls, dioxins, herbicides	Folliculitis and acneiform dermatosis
Ultraviolet radiation	Cancer

6.2 Chemical and Drug Injury

Everyone is exposed to different chemicals and drugs. These chemicals and drugs are broadly divided into the following two categories: (1) therapeutic (iatrogenic) agents, for example, drugs, which when administered indiscriminately are associated with adverse effects, and (2) nontherapeutic agents, for example, alcohol, lead, carbon monoxide, and drug abuse.

6.2.1 Therapeutic (Iatrogenic) Drug Injury

Drugs are often indispensable for the treatment of diseases. Of course, if used inappropriately, it can also cause many injuries. These injuries can contribute substantially to patient morbidity, cost of therapy, and length of stay. Although patient management is based on rational drug treatment, 2–5% of patients do have adverse drug reactions. Generally speaking, the risk of adverse drug reactions increases with the number of drugs used. Fortunately, most of these injuries can be prevented with appropriate techniques and preventive precautions. If a therapeutic (iatrogenic) drug injury does occur, prompt recognition and timely, appropriate treatment may prevent further tissue injury, pain, or even function loss.

Drug-induced injury refers to the physiological and biochemical disorders, structural changes, and other diseases occurring in the course of drug use, which is the consequence of adverse drug reactions. Drug-induced injuries can be divided into two categories. The first type of injuries is due to side effects of drugs, excessive doses, or injury caused by drug interactions. Such injuries can be prevented by rational use of drugs, with little harm. The second type is allergic reactions or specific reactions. This kind of injuries are difficult to prevent; though their incidence is low, it is very harmful. It often leads to death. Drug damage occurs in almost every organ and causes a variety of diseases or injuries, such as arrhythmia, diffuse pneumonia, pulmonary fibrosis, violent hepatitis, chronic active hepatitis, nephrotic syndrome or renal failure, dermatitis, aplastic anemia, hemolytic anemia, mental disorder, gastrointestinal bleeding, spinal cord injury, and cancer.

6.2.1.1 Cause of Iatrogenic Drug Injury

Drug-induced injury involves a variety of reasons, usually thought to be related to the following factors: (1) overdose, (2) genetic predisposition, (3) exaggerated pharmacologic response, (4) interaction with other drugs, (5) unreasonable prescription, and (6) unknown factors. Some of the common forms of iatrogenic drug injuries and the offending drugs are listed in Table 6.6.

Table 6.6 Iatrogenic drug injury

Adverse Effect	Offending drug
Gastritis, peptic ulcer Pancreatitis	Aspirin, nonsteroidal anti-inflammatory drugs (NSAIDs) Azathioprine, mercaptopurine, didanosine
Cholestatic jaundice Hepatitis	Amoxicillin/clavulanate, Trimethoprim/ sulfamethoxazole Halothane, isoniazid, acetaminophen
Cerebrovascular accidents Spinal cord injury Peripheral neuropathy eight nerve deafness	Anticoagulants Intrathecal chemotherapy Vincristine, antimalarials, streptomycin
Exfoliative dermatitis	Penicillin, sulfonamides
Arrhythmias Cardiomyopathy	Idarubicin, mitoxantrone, epirubicin, etc. Cyclophosphamide, anthracyclines, trastuzumab, etc.
Aplastic anemia Extravasation	Chloramphenicol, flucloxacillin Vasoconstrictive substances, cytotoxic agents
Alveolitis, interstitial pulmonary fibrosis Asthma	Antineoplastic drugs Aspirin, indomethacin
Acute tubular necrosis Nephrotic syndrome	Gentamycin, kanamycin Heroin, interferon alfa, lithium
……	……

6.2.2 Alcoholism

Alcoholism, usually known as drunkenness, refers to the abnormal state of the body after a large amount of alcohol, which causes the most serious damage to the nervous system and liver.

When alcohol is absorbed, its metabolism goes through three steps: first, ethanol is converted into acetaldehyde by ethanol dehydrogenase; then acetaldehyde is oxidized into acetic acid by acetaldehyde dehydrogenase; and finally, the acetic acid will be decomposed into carbon dioxide and water. Acetaldehyde can cause the secretion of adrenaline and norepinephrine. At this point, the patient's characteristics are flushing and rapid heartbeat. Alcohol has direct neurotoxicity, cardiotoxicity, and hepatotoxicity, so patients have a series of neurological abnormalities and even coma and shock after poisoning. In addition, alcohol can cause heart diseases, hypoglycemia, and metabolic acidosis.

It is divided into acute poisoning and chronic poisoning. The former can cause serious injuries to the patient in a short time and can even lead to death directly or indirectly. The latter causes cumulative damage to patients, such as alcohol dependence, mental disorders, alcoholic cirrhosis, and the induction of certain cancers (oral and tongue cancer, esophageal cancer, cancer of the esophagus, and liver cancer).

Alcohol dependence is characterized by higher tolerance. The usual dose is insufficient to produce a comfortable effect, and more and more doses are needed to experience the expected effects of alcohol. Dependence can also trigger symptoms of abstinence, such as nausea, sweating, tremor, and anxiety, when patients attempt to abstain from alcohol.

Acute and chronic adverse reactions are associated with the amount and duration of alcohol intake. Generally speaking, 10 g of ethanol exists in (1) a can of beer (or half bottle of beer), (2) 120 mL of pure wine, and (3) 30 mL of 43% liquor. It may be harmful to take 40 g of ethanol per day, and it is undoubtedly dangerous to take 100 g or more per day.

6.2.2.1   Acute Alcoholism

After drinking, about 20% of alcohol was absorbed by the stomach and 80% by the duodenum and small intestine. The toxic dose and lethal dose of alcohol vary from person to person. The toxic dose is generally 70–80 g, and the lethal dose is 250–500 g. Whether alcohol intoxication occurs after drinking is related to the following factors: (1) limosis (when drinking on an empty stomach, alcohol absorbs quickly), (2) fat intake (fat food can slow down alcohol absorption), and (3) the gastrointestinal function and ability to metabolize alcohol.

1. Central Nervous System

    Alcoholic brain injury refers to chronic damage of frontal lobe and limbic system-related memory and advanced mental functions caused by alcohol consumption. The clinical manifestations of alcoholism include alcohol-induced amnesia syndrome and frontal lobe syndrome. The main manifestations of amnesia syndrome are short-term memory impairment, and the manifestation of frontal lobe damage syndrome includes the defects of abstract thinking, concept formation, planning, and complex information processing, while other cognitive functions are relatively intact and clear consciousness.

    Alcohol acts as a central nervous system inhibitor; the intensity of alcohol's influence on the central nervous system is related to the amount and duration consumed, which can be reflected by the level of alcohol in the blood:

    (a) Initial effect of alcohol on subcortical structures is followed by disordered cortical function, motor ataxia, and behavioral changes. These changes are evident when blood alcohol content does not exceed 100 mg/dL, which is the upper limit of alcohol sobriety defined by most national law enforcement agencies in dealing with drunk driving cases.
    (b) Blood level of 100–200 mg/dL is associated with depression of cortical centers, lack of coordination, impaired judgement, and drowsiness.
    (c) Stupor and coma supervene when blood alcohol level is about 300 mg/dL.
    (d) Blood level of alcohol above 400 mg/dL can cause anesthesia, depression of medullary center, and death from respiratory arrest.

    However, chronic alcoholics develop the tolerance and adaptability of the central nervous system, so they can tolerate higher levels of alcohol in the blood without such a serious impact.
2. Stomach

Acute alcohol poisoning can cause gastric ulcer, gastric bleeding, and acute gastritis.
3. Liver

Severe alcoholism can cause extensive necrosis of liver cells and even liver failure, namely, alcoholic hepatitis. Alcoholic hepatitis is characterized by fatigue, anorexia, weight loss, hepatomegaly, and more seriously fever, jaundice, and vomiting. Alcoholic hepatitis represents an acute presentation of alcoholic liver disease. Acute alcoholic injury to the liver is explained in Chap. 9.

6.2.2.2 Chronic Alcoholism

Chronic alcoholism produces widespread injury to organs and systems. Contrary to the earlier view that chronic alcoholic injury is caused by nutritional deficiency, it is now known that most alcohol-related injuries to different organs are caused by the toxicity of alcohol and the accumulation of acetaldehyde, the main toxic metabolite in the blood. Other proposed mechanisms of tissue injuries in chronic alcoholism are free radical-mediated injuries and genetic susceptibility to alcohol dependence and tissue damage.

Some of the more important organ effects in chronic alcoholism are as follows:

1. Nervous System

Long-term alcoholism can cause serious brain damage. The main lesion is the loss of neurons, affecting the whole brain, especially in the cerebellum. Neuronal loss in alcoholics occurs in discrete anatomical regions. In the cerebral cortex, the loss is restricted to the superior frontal cortex, but the magnitude of this loss is too small to be reliably detected by routine nonquantitative evaluation.

No loss of neurons occurs in the primary motor cortex or hippocampus. In subcortical regions, there is neuronal loss from the supraoptic and paraventricular nuclei of the hypothalamus, but not from the mamillary bodies, anterior and dorsomedial nuclei of the thalamus, serotonergic dorsal raphe, basal forebrain, or cerebellum. Neuronal loss from the hypothalamus is related to maximum daily alcohol consumption.

Autopsy and brain imaging studies indicate reductions and abnormalities in overall brain size and shape, specifically in structures such as the cerebellum, basal ganglia, and corpus callosum.
2. Liver

Alcoholic liver disease is a disorder of liver cell structure and/or dysfunction due to excessive consumption of alcohol. There is a wide spectrum of histology for alcoholic liver disease that includes steatosis, steatohepatitis, alcoholic hepatitis (without steatosis), alcoholic foamy degeneration, cholestasis, veno-occlusive disease, central hyaline sclerosis, and micronodular cirrhosis. Initially, it is usually manifested as fatty liver, which can develop into alcoholic hepatitis, alcoholic liver fibrosis, and alcoholic cirrhosis (severe scarring), which is associated with liver failure, intestinal hemorrhage, and liver cancer.

3. Pancreas

   Chronic calcifying pancreatitis and acute pancreatitis are serious complications of chronic alcoholism.

4. Alimentary tract

   Alcohol abuse is associated with gastritis, gastric and esophageal ulcers, fatal intestinal bleeding, and increased risk for cancers of the mouth and esophagus. Gastritis, peptic ulcer, and esophageal varices associated with fatal massive bleeding may occur.

5. Cardiovascular system

   Alcohol abuse is associated with increased risk of dilated cardiomyopathy and increased risk of cardiovascular diseases. By contrast, moderate social consumption of alcohol has some protective effect. Level of HDL (atherosclerosis-protective lipoprotein) has been shown to increase with moderate consumption of alcohol.

6. Endocrine system

   In men, testicular atrophy, feminization, loss of libido and potency, and gynecomastia may develop. Although these effects appear to be associated with decreased testosterone levels, this is due to impaired ability of the liver to inactivate estrogen in alcoholic cirrhosis.

7. Blood

   Hematopoietic dysfunction may occur due to secondary megaloblastic anemia and increased red blood cell volume. This is related to atrophic gastritis caused by alcohol, which leads to vitamin B_{12} absorption disorder.

8. Immune system

   Alcoholics are more vulnerable to various infections, which is due to liver damage and the decline of the function of synthetic albumin.

9. Pregnancy

   Even small amounts of alcohol have adverse effects on the fetus. Fetal alcohol syndrome is a permanent birth defect caused by maternal alcohol abuse during pregnancy. The severity of the disease was related to the amount, frequency, and duration of alcohol intake by the pregnant women. Alcohol can inhibit fetal growth, resulting in low birth weight. More seriously, alcohol exposure during pregnancy will hinder the development of fetal nerve cells and brain structures, which will lead to a series of primary cognitive and functional disorders after birth, including weak memory, insufficient attention, impulsive behavior, and poor understanding.

10. Cancer

    Alcohol is considered a clear risk factor for cancer. Alcohol is directly responsible for 5.5% of new cancer cases and 5.8% of cancer deaths worldwide. Alcohol has a causal relationship with pharyngeal cancer, esophageal cancer, hepatocellular carcinoma, female breast cancer, and colorectal cancer. Alcohol may also be associated with pancreatic cancer, gastric cancer, and lung cancer.

As more evidence emerges, the list of alcohol-related cancers will continue to grow.

6.2.3   Drug Abuses

Drug abuse refers to drug addicts indulging in substances other than alcohol and nicotine and overusing these substances to change their mood. This use has nothing to do with medical purposes, and the result is that the abusers rely on this substance, forcing them to continue to use it indefinitely. Once they become addicted, they can't extricate themselves. Drug abuse is associated with suicide, homicide, assaults, motor-vehicle injury, HIV infection, pneumonia, mental illness, hepatitis, and sudden death from cardiac disease or coma.

Addiction can be psychological, physical, or both. In physical addiction, corporeity becomes physiologically dependent on the drug, and withdrawal may produce physical effects ranging from anxiety to seizures and death. Psychological addiction is self-explanatory. Tolerance comes from long-term drug use. Drug detoxification reduces tolerance, which is a dangerous feature.

Drug abuse in pregnancy is dangerous for the fetus. Like maternal alcohol use, maternal drug use may be associated with prematurity and birth defects. Some babies are also addicted, and these babies must be slowly detoxification.

Drugs can be ingested, sniffed or inhaled, or injected. Greater effect is produced by injection or by sniffing or inhaling (including smoking) for quicker absorption into the bloodstream. Because intestinal absorption is slow, intake has less expected effect.

Illegal drugs fall into four main categories: depressants, stimulants, narcotics, and hallucinogens. Some commonly abused drugs and substances will be discussed here.

6.2.3.1   Derivatives of Opium

Alkaloids extracted from opium and its derivatives can interact with the specific receptors of central neurons, relieve pain, and produce happiness. Opioid and its semisynthetic derivatives include morphine, two ethyl morphine (heroin), hydrogenated morphine, codeine, and oxycodone. The synthetic morphine substances include propiophenone, fentanyl, methadone, pethidine, and agonist antagonist tazocin. Large doses of opioids can cause stupor, coma, and respiratory depression.

Repeated use of opioids will cause tolerance and neural adaptation, which is related to withdrawal response. Opioid withdrawal syndrome includes craving, anxiety, bad mood, yawning, sweating, goose bumps, tears, runny nose, nausea or vomiting, diarrhea, pain spasm, muscle pain, fever, and insomnia.

Intravenous morphine abuse has many physical consequences, including hepatitis B, hepatitis C, HIV infection, septicemia, endocarditis, pneumonia and lung abscess, thrombotic phlebitis, and rhabdomyolysis, as well as psychological and social damage.

Heroin is a semisynthetic drug derived from morphine alkaloids. It is commonly known as white powder or white flour. It is the essential product of the opioid drug series. Generally, it includes heroin (two acetyl morphine), heroin salt (heroin hydrochloride, nitrate, tartrate, and citrate), and the hydrate of heroin salt.

Heroin has great harm to the physical and mental health of human beings. Long-term heroin abuse can lead to disintegration of personality, psychopathy, and reduced life expectancy, especially the most obvious injury to the nervous system.

Heroin has a wide range of adverse physical effects, which can be classified according to: (1) pharmacological effects of pharmaceuticals, (2) reactions to cutting agents or contaminants, (3) allergic reactions to drugs or their adulterants, and (4) diseases transmitted by sharing needles.

Some of the most important adverse effects of heroin are the following:

1. Sudden death

    Sudden death refers to a person who is healthy or seemingly healthy dies in a short time due to natural illness. Heroin-induced death is usually caused by overdose. However, sudden death sometimes is due to a loss of tolerance for the drug, such as after a period of incarceration. The mechanisms of death include profound respiratory depression, arrhythmia and cardiac arrest, and pulmonary edema.

2. Pulmonary disease

    Pulmonary complications induced by heroin abuse include edema, septic embolism, pulmonary abscess, opportunistic infections, talcum, and other foreign body granulomas with adulteration. Although granulomas occurs mainly in the lungs, they are sometimes found in the spleen, liver, and lymph nodes to drain the upper limbs. Examination under polarized light often highlights the trapped talc crystals, sometimes wrapped in foreign body giant cells.

3. Infections

    Infection is a common complication of heroin abuse. The most common sites are skin and subcutaneous tissue, heart valves, liver, and lungs. More than 10% of the patients hospitalized with heroin abuse have infective endocarditis, which often takes a distinctive form involving right-sided heart valves, particularly the tricuspid. Endocarditis can be caused by infection with a variety of pathogenic microorganisms, such as bacteria, fungi, and viruses. Bacterial endocarditis often has valve perforation or valve vegetations. If the vegetation falls off, it can cause embolic abscess of distant organs or sepsis. Most cases are caused by *Staphylococcus aureus*. Viral hepatitis is the most common infection among drug addicts and is obtained through sharing dirty needles, which has also led to a very high incidence of human immunodeficiency virus (HIV) infection among intravenous drug users.

4. Skin and subcutaneous lesions

    Skin damage may be the most common symptom of intravenous drug abuse. Acute lesions include abscesses, cellulitis, and ulcers caused by subcutaneous injection. Scarring at injection sites, hyperpigmentation over commonly used veins, and thrombosed veins are the usual sequelae of repeated intravenous inoculations.

5. Renal problems

Drug users are prone to suffer from various kidney diseases. If the one-time dose of drug users exceeds the body's tolerance, it will lead to drug poisoning, and acute renal failure is easy to occur during the protracted coma. Sometimes, drug addicts will suffer from a variety of urinary infectious diseases. The common ones are glomerulonephritis, acute and chronic urinary tract infection, pyelonephritis, cystitis, renal abscess, etc.

6.2.4   Lead Poisoning

Lead can cause a series of physiological and biochemical changes and affect the functions of the central and peripheral nervous system, the cardiovascular system, the reproductive system, and the immune system and cause the diseases of the gastrointestinal tract, the liver, the kidney, and the brain. Children and pregnant women are especially vulnerable to lead. Lead poisoning reduces children's intelligence, learning ability, perception, and understanding ability, leading to attention deficit hyperactivity disorder, impulse, and language learning disorders. Lead poisoning occurs when the venous blood lead level exceeds 200 μg/L for two consecutive times. According to the blood lead level, it can be divided into mild, moderate, and severe lead poisoning. A blood lead level of 200–249 μg/L is considered as mild lead poisoning. If the blood lead level is 250–449 μg/L, it is a moderate lead poisoning. If the blood lead level is higher than 450 μg/l, it means severe lead poisoning.

6.2.4.1   The Effects of Lead Exposure

1. The neurotoxic effects of lead are attributed to the inhibition of neurotransmitters caused by the disruption of calcium homeostasis.
2. Lead interferes with the normal remodeling of cartilage and primary bone trabeculae in the epiphyses in children. This causes increased bone density detected as radiodense lead lines.
3. Lead inhibits the healing of fractures by increasing chondrogenesis and delaying cartilage mineralization.
4. Lead affects the synthesis of hemoglobin. Porphyrin metabolism disorder is one of the important and early changes in the mechanism of lead poisoning. Porphyrin is an intermediate in the process of hemoglobin synthesis, and it is affected by a series of sulfhydryl enzymes in the process of hemoglobin synthesis. It has been proved that lead at least inhibits the delta aminolevulinic acid dehydrase (delta aminovalerate dehydrase, ALAD), fecal porphyrin oxidase, and ferrous complex enzyme. ALAD is a metalloenzyme that consists of eight identical subunits and

eight zinc ions. Zinc ions play an important role in the activity and stability of enzymes. Lead can replace the zinc ions of the active site, inhibit the activity of ALAD, and inhibit the formation of ALA, which leads to the increase of ALA in the blood. Of course, the ALA from urine is also increased. In addition, lead can also inhibit the activity of fecal porphyrin oxidase and prevent the oxidation of fecporphyrin III to protoporphyrin IX, which results in the increase of fecal porphyrin in blood and increased excretion of fecal porphyrin in urine. Lead can also inhibit the ferrous complex enzyme, so that protoporphyrin IX cannot be combined with two valent iron to be heme. Protoporphyrin in erythrocytes can bind with abundant zinc in mitochondria of red blood cells, resulting in increased zinc protoporphyrin. Therefore, urinary ALA, fecal porphyrin, and hematoporphyrin or zinc protoporphyrin are all diagnostic indicators of lead poisoning.

6.2.4.2 Morphological Changes of Lead Poisoning

The major anatomic targets of lead toxicity include the bone marrow and blood, nervous system, gastrointestinal tract, and kidneys.

1. Blood and marrow changes occur fairly rapidly and are characteristic. Lead poisoning can lead to anemia. Compensatory erythrocyte proliferation in bone marrow occurs due to disordered hemoglobin synthesis. The dot cells, reticulocytes, and erythrogranulocytes in the blood increased. The basophilic substances of these three red blood cells contain mitochondria and microsomal fragments and RNA. The inhibition of ferrochelatase by lead may result in the appearance of a few ring sideroblasts, red cell precursors with iron-laden mitochondria that are detected with a Prussian blue stain. In peripheral blood, hemoglobin synthesis deficiency is characterized by erythrocytic, hypochromic anemia, usually accompanied by mild hemolysis. More obvious is the dotted basophilic dotting of red blood cells.

2. Brain damage is prone to occur in children. Lead can easily pass through the placenta and the immature blood-brain barrier. Lead poisoning may cause sensory, motor, intellectual, and psychological disorders in young children, including reduced IQ, learning disabilities, mental and motor retardation, blindness, and, in more severe cases, psychosis, epilepsy, and coma. Lead toxicity in the mother may impair brain development in the prenatal infant. At the more severe end of the spectrum lies marked brain edema, demyelination of the cerebral and cerebellar white matter, and necrosis of cortical neurons accompanied by diffuse astrocytic proliferation. In adults, the CNS is less often affected, but frequently, a peripheral demyelinating neuropathy appears, typically involving the motor nerves of the most commonly used muscles. Thus, the extensor muscles of the wrist and fingers are often the first to be affected (causing wrist-drop), followed by paralysis of the peroneal muscles (causing foot drop).

3. Lead poisoning also often causes gastrointestinal symptoms. The mucous membrane of the digestive tract has the ability to secrete lead. In the process of lead

poisoning, lead directly acts on the gastric mucosa, destroys the regeneration ability of the gastric mucosa, and causes the inflammation of the gastric mucosa. Studies have shown that the detection rate of pathological damages of gastric mucosa in patients with chronic lead poisoning is 96.7%, and there may be atrophic gastritis. It has been reported that patients with chronic moderate or severe lead poisoning are initially diagnosed as superficial gastritis, and 91% turn to atrophic gastritis after 3 years. Lead "colic" is characterized by extreme severity and local abdominal pain.

4. Kidney lesions. Lead can affect the function of mitochondria in renal tubular epithelial cells, inhibit the activities of Na^+ and K^+-ATP enzymes, and cause renal tubular dysfunction and even damage. Acute poisoning mainly affects the proximal convoluted tubules, resulting in cell membrane damage, cell swelling, mitochondrial swelling and rupture, and matrix granule loss. In addition to damaging the renal tubules, chronic poisoning mainly manifested as progressive interstitial fibrosis, which first appeared around the renal tubules and then gradually expanded outward. Microtubule atrophy and fibroblast proliferation can be seen simultaneously.

5. The cardiac lesions. Long-term lead exposure can cause elevated blood pressure, toxic myocarditis, and myocardial damage. Lead exposure can increase the oxygen free radicals in the body, which results in lipid peroxidation damage, including myocardial cell membrane and myocardial microsomal membrane, and affects the cation transfer enzyme of myocardial microsomal membrane, with overloading the Ca^{2+} ion in the aorta and other vascular cells and makes the myocardial cell work disorder.

6. To those poor oral hygiene, blue black lead line can be visible on the incisors, canine gingival margin.

6.3 Injury by Physical Agents

Physical injury refers to human injuries caused by various physical factors, including radiation injury, high and low temperature injury, and electric injury.

6.3.1 Injury by Radiation

Radiation is divided into two categories: ionizing and nonionizing radiation. Ionizing radiation includes cosmic rays, X-rays, and radiation from radioactive substances. Nonionizing radiation includes ultraviolet, thermal radiation, radio waves, and microwaves. The energy of nonionizing radiation, such as ultraviolet (UV) and infrared light, microwaves, and sound waves, can move atoms in a molecule or cause them to vibrate but is not sufficient to displace electrons from atoms. By contrast, ionizing radiation has sufficient energy to remove tightly bound electrons.

Collision of these free electrons with other atoms releases additional electrons, in a reaction cascade referred to as ionization.

Radiation injury refers to acute, delayed, or chronic tissue damage caused by ionizing radiation. Acute, chronic, and long-term effects were observed, among which acute injury was most commonly seen in nuclear radiation accidents.

The main sources of ionizing radiation are (1) X-rays and gamma rays, which are electromagnetic waves of very high frequencies, and (2) high-energy neutrons, alpha particles (composed of two protons and two neutrons), and beta particles, which are essentially electrons. At equivalent amounts of energy, alpha particles induce heavy damage in a restricted area, whereas X-rays and gamma rays dissipate energy over a longer, deeper course and produce considerably less damage per unit of tissue.

6.3.1.1 Main Determinants of the Biologic Effects of Ionizing Radiation

In addition to the physical properties of the radiation, its biologic effects depend heavily on the following variables:

1. Radiation sources and radiographic conditions: The type of radiation, the dose of radiation, the location and area of the radiation, the fractionation, and the single irradiation all have different damage effects. In general, the radiation dose is the main factor of the image damage effect. The greater the dose, the more significant the effect, but not the linear relationship.
2. Different Parts of the Body Have Different Sensitivity to Radiation. The abdomen is most sensitive to radiation, in turn, the pelvic cavity, head, chest, and limbs. When the other conditions are the same, the larger the irradiated area is, the greater the damage effect. Because ionizing radiation damages DNA, rapidly dividing cells are more vulnerable to injury than quiescent cells. In addition to damaging DNA transcription at very high doses, DNA damage is compatible with the survival of non-differentiated cells such as neurons and muscle cells. Therefore, it is understandable that tissues with high rate of cell turnover, such as gonads, bone marrow, lymphoid tissues, and gastrointestinal mucosa, are highly vulnerable to radiation and show early damage after exposure.
3. Rate of delivery. When the total radiation dose is the same, the damage effect of multiple radiation is lower than the damage effect of a full dose of radiation, and the more the division times, the longer the interval, the smaller the damage effect, which may be related to the compensation and repair of the body.
4. Hypoxia. Radiation decomposition of water to produce reactive oxygen species (ROS) is an important mechanism of DNA damage induced by ionizing radiation. Tissue hypoxia, for example, may exist in the center of rapidly growing, poorly vascularized tumors, thus reducing the degree of damage and effectiveness of radiation therapy for tumors.

6.3.1.2 DNA Damage and Carcinogenesis

The ionizing radiation damage DNA can be divided into two types: direct and indirect effects. The direct effect is DNA directly absorbed radiation energy and is damaged. The indirect effect is that other molecules around DNA (mainly water molecules) absorb ray energy to produce highly reactive free radicals and then damage DNA. Ionizing radiation can cause many types of damage in DNA, including single-base damage, single- and double-strand breaks, and cross-links between DNA and protein. In surviving cells, simple defects may be reparable by various enzyme repair systems. However, double-strand breaks may persist without repair, or the repair of lesions may be imprecise (error prone), creating mutations. If cell-cycle checkpoints are not functioning (for instance, because of mutations in p53 gene), cells with abnormal and unstable genomes survive and may expand as abnormal clones to form tumors eventually.

6.3.1.3 Morphological Changes of Radiation Injury

When the radiation dose is large, vessels may show dilation only during acute injury stage. Necrosis is the most important lesion, followed by inflammatory reaction. Subsequently, endothelial cell proliferation and collagen transparency were observed in the irradiated blood vessels. With the thickening of the media layer, the vascular lumen was obviously narrow or occluded. At this time, the increase of collagen in the irradiation field leads to scar formation and contraction, usually becoming obvious. Later, or higher doses, a variety of changes, including endothelial cell swelling and vacuolization, and even necrotic small blood vessels, such as capillaries and small veins. The affected vessels may rupture or develop thrombosis.

When the radiation dose is relatively small, cells can survive radiation damage, showing extensive changes in the structure of chromosomes, including deletion, fragmentation, and translocation. Mitotic spindles are often disordered, and polyploidy and aneuploidy may be encountered. The nuclear swelling, condensation, and agglutination of chromatin may occur, and the breakage in the nuclear membrane can also be noticed. In addition to affecting DNA and nucleus, radiative energy may cause multiple cytoplasmic changes, including cytoplasmic swelling, mitochondrial aberration, and endoplasmic reticulum degeneration. The rupture of the plasma membrane and the focal defect may occur. Of course, apoptosis may also occur.

The morphology of tissues and cells after radiation injury is similar to that of cancer. It is characterized by cell pleomorphism, giant cell formation, nuclear change, and mitosis. Cells with abnormal nuclear morphology can be produced for several years, including multinucleated giant cells or multinucleated giant cells.

6.3.1.4 Effects on Organ Systems

The most sensitive organs and tissues are gonads, hematopoietic and lymphatic systems, and the lining of the gastrointestinal tract. Changes in hematopoietic and lymphatic systems, as well as cancer caused by environmental or occupational exposure to ionizing radiation, are summarized below:

1. Hematopoietic and lymphoid systems

   It is noteworthy that hematopoietic and lymphatic systems are extremely sensitive to radiation damage. Hematopoietic precursors in bone marrow are also sensitive to radiant energy, resulting in dose dependent bone marrow aplastic anemia. Radiation directly destroys lymphocytes, both in circulating blood and tissues (lymph nodes, spleen, thymus, and intestines). High dose and large area of exposure and severe lymphocytic reduction may occur in the time of irradiation, along with shrinkage of the lymph nodes and spleen. With sublethal doses of radiation, the regeneration of surviving progenitor cells is rapid, leading to a return to normal lymphocyte counts. The acute influence of bone marrow irradiation on the peripheral blood count reflects the turnover kinetics of granulocytes, platelets, and red cells of the components, and their half-life is less than 1 day, 10 days, and 120 days, respectively.

   Neutropenia occurred within a few days after circulating neutrophil counts increased briefly. Neutrophil counts reached the lowest point, usually at near zero counts, at the second week. If the patient survives, a complete recovery of granulocytes may take 2–3 months.

   Thrombocytopenia occurs at the end of the first week. The lowest point of platelet counts occurs behind granulocytes, and the recovery of platelets is later than the recovery of granulocytes. Anemia occurs after 2–3 weeks of radiation exposure and may last for several months. High doses of radiation will result in more severe cell loss and longer recovery. Very high doses kill marrow stem cells and induce permanent aplasia (aplastic anemia) marked by a failure of blood counts recovery, whereas with lower doses, the aplasia is transient.

2. Gonads

   Gonad is one of the highly sensitive organs to ionizing radiation, which often causes infertility and menstrual disorders under radiation accidents and occupational exposure conditions. Ionizing radiation can damage the ovary and endometrium or cause endometrial atrophy, leading to menopause. When menopause lasts for more than 3 months, it is called radioactive amenorrhea. Testicular atrophy in males and destruction of ovaries in females can cause infertility. Depending on the radiation dose, the damage ranges from temporary infertility to permanent infertility.

3. Gastrointestinal injury

   Radiation injury to the intestine can be divided into three periods, namely, acute phase, subacute stage, and chronic stage. The acute phase occurs in the early stage of radiation, subacute phase occurs 2–12 months after radiation, and chronic phase occurs after 12 months of radiation.

Acute injury is most obvious in cells with strong metabolic ability and active mitosis, especially in basal cells and crypt cells of intestinal mucosa. The intestinal mucosa decays, the villi of the intestinal wall became shorter, and the surface area of the intestinal epithelium became smaller. When the cell regeneration system is further damaged, tiny ulcers are formed, and as time goes on, small ulcers fuse with each other to form an eye-visible ulcer. At the same time, there is edema, inflammatory cell infiltration, telangiectasia, and even bleeding in the submucosa. The endothelial cells of the submucosal arterioles can be swollen and separate from the basement membrane. Progressive vascular and connective tissue lesions can cause obliteration arterio-phlebitis and microvascular insufficiency. The intestinal mucosa may be ulcerated by plaque like ischemia. Large foam cells can be seen under the intima of the vessels. Because of ischemia, there are fibrous tissue hyperplasia in the submucosa, and large gigantic shape fibroblasts appear.

Chronic injury is caused by the indirect effect of radiation, mainly due to progressive occlusive arterioles and extensive collagen deposition and fibrosis. With the decrease of intestinal wall blood vessels, intestinal wall blood supply decreased, and intestinal wall ischemia occurred, resulting in intestinal mucosal atrophy and telangiectasia. As the vasculitis progressively worsens, necrosis, ulceration, and perforation of the intestinal wall can occur. Among them, ulcer is the most common, which can cause perforation of intestinal wall and cause peritonitis or abdominal abscess. Healing and repair of ulcers can lead to fibrosis and scarring, resulting in intestinal stenosis and intestinal obstruction. Some patients can form fistula. Cancer can also be induced by radiation in the later period.

4. Heart

The damaged sites include pericardium, epicardium, endocardium, and even heart valves, conduction systems, and coronary arteries. General damage is closely related to the area of radiotherapy, and pericardium and myocardial damage are most common. Pericardial effusion and thickening can be seen in cases of radiation heart injuries. Myocardial fibrosis can be seen in all layers of the ventricular wall. The changes can be diffuse or focal, but the right ventricle is more obvious, which may be related to the close proximity of the right ventricle to the chest wall. At the end of the course, some people have left ventricular contraction with one or more valvular thickening, among which tricuspid and aortic valve injuries are most common. However, mitral atresia and mural thrombosis are rare. In about 40% of patients, severe coronary stenosis can be found. The main reason is the formation of atherosclerotic plaque. The plaque is mainly fibrous lesions, and the proximal end is more significant than the distal end. Under the light microscope, transient granulocyte infiltration and edema around the blood vessels can be observed during the acute stage of radiological heart damage, and cardiomyocyte edema, hyaline change, and fatty degeneration can be seen later. Subsequently, scattered spots of fibrosis and necrosis would be found. Fibroplasia can be seen in the interstitium and surrounding vessels, and the number of smooth muscle cells in the blood vessels decreased. Under electron microscope, cardiomyocytes arranged in disorder, broken and atrophied cardiac muscle fibers, nuclei deformed, mitochondrial sarcoplasmic reticulum, and nuclear structure destroyed.

5. Kidney

Early lesions may be not obvious. Four to six days after irradiation, epithelial cell degeneration and fusion of foot processes could be seen. The lesions will continue to develop, characterized by endoplasmic reticulum dilatation, a large number of lipid particles, autophagic vacuoles, extensive fusion of podocytes, and swelling of endothelial cells. The lesion of renal tubules mainly appears in the proximal convoluted tubules, showing atrophy and necrosis of the renal tubules. The basement membrane of the renal tubules will take on focal diffuse thickening, distortion, and widening of the intercellular space.

6. Eyes

Radiation can cause radiation cataract, radiation retinopathy or optic neuropathy, keratitis, or iridocyclitis. Neurofibrillar infarction, retinal hemorrhage, microaneurysm, telangiectasis and exudation, and neovascularization can be seen by funduscopy.

7. Skin

Acute radiation skin injury refers to acute radiation dermatitis and radiation skin ulcer caused by multiple or a large doses of X-ray, gamma-ray, and beta-ray in a short period of time (several days). The incubation period is several days, and it is divided into three degrees according to the severity of the damage. If the dose exceeds the threshold dose, local temporary inflammatory reaction can be seen, which is characterized by hair follicle papules and temporary hair loss, that is, the first-degree damage. The first grade is marked erythema, most obvious in 2–6 weeks, with pruritus, alopecia, and pigmentation after erythema subsided. With the increase of radiation dose, the symptoms develop from dry dermatitis (erythema) to exudative reaction, which is the second-degree injury. Within 1–3 weeks, local flushing, swelling, and blisters will appear in the injured area, followed by surface erosion, erythema, pain, and finally repaired, resulting in scab formation, pigmentation healing, permanent alopecia, and so on. Severe lesions can involve deep or subcutaneous tissue of the dermis and form carrion and necrotic ulcers, which is the third-degree injuries.

Chronic radiation-induced skin injuries can be also divided into three degrees. In the first degree, there are dry skin, pigmented or lost, rough, nail dark, or longitudinal ridge color bar; in the second degree, there is skin hyperkeratosis, chapped or atrophy, capillary dilatation, nail thickening deformation, and so on; in the third degree, there are necrotic ulcer, horny protuberance, finger end keratinization fusion, joint deformation, dysfunction, and so on. Chronic radiation damage can cause verrucous hyperplasia of the skin, or form intractable ulcers, which can be secondary to basal cell carcinoma or squamous cell carcinoma.

6.4   Nutritional Diseases

Nutritional disease is a kind of disease caused by excessive or too little nutrients in the body or imbalances that cause excess nutrition or nutrition deficiency and abnormal nutrition metabolism. Nutritional diseases include nutritional disorders, obesity, vitamin deficiency, and hypervitamin.

Too much energy is often stored in the form of fat in our subcutaneous tissue, around the internal organs, and on the abdominal omentum. The excess fat not only ruins our body out of shape but also increases the burden of the body, reduces the function of the heart and lungs, and causes great pressure on the body, especially the joints of the lower extremities, which may induce the degenerative arthropathy. In addition, too much fat can also interfere with the absorption of other nutrients such as protein, calcium, and iron.

Excessive intake of certain nutrients, which cannot be promptly metabolized in the body, may cause poisoning. Vitamin A, vitamin D, vitamin E, and vitamin K, which are fat-soluble nutrients, are not easily expelled from the body and can cause poisoning. Too much protein intake will also increase the liver and kidney metabolic burden and prevent iron absorption.

6.4.1   Pathogenesis of Deficiency Diseases

Nutritional deficiency is a kind of malnutrition due to inadequate intake of nutrients with various clinical manifestations. In recent years, various subclinical nutritional deficiencies have been found due to the increasingly perfect nutritional function tests. Therefore, nutritional deficiency also includes this part. The nutritional deficiency may be of two types: (1) Primary nutritional deficiency is due to a diet that lacks or reduces essential nutrients. The most common reason is inadequate intake of food, which can be either primary or secondary. The deficiency of food caused by social factors such as disaster or war is often lacking in primary intake of food, but it is mainly characterized by malnutrition of heat energy protein. Dietary bias can cause a deficiency of nutrients. Sometimes, a group's habit bias can even lead to the prevalence of a kind of nutritional deficiency disease. In addition, unreasonable cooking of food can also lead to loss of nutrition and malnutrition. For example, vegetables are first cut and then washed, and most vitamin C is destroyed by hot bleaching and extrusion. (2) Secondary or conditional deficiency is a variety of factors contributing to malnutrition. The secondary causes of inadequate food intake are inanorexia, coma, insanity or anorexia, oral and maxillofacial surgery, esophagus cancer, and cardia cancer, which cause obstruction of the esophagus and stomach. Nasogastric feeding or parenteral nutrition is often used in these diseases, but if the amount of recharge cannot meet the needs of patients, there will still be symptoms of nutritional deficiency.

6.4.2 Obesity

Obesity is a common group of metabolic disorders. When the body eats more calories than the consumption of heat, the excess heat is stored in the body in the form of fat, which tends to be more than normal physiological needs and then becomes obesity. The weight of adipose tissue in normal male adults is 15–18% of body weight and 20–25% for females. The proportion of body fat increased with age. The most widely used method to gauge obesity is body mass index (BMI), a measure of the ratio of body weight to height as expressed by body weight in kilograms divided by the square of the height in meters. Whether male or female, the threshold of BMI is 30.

6.4.2.1 Etiology

The main external cause is too much diet and too little activity. Calorie intake is more than caloric consumption, and fat synthesis is the material basis of obesity. The internal factor is the disorder of fat metabolism.

1. Genetic factors
    There is a certain genetic background in the pathogenesis of simple obesity in humans. Studies have shown that one side of the parents is obese, and their children's obesity rate is about 50%. If both parents are obese, the obesity rate of their children will rise to 80%. Human obesity is generally considered to be a polygenic inheritance, and heredity plays an important role in its pathogenesis.
2. Neuropsychic factors
    It is known that there are two pairs of neural nuclei related to feeding behavior in the hypothalamus. One pair is the ventral contralateral nucleus, also known as the satiety center; the other is the ventral lateral nucleus, also known as the hunger center. When satiety centers are excited, they refrain from eating. When they are destroyed, appetite is stimulated. The hungry center works opposite to the satiety center. The two factors regulate each other, restrict each other, and are in a dynamic balance under physiological conditions, so that the appetite is regulated in normal range and maintains normal body weight.
    If the lateral abdominal nucleus is destroyed, the abdominal nucleus function is relatively active, and overeating is not tired, leading to obesity. On the contrary, when the ventrolateral nucleus is destroyed, the ventromedial nucleus is relatively hyperfunctional and anorexia, resulting in emaciation.
3. Endocrine factors
    Many hormones, such as thyroxine, insulin, and glucocorticoids, can regulate the intake of food. Therefore, it is assumed that these hormones may be involved in the pathogenesis of simple obesity. Obese people have insulin resistance and cause hyperinsulinemia, which can regulate the insulin receptor and increase insulin resistance. Increased insulin secretion stimulates food intake and inhibits

lipolysis, resulting in fat accumulation in the body. Sex hormones may play a role in the pathogenesis of simple obesity.

Too much intake can produce an excess of gastric inhibitory polypeptide (GIP) by stimulating the small intestine, which stimulates the release of insulin from islet beta cells. When the pituitary function is low, the secretion of growth hormone, gonadotropin, and thyroid-stimulating hormone is reduced, which causes the hypofunction of the gonadal and thyroid glands, and a special type of obesity can occur at this time. This may be related to decreased fat mobilization and increased synthesis. Clinically, obesity is more likely to occur in women, especially by parturients or by women who take oral contraceptives, suggesting that estrogen is related to the metabolism of fat. When adrenocortical hyperfunction occurs, cortisol secretion increases, and plasma glucose and blood sugar increase, which in turn stimulates insulin secretion and increases fat synthesis.

4. Other factors

Lifestyle or hobbies, feeding behavior, climate, and social psychological factors may lead to obesity. For example, a sedentary lifestyle is likely to lead to obesity.

6.4.2.2 Pathogenesis

Adipocytes constitute adipose tissue, which exists around blood vessels and in the matrix chamber. Besides the generally accepted role of adipocytes for fat storage, these cells also release endocrine-regulating molecules. These molecules include energy regulatory hormone (leptin), cytokines (TNF-α and interleukin-6), insulin sensitivity-regulating agents (adiponectin and resistin), prothrombotic factors (plasminogen activator inhibitor), and blood pressure regulating agent (angiotensinogen).

Adipose mass is increased due to the enlargement of adipose cells due to excess of intracellular lipid deposition as well as due to increase in the number of adipocytes.

The most important environmental factor is excessive intake of nutrients, which can lead to obesity. However, according to observations that obesity is familial and also found in identical twins, the underlying molecular mechanisms of obesity are beginning to emerge. At present, more than five genes are considered to be related to obesity. Among them, two obesity genes have been studied most, that is, ob gene and its protein product leptin and db gene and its protein product leptin receptor.

6.4.2.3 Morphologic Features

In obese people, a large amount of adipose tissue is deposited in subcutaneous tissue, skeletal muscle, and internal organs such as kidney, heart, and omentum, causing the volume of these organs to increase, which is due to the increase in the size and number of fat cells, that is, there is hypertrophy and proliferation. In addition, fatty liver is more common in obese people. Fatty liver, also known as liver steatosis, is the appearance of fat droplets in the cytoplasm of liver cells. Lipids are mainly neutral fats, namely, triglycerides and fatty acids, while phospholipids, cholesterol, and cholesterol lipids are less. In severe cases, it can be seen that the liver is swollen, soft, and light yellow to yellowish in color, and the section structure is vague and greasy, and some are even as fragile as mud. Microscopically, vacuoles of different sizes can be seen in the cytoplasm of liver cells. At first, they are mostly seen around the nucleus and then become larger and more densely scattered in the whole cytoplasm. In severe cases, they can fuse into a large vacuole, which looks like fat cells.

6.4.2.4 Consequences of Obesity

Being obese is bad for the health (Table 6.7). Due to discrimination in housing and work, poor physical image, and low self-esteem, obese people are at greater risk for social, economic, and psychological problems, including diagnosed mental illness. For example, obese people are more often poor and less likely to be employed. What's more, the ill effects of obesity are pervasive and appear in unexpected ways. Obese people are more likely to have family accidents because they are not healthy enough to control their weight. On average, they are twice as likely to suffer from hearing loss, poor vision, and movement disorders in their arms or legs as normal people. As for the reason why obesity is prone to these disorders, it is currently believed that this is related to the fact that obese people are prone to atherosclerosis, which may lead to decreased blood flow of important organs, such as the cochlea, fundus, and brain.

Obesity can cause the following abnormalities:

1. Hyperinsulinemia. Increased volume and quantity of adipocytes lead to increased or decreased expression of secretory hormones, thus affecting insulin levels, resulting in insulin resistance and hyperinsulinemia. The excessive storage of fat leads to the enhancement of lipid degradation, resulting in the formation of a large number of free fatty acids (FFA), which are transported to the liver and peripheral tissues, leading to the disorder of liver sugar utilization and new sugar generation. In addition, the activity of insulin receptor tyrosine kinase is inhibited in the high FFA environment, which inhibits the expression and activity of the insulin receptor substrate-1 (IRS-1) and leads to insulin resistance

Table 6.7 Relative risk for cancer, cardiovascular disease, and diabetes in people with various body mass indices (BMI)

Body mass index classification	Increased cardiovascular risk	Increased diabetes risk	Increased cancer risk
18.5–24.9 Normal	None	None	None
25–29.9 Overweight	20%	100%	10%
30–40 Obese	90%	250%	30%
≥40 Morbidly obese	140%	550%	70%

2. Type 2 diabetes mellitus. Studies showed that there are a positive correlation between obesity and diabetes; the more obese people, the higher the incidence of type 2 diabetes; and 90% of patients with type 2 diabetes has showed a state of obesity. Obesity tends to aggravate the state of diabetes, and in many cases, weight loss tends to lead to improvements in diabetes.
3. Hypertension. In a large sample survey, it was found that if body mass index increased by 10%, systolic blood pressure increased by 2–6 mmHg, while diastolic blood pressure increased by 1–3 mmHg on average. A strong association between hypertension and obesity is observed, which is perhaps due to increased blood volume.
4. Hyperlipoproteinemia. Obesity is strongly associated with VLDL and mildly with LDL. Also, obesity is associated with low blood HDL cholesterol and high triglyceride; both are risk factors in coronary artery disease.
5. Atherosclerosis. Obese patients are often associated with lipid metabolism disorders. The most common result of lipid metabolism disorders is lipoid deposition, mainly manifested by low-density lipoprotein (LDL) elevation and high-density lipoprotein (HDL) level. Obesity predisposes to development of atherosclerosis. As a result of atherosclerosis and hypertension, there is increased risk of myocardial infarction and stroke in obese individuals.
6. Hypoventilation syndrome. The bulge of abdominal fat and the thick pelt of fat on the chest and breasts limit ventilation. This is characterized by hypersomnolence, both at night and during day in obese individuals along with carbon dioxide retention, hypoxia, polycythemia, and eventually right-sided heart failure. Chronic hypoxia causes vasoconstriction in the systemic and pulmonary circulation, causing high blood pressure, pulmonary hypertension, and right heart failure.
7. Osteoarthritis. Obese people are more likely to develop degenerative joint disease, which is related to the fact that the patient's joints need to bear more body weight.
8. Cancer. Obesity can promote the development of cancer by changing the metabolic state of the body, releasing a large amount of inflammatory factors, changing the fat factor, and damaging the immune system. Overweight and obesity significantly increase the risk of at least 12 cancers, such as endometrial cancer, breast cancer, kidney cancer, colorectal cancer, and prostate cancer.

6.4.3   Starvation

Generally, starvation mainly refers to insufficient heat. However, the broad sense of starvation also includes oxygen hunger and water starvation. Protein starvation, calcium starvation, and vitamin starvation are synonymous scientifically with protein deficiency, calcium deficiency, and vitamin deficiency. The causes of starvation are varied, such as natural disasters, poverty, captives, imprisons, hunger, specific religious activities, and food or physical function. What is more noteworthy is the nutritional imbalance caused by inadequate food or partial insufficiency, due to economic conditions, cultural scientific literacy, religious and custom concepts, and special physiological or pathological reasons. These people are in a semi-starvation of certain nutrients, which are not uncommon in the poor, backward areas and classes. Some diseases affect the absorption of nutrients, such as wasting diseases (infections, inflammatory conditions, and liver disease) and cancer, leading to starvation. A variety of diseases or nutritional deficiencies can cause cachexia, which is mainly manifested as emaciation, anemia, debility, systemic failure, etc. The cachexia caused by cancer is related to the synthesis and release of various cytokines by cancer cells, such as tumor necrosis factor a, IL-1, and IL-6.

6.4.3.1 Metabolic Changes

Under the regulation of insulin reduction and glucagon increase, the metabolic characteristics of starvation process are as follows:

1. Muscle breakdown increases, and most of the amino acids released are converted to alanine and glutamine.
2. The effect of sugar isogenesis is enhanced. Under normal circumstances, the body's glucose storage can only meet one day's metabolic needs. During fasting state, insulin-independent tissues such as the brain, blood cells, and renal medulla continue to utilize glucose, while insulin-dependent tissues like muscle stop taking up glucose. It requires the liver to release glycogen to maintain normal blood glucose level. Subsequently, hepatic gluconeogenesis from other sources such as breakdown of proteins takes place. It can be seen that gluconeogenesis is mainly carried out in the liver during the starvation process. Proteins break down to release amino acids, which are used as fuel for hepatic gluconeogenesis so as to maintain glucose needs of the brain.
3. Fat decomposition accelerated, glycerol and fatty acid content increased in plasma, and the result was gluconeogenesis. Then, starvation can continue until all body fat is stored up and death occurs. After about one week of starvation, protein breakdown is decreased, while triglycerides of adipose tissue are broken down to form glycerol and fatty acids. Glycerol can produce sugar directly, and fatty acid not only can provide the energy of sugar isogenesis but also can produce acetyl coenzyme A and promote the isogenesis of amino acid, pyruvic acid, lactic acid, and so on. About 1/4 of fatty acids decomposed by fat can transform into ketone bodies in the liver, so the plasma ketone bodies can increase several hundred times when starving.

6.4.3.2 Clinical Presentation

The clinical manifestations of hunger include weakness, fatigue, depression, haggard, emaciation, and apathy. The eyes are dull, and the skin is rough and hangs in fold due to the loss of fat. The face is often pigmented. The bones protrude. A shrunken limb resembles a stick. On the face and abdomen, there is an incongruous edema. If the starvation be continued for long time, it will cause hair to fall off. For children, hunger leads to growth retardation. For women, hunger often leads to amenorrhea.

6.4.3.3 Morphological Changes

Most organs shrink due to atrophy. The fat disappeared, and the muscles were wasted. The heart is also smaller, and the blood pressure and heart output are low. The pulse is slow. The liver first appears fatty degeneration, but long starved, the fat and protein stored in the liver were decomposed, and the liver cells became very small, with few organelles. Hypoalbuminemia developed later only in the disease. The pancreas and other exocrine glands atrophy. Edema is common in the face and extremities, and ascites are also common. The wall of the intestines is so thin that it is transparent. The erythrocyte of the patient is reduced, and anemia appears. Hemosiderosis is common. Lymphoid tissue atrophies. T-cell function and neutrophil phagocytosis become impaired, but B-cell function is preserved. Opportunistic infections or pneumonia occurs and often lead to death.

6.4.4 Disorders of Vitamins

Vitamins are essential nutrients to maintain human health, including vitamin A, vitamin B, vitamin C, vitamin D, and vitamin E. They are all organic compounds with small molecular weights. Most of them cannot be synthesized in the body, or the amount of synthesis is difficult to meet the needs of the body. Therefore, they must be supplied by food. Such substances have the following common characteristics: (1) It is found in natural food, and the overwhelming majority cannot be synthesized in the body (vitamins D and K and other few vitamin exceptions). (2) It is not the structural component of the body and does not provide energy, but it plays an important role in regulating the metabolic process of a certain substance. (3) Although the body only needs a small amount of vitamins every day to meet the needs of metabolism, it must not be lacking. Otherwise, it will cause vitamin deficiency.

Vitamins can be classified into two groups according to their solubility: fat-soluble and water-soluble vitamins. Fat-soluble vitamins include vitamin A, vitamin D, vitamin E, and vitamin K, which dissolve in fat and go into the body with fat; water-soluble vitamins include vitamin C and vitamin B (B1, B2, B6, B12, niacin, pantothenic acid, folic acid, biotin, etc.), which can dissolve in water. Because of the different solubility of the two kinds of vitamins, their absorption, excretion, and accumulation in the body are different, resulting in different symptoms of different vitamin deficiency. Table 6.8 sums up various clinical disorders that can be produced by vitamin deficiencies.

6.4.4.1 Etiology of Vitamin Deficiencies

There are many reasons for vitamin deficiency, which are common as follows:

1. The supply of vitamins is insufficient. It includes not only the inadequacy of food itself but also the inadequacy of food intake. It may also cause the destruction and loss of vitamins due to improper use of cooking methods in food processing, resulting in inadequate dietary supply.

Table 6.8 Disorders caused by vitamin deficiencies

Vitamins	Deficiency disorders
Fat-soluble vitamins	
Vitamin A (Retinol)	Nyctalopia, xerophthalmia, dry skin, metaplasia and keratinization of mucosal epithelial surfaces
Vitamin D (Calcitriol)	Rickets in growing children Osteomalacia, osteoporosis, hypocalcemic tetany, type 1 diabetes, cancer, and multiple sclerosis
Vitamin E	Ataxia, degeneration of neurons, retinal pigments, axons of peripheral nerves; denervation of muscles
Vitamin K	Hypoprothrombinemia (severe bleeding)
Water-soluble vitamins	
Vitamin C (Ascorbic acid)	Scurvy, poor wound healing, vasomotor instability and connective tissue disorders
Vitamin B complex	
Thiamin (Vitamin B1)	Neurological and cardiovascular disorders
Riboflavin (Vitamin B2)	Degenerative changes in the nervous system, endocrine dysfunction, skin disorders, and anemia
Niacin (Vitamin B3)	Pellagra (a disease that is characterized by inflammation of mucous membranes, skin lesions, and diarrhea)
Pantothenic acid (Vitamin B5) Pyridoxine (Vitamin B6)	Unknown Neurological disorders and anemia
Folate (Vitamin B9)	Megaloblastic anemia, growth retardation, congenital (neural tube) defects
Cyanocobalamin (Vitamin B12)	Pernicious anemia
Biotin (vitamin H)	Growth retardation, dermatological abnormalities, and neurological disorders

2. The ability of the body to absorb and utilize vitamins is reduced. Excessive dietary fiber intake and other factors can lead to reduced vitamin uptake. In addition, gastrointestinal dysfunction leads to a decline in vitamin uptake and utilization.
3. The physiological needs of vitamins are relatively increased. Demand for multivitamins increased during pregnancy, lactation, and growth. In the cold, hot, and other special environmental conditions or diseases, the body will also increase the demand for vitamins.
4. Vitamin excretion is increased. Vomiting, diarrhea, and other conditions may lead to an increase in the elimination of multiple vitamins, especially water-soluble vitamins.

6.4.5   Vitamin A (Retinol) Deficiency

Vitamin A is necessary to maintain the integrity of all epithelial tissues. If vitamin A is deficient, the epithelial cells of the eye, respiratory, digestive, urethra, and reproductive organs are significantly affected, showing in epithelial xerosis, hyperplasia, and desiccation. In addition, vitamin A can promote growth and development. When vitamin A is deficient, human skeletal growth is poor, growth and development are hindered, and even reproductive capacity is reduced. Vitamin A is a component of the photoreceptor in visual cells. When vitamin A is deficient, the sensitivity to weak light will be reduced, which leads to dark adaptation disorder, and even night blindness occurs in severe cases.

6.4.5.1   Lesions in Vitamin A Deficiency

Consequent to vitamin A deficiency, following pathologic changes are seen:

1. Ocular lesions
   Eye lesions are the most obvious. Ocular lesions are characterized by dry conjunctiva and cornea, loss of luster, and tear loss. Night blindness is usually the first sign of vitamin A deficiency. As a result of replacement metaplasia of mucus secreting cells by squamous cells, there is dry and scaly scleral conjunctiva. The lacrimal duct also shows hyperkeratosis. Keratinized epithelium accumulating into foam white spots, called conjunctival plaques or Bitot's spots, is the turbid area of the focal triangle. Then the cornea is dry, cloudy, softened, conscious photophobia, impeding transmission of light. Corneal ulcers may occur, which may cause infection and cause keratomalacia. Serious cases even have corneal perforation, leading to iris and lens prolapse. Ultimately, infection, scarring, and opacities lead to blindness.
2. Cutaneous lesions

Skin lesions are characterized by dry skin, desquamation, epithelial keratosis, and keratosis filling hair follicles to form hair follicles and papules. When touching the skin, it feels rough and sandy. With the extension of limbs and shoulders, it can develop to the neck, back, and even face. The hair follicle keratinization causes the hair to be dry, losing luster and being easy to fall off. Nails and toenails are brittle and easy to break.

3. Other lesions

   (a) Squamous metaplasia of respiratory epithelium of bronchus and trachea may predispose to respiratory infections.
   (b) Squamous metaplasia of pancreatic ductal epithelium may lead to obstruction and cystic dilatation.
   (c) Squamous metaplasia of urothelium of the pelvis of kidney may predispose to pyelonephritis and perhaps to renal calculi.
   (d) Long-standing metaplasia may cause progression to anaplasia under certain circumstances.
   (e) Bone growth in vitamin A deficiency is retarded.
   (f) Immune dysfunction may be caused by impaired barrier epithelium.
   (g) Pregnant women may have increased risk of maternal infection, mortality, and impaired embryonic development.

6.4.6 Hypervitaminosis A

Vitamin A poisoning (vitamin A toxicity) is a toxic syndrome caused by excessive intake of vitamin A. According to the study, infants and young children take vitamin A, such as a dose of more than 300,000 international units (a gram of common cod liver oil containing vitamin A 850 international units), which can cause acute poisoning. Taking 50,000–100,000 units per day for 6 months or so can cause chronic poisoning. In addition, children's sensitivity and tolerance to vitamin A can be significantly different, and some children may also have mild symptoms of poisoning even if they do not exceed the above range.

6.4.6.1 Acute Toxicity

Usually, vitamin A injected with 300,000 IU can produce toxic symptoms within a few days. It is manifested as loss of appetite, irritability or lethargy, vomiting, anterior fontanelle enlargement, head circumference enlargement, craniofacial dehiscence, papillary edema, etc. Increased intracranial pressure is common in acute type, which is due to increased cerebrospinal fluid volume or absorption disorders.

6.4.6.2 Chronic Toxicity

Toxicity usually disappears when excessive vitamin A intake is prevented. When the dose of vitamin A reaches tens of thousands of units per day, such as 1500 IU vitamin A per kilogram weight of infants and young children, chronic poisoning will occur.

Early symptoms include irritability, anorexia, low fever, sweating, and hair loss, followed by typical bone pain, metastatic pain, and soft tissue swelling. There is no inflammation such as redness, swelling, heat, and pain in the painful part, which is common in the long bones of the limbs. Because of long bone epiphysis involvement, it can cause short stature.

Some cases have swelling and pain in the temporal and occipital regions, which can be misdiagnosed as cranial osteomalacia. The symptoms of increased intracranial pressure such as headache, vomiting, wide anterior fontanelle, separation of cranial seams, and strabismus in the eyes, nystagmus, and diplopia are another feature of this disease. In addition, there are skin pruritus, desquamation, rash, chapped lips, dry hair, hepatomegaly, splenomegaly, abdominal pain, myalgia, bleeding, kidney disease, and regenerative anemia with leukocyte reduction. The blood alkaline phosphatase increased. It has been reported that chronic hepatomegaly and splenomegaly can cause cirrhosis, increased portal hypertension, and even death.

6.4.7 Vitamin D (Calcitriol) Deficiency

Vitamin D deficiency is a disease characterized by abnormal calcium and phosphorus metabolism and poor calcification of bone like tissue. The clinical manifestations of vitamin D deficiency are rickets in children, tetany, and osteomalacia in adults.

6.4.7.1 Etiology

1. Lack of sunlight. In China, due to the long winter in the north and the long rainy season in the south, people's outdoor activities are reduced, resulting in a significant reduction in the chances of people receiving sunlight.
2. Low intake of vitamin D. Low vitamin D content in breast milk or other dairy products cannot meet the growth and development needs of infants, which can lead to vitamin D deficiency in infants. Also, unreasonable diet can lead to vitamin D deficiency.
3. Fast growth rate of the baby. Since the bones of babies grow very fast, vitamin D is in great demand. Premature infants and multiple births have insufficient vitamin D reserves.

4. The influence of some diseases. Severe chronic kidney disease, hepatic disease, and gastrointestinal tract disorders can affect the absorption of vitamin D or affect the synthesis of 1,25-dihydroxyvitamin D3.
5. Effect of some drugs. If vitamin D metabolism is interfered with by phenobarbital and glucocorticoids, vitamin D deficiency may occur.

6.4.7.2 Lesions in Vitamin Deficiency

Lesions in vitamin D deficiency from any of the above mechanisms includes: (1) rickets in growing children, (2) osteomalacia in adults, and (3) hypocalcemic tetany due to neuromuscular dysfunction.

6.4.7.3 Rickets

Rickets is a heterogeneous group of acquired and inherited diseases resulting in disturbances in calcium and/or phosphate homeostasis, affecting the growing skeleton. Rickets is characterized by impaired apoptosis of hypertrophic chondrocytes, resulting in widening of the growth plates in bones and is usually associated with osteomalacia. It is caused by vitamin D deficiency in infants and adolescents. Rickets mainly occur in infants under 2 years of age, especially within 3–18 months, and can be prevented by taking adequate vitamin D.

The main characteristics of rickets are as follows: (1) Irregular overgrowth of small blood vessels in disorganized and weak bone; (2) proliferation of cartilage cells at the epiphyses followed by inadequate provisional mineralization; (3) persistence and overgrowth of epiphyseal cartilage, deposition of osteoid matrix on inadequately mineralized cartilage resulting in enlarged, and expanded costochondral junctions; (4) deformed bones due to lack of structural rigidity; and (5) mesenchymal cells differentiate into osteoblasts with laying down of osteoid matrix, which fails to get mineralized resulting in soft and weak flat bones.

Clinical presentation depends on age at onset, duration of disease, and underlying pathophysiology. Common features are thickened wrists and ankles due to widened metaphyses, growth failure, bone pain, muscle weakness, waddling gait, and leg bowing. Severe rickets in infancy may also include delayed closure of the fontanelles, parietal and frontal bossing, and craniotabes (soft skull bones). The main clinical manifestations are listed below:

1. Craniotabes. In 6-month-old infants, rickets is mainly characterized by skull changes. The front fontanelle margin is soft and thin. At 6 months of age, table tennis-like toughness is felt by touching the bone seams, but the central frontal and parietal bones tend to thicken gradually. At 7–8 months, the head will become a "square skull" appearance, and the head circumference was larger than that of the normal skull.

2. Thoracic deformity. (1) Rachitic rosary is a thoracic deformity caused by excessive cartilage growth at costochondral junction. The epiphyseal end is enlarged by the accumulation of bone-like tissue. Along the ribs, the rounded protuberance can be touched at the junction of ribs and costal cartilage, from upper to lower like beaded protuberances, and the most obvious seventh to tenth ribs are called rachitic rosary. (2) Harrison's sulcus appears due to indrawing of soft ribs on inspiration. In children with severe rickets, a horizontal depression is formed at the lower edge of the thorax, namely, the costal phrenic groove or Harrison's sulcus. (3) Pigeon-chest deformity is the anterior protrusion of sternum and adjacent cartilage due to action of respiratory muscles. (4) The funnel chest is the inward depression of the sternum and looks like a deep hopper.
3. Limb deformity. (1) Lower limb deformity. As a result of osteomalacia and muscle and joint relaxation, the bone is not enough to support the weight of the body, resulting in femoral and tibiofibular bending deformation, which forms a serious genu varus ("O") or genu valgus ("X") deformity. (2) The hand and foot bracelet. In patients with serious diseases, wrists, ankles, and feet can also form a blunt round protuberance, which looks like a bracelet is wrapped around the wrists or ankles.
4. Lumbar lordosis is due to spinal and pelvic involvement.

Further Reading

Bhardwaj JR. Boyd's textbook of pathology. 10th ed. New Delhi: Wolters Kluwer; 2013.

Chalasani NP, Maddur H, Russo MW, et al. ACG clinical guideline: diagnosis and management of idiosyncratic drug-induced liver injury. Am J Gastroenterol. 2021;116:878–98.

Curigliano G, Cardinale D, Dent S, et al. Cardiotoxicity of anticancer treatments: epidemiology, detection, and management. CA Cancer J Clin. 2016;66:309–25.

Hafner D, Leifheit-Nestler M, Grund A, et al. Rickets guidance: part I—diagnostic workup. Pediatr Nephrol. 2022;37:2013–36.

Karch SB. Karch's pathology of drug abuse, 4th ed. CRC Press; 2009.

Kodner C. Diagnosis and management of nephrotic syndrome in adults. Am Fam Physician. 2016;93(6):479–85.

Kumar V, Abbas AK, Aster JC. Robbins and Cotran pathologic basis of disease, 9th ed. Saunders, an imprint of Elsevier Inc.; 2014.

Kumar V, Abbas AK, Aster JC. Robbins basic pathology. 10th ed. Elsevier; 2017.

Love S, Louis DN, Ellison DW. Greenfield's neuropathology. 8th ed. London: Hodder Arnold; 2008.

McConnell TH. The nature of disease : pathology for the health professions. 2nd ed. Philadelphia: Wolters Kluwer Health. Lippincott Williams & Wilkins; 2013.

Mohan H. Textbook of pathology, 7th ed. Jaypee Brothers Medical Publishers; 2015.

Said HM. Intestinal absorption of water-soluble vitamins in health and disease. Biochem J. 2011;437(3):357–72. https://doi.org/10.1042/BJ20110326.

Strayer DS. Rubin's pathology: clinicopathologic foundations of medicine. 7th ed. Philadelphia: Wolters Kluwer Health; 2014.

Weissman S, Aziz M, Perumpail RB, et al. Ever-increasing diversity of drug-induced pancreatitis. World J Gastroenterol. 2020;26(22):2902–15.

Wilson LR, Tripkovic L, Hart KH, et al. Vitamin D deficiency as a public health issue: using vitamin D2 or vitamin D3 in future fortification strategies. Proc Nutr Soc. 2017;76:392–9. https://doi.org/10.1017/S0029665117000349.

Chapter 7
The Blood Vessel and Heart

Xiaomin Hou

Contents

X. Hou (✉)
School of Basic Medical Science, Shanxi Medical University, Taiyuan, China

© Zhengzhou University Press 2024
K. Chen et al. (eds.), *Textbook of Pathologic Anatomy*,
https://doi.org/10.1007/978-981-99-8445-9_7

Objectives

This chapter introduces the pathoanatomical knowledge of cardiovascular diseases in detail to provide a basis for the research and treatment of related diseases.

Key Concepts

1. Atheromas
2. Ischemic heart disease
3. Primary hypertension
4. Aschoff body
5. Stenosis
6. Insufficiency
7. Primary cardiomyopathy
8. Myocarditis
9. Aneurysm

Introduction

Cardiovascular disease is a class of diseases that refers to any condition affecting the cardiovascular system, principally cardiac disease, vascular diseases of the brain and kidney, and peripheral arterial diseases. The causes of cardiovascular diseases are diverse, but atherosclerosis and hypertension are the most common. A number of physiological and morphological changes that occur with aging alter cardiovascular function and lead to an increased risk of cardiovascular disease, even in healthy asymptomatic individuals.

7.1   Atherosclerosis

Atherosclerosis is a generic, inclusive term that describes the thickening and hardening of the arterial wall. Included in this term are three pathologic entities: atherosclerosis, arteriosclerosis, and arterial medical calcification.

Atherosclerosis is a multifactorial degenerative disease characterized by the formation of plaques on the intima. These lesions are called atheromas. Atheromatous plaques are raised lesions composed of soft, gummous lipid cores covered by fibrous caps. Large and medium arteries are usually involved.

7.1.1 Etiology and Pathogenesis

7.1.1.1 Etiology

So far, the causes of atherosclerosis are not very clear. The prevalence and severity of atherosclerosis are correlated with a number of risk factors.

Constitutional Risk Factors

1. Age: Atherosclerosis usually remains clinically silent until lesions reach a critical threshold in middle age or later.
2. Genetics: Family history is the most important independent risk factor for atherosclerosis. Certain Mendelian disorders are strongly associated with atherosclerosis, but these account for only a small percentage of cases.

Modifiable Major Risk Factors

1. Hypertension
    Hypertension, especially increased diastolic blood pressure, is a major risk factor for the development of atherosclerosis.
2. Diabetes Mellitus
    Diabetes mellitus is associated with raised circulating cholesterol levels and markedly increases the risk of atherosclerosis.
3. Hyperlipidemia
    Hypercholesterolemia is another major risk factor for the development of atherosclerosis and is sufficient to induce lesions in the absence of other risk factors.
4. Cigarette Smoking
    It is a well-established risk factor in men and probably accounts for the increasing incidence and severity of atherosclerosis in women. Prolonged (years) smoking of one or more packs of cigarettes a day doubles the rate of ischemic heart disease-related mortality, while smoking cessation reduces the risk.

Additional Risk Factors

Other factors that contribute to risk include the following:

1. Metabolic Syndrome
    It is associated with central obesity and is characterized by dyslipidemia, hypertension, hypercoagulability, insulin resistance, and a pro-inflammatory state, which may be triggered by cytokines released from adipocytes. Hypertension, dyslipidemia, and hyperglycemia are all cardiac risk factors,

while the systemic hypercoagulable and pro-inflammatory state may contribute to endothelial dysfunction and/or thrombosis.

2. C-Reactive Protein (CRP) Levels

   CRP levels predict the risk of stroke, peripheral artery disease, myocardial infarction, and sudden cardiac death, even among apparently healthy persons.

3. Inflammation

   Inflammatory cells are present during all stages of atheromatous plaque formation and are intimately linked with plaque progression and rupture. With increasing recognition of the role of inflammation, measures of systemic inflammation have become important in risk stratification.

4. Elevated Levels of Procoagulants

   Excessive activation of thrombin, which can initiate inflammation through cleavage of protease-activated receptors (PARs) on endothelium, leukocytes, and other cells, may be particularly atherogenic.

5. Hyperhomocysteinemia

   Serum homocysteine levels correlate with stroke, coronary atherosclerosis, venous thrombosis, and peripheral vascular disease. Homocystinuria, due to rare inborn errors of metabolism, causes elevated circulating homocysteine (more than 100 μmol/L).

6. Other Factors

   They are associated with difficult-to-quantify risks, including lack of exercise and a competitive, stressful lifestyle ("type A personality").

7.1.1.2 Pathogenesis

Response to Injury Hypothesis (Endothelial Injury)

Endothelial cell injury is the cornerstone of the response to injury hypothesis. Endothelial cell loss due to any kind of injury—induced experimentally by irradiation, immune complex deposition, hemodynamic forces, mechanical denudation, or chemicals—results in intimal thickening in the presence of high-lipid diets. The importance of hemodynamic factors in atherogenesis is illustrated by the observation that plaques tend to occur at the ostia of exiting vessels, at branch points, and along the posterior wall of the abdominal aorta, where there is turbulent blood flow. Suspected triggers of early atheromatous lesions include toxins from cigarette smoke, homocysteine, hypertension, and hyperlipidemia, which can stimulate pro-atherogenic patterns of endothelial cell gene expression.

Lipid Infiltration

Common lipoprotein abnormalities in the general population include (1) decreased high-density lipoprotein cholesterol levels, (2) increased low-density lipoprotein (LDL) cholesterol levels, and (3) increased levels of lipoprotein.

The mechanisms by which dyslipidemia contributes to atherogenesis include the following:

1. Chronic hyperlipidemia, particularly hypercholesterolemia, can directly impair endothelial cell function by increasing local oxygen free radical production; among other things, oxygen free radicals accelerate nitric oxide decay, dampening its vasodilator activity.
2. With chronic hyperlipidemia, lipoproteins accumulate within the intima, where they are hypothesized to generate two pathogenic derivatives, cholesterol crystals and oxidized LDL. LDL is oxidized through the action of oxygen free radicals generated locally by macrophages or endothelial cells and ingested by macrophages through the scavenger receptor, resulting in foam cell formation.

Smooth Muscle Cell Proliferation and Extracellular Matrix Synthesis

Intimal smooth muscle cell proliferation and extracellular matrix deposition lead to the conversion of the earliest lesion, a fatty streak, into a mature atheroma, thus contributing to the progressive growth of atherosclerotic lesions. Several growth factors are implicated in smooth muscle cell proliferation and extracellular matrix synthesis, including platelet-derived growth factor (PDGF), fibroblast growth factor (FGF), and transforming growth factor alpha (TGF-α). The recruited smooth muscle cells synthesize extracellular matrix, which stabilizes atherosclerotic plaques. However, smooth muscle cell apoptosis and breakdown of matrix can lead to the development of unstable plaques.

Inflammation

Inflammation contributes to the initiation, progression, and complications of atherosclerotic lesions. Monocytes differentiate into macrophages and avidly engulf lipoproteins, including small cholesterol crystals and oxidized LDL. Activated macrophages also produce toxic oxygen species that drive LDL oxidation and elaborate growth factors that stimulate smooth muscle cell proliferation. Cholesterol crystals appear to be particularly important instigators of inflammation through activation of the inflammasome and subsequent release of IL-1.

Infection

There is circumstantial evidence linking infections to atherosclerosis. Cytomegalovirus, Chlamydia pneumoniae, and herpesvirus have all been found in atherosclerotic plaque, and seroepidemiologic studies increased antibody titers to Chlamydia pneumoniae in patients with more severe atherosclerosis. Infections with these organisms, however, are exceedingly common (as is atherosclerosis), making it difficult to draw conclusions about causality.

It is important to recognize that atherosclerosis can be induced in germ-free mice, indicating that there is no obligate role for infection in the disease process.

7.1.2 Morphology

According to the development of the plaque, the basic pathological changes are as follows:

1. Fatty Streaks

    Fatty streaks begin as flat, minute yellow macules that coalesce into elongated lesions, 1 cm or more in length, which are composed of lipid-filled foamy cells but are only minimally raised and do not cause any significant flow disturbance. In electron microscopy, there are two kinds of foamy cells: one is smooth muscle-derived foam cell and the other is macrophage-derived foam cell. The link between fatty streaks and atherosclerotic plaques is uncertain; although fatty streaks may evolve into plaques, not all are destined to progress. Nevertheless, it is notable that coronary fatty streaks form during adolescence at the same anatomic sites that are prone to plaques later in life.

2. Fibrous Plaque

    The term "fibrous plaque" refers to the gross morphologic appearance of the lesion that is the hallmark of atherosclerosis. Grossly, the lesions are raised, pearly white to gray, smooth-surfaced, plaque-like structures in the intima that vary in diameter from a few millimeters to over a centimeter. Microscopically, the surface of the plaque is composed of hyperplastic collagen fibers. The fibrous cap is made up of elongated smooth muscle cells and avascular connective tissues. The connective tissue contains glycosaminoglycans, elastic fibers, and collagen fibers synthesized by the smooth muscle cells. The collagen in the fibrous cap may be dense and hyalinized. The central plaque consists of some smooth muscle cells, extracellular matrix, inflammatory cells, and foam cells.

3. Atheromatous Plaque (Atheroma)

    Atheromatous plaques are white to yellow raised lesions; they range from 0.3 to 1.5 cm in diameter but coalesce to form larger masses. Atherosclerotic plaques are patchy, usually involving only a portion of any given arterial wall; on cross-section, therefore, the lesions appear "eccentric." The focal nature of atherosclerotic lesions may be related to the vagaries of vascular hemodynamics.

    The proportion and configuration of each component vary from lesion to lesion. Most commonly, plaques have a superficial fibrous cap composed of dense collagen and smooth muscle cells. Where the cap meets the vessel wall is a more cellular area containing smooth muscle cells, T cells, and macrophages. Deep to the fibrous cap is a necrotic core, containing variably organized thrombus, lipid, necrotic debris, lipid-laden macrophages and smooth muscle cells (foam cells), fibrin, and other plasma proteins. The extracellular cholesterol frequently takes the form of crystalline aggregates that are washed out during routine tissue processing, leaving behind empty "cholesterol clefts." The periphery of the lesions shows neovascularization (proliferating small blood vessels). The media deep to the plaque may be attenuated and exhibit fibrosis secondary to smooth muscle atrophy and loss.

7.1.3 Clinical Consequences

The principal pathophysiologic outcomes depend on the size of the affected vessel, the stability and size of the plaques, and the degree to which plaques disrupt the vessel walls.

7.1.3.1 Atherosclerotic Stenosis

At early stages, remodeling of the media tends to preserve the luminal diameter by increasing the vessel circumference. Owing to limits on remodeling, however, eventually, the expanding atheroma may impinge on blood flow.

7.1.3.2 Acute Plaque Changes

Plaque erosion or rupture typically triggers thrombosis, leading to partial or complete vascular obstruction and often tissue infarction. The causes of acute plaque changes are complex and include both intrinsic (e.g., plaque structure and composition) and extrinsic (e.g., blood pressure) factors. These factors combine to weaken the integrity of the plaque, making it unable to withstand vascular shear forces. Certain types of plaques are believed to be at particularly high risk of rupturing.

Atherosclerotic plaques are susceptible to several clinically important changes:

1. Rupture, ulceration, or erosion
2. Hemorrhage into a plaque
3. Atheroembolism
4. Aneurysm formation

7.2 Ischemic Heart Disease

Ischemic heart disease (IHD) refers to a group of closely related syndromes caused by an imbalance between myocardial oxygen demand and cardiac blood supply. Although ischemia can result from increased demand (e.g., increased heart rate or hypertension) or diminished oxygen-carrying capacity (e.g., anemia, carbon monoxide poisoning), in the vast majority of cases, IHD is due to a reduction in coronary blood flow caused by obstructive atherosclerotic disease. Thus, IHD is often called coronary artery disease (CAD) or coronary heart disease.

7.2.1 Angina Pectoris

Angina pectoris is intermittent chest pain caused by transient, reversible myocardial ischemia. There are three variants.

7.2.1.1 Unstable Angina

Unstable angina, also called crescendo angina, is characterized by an increasing frequency of pain, precipitated by progressively less exertion. It is more intense and longer lasting than stable angina. This is due to plaque disruption and superimposed partial thrombosis, distal embolization of the thrombus, and/or vasospasm.

7.2.1.2 Typical Angina

Typical angina, also called stable angina, is episodic chest pain associated with less exertion or some other form of increased myocardial oxygen demand (e.g., tachycardia or hypertension due to fever, anxiety, or fear). The pain is classically described as a crushing or squeezing substernal sensation, which can radiate down the left arm or to the left jaw (referred pain).

7.2.1.3 Prinzmetal Angina

Prinzmetal angina, also called variant angina, usually occurs at rest due to coronary artery spasm. The etiology is not clear, but Prinzmetal angina typically responds promptly to the administration of vasodilators such as nitroglycerin or calcium channel blockers.

7.2.2 Myocardial Infarction

Myocardial infarction (MI), popularly called heart attack, is necrosis of heart muscle resulting from ischemia. The major underlying cause of MI is atherosclerosis; therefore, the frequency of MIs rises progressively with increasing age and the presence of other risk factors such as smoking, diabetes, and hypertension.

MIs are caused by acute coronary artery thrombosis. In most cases, disruption of an atherosclerotic plaque results in the formation of a thrombus. Clinical features include chest pain accompanied by breathlessness, vomiting, and collapse or syncope.

7.2.2.1 Morphology

Nearly all transmural infarcts affect at least a portion of the left ventricle and/or ventricular septum. Roughly 15–30% of MIs affecting the posterior or posteroseptal wall also extend into the adjacent right ventricular wall. Other coronary occlusions are occasionally encountered. These include the left main coronary artery or secondary branches, such as the diagonal branches of the left anterior descending (LAD) artery or marginal branches of the left circumflex (LCx) artery. In contrast, significant atherosclerosis or thrombosis of penetrating intramyocardial branches of coronary arteries rarely occurs. Severe coronary occlusion without associated myocardial damage suggests the prior formation of protective collateral connections.

According to the area and depth of MI, it can be divided into two types: subendocardial and transmural MI:

1. MIs less than 12 h old are usually not grossly apparent.
2. For infarcts more than 3 h old, an infarcted area is revealed as an unstained pale zone.
3. By 12–24 h after MI, an infarct can usually be grossly identified by a reddish-blue discoloration caused by stagnant, trapped blood. The infarct then becomes more sharply delineated as a yellow-tan, softened area.
4. By 10–14 days, infarcts become rimmed by hyperemic granulation tissue.
5. Over the succeeding weeks, the MI evolves into a fibrous scar.

The microscopic appearance also undergoes a characteristic sequence of changes:

1. Within 4–12 h, typical features of coagulative necrosis become detectable.
2. 1–3 days after MI, necrotic myocardium elicits acute inflammation.
3. 5–10 days after MI, a wave of macrophages removes necrotic myocytes, and neutrophil fragments are present.
4. 2–3 weeks after MI, the infarcted zone is progressively replaced by granulation tissue.
5. By the end of the sixth week, a dense collagenous scar is formed.

7.2.2.2 Clinical Features

MI is usually heralded by severe, crushing substernal chest pain or discomfort that can radiate to the neck, jaw, epigastrium, or left arm. The pain of an MI typically lasts from 20 min to several hours and is not significantly relieved by nitroglycerin or rest. In a substantial minority of patients, MIs can be entirely asymptomatic.

Electrocardiographic abnormalities include Q waves (indicating transmural infarcts) and ST segment abnormalities and T-wave inversion (representing abnormalities in myocardial repolarization).

Laboratory evaluation of MI mainly includes the blood levels of creatine kinase (CK and CK-MB), cardiac troponins T and I (TnT, TnI), myoglobin, lactate dehydrogenase, and others. Troponins and CK-MB have high specificity and sensitivity for myocardial damage.

TnI and TnT are not normally detectable in the circulation, but after acute MI, both troponins become detectable after 2–4 h and peak at 48 h.

CK-MB is the second-best marker after the cardiac-specific troponins. CK-MB activity begins to rise within 2–4 h of MI, peaks at 24–48 h, and returns to normal within approximately 72 h.

7.2.2.3 Consequences and Complications of MI

Nearly three-fourths of patients have one or more complications after acute MI, illustrated as follows:

1. Pericarditis
   A fibrinous or hemorrhagic pericarditis usually develops within 2–3 days of a transmural MI and typically spontaneously resolves with time.
2. Mural Thrombus
   With any infarct, the combination of a local loss of contractility with endocardial damage can foster mural thrombosis and, potentially, thromboembolism.
3. Contractile Dysfunction
   Typically, there is some degree of left ventricular failure, with hypotension, pulmonary vascular congestion, and fluid transudation into the pulmonary interstitial and alveolar spaces. Severe "pump failure" occurs in 10–15% of patients after acute MI, generally with a large infarct (often >40% of the left ventricle). Cardiogenic shock has a nearly 70% mortality rate and accounts for two-thirds of in-hospital deaths.
4. Ventricular Aneurysm
   Aneurysms of the ventricular wall most commonly result from a large transmural anteroseptal infarct that heals with the formation of thin car tissue. Complications of ventricular aneurysms include mural thrombus, arrhythmias, and heart failure, but rupture of the fibrotic wall does not occur.
5. Arrhythmias
   Following MI, many patients develop arrhythmias, which are undoubtedly responsible for many of the sudden deaths. MI-associated arrhythmias include heart block, ventricular premature contractions, ventricular fibrillation, tachycardia, or sinus bradycardia.
6. Myocardial Rupture
   Rupture complicates somewhere between 1 and 5% of MIs but is a frequent cause of MI-associated death.

7. Papillary Muscle Dysfunction

   More frequently, postinfarct mitral regurgitation results from ischemic dysfunction of a papillary muscle and underlying myocardium and later from papillary muscle fibrosis and shortening, or ventricular dilation.

8. Infarct Expansion

   Because of the weakening of necrotic muscle, there may be disproportionate stretching, thinning, and dilation of the infarct region (especially with anteroseptal infarcts); this is often associated with mural thrombus.

7.2.3 Chronic IHD

7.2.3.1 Morphology

Hearts from patients with chronic IHD are usually enlarged and heavy due left ventricular dilation and hypertrophy. There is moderate to severe atherosclerosis of the coronary arteries, sometimes with total occlusion. Discrete, gray-white scars of healed infarcts are usually present. The endocardium generally shows patchy, fibrous thickening, and mural thrombi may be present. The major microscopic findings include myocardial hypertrophy, diffuse subendocardial myocyte vacuolization, and fibrosis from previous infarcts.

7.2.3.2 Clinical Features

Chronic IHD is characterized by the development of severe, progressive heart failure, sometimes punctuated by episodes of angina or MI. Arrhythmias are common and, along with congestive heart failure and intercurrent MI, account for many deaths.

7.2.4 Sudden Cardiac Death

Coronary artery disease is the most common underlying cause, and in many adults, sudden cardiac death (SCD) is the first clinical manifestation of IHD. In younger victims, the following non-atherosclerotic causes of SCD become more common: hereditary or acquired abnormalities of the cardiac conduction system, congenital coronary arterial abnormalities, isolated hypertrophy, hypertensive or unknown causes, mitral valve prolapse, aortic valve stenosis, dilated or hypertrophic cardiomyopathy, myocarditis or sarcoidosis, and pulmonary hypertension. The most important cause is the autosomal dominant long QT syndrome, due to mutations in various cardiac ion channels. Increased cardiac mass is an independent risk factor for SCD; thus, some young individuals who die suddenly (including athletes) have unsuspected hypertrophic cardiomyopathy, myocarditis, or congenital abnormalities of coronary arteries.

The ultimate mechanism of SCD is most often a lethal arrhythmia, such as ventricular fibrillation.

Subendocardial myocyte vacuolization indicative of severe chronic ischemia is common. Only 10–20% of cases of SCD are of non-atherosclerotic origin.

7.3   Hypertension

Hypertension is one of the most common serious cardiovascular diseases. Arterial hypertension is defined as a sustained rise in systemic blood pressure above 140 mmHg (18.4 kPa) systolic and/or 90 mmHg (12.0 kPa) diastolic. Hypertension can be classified into two main types according to its etiology. Primary (essential or idiopathic) hypertension refers to the elevation of blood pressure with age but with no apparent cause, while secondary hypertension elevated blood pressure due to an identifiable cause. It is also called symptomatic hypertension. "Hypertension disease" refers to primary hypertension.

With the development of economic conditions, cardiovascular diseases incidence has increased due to changes in lifestyle, mental stress, and malfunction.

Primary hypertension is one of the most common cardiovascular diseases in China. Unfortunately, in the vast majority of patients with systemic hypertension, the underlying cause is unknown. Primary hypertension tends to be familiar and develops in later adult life. The incidence of hypertension and its complications varies by gender and race. Males tend to have higher blood pressure than females at the same age before 55 years old. However, at the age of 75, females tend to have a higher incidence of hypertension.

Secondary hypertension accounts for 5–10% of hypertension cases. It is precipitated by an identifiable background abnormality. These conditions may be divided into renal, endocrine, cardiovascular, and neurologic types. Special types of hypertension include intracranial hemorrhage, unstable angina pectoris, acute MI, pregnancy-associated hypertension, and hypertension crisis due to hypertensive encephalopathy, aortic dissection, and eclampsia.

7.3.1   Etiology and Pathogenesis

The causes of hypertension are not yet fully understood. It is believed that hypertension results from the interaction of genetic and environmental factors.

7.3.1.1 Risk Factors

1. Dietary Factors

   (a) Alcohol consumption is one of the factors in the pathogenesis of hypertension. Alcohol can activate the sympathetic nervous-catecholamine (CA) system.
   (b) Obesity is a major medical and public health problem worldwide and can lead to hypertension and other cardiovascular diseases.
   (c) Mutations in proteins may affect sodium resorption. Additionally, heavy consumption of salt has been implicated as an exogenous factor in hypertension.

2. Environmental Factors

   Environmental factors can modify the expression of the genetic determinants of increased pressure.

3. Genetic and Familial Aggregation

   Genetic factors play a role in determining pressure levels, as evidenced by studies comparing blood pressure in monozygotic and dizygotic twins and by studies of familial aggregation of hypertension.

7.3.1.2 Pathogenesis

Although the specific triggers are unknown, it appears that both altered renal sodium handing and increased vascular resistance contribute to essential hypertension.

1. Genetic Factors

   Genetic factors play an important role in determining blood pressure. Polymorphisms of the renin-angiotensin system may also contribute to the known racial differences in blood pressure regulation. Hypertension has been linked to specific angiotensinogen polymorphisms and angiotensin II receptor variants.

2. Environmental Factors

   These include physical inactivity, obesity, stress, smoking, and high levels of salt consumption, which modify the impact of genetic determinants.

3. Increased Vascular Resistance

   This may stem from vasoconstriction or structural changes in vessel walls. These are not necessarily independent factors, as chronic vasoconstriction may result in permanent thickening of the walls of affected vessels.

4. Reduced Renal Sodium Excretion

   This is probably a key pathogenic feature. Decreased sodium excretion causes an obligatory increase in fluid volume and increased cardiac output, thereby elevating blood pressure.

7.3.2   *Morphology*

According to pathological features, primary hypertension can be divided into two types: benign and malignant hypertension.

7.3.2.1   Benign Hypertension

Benign hypertension, also called chronic hypertension, is encountered in about 95% of hypertensive subjects. The patients are usually asymptomatic, and the diagnosis is often made incidentally. The course is stable, with a slow rise in blood pressure. With regard to the development of the diseases, these patients fall into three phases:

1. Dysfunction
   This is the early stage of hypertension. Arterioles show intermittent spasm. Blood pressure is increased, accompanied by headaches. However, blood pressure can return to normal after the spasm eases.
2. Artery Lesion
   This is the characteristic lesion of hypertension. Under microscopy, the intimal smooth muscle cells proliferate, and elastic tissue fibers are deposited between the proliferating cells. Such intimal proliferation is most often observed in renal arteries over 50 μm in diameter and occasionally in arterioles. One of the earliest changes noted in the wall of arteries and arterioles in hypertension is medial hypertrophy due to the work hypertrophy of medial smooth muscle cells and the accumulation of acid mucopolysaccharides, water, and electrolytes, which also contributes to medical thickening. These lesions can be referred to as hyaline arteriolosclerosis.
3. Viscera Lesion

   (a) Brain
       There are three main types:

       • Softening of the brain: Small artery arteriosclerosis leads to cerebral ischemia and results in microinfarct formation.
       • Hypertensive encephalopathy: The clinical features of acute hypertensive encephalopathy include reduced consciousness, headache, vomiting, nausea, and variable neurological signs. It is also called hypertensive crisis.
       • Cerebral hemorrhage: This is a serious and life-threatening complication.

       Hypertensive intraparenchymal hemorrhages typically occur in the basal ganglia, thalamus, pons, and cerebellum. Intracerebral hemorrhage can be clinically devastating when it affects large portions of the brain and extends into the ventricular system. Rupture of small artery arteriosclerosis or microaneurysm leads to cerebral hemorrhage with increased blood pressure.

   (b) Heart

An increase in myocardial wall tension results in increased myocardial oxygen consumption, initiating a series of biochemical events that lead to left ventricular hypertrophy. The essential feature of hypertensive heart disease is left ventricular hypertrophy. The musculi papillares and adductors are thickened.

(c) Retina

The central artery of the retina may exhibit arteriosclerosis. In severe cases, this can lead to papilledema, retinal hemorrhage, and significant vision loss.

4. Kidney

Hyaline and fibrous changes in blood vessels cause vascular narrowing, which leads to glomerular ischemic atrophy and fibrosis, as well as tubular atrophy and interstitial fibrosis. Additionally, glomerular compensatory hypertrophy and renal tubular compensatory enlargement may be observed. The kidneys may appear normal in size or moderately reduced. The cortical surfaces show fine, even granularity, also called primary granular atrophy of the kidney.

7.3.2.2 Malignant Hypertension

Malignant hypertension is characterized by severe hypertension of rapidly increasing severity. Blood pressure often exceeds to 230/130 mmHg. The malignant phase may be superimposed on benign hypertension; less commonly, it may start de novo. The pathognomonic lesions of malignant hypertension are proliferative endarteritis and fibrin "onion skin" lesions.

7.4 Rheumatism

Rheumatism, also called rheumatic fever (RF), is an acute, immunologically mediated, multisystem inflammatory disease. It usually occurs a few weeks after an episode of group A β-hemolytic streptococcal pharyngitis. The connective tissue is mainly involved in blood vessels and the heart and skin. It is also referred to as connective tissue disease or collagen disease.

Acute RF is closely related to infection with group A streptococci. It is a hypersensitivity reaction induced by host antibodies elicited by group A streptococci.

7.4.1 Basic Pathological Changes

According to the pathological features, the course of disease development is divided into three phases described as follows:

1. Alterative and Exudative Phase

   This is the early stage of acute RF. Discrete inflammatory lesions are found in the connective tissues throughout the body, including the skin, synovium, the heart, and joints. There is focal mucoid degeneration and fibrinoid necrosis of the collagen with surrounding polymorphs, plasma cells, and monocytes. This phase can last for 1 month.

2. Proliferative or Granulomatous Phase

   The hallmark of acute RF is the presence of multiple foci of inflammation called Aschoff bodies. Aschoff bodies consist of a central zone of degenerating, hypereosinophilic extracellular matrix infiltrated by lymphocytes (primarily T cells), occasional plasma cells, and plump activated macrophages called Anitschkow cells. These activated macrophages can also fuse to form giant cells. This phase lasts for 2–3 months.

3. Fibrous Phase or Healed Phase

   Progressive fibrosis of Aschoff body results in the formation of small scars. This phase lasts for 2–3 months.

7.4.2 Rheumatic Heart Disease

Rheumatic heart disease (RHD) can involve myocardium, endocardium, and pericardium. Chronic valvular deformities are the most important consequences of RHD, characterized by diffuse and dense scarring of valves resulting in permanent dysfunction.

7.4.2.1 Rheumatic Endocarditis

Of the heart valves, the mitral valve is most often affected, followed by the combined involvement off mitral and aortic, aortic, and tricuspid valves, while the pulmonary valve is rarely involved.

Valve involvement results in fibrinoid necrosis along the lines of closure, forming 1–2 mm vegetations (verrucae) that have little effect on cardiac function. These irregular, warty projections probably arise from the precipitation of fibrin at sites of erosion caused by underlying inflammation and collagen degeneration. The vegetations consist of platelets and fibrin, with fibrinoid necrosis and occasional small Aschoff bodies.

7.4.2.2 Rheumatic Myocarditis

Rheumatic myocarditis is characterized by scattered Aschoff bodies within the interstitial connective tissue of the left atrial and left ventricular myocardium, particularly in the subendocardial fibrous tissues and around blood vessels in the intermuscular fibrous septa.

7.4.2.3 Rheumatic Pericarditis

Rheumatic pericarditis is characterized by fibrinous or serous exudation. Hydropericardium results from serous exudation, while "shaggy heart" or "cor villosum" refers to fibrinous exudates that are heavily layered on the epicardial surface of the heart. A large amount of fibrinous exudation that cannot be completely absorbed can lead to organization and result in constrictive pericarditis.

7.4.2.4 Clinical Features

Typically, symptoms occur 2–3 weeks after an episode of streptococcal pharyngitis. The predominant clinical manifestations are carditis and arthritis. Clinical features of carditis include pericardial friction rubs and arrhythmias. Myocarditis can be so severe that it causes cardiac congestive dilation, leading to functional mitral insufficiency and even heart failure. Nevertheless, fewer than 1% of patients die of acute rheumatic endocarditis.

7.4.3 Rheumatic Arthritis

About 75% of patients with rheumatic fever in the early stages exhibit clinical manifestations of rheumatic arthritis. Large joints such as the ankle, shoulder, wrist, elbow, and knee are most frequently involved. Redness, swelling, heat, pain, and dysfunction are observed in these joints. Fibrinous and serous exudation fills the articular cavity. Atypical Aschoff bodies can be seen in the adjacent soft tissue.

7.4.4 Rheumatic Arteritis

Any arteries can be involved, with the small artery involvement being common, such as the renal artery, cerebral artery, coronary artery, pulmonary artery, mesenteric artery, etc. In the acute phase, fibrinoid necrosis, mucous degeneration, and lymphocyte infiltration of the vascular wall are prominent, accompanied by the formation of Aschoff bodies. In the late stage, the vascular wall changes to thickening with fibrosis, and the vascular cavity narrows, accompanied by thrombosis.

7.5 Infective Endocarditis

Infective endocarditis (IE) is a serious infection requiring prompt diagnosis and intervention.

7.5.1 Etiology and Pathogenesis

Infection occurs when organisms are implanted on the endocardial surface during episodes of bacteremia. In some instances, the cause of the hematogenous infection is obvious, as in the case of intravenous drug abusers who inject contaminated materials directly into bloodstream; an infection elsewhere or a previous dental, surgical, or other interventional procedure may also seed the bloodstream. In other cases, however, the source of bacteremia is occult and presumably related to trivial injuries to the skin or mucosal surfaces.

IE is traditionally classified into acute and subacute forms, primarily based on clinical tempo and severity. The distinctions are attributable to intrinsic microbial virulence and whether underlying cardiac disease is present. The disease typically appears insidiously and follows a protracted course of weeks to months, with most patients recovering after appropriate antibiotic therapy.

7.5.2 Acute IE

Acute IE usually suggests a destructive, tumultuous infection, frequently involving a highly virulent organism attacking a previously normal valve and causing death within days to weeks in more than 50% of patients despite antibiotics and surgery.

In acute IE, bulky, friable, and potentially destructive vegetations containing fibrin, inflammatory cells, and microorganisms are present on the heart valves. The mitral and aortic valves are the most common sites of infection.

Systemic emboli may occur at any time because of the friable nature of the vegetations, and they may cause infarcts in the brain, myocardium, kidney, and other tissues. Abscesses often develop at the site of such emboli.

7.5.3   Subacute IE

Subacute IE refers to infections by organisms of low virulence colonizing of a previously abnormal heart, especially when there are deformed valves.

The vegetations of subacute IE tend to be somewhat firmer and are associated with less valvular destruction than those of acute IE. The aortic and mitral valves are the most common sites of infection. Microscopically, the vegetations of typical subacute IE are distinguished from those of acute disease by the presence of granulation tissue at their bases, suggesting chronicity. As time passes, calcification, fibrosis, and a chronic inflammatory infiltrate may develop.

7.6   Valvular Heart Disease

Valvular heart disease is caused by congenital disorders or a variety of acquired diseases that lead to valve abnormalities. These abnormalities can result in stenosis or insufficiency (regurgitation or incompetence), or both. Stenosis is the failure of a valve to open completely, obstructing forward flow. Valvular stenosis is almost always a chronic process caused by a primary cuspal abnormality. Insufficiency results from the failure of a valve to close completely, allowing reversed flow. Valvular insufficiency may result from either intrinsic disease of the valve cusps (e.g., valve destruction) or distortion of the supporting structures without primary changes in the cusps. It can appear acutely, as with chordal rupture, or chronically due to leaflet scarring and retraction.

7.6.1   Mitral Stenosis

The most common causes of mitral stenosis (MS) are postrheumatic or postinflammatory diseases, accounting for 99% of cases. Rare causes include congenital valvular or supravalvular stenosis, SLE (systemic lupus erythematosus), Whipple endocarditis, and extensive calcification of the mitral annulus.

7.6.1.1   Morphology

The mitral valve leaflets are swollen, and tiny flat vegetations can be seen along the lines of closure at the acute stage. Microscopically, chronic inflammation, edema, and platelet-fibrin thrombi are present; Aschoff bodies are found in a minority of cases. In chronic RHD, the valve leaflets exhibit thickening, retraction, and calcification, with chordae fused and shortened, reducing the orifice to an oval, narrow "fish mouth" opening. Microscopically, typical findings include fibrosis, calcification with or without ossification, neovascularization, and a variable chronic inflammatory cell infiltrate containing lymphocytes, monocytes, and mast cells.

7.6.1.2 Hemodynamic Phenomena

When the mitral orifice is reduced, blood can flow from the left atrium into the left ventricle only if it is propelled by a pressure gradient. To double the flow through the stenotic mitral valve, the pressure gradient has to be increased four times. The left atrium progressively dilates and may harbor mural thrombi. Such an increase in atrial pressure will be transmitted to the pulmonary veins and capillaries, producing dyspnea; if severe enough, it may give rise to frank pulmonary edema. Left atrial hypertension, resulting in dilatation, hypertrophy, and fibrosis of the atrial wall, can lead to atrial fibrillation.

7.6.1.3 Clinical Features

In patients whose mitral orifices are large enough to accommodate normal blood flow with only mild elevations of left atrial pressure, marked elevations of pressure leading to dyspnea and cough may be precipitated by sudden changes in heart rate, volume status, or cardiac output. As mitral stenosis progresses, patients become limited in daily activities and may experience orthopnea and paroxysmal nocturnal dyspnea. The characteristic physical finding in mitral stenosis is a rumbling diastolic murmur, which is heard at the cardiac apex. Liver congestion, jugular filling, serous cavity effusion, and edema of the lower limbs can also be present. Chest X-rays may show an enlarged left atrium and a pear-shaped heart.

7.6.2 Mitral Insufficiency

A variety of different pathological entities can lead to regurgitation or incompetence of the mitral valve. RHD is still the most common cause of mitral regurgitation, followed by myxomatous degeneration, mitral valve prolapse, IHD, IE, postinflammatory disease, ruptured chordae tendineae, and annular calcification.

7.6.2.1 Morphology

The valve associated with mitral insufficiency from RHD is often shortened and thickened and has fused chordae, anchoring the posterior cusp of the mitral valve to the wall of the left ventricle.

7.6.2.2 Hemodynamic Phenomena

In patients with mitral regurgitation, the left ventricle ejects blood via two routes: (1) through the aortic valve to the systemic circulation and (2) through the mitral valve into the left atrium. To maintain normal systemic cardiac output, more complete systolic emptying of the left ventricle is required. Over many years, the additional burden imposed by mitral regurgitation on the myocardium leads to dilatation and hypertrophy of the left ventricle. After many decades, left ventricular failure develops, eventually leading to dilatation and hypertrophy of the right ventricle and right atrium, or right ventricular failure.

7.6.2.3 Clinical Features

Patients with chronic mild-to-moderate isolated mitral insufficiency are usually asymptomatic. Orthopnea, exertional dyspnea, and fatigue are the most prominent complaints in patients with chronic severe mitral insufficiency. In patients with chronic severe mitral insufficiency, a whiffing systolic murmur is often heard at the cardiac apex. The chest X-ray may show a ball-shaped heart.

7.6.3 Aortic Stenosis

Aortic stenosis is usually the result of rheumatic valvular injury and a congenital abnormality of the valve that leads to fibrosis and calcification of the valve. Other less common causes include unicuspid or unicommissural valves, dysplastic valvular formation, and a group of rare metabolic disorders.

7.6.3.1 Morphology

The classic gross appearances of aortic valve leaflets are thickening, rigidity, calcification, and fusion, with the orifice reduced to a narrow opening. The histopathologic changes reflect the extensive calcific alterations that begin in the fibrosa layer and expand into the sinuses.

7.6.3.2 Hemodynamic Phenomena

The hemodynamic burden imposed by critical aortic valve stenosis is systolic overloading of the left ventricle. To overcome the impedance to emptying imposed by the stenotic valve, the left ventricle increases its contractile mass. Dilatation of the left ventricular chamber and hypertrophied muscle result in elevated left ventricular end diastolic pressure. This leads to left ventricular failure and, eventually, to dilatation and hypertrophy of the right ventricle and right atrium, or right ventricular failure.

7.6.3.3 Clinical Features

In severe aortic stenosis, the resulting left ventricular outflow obstruction can lead to left ventricular pressures as high as 200 mmHg or more. The hypertrophied myocardium tends to be relatively ischemic, and angina can develop. Syncope may occur due to poor perfusion of the brain. The characteristic physical finding in aortic valve stenosis is an ejecting systolic murmur, usually heard at the aortic valve area. The chest X-ray may show a boot-shaped heart.

7.6.4 Aortic Insufficiency

Regurgitation of blood back through the aortic valve during diastole occurs as a result of a number of different pathological entities: bicuspid aortic valve, RHD, myxomatous degeneration, IE, Marfan syndrome, or septal myomectomy.

7.6.4.1 Morphology

In root dilatation associated with Marfan syndrome, the leaflets display a range of myxomatous expansion of the spongiosa layer, highlighted by stains for glycosaminoglycans such as colloidal iron. Aortic insufficiency caused by postrheumatic lesions contains limited amounts of calcific deposits and fibrosis. In age-related aortic root dilatation, the leaflets show minimal or no degenerative features under the microscope.

7.6.4.2 Hemodynamic Phenomena

Aortic regurgitation produces a volume overload of the ventricle. Left ventricular end diastolic volume is increased by the quantity of regurgitant blood flow. Initially, compensatory left ventricular hypertrophy and dilatation maintain cardiovascular homeostasis. Over a period of years, left ventricular function deteriorates, and signs and symptoms of heart failure ensue. Left ventricular diastolic pressure rises and is transmitted to the pulmonary capillaries, producing pulmonary congestion and the sensation of dyspnea. Thus, it leads to dilatation and hypertrophy of the right ventricle and right atrium, or right ventricular failure.

7.6.4.3 Clinical Features

Symptomatic aortic valve regurgitation is characterized by the presence of one or more of a triad of symptoms: orthopnea, dyspnea, and angina. The characteristic physical finding in aortic valve regurgitation is a whiffing diastolic murmur, which is usually best heard along the left sternal border. The carotid pulsation, Corrigan's pulse, capillary pulsation, and vascular shot sound phenomenon can be present.

7.7 Cardiomyopathy

Myocardial diseases are termed cardiomyopathies. In many cases, cardiomyopathies are of unknown etiology (idiopathic); however, several previously "idiopathic" cardiomyopathies have been shown to be caused by specific genetic abnormalities in cardiac energy metabolism or structural and contractile proteins.

Cardiomyopathies can be subdivided by a variety of criteria into two major groups: (1) primary, which includes those entities in which the disease is solely or predominantly confined to the heart muscle, and (2) secondary, in which the heart is involved as a part of a generalized multiorgan disorder. A more clinical and functional classification of cardiomyopathies is as follows: dilated cardiomyopathy (90% of cases), hypertrophic cardiomyopathy, restrictive cardiomyopathy, Keshan disease, and arrhythmogenic right ventricular cardiomyopathy.

7.7.1 Dilated Cardiomyopathy

Dilated cardiomyopathy (DCM), also called congestive cardiomyopathy, is defined clinically by global left ventricular systolic dysfunction and increased left ventricular cavity diameter in the absence of hypertension, valvular disease, or significant coronary artery disease. Some DCM cases have a familial (genetic) basis. Others result from a variety of acquired myocardial insults, including toxic exposures (e.g., chronic alcoholism), myocarditis, and pregnancy-associated changes.

7.7.1.1 Morphology

Because of the wall thinning that accompanies dilation, the ventricular thickness may be less than, equal to, or greater than normal. Mural thrombi are common and may be a source of thromboembolism. The heart in DCM is characteristically enlarged (increase two to three times compared with its normal weight) and flabby, with dilation of all chambers.

The histologic abnormalities in DCM are nonspecific. Microscopically, most myocytes are hypertrophied with enlarged nuclei, but many are irregular, stretched, and attenuated. There is variable interstitial and endocardial fibrosis; scattered scars are also often present, probably marking previous myocyte ischemic necrosis caused by reduced perfusion and increased demand.

7.7.1.2 Pathogenesis

The causes of DCM can be grouped into four broad categories: family history of cardiomyopathy, idiopathic dilated cardiomyopathy (no known association), acquired DCM (including viral coxsackievirus B and other enteroviruses, peripartum cardiomyopathy, alcohol or other toxic exposure), and secondary DCM (including IHD, autoimmune diseases, endocrine, metabolic, and nutritional disorders).

7.7.1.3 Clinical Features

DCM can occur at any age, but it most commonly occurs between the ages of 20 and 50. The fundamental defect in DCM is ineffective contraction. Fifty percent of patients die within 2 years, and only 25% survive longer than 5 years. Death is usually due to progressive cardiac failure or arrhythmia. In most cases, cardiac transplantation is the only definitive treatment.

7.7.2 Hypertrophic Cardiomyopathy

Hypertrophic cardiomyopathy (HCM) (also known as idiopathic hypertrophic subaortic stenosis) is characterized by abnormal diastolic filling, myocardial hypertrophy, and, in a third of cases, ventricular outflow obstruction. The heart is heavy, thick-walled, and hypercontracting, in striking contrast to the flabby, poorly contractile heart in DCM.

7.7.2.1 Morphology

The essential gross feature of HCM is massive myocardial hypertrophy without ventricular dilation. The classic pattern of HCM involves disproportionate thickening of the ventricular septum relative to the left ventricle free wall. The characteristic histologic features in HCM are severe myocyte hypertrophy, myocyte (and myofiber) disarray, and interstitial and replacement fibrosis.

7.7.2.2 Pathogenesis

Almost all cases of HCM are caused by missense point mutations in one of several genes encoding the sarcomeric proteins that form the contractile apparatus of striated muscle. More than 100 causal mutations have been identified in at least 12 sarcomeric genes, with the β-myosin heavy chain being most frequently affected, followed by myosin-binding protein C and troponin T. These three genes account for 70–80% of all cases of HCM.

7.7.2.3 Clinical Features

High left ventricular pressures, massive hypertrophy, and compromised intramural coronary arteries often lead to myocardial ischemia (with angina), even without coronary artery disease. Major clinical problems include atrial fibrillation with mural thrombus, IE of the mitral valve, congestive heart failure, arrhythmias, and sudden death. Most patients improve with therapy that promotes ventricular relaxation; occasionally, partial surgical excision of septal muscle is necessary to relieve outflow tract obstruction.

7.7.3 Restrictive Cardiomyopathy

Restrictive cardiomyopathy (RCM) is characterized by a primary decrease in ventricular compliance, resulting in impaired ventricular filling during diastole. RCM can be idiopathic or associated with systemic diseases that also affect the myocardium, such as radiation fibrosis, amyloidosis, hemochromatosis, sarcoidosis, or products of inborn errors of metabolism.

7.7.3.1 Morphology

In idiopathic restrictive cardiomyopathy, the ventricles are of approximately normal size or slightly enlarged, the cavities are not dilated, and the myocardium is firm. Biatrial dilation is commonly observed. Microscopically, there is interstitial fibrosis, varying from minimal and patchy to extensive and diffuse. RCM of different causes may have similar gross morphology. However, endomyocardial biopsy can reveal disease-specific features.

7.7.3.2 Clinical Features

Due to increased ventricular stiffness, there is reduced diastolic ventricular volume with normal or near-normal systolic function (ejection fraction) and wall thickness. The atria are dilated, with non-hypertrophied, nondilated ventricles.

7.7.4 Keshan Disease

Keshan disease is an endemic cardiomyopathy found in Keshan, northeast China. The first patient was identified in 1935. This disease is characterized by a blood circulation disorder, endocardial abnormality, and myocardial necrosis. Selenium deficiency is thought to be a major factor, according to Chinese scientists.

Scar lesions are scattered on the ventricular wall, and mural thrombus are found on ventricular trabecular muscles or the left or right atrial appendage. Microscopically, different degrees of cellular granular degeneration, vacuole degeneration, fatty degeneration, and coagulation necrosis can be seen. Scar formation is prominent in chronic cases. The pathological features of Keshan disease include severe myocardial degeneration, necrosis, and scarring. Grossly, the heart is large, spherical, and increased in weight. Both sides of the heart chambers are expanded with ventricular wall thinning, especially in the apex area.

7.7.5 Arrhythmogenic Right Ventricular Cardiomyopathy

Arrhythmogenic right ventricular cardiomyopathy is a unique (albeit uncommon) entity with a clinical presentation involving right-sided heart failure and various rhythm disturbances (including SCD). Morphologically, the right ventricular wall is severely thinned as a result of myocyte replacement by massive fatty infiltration and lesser accounts of fibrosis. Most cases are sporadic, but familial forms do occur with gene defects localized to chromosome 14 (autosomal dominant inheritance with variable penetrance). Most of the mutations seem to involve desmosomal junctional proteins.

7.8 Myocarditis

Myocarditis refers to generalized inflammation of the myocardium associated with necrosis and degeneration of myocytes. The term is generally used only for diffuse myocardial inflammation that results in symptoms, although focal myocarditis has been described. In myocarditis, there is inflammation of the myocardium with resulting injury. According to etiology, myocarditis can be divided into infective and noninfective types. The causes of the former include virus, bacteria, parasites, fungi, etc. The latter type is usually caused by allergic reaction, drugs, or physical or chemical factors. However, most cases are caused by viral infection.

7.8.1   Viral Myocarditis

Viral myocarditis, also called lymphocytic and idiopathic myocarditis, is most common. The most frequently implicated agents are coxsackie A and B, ECHO, Polio, and influenza viruses. The pathogenesis of myocyte injury is complex, involving myocyte receptors specific for enteroviruses and adenoviruses and immune-regulated myocyte cell death.

Grossly, myocarditis reveals nonspecific findings ranging from normal to dilatation of all four chambers, depending on the specific agents and the duration of illness. Multinucleated giant cells often surround individual myocytes, and there is focal or patchy acute myocyte necrosis associated with inflammatory cell infiltrates. In the early stage, necrosis and accumulation of interstitial proteinaceous material are prominent, whereas during the resolving phase, fibroblast proliferation and interstitial collagen deposition become predominant.

Most viruses that cause myocarditis also cause pericarditis. Most patients present with symptoms of concomitant pericardial inflammation, chest pain, and palpitations. The clinical diagnosis is made on the basis of electrocardiographic changes, myocardial damage evidenced by elevated troponin, evidence of systemic inflammation, lack of epicardial coronary artery disease, and generally mild and transient ventricular dysfunction.

7.8.2   Bacterial Myocarditis

Myocarditis involvement is most commonly encountered with meningococcal, diphtheritic, and leptospiral infections among bacteria. The exotoxin of the diphtheria bacillus causes cardiac damage by inhibiting protein synthesis, as it specifically interferes with a transferase involved in the delivery of amino acids to elongating polypeptide chains. The myocardium may be affected in any bacterial infection when the causative agent is blood-borne, for example, with salmonellosis or any form of bacterial endocarditis.

Bacterial myocarditis is characterized by a different mixed inflammatory cell infiltrate, with neutrophils as the major component. Organisms can often be marked by special stains. Microabscesses can occur when septic emboli lodge in the coronary circulation, often as a consequence of infective endocarditis.

7.8.3   Isolated Myocarditis

Isolated myocarditis refers to sporadic cases of myocarditis that occur without a discernible cause in previously healthy individuals. A fatal type of isolated myocarditis is known as idiopathic giant cell myocarditis of Fiedler's myocarditis. This type affects males more than females.

Regardless of the histological changes during the acute phase of the disease, all inflammatory lesions either resolve, leaving no residuals, or undergo progressive fibrosis. Often, the persistent connective tissue is sufficiently scattered and subtle to be virtually imperceptible at a later date. With more severe damage, focal minute scars may remain. Transient ECG abnormalities may be the only indication of its presence, particularly abnormalities of the ST segment and T wave, which point to a diffuse myocardial lesion.

Acute symptomatic myocarditis often manifests itself as malaise, dyspnea, and low-grade fever, with tachycardia more pronounced than the fever alone would warrant. Most patients recover from acute myocarditis, although a few may die of congestive heart failure or arrhythmias. Patients with isolated myocarditis usually die from congestive heart failure or sudden death from arrhythmias.

7.9   Aneurysm and Dissections

Aneurysm is congenital or acquired localized abnormal dilations of blood vessels or the heart. The two most important causes of aneurysms are atherosclerosis and cystic medial degeneration of the arterial media. Arterial aneurysms can also be caused by systemic diseases, such as vasculitis. Other causes that weaken vessel walls include trauma, congenital defects, infections (mycotic aneurysms), and syphilis.

Chapter 8
Respiratory Diseases

Zhihong Yang and Yingying Zhang

Contents

Z. Yang (✉) · Y. Zhang
School of Basic Medical Science, Kunming Medical University, Kunming, China

© Zhengzhou University Press 2024
K. Chen et al. (eds.), *Textbook of Pathologic Anatomy*,
https://doi.org/10.1007/978-981-99-8445-9_8

Objectives

The objective is to comprehend the pathological alterations, clinicopathologic relationships, and distinctions between lobular and lobar pneumonia, as well as the pathological alterations associated with silicosis and lung cancer metastasis.

Additionally, you should be familiar with the pathological changes of emphysema and pulmonary heart disease, the etiology and clinicopathologic correlations of lung cancer, the etiology and complications of silicosis, the pathological changes and progression of nasopharyngeal carcinoma, as well as the characteristics of lobar and lobular pneumonia.

Furthermore, it is important to understand the causes, mechanisms, and symptoms of silicosis and pulmonary arterial hypertension.

Key Concepts

 1. Chronic bronchitis
 2. Emphysema
 3. Asthma
 4. Bronchiectasis
 5. Chronic obstructive pulmonary disease
 6. Cor pulmonale
 7. Acute respiratory distress syndrome
 8. Respiratory distress syndrome of the newborn
 9. Lung cancer
10. Adenocarcinoma in situ

Introduction

The respiratory system is comprised of the nostrils, pharynx, larynx, trachea, bronchi, and lungs. Typically, the laryngeal cricoid functions as the boundary between the upper and lower respiratory tracts. The nose, pharynx, and larynx are components of the upper respiratory system, whereas the trachea and bronchi are components of the lower respiratory system. The right lung has three lobes, while the left lung has only two. The principal right and left bronchi emanate from the trachea and then divide in opposite directions to form airways of decreasing size. Bronchi lack submucosal glands and cartilage in their wall structures, whereas bronchioles possess these glands. Terminal bronchioles, which have a diameter of less than 2 mm, are the result of bronchiole branching. The distal terminal bronchiole or pulmonary acini of the lung contains alveoli, the primary site for gas exchange. The microscopic structure of the alveolar walls (also known as alveolar septa) consists of the following components: The first three structures are the capillary endothelium, the basement membrane, and the interstitial tissue that surrounds it. The second structure is the alveolar epithelium, which consists of a continuous layer of two primary cell types: type I pneumocytes (95%), type II pneumocytes (5%), and a few alveolar macrophages. The primary functions of the lungs are to discharge carbon dioxide and replace oxygen in the circulation. The circumstance is hazardous to the respiratory system. As a result, respiratory ailments can be caused by inhaling a variety of hazardous substances, such as toxic gases, particles, and microorganisms, into the airways. Due to immunological and nonimmune defense mechanisms, the respiratory system is able to repel the invasion of hazardous microorganisms. Respiratory maladies develop when the immune system is compromised or when the pathogens are highly potent [1, 2].

8.1 Respiratory Tract and Pulmonary Infections

The respiratory system has direct communication with the outside environment. Harmful chemicals and pathogenic bacteria can enter the respiratory system and cause respiratory inflammatory disorders. Infections are the most prevalent respiratory illnesses. This section discusses acute tracheobronchitis, acute bronchiolitis, and pneumonia.

8.1.1 Acute Tracheobronchitis

A common respiratory condition, acute tracheobronchitis often affects youngsters and the elderly. It occurs after a cold and an upper respiratory infection and is more common during the colder months.

8.1.1.1 Etiology

This condition is mostly caused by viral infections, such as rhinovirus, influenza virus, respiratory syncytial virus, and adenovirus, as well as secondary bacterial infection, such as influenza bacillus and Streptococcus pneumoniae. Furthermore, inhaling toxic dust or gas (such as sulfur dioxide, chlorine, etc.) might induce acute tracheobronchitis in a few situations.

8.1.1.2 Morphology

In severe instances, the tracheobronchial mucosa is red and swollen, with white or yellowish mucus discharges on the surface, adhering to areas of necrosis and ulcers on the mucosa. It is classified based on the characteristics of the disease:

1. Acute catarrhal tracheobronchitis: mucosal and submucosal congestion and edema, with neutrophil infiltration; the luminal surface coated with yellow, sticky secretions. The secretions might induce coughing or blockage.
2. Acute suppurative tracheobronchitis: characterized by suppurative inflammation. A considerable number of neutrophils infiltrate the mucosa and submucosa, and the inflammation might spread to the alveoli around the bronchioles. Patients frequently cough up yellow, sticky purulent sputum.
3. Acute ulcerative tracheobronchitis: a dangerous condition typically brought on by a combination of a virus and pyogenic bacteria. The mucosa of the lumen develops superficial necrosis, erosion, and ulcers in the early stages. The trachea and bronchial tubes' shape and function revert to normal once the inflammation subsided and the mucus epithelium is healed by basal cell growth.

8.1.2   Acute Bronchiolitis

An acute inflammation of the bronchioles with a diameter of less than 2 mm is known as acute bronchiolitis. Infants under the age of 4 frequently have it, particularly those under 1 year, who make up roughly 90% of cases.

8.1.2.1   Etiology

Winter viral infections, particularly respiratory syncytial virus, adenovirus, and parainfluenza virus, are to blame for the majority of cases.

8.1.2.2   Morphology

Hyperemia and edema are caused by bronchial mucus. Pseudostratified ciliated columnar epithelium, which is necrotic, is replaced by columnar epithelium and simple squamous epithelium. There are more goblet cells and they secrete more mucus. Mononuclear cells and lymphocytes infiltrate. Acute focal atelectasis or obstructive emphysema results from the lumen being fully or partly obstructed by inflammatory exudates. Because the bronchioles are so thin, peribronchiolitis or localized pneumonitis can readily result from the inflammation spreading to the alveoli and nearby pulmonary interstitium.

8.1.3 Pneumonia

As the most prevalent illness of the respiratory system, pneumonia is commonly described as any infection of the lung. A World Health Organization survey found that 75% of acute respiratory illness mortality was attributable to pneumonia. Pneumonia ranks sixth among all causes of mortality in China. Pneumonia can present as an acute, fulminant, or chronic illness with a protracted course of symptoms.

Numerous pathogenic causes can cause pneumonia. According to the etiology, radiation pneumonia, aspiration pneumonia, lipoid pneumonia, and allergy pneumonia are produced by physicochemical causes. In contrast, bacterial pneumonia, viral pneumonia, mycoplasma pneumonia, fungal pneumonia, and parasitic pneumonia are caused by biological factors. Alveolar pneumonia, which makes up the majority of pneumonia cases, and interstitial pneumonia can be distinguished based on the location of the lesion. There are three types of pneumonia: lobar pneumonia, lobular pneumonia, and interstitial pneumonia, depending on the severity of the lesion.

8.1.3.1 Bacterial Pneumonia

Bacterial pneumonia, which makes up roughly 80% of all cases of pneumonia, is the most prevalent form. Lobar pneumonia and lobular pneumonia (also known as bronchopneumonia) are two anatomical and radiological patterns of acute bacterial pneumonias.

1. **Lobar Pneumonia**

   Pneumococcal infection, characterized by acute lung fibrinous inflammation, is the primary cause of lobar pneumonia. The symptoms often arise all at once and include a high fever, trembling chills, pleuritic chest discomfort, and a productive mucopurulent cough; patients may also occasionally display hemoptysis. Pneumococcus is the primary culprit in more than 90% of cases of lobar pneumonia, particularly the most potent type III strains.

   (a) **Morphology and Clinical Features**: Most frequently, the lower lobes or right middle lobe are involved. Prior to the discovery of antibiotics, pneumococcal pneumonia affected entire or nearly entire lobes and progressed through four stages: obstruction, red hepatization, gray hepatization, and resolution. This common occurrence is altered or halted by antibiotic treatment administered early on. In the initial stage, the affected lobes are heavy, red, and sluggish; histologically, the vessels are congested, with proteinaceous fluid, dispersed neutrophils, and numerous bacteria in the alveoli. Toxemia results in shivering, fever, and an increase in the number of white blood cells. Auscultation reveals damp rales, and the chest film exhibits a scaly trace. After 1 or 2 days, the second stage of red hepatization begins to develop. At this juncture, the affected lobe is still crimson and has a liver-

Fig. 8.1 Lobar
pneumonia. The affected
lobe becomes gray, dry,
and firm

like consistency, with neutrophils, red blood cells, and fibrin occupying the
alveolar spaces (Fig. 8.1). There are a large number of microorganisms in
the exudate. As PaO_2 levels decrease, hypoxia can cause dyspnea and cyanosis in patients. Alveolar macrophages consume the exudate's erythrocytes,
causing the sputum to become rusty. This is diagnostically significant. A
pleural lesion causes fibrinous pleurisy and chest pain, which is exacerbated
by respiration and wheezing. On the X-ray, a significant region of dense
shadow is visible. Gray hepatization is the next stage. Due to the dissolution
of red blood cells, the affected lobe becomes gray, desiccated, and solid
(similar to the consistency of the liver), but fibrinous suppurative exudate is
still visible in the alveoli (Fig. 8.2). Ischemia is characterized by the compression of alveolar capillaries, which reduces oxygenation, despite the
presence of pulmonary consolidation signs. When the sputum changes to
purulent mucus, the remaining symptoms subside. In uncomplicated cases,
resolution follows the final phase. Enzymes within the alveoli break down
exudates into granular, semifluid fragments that can be reabsorbed, consumed by macrophages, expelled, or organized by fibroblasts growing into
them [1, 2].

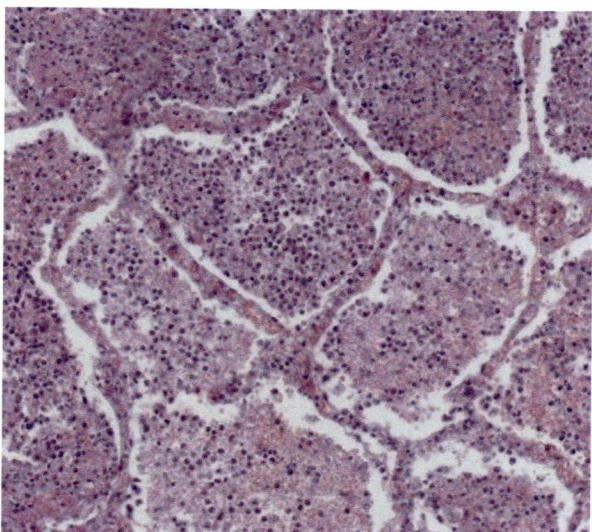

Fig. 8.2 Lobar pneumonia

(b) **Complications**: Proper treatment can restore lung structure and function; however, difficulties may sometimes arise: (1) organization of the intra-alveolar exudate may transform some areas of the lung into solid fibrous tissue, called pulmonary carnification; (2) an abscess may form due to tissue destruction and necrosis; (3) infection may spread to the pleural cavity, causing an empyema; (4) bacteremic dissemination to the pericardium, heart valves, brain, and joints may cause meningitis, suppurative arthritis, or infective endocarditis; (5) septic shock is the most serious complication and can cause death.

2. **Lobular Pneumonia**

It is mostly caused by pyogenic infections such as Staphylococcus aureus and hemolytic streptococci. The lesions start in the bronchioles and spread to the peripheral lung tissues, resulting in a pyogenic inflammation throughout the lung lobules as a whole. Because of the lesion in the bronchioles, lobular pneumonia is also known as bronchopneumonia. It primarily affects youngsters and the elderly.

(a) **Morphology and Clinical Features**: Bronchopneumonia is distinguished by patchy consolidation of the lung that is dispersed throughout one or more lobes, most usually bilateral and basal. Well-developed lesions are typically 3–4 cm in diameter, somewhat raised, dry, granular, and gray-red to yellow in color. These foci may confluence in severe cases, resulting in the appearance of a lobar consolidation (confluent bronchopneumonia). The lung tissue adjacent to the consolidation area is frequently congested and edematous,

Fig. 8.3 Lobular
pneumonia. Foci of
inflammatory consolidation
are distributed in patches,
gray-red to yellow

Fig. 8.4 Lobular
pneumonia

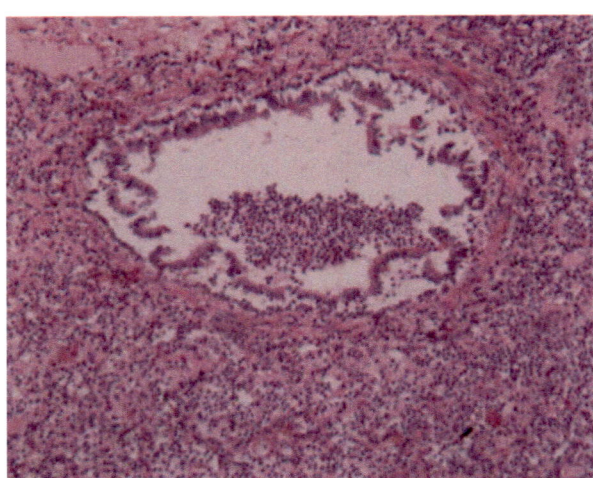

while the broad intervening areas are usually normal. Pleural involvement is less common than in lobar pneumonia. Suppurative exudates fill the bronchi, bronchioles, and surrounding alveolar spaces histologically [1, 2](Figs. 8.3 and 8.4).

(b) **Clinical Manifestations**: Although lobular pneumonia can be a consequence of various diseases, its clinical manifestations are sometimes obscured by primary disorders. Coughing, expectoration, and fever are the most common symptoms. Coughing and sputum production are caused by bronchial mucosal inflammation, and the phlegm is purulent. The lesions appear to be focally distributed, and physical signs of the lung (excluding confluent bronchopneumonia) are often not visible. An X-ray of the chest reveals irregular tiny flaky or speckled hazy shadows. If the condition is detected early and treated properly, the exudate in the lungs can be entirely absorbed, and the patients can recover. However, the outlook is generally bleak for young children and the elderly. Complications of lobular pneumonia are far more likely than those of lobar pneumonia and include heart failure, respiratory failure, sepsis, lung abscess, and empyema.

3. **Additional Important Bacterial Pneumonias**

(a) **Haemophilus influenzae**: In children and adults with chronic lung illnesses such as chronic bronchitis, cystic fibrosis, and bronchiectasis, Haemophilus influenzae can produce a potentially life-threatening form of pneumonia, which commonly occurs after a respiratory virus infection. The most common bacterial cause of acute chronic obstructive pulmonary disease (COPD) exacerbation is Haemophilus influenzae. Although encapsulated Haemophilus influenzae type b was a major cause of epiglottitis and suppurative meningitis in infants in the past, the use of vaccines against this pathogen in infancy has considerably reduced the risk [1, 2].

(b) **Klebsiella pneumoniae**: The most prevalent cause of gram-negative bacterial pneumonia is Klebsiella pneumoniae. Klebsiella-related pneumonia is common among the elderly and malnourished, particularly chronic alcoholics. Because the bacterium produces an abundance of viscid capsular polysaccharide, thick and viscous sputum is characteristic of this variety, making coughing difficult for the patient [1, 2].

(c) **Legionella pneumophila**: Legionella pneumophila is the causative agent of Legionnaires' disease, which can be both epidemic and sporadic. People with specific predisposing diseases, such as cardiac, renal, immunologic, or hematologic disorders, are more likely to develop Legionnaires' disease. Organ transplant recipients are especially susceptible to infection. Legionnaires' disease can be highly severe, often necessitating hospitalization, and the death rate in immunocompromised individuals ranges from 30 to 50%. Rapid diagnosis can be aided by identifying Legionella antigens in urine or a positive fluorescent antibody test in sputum samples; however, culture remains the gold standard [1, 2].

8.1.3.2 Interstitial Pneumonia (Atypical Pneumonia)

The term "primary atypical pneumonia" was first used to describe an acute febrile respiratory disease characterized by patchy inflammatory alterations in the lungs, mostly in the alveolar septa and pulmonary interstitium. The term "atypical" refers to the presence of moderate volumes of sputum, the absence of physical signs of consolidation, and the absence of exudates in the alveoli [1, 2].

1. **Pathogenesis and Etiology**: Atypical pneumonia is caused by a variety of infections. The most frequent is Mycoplasma pneumoniae, which is especially prevalent in children and young people. Other causes include viruses such as influenza A and B, respiratory syncytial viruses, human metapneumovirus, and adenovirus.

   Organism adhesion to the respiratory epithelium is a typical etiology, followed by cell necrosis and an inflammatory reaction.

   When this process extends to the alveoli, interstitial inflammation typically occurs, although some fluid may also leak into alveolar spaces, resulting in abnormalities on the chest X-ray that are similar to those seen in bacterial pneumonia.

   Because respiratory epithelial injury inhibits mucociliary clearance, viral respiratory infections are easily worsened by bacterial infection.

   Infants, the elderly, malnourished patients, alcoholics, and immunocompromised individuals are more susceptible to developing severe lower respiratory tract infections.
2. **Morphology (Secondary)**: Atypical pneumonias have the same morphologic appearance regardless of the source. The process can be spotty or involve entire lobes bilaterally or unilaterally. The affected lobes appear red-blue and congested. Histologically, the inflammatory reaction is primarily restricted to the alveolar walls. The alveolar walls become enlarged and edematous, with lymphocytes, histiocytes, and occasionally plasma cells infiltrating. In contrast to bacterial pneumonias, atypical pneumonias have almost no cellular exudate in the alveolar spaces. In extreme situations, diffuse alveolar damage with pink hyaline membranes may occur [1, 2].

   Reconstruction of the original structure occurs after disease remission in milder and less complex situations. Bacterial co-infection alters the histologic appearance by inducing ulcerative bronchitis and bronchiolitis.
3. **Clinical Characteristics**: Primary atypical pneumonia has a wide range of clinical manifestations. The typical initial manifestation is an acute, nonspecific febrile sickness marked by fever, headache, and malaise, followed by a cough with little sputum. In clinical settings, it can present as a severe upper respiratory tract infection or as a fulminant, life-threatening infection in immunocompromised patients. Because edema and exudation are key sites where alveolar capillaries can become clogged, respiratory discomfort may occur, which appears to be disproportionate to physical and X-ray findings. Determining the cause may be challenging. Mycoplasma antigen detection and polymerase chain reaction (PCR) detection of mycoplasma DNA can be used for diagnosis. Patients with

community-acquired pneumonia who do not have bacterial infections can be treated with a macrolide antibiotic effective against Mycoplasma and Chlamydia pneumoniae, which are the most common pathogens causing treatable illness.

8.1.3.3 SARS (Severe Acute Respiratory Syndrome)

Severe acute respiratory syndrome coronavirus (SARS-CoV) is a zoonotic viral respiratory disease caused by the SARS coronavirus. The primary mode of transmission for SARS is contact of the mucosal membranes with respiratory droplets or fomites. Early signs include fever, myalgia, malaise, cough, sore throat, and other nonspecific symptoms. The single symptom shared by all patients appears to be a fever higher than 38 °C (100 °F). The disease primarily causes abnormalities in the lung and immune system. The lungs become patchy or completely consolidated, turning dark red; histologically, the lungs of dying patients frequently show diffuse alveolar destruction with hyperemia, hemorrhage, and edema. Many cells populate the alveolar spaces, including alveolar epithelial cells, mononuclear cells, lympho-cytes, and plasma cells, as well as the formation of hyaline membranes. Some alve-olar epithelial cells contain virus inclusion bodies in their cytoplasm. Fibrinoid necrosis occurs in the walls of pulmonary small blood vessels, where thrombus may occur.

8.2 Chronic Obstructive Pulmonary Disease (COPD)

COPD is a form of obstructive lung disease that includes chronic bronchitis, emphy-sema, asthma, and bronchiectasis. It is characterized by long-term breathing issues and inadequate airflow. Shortness of breath and cough with sputum production are the most common symptoms. COPD is a progressive condition, meaning that it usu-ally worsens over time.

8.2.1 Chronic Bronchitis

Chronic bronchitis is a common condition characterized by prolonged nonspecific inflammation of the tracheal mucosa and its surrounding tissues. It is widespread among cigarette smokers and residents of smog-filled cities. Chronic bronchitis is diagnosed clinically; the presence of a continuous productive cough for at least three consecutive months in at least two consecutive years is required [1, 2].

8.2.1.1 Pathogenesis and Etiology

Chronic bronchitis results from a prolonged and complex interaction of several causes. While smoking is the most common cause, other air pollutants such as sulfur dioxide and nitrogen dioxide also play a role. These environmental irritants cause mucus gland enlargement in the trachea and main bronchi, leading to an increase in the number of mucin-secreting goblet cells in the surface epithelium of smaller bronchi and bronchioles. These irritants also induce inflammation, which includes the invasion of CD8+ cells, macrophages, and neutrophils. Recurrent viral infections and co-bacterial infections are additional key factors in the development and exacerbation of chronic bronchitis.

8.2.1.2 Morphology

The hallmark of chronic bronchitis is mucus hypersecretion, which begins in the major airways. The mucosal lining of the airways, including the large airways, smaller bronchi, and bronchioles, is usually hyperemic and swollen with edema fluid and is frequently covered with mucinous or mucopurulent secretions. Histologically, chronic bronchitis is characterized by enlargement of the mucus-secreting glands in the trachea and larger bronchi, as well as an increase in the number of goblet cells. Inflammatory cells, including lymphocytes, plasma cells, and occasionally neutrophils, invade the area. Squamous metaplasia and dysplasia of the bronchial epithelium can occur in chronic bronchiolitis. Goblet cell metaplasia, mucus plugging, inflammation, and fibrosis all contribute to bronchiole narrowing. In the most severe forms, the lumen may be completely obliterated [1, 2].

8.2.1.3 Clinical Characteristics

In the early stages of the disease, the productive cough produces mucoid sputum but does not impede airflow. Some patients with chronic bronchitis may have hyperresponsive airways with intermittent bronchospasm and wheezing. Some individuals develop severe COPD with airflow obstruction, leading to hypercapnia, hypoxemia, and, in extreme cases, cyanosis. Particularly heavy smokers may acquire chronic airflow restriction, which is generally accompanied by evidence of emphysema. Chronic bronchitis can progress to pulmonary hypertension and heart failure.

8.2.2 Emphysema

Emphysema is a chronic obstructive airway illness characterized by permanent enlargement of air spaces distal to terminal bronchioles, along with wall damage but no visible fibrosis.

8.2.2.1 Pathogenesis

Emphysema is frequently caused by smoking; however, it can also result from other conditions such as chronic bronchitis. It is associated with obstructive ventilation function disturbance, decreased elasticity of respiratory bronchioles and alveolar walls, and a lack of $\alpha 1$-antitrypsin.

8.2.2.2 Emphysema Types

Emphysema is classified into four categories based on its anatomic distribution inside the lobule: centriacinar, panacinar, distal acinar, and irregular. Only certriacinar and panacinar emphysema result in clinically substantial airway obstruction [1, 2].

1. Centriacinar (Centrilobular) Emphysema: This form is most common in heavy smokers who do not have congenital $\alpha 1$-antitrypsin deficiency. It is approximately 20 times more prevalent than panacinar emphysema. The pattern of involvement in the lobules is distinctive: the middle or proximal parts of the acini, created by respiratory bronchioles, are damaged, while the distant alveoli are spared. Upper lobe lesions are more common and severe, especially in the apical portions.
2. Panacinar (Panlobular) Emphysema: The acini are uniformly enlarged, beginning at the level of the respiratory bronchiole and progressing to the terminal blind alveoli. Panacinar emphysema occurs more frequently in the lower lung sections than centriacinar emphysema. This type is linked to a lack of $\alpha 1$-antitrypsin.
3. Distal Acinar Emphysema: The etiology of this type is unknown. The distal section of the acinus is primarily affected, although the proximal portion is normal. Emphysema is particularly noticeable in the pleura, along the lobular connective tissue septa, and at the lobule borders. It is typically worse in the upper portion of the lungs. Multiple, continuous, expanded air gaps, sometimes forming cystic structures, are typical observations. This type is most likely responsible for many cases of spontaneous pneumothorax in young adults.
4. Irregular Emphysema: When the acinus is irregularly involved, it is referred to as irregular emphysema and is nearly always associated with scarring, such as that caused by healed inflammatory disorders. Clinically, this type is asymptomatic; however, it is most likely the most common variety of emphysema.

8.2.2.3 Morphology

The macroscopic appearance of the lung is used to diagnose and classify emphysema. When the pathologic process is fully developed, panacinar emphysema produces voluminous lungs that often overlap the heart and obscure it when the anterior chest wall is removed. Centriacinar emphysema has less remarkable macroscopic features than panacinar emphysema. Unless the disease is advanced, the lungs in centriacinar emphysema are less pale and less massive. Generally, centriacinar emphysema affects the upper two-thirds of the lungs more severely than the lower portions. Histologically, alveolar walls are destroyed, resulting in increased air space. Due to alveolar injury, the number of alveolar capillaries is reduced. As the disease progresses, adjacent alveoli merge, leading to even larger aberrant air spaces and the formation of blebs or bullae. The loss of elastic tissue in the surrounding alveolar walls reduces radial tension on the small airways. Bronchial and bronchiolar inflammation is persistent. As a result, during expiration, the airways compress, leading to chronic airflow obstruction in severe emphysema (Figs. 8.5 and 8.6) [1, 2].

8.2.2.4   Clinical Characteristics

Typically, dyspnea is the first symptom, appearing gradually but worsening over time. Coughing and wheezing may be the initial symptoms in some patients with chronic bronchitis or chronic asthmatic bronchitis. The patient is typically barrel-chested and dyspneic, with noticeably protracted expiration, and may sit forward in a hunched-over position, trying to expel air from the lungs with each expiratory effort. In pulmonary function tests, FEV1 is reduced while FVC is normal or near-normal, resulting in a decreased FEV1 to FVC ratio. As the condition progresses, pulmonary hypertension develops due to the decreased number of alveolar capillaries caused by alveolar destruction, leading to cor pulmonale. Emphysema patients may die from respiratory failure, respiratory acidosis, hypoxia, unconsciousness, or right heart failure (cor pulmonale).

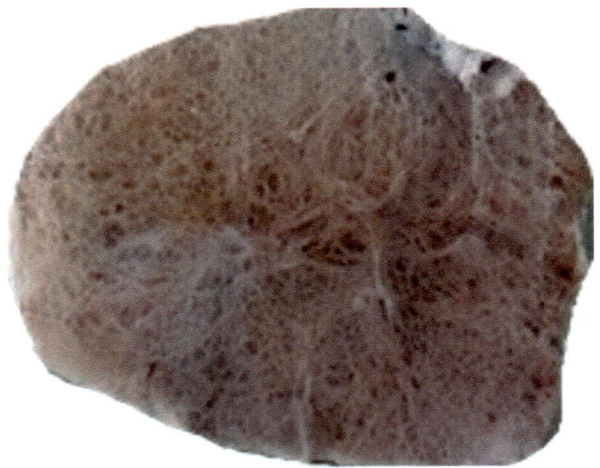

Fig. 8.5 Emphysema

Fig. 8.6 Emphysema

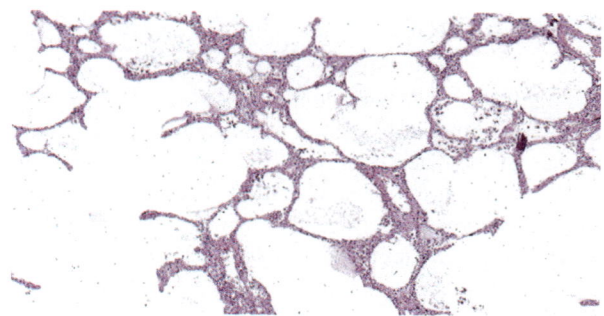

8.2.3 Asthma

Asthma is a chronic inflammation of the airways that can produce recurring episodes of wheezing, shortness of breath, chest tightness, and coughing, often at night and/or early in the morning. Airway obstruction is sporadic and reversible, with chronic bronchial inflammation, eosinophil infiltration, bronchial smooth muscle cell hypertrophy and hyperreactivity, and increased mucus production. The etiology of this condition is complex, and the pathophysiology is not fully understood. Asthma is caused by a genetic predisposition to type I hypersensitivity (atopy), acute and chronic airway inflammation, and bronchial hyperresponsiveness to a variety of stimuli. The disease involves multiple cell types and inflammatory mediators. Type-2 helper T (TH2) cells may play a significant role in asthma etiology [1].

8.2.3.1 Morphology

The lungs are grossly overextended due to overinflation, and there may be bronchiectasis and small patches of atelectasis. Thick, persistent mucus plugs form in the bronchi and bronchioles, causing blockage. The mucus plugs have whorls of shed epithelium on histology. There is also an eosinophil influx and Charcot-Leyden crystals. Other morphological changes associated with asthma, known as "airway remodeling," include (1) thickening of the basement membrane of the airway wall, (2) increased vascularity in the submucosa, (3) an increase in the size of the submucosal glands, and (4) hypertrophy and/or hyperplasia of the bronchial muscle [1, 2].

8.2.3.2 Clinical Characteristics

Asthma attacks are characterized by acute dyspnea and wheezing, particularly during expiration. Most attacks last one to several hours and resolve spontaneously or with treatment, usually involving bronchodilators and corticosteroids. Asthmatic patients are at risk of dying from complications such as hypercapnia, acidosis, and severe hypoxia.

8.2.4 Bronchiectasis

Bronchiectasis is a persistent necrotizing infection that causes permanent dilatation of the bronchi and bronchioles, as well as muscle and elastic tissue loss. The main causes of the disease are persistent infection or blockage. Patients experience frequent coughing and expectoration of copious volumes of purulent sputum as the disease progresses. A thorough history and radiographic demonstration of bronchial dilatation aid in diagnosis.

8.2.4.1 Morphology

Bronchiectasis typically affects the lower lobes bilaterally, particularly the most vertical air channels, with the most severe involvement occurring in the more distal bronchi and bronchioles. The airways may be dilated to four times their normal diameter, and the lung can be traced practically to the pleural surfaces. Due to these dilations, the airways may appear cylindroid or saccular. Histological results vary according to disease activity and chronicity. Extensive areas of ulceration are caused by an intense acute and chronic inflammatory exudate in the walls of the bronchi and bronchioles, as well as desquamation of the lining epithelium. In more severe cases, bronchial and bronchiolar wall fibrosis, as well as peribronchiolar fibrosis, develop. Necrosis can sometimes destroy the bronchial or bronchiolar walls, resulting in the formation of an abscess [1, 2] (Fig. 8.7).

8.2.4.2 Clinical Features

Patients present with a severe, persistent cough that is accompanied by mucopurulent, occasionally foul-smelling sputum. Sputum may include blood, and hemoptysis may result. Patients may develop finger clubbing. Severe cases are frequently associated with extensive bronchiectasis, significant obstructive ventilation abnormalities, hypoxemia, hypercapnia, pulmonary hypertension, and possibly cor pulmonale.

8.3 Pneumoconiosis

Pneumoconiosis was originally used to characterize the non-neoplastic response of the lungs to mineral dust intake. Pneumoconiosis now covers organic and inorganic particle disorders, as well as several non-neoplastic lung diseases induced by chemical fumes and vapors. The three most prevalent mineral dust pneumoconiosis causes are exposure to coal dust, silica, and asbestos, all of which are almost always caused by occupational exposure. Furthermore, the elevated cancer risk associated

Fig. 8.7 Bronchiectasis

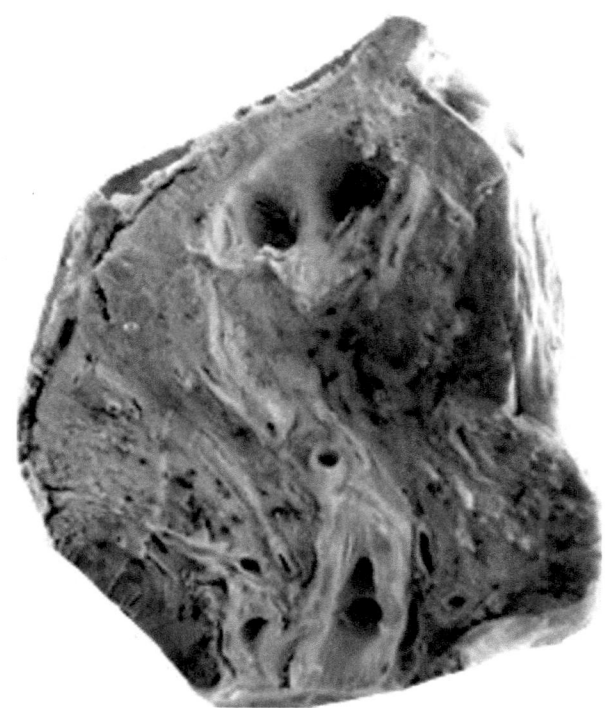

with asbestos exposure extends to family members of asbestos workers as well as anyone who is exposed to asbestos outside of the workplace.

8.3.1 Pathogenesis

The lung's sensitivity to mineral dusts is influenced by a variety of parameters, including particle size, shape, solubility, and reactivity. Particles greater than 5–10 μm in diameter are unlikely to reach distal airways, but particles smaller than 0.5 μm flow into and out of alveoli with little accumulation and harm. Particles 1–5 μm in diameter are the most harmful, becoming lodged at the distal airway bifurcation. Coal dust is generally inert, and a significant amount of coal dust must be deposited in the lungs before clinical lung illness is visible. At lesser quantities, silica, asbestos, and beryllium are more reactive than coal dust and can trigger fibrotic reactions in the lungs. The majority of dusts inhaled become caught in mucus layers and are promptly removed from the lungs by ciliary motions. Some particles, however, become lodged at the bifurcation of the alveolar ducts, where macrophages congregate and ingest the particles. The pulmonary alveolar macrophage is an important biological component in the development and maintenance of lung damage and fibrosis. The particles inhaled cause macrophages to release high amounts of cytokines, which mediate an inflammatory response and stimulate fibroblast proliferation and collagen deposition. Some of the inhaled particles may reach the lymphatics via direct drainage or within migrating macrophages, triggering an immune response to particle components and/or self-proteins modified by the particles, resulting in an amplification and extension of the local reaction [1, 2].

8.3.2 Silicosis

Silicosis is the most frequent pneumoconiosis in the world, with crystalline silica being the most common cause, primarily found in work settings. Workers in sandblasting and hard-rock mining are particularly vulnerable. Silica comes in two varieties: crystalline and amorphous. Crystalline forms are far more toxic, including quartz, which is most closely associated with the prevalence of silicosis. After inhalation, particles, particularly those with diameters of 1–5 μm, interact with epithelial cells and macrophages. As silica particles are ingested, pulmonary macrophages activate and release mediators such as interleukin-1 (IL-1), tumor necrosis factor (TNF), fibronectin, lipid mediators, and oxygen-derived free radicals. These mediators contribute to the formation of silicotic nodules and pulmonary fibrosis.

8.3.2.1 Morphology

The primary pathological alterations of silicosis are the production of silicotic nodules and widespread pulmonary fibrosis. Silicotic nodules are small, 3–5 mm in diameter, round or oval, distinct, and pale-to-blackened (if coal dust is present), with a solid appearance. Nodules are typically found in the upper zones of the lungs (Fig. 8.8). As the disease progresses, the nodules may merge into a hard mass. Necrosis occurs due to ischemia and hypoxia, leading to the formation of a silicotic cavity. Silicotic nodule formation includes cellular silicotic nodule, fibrous silicotic nodule, and hyaline silicotic nodule. Histologically, a typical silicon nodule is composed of concentrically organized hyaline collagen fibers. The "whorled" appearance of the collagen fibers is characteristic of silicosis (Fig. 8.9). Polarized microscopy reveals weakly birefringent silica particles, primarily in the centers of the nodules. In addition to the silicotic nodules, diffuse pulmonary fibrosis is present in the lung. The pleura may also become thickened due to diffuse fibrosis. Fibrotic lesions in the hilar lymph nodes are possible as well. On X-ray, the lymph nodes may develop thin sheets of calcification with "eggshell" calcification [1, 2].

Fig. 8.8 Silicosis

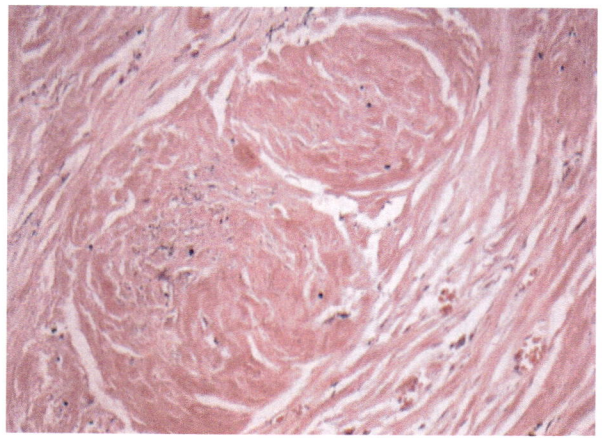

Fig. 8.9 Silicosis (silicotic nodule)

8.3.2.2 Clinical Characteristics

Silicosis is classified into three stages based on the number, size, and location of silicotic nodules, as well as the degree of pulmonary fibrosis. The number of silicotic nodules increases progressively, and the pleura thickens due to fibrosis. Silicosis is typically detected on routine chest radiographs, which show small nodules in the upper zones of the lung in asymptomatic individuals, but pulmonary function may be normal or only slightly altered.

Patients with progressive massive fibrosis (PMF) may experience shortness of breath. Even if the patient is no longer exposed to silica, the disease continues to advance slowly and severely compromises lung function. Because silicosis causes a decrease in cell-mediated immunity, and crystalline silica may impair the ability of pulmonary macrophages to destroy phagocytosed mycobacteria, it is often associated with an increased susceptibility to tuberculosis (silico-tuberculosis) and other diseases. Pneumovascular fibrosis can cause pulmonary hypertension and cor pulmonale in approximately 60–75% of individuals. Emphysema and spontaneous pneumothorax may also occur in patients [1, 2].

8.3.3 Coal Worker's Pneumoconiosis (CWP)

Asymptomatic anthracosis, simple CWP, and severe CWP are all types of CWP. Coal miners' lung lesions are diverse. In silent anthracosis, there is no visible cellular reaction to pigment buildup. Accumulations of macrophages occur with little or no pulmonary impairment in simple CWP. In severe CWP, which progresses to PMF, fibrosis is widespread and lung function is substantially reduced. According to statistics, only about 10% of cases with simple CWP progress to severe fibrosis [1, 2].

8.3.3.1 Morphology

Pulmonary anthracosis is the most common and innocuous coal-induced pulmonary lesion in coal miners, as well as in urban populations and smokers. Carbon pigment inhaled is absorbed by pulmonary macrophages. They then accumulate in connective tissue along lymphatics or in lymph nodes. Coal macules and slightly larger coal nodules distinguish simple CWP. The coal macules are composed of dust-laden macrophages, and the nodule contains a small network of collagen fibers. Lesions are found throughout the lung, but the upper lobes and upper zones of the lower lobes are most affected (Fig. 8.10). Pneumoconiosis in coal miners is typically marked by many, deep-black scars greater than 2 cm, and sometimes up to 10 cm. Dense collagen and pigment can be found histologically in the lesions [1, 2].

Fig. 8.10 Coal worker's pneumoconiosis

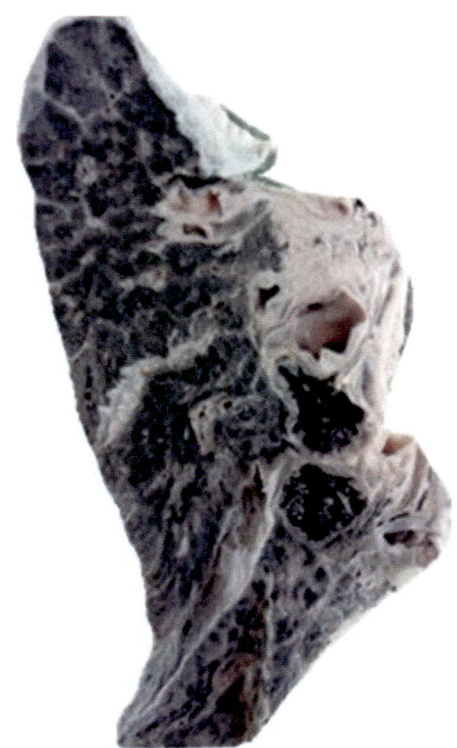

8.3.3.2 Clinical Features

CWP is typically a benign condition with no impact on lung function. If patients develop progressive massive fibrosis, they will experience pulmonary dysfunction, pulmonary hypertension, and cor pulmonale. The transition from CWP to PMF is influenced by a number of factors, including the amount of coal dust exposure and the total dust burden. Even when patients are no longer exposed to coal dust, severe fibrosis can eventually emerge.

8.3.4 Asbestosis

Asbestos exposure in the workplace has been linked to the following diseases, according to epidemiological studies: (1) parenchymal interstitial fibrosis (asbestosis); (2) localized fibrous plaques or, in rare cases, diffuse fibrosis in the pleura; (3) pleural effusions; (4) lung cancer; (5) malignant pleural and peritoneal mesotheliomas; and (6) laryngeal carcinoma [1, 2].

8.3.4.1 Pathology

Asbestos is a type of crystalline hydrated silicate with a fibrous structure. The concentration, size, shape, and solubility of various types of asbestos affect whether inhalation of the substance causes illness. Asbestos comes in two varieties: serpentine and amphibole. The fiber in serpentine is curled and flexible, whereas the fiber in amphibole is straight, hard, and brittle. Both forms can cause asbestosis, lung cancer, and mesothelioma. Chrysotile, a type of serpentine, is the most common type of industrial asbestos. Although amphiboles are less frequent, they are more pathogenic than chrysotile serpentine.

8.3.4.2 Morphology

Asbestosis is characterized by diffuse pulmonary interstitial fibrosis, thickening of the visceral pleura, pleural plaques on the parietal pleura, and asbestos bodies. Asbestosis typically begins in the lower lobes and subpleural space, but as fibrosis progresses, the middle and upper lobes of the lungs also become affected. The lungs eventually shrink and stiffen in texture. Fibrosis thickens the pleura. Pleural plaques, the most prevalent manifestation of asbestosis, are well-circumscribed plaques of thick collagen, typically including calcium, in the parietal pleura. Early lesions consist of interstitial pneumonia induced by asbestos exposure, as indicated by histology. Type I alveolar epithelial cells are destroyed by type II alveolar epithelial hyperplasia, and macrophages accumulate in the pulmonary alveoli. Lymphocytes and monocytes proliferate in the pulmonary mesenchyme. Asbestos bodies, which are an important diagnostic feature of asbestosis, develop when macrophages attempt to ingest asbestos fibers. These bodies are composed of iron from phagocyte ferritin and appear as golden brown, fusiform, or beaded rods with a translucent center. Asbestos bodies can occasionally be seen in the lungs of healthy individuals, though in much lower numbers and without interstitial fibrosis [1, 2].

8.3.4.3 Clinical Features

The clinical symptoms of asbestosis are similar to those of other chronic interstitial lung disorders. Dyspnea often worsens 10–20 years after exposure, accompanied by coughing and expectoration. Scarring caused by fibrosis can trap and constrict pulmonary arteries and arterioles, leading to pulmonary hypertension, cor pulmonale, and even mortality. Pleural plaques are frequently asymptomatic and appear as confined densities on radiography. The combined incidence of asbestosis and tuberculosis is about 10%, which is lower than the rate for silicosis. Patients with asbestosis are more likely to develop lung cancer and other malignant tumors. Asbestos workers are five times more likely to develop lung cancer.

8.4   Pulmonary Hypertensive Heart Disease: Cor Pulmonale

Pulmonary heart disease is characterized by right ventricular hypertrophy and dilatation, often accompanied by right heart failure, resulting from pulmonary hypertension induced by a primary illness of the pulmonary parenchyma or pulmonary vascular system.

Chronic cor pulmonale is characterized by right ventricular hypertrophy (frequently accompanied by right atrial enlargement). It is caused by pulmonary hypertension, which arises from primary parenchymal or vascular lung disease. Simultaneous enlargement of the right ventricle and right atrium is typical. When a ventricular failure develops, the right ventricle and atrium often dilate. Because chronic cor pulmonale occurs in the context of pulmonary hypertension, the pulmonary arteries frequently contain atherosclerotic plaques and other lesions that reflect prolonged elevated stress [1, 2] (Fig. 8.11).

Diseases of the pulmonary parenchyma and conditions that increase the risk of developing cor pulmonale are as follows [1, 2]:

- Chronic obstructive pulmonary disease (COPD)
- Diffuse interstitial fibrosis of the lungs
- Pneumoconiosis
- Chronic pulmonary fibrosis
- Bronchiectasis
- Pulmonary vascular disease

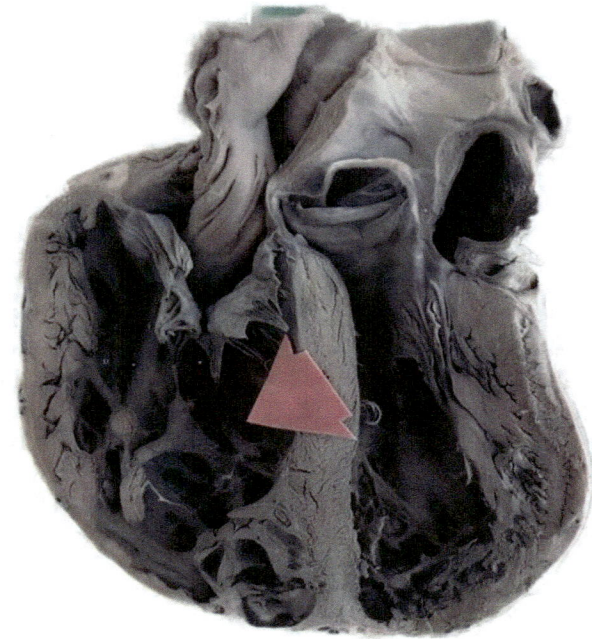

Fig. 8.11 Chronic cor pulmonale. The right ventricle (on the left side of this image) is significantly dilated and hypertrophied, with a thicker free wall and hypertrophied trabeculae. The enlarged right ventricle has deformed the shape and volume of the left ventricle

- Repeated pulmonary embolism
- Pulmonary hypertension of unknown cause
- Inflammation of several arteries in the lungs (like Wegener's granulomatosis)
- Vascular occlusion due to medication, poison, or irradiation
- Microembolization of tumor cells throughout the lungs

Disorders affecting breathing are as follows:

- Kyphoscoliosis
- Pickwickian syndrome (extreme obesity)
- Muscle and nerve disorders
- Diseases causing pulmonary artery constriction
- Metabolic acidosis
- Hypoxemia
- Major airway obstruction
- Alveolar hypoventilation of unknown cause

8.5 Acute Respiratory Distress Syndrome (ARDS)

ARDS is a clinical syndrome induced by injury to diffuse alveolar capillaries and epithelial cells. The typical progression of the disease is a rapid onset of life-threatening respiratory insufficiency, cyanosis, and severe arterial hypoxemia that is resistant to oxygen therapy and may progress to multisystem organ failure. Diffuse alveolar injury is the histologic manifestation of pulmonary ARDS.

Note that respiratory distress syndrome in neonates is pathologically distinct; it is caused by a lack of surface surfactant.

8.5.1 *Morphology*

During the acute phase of ARDS, the lungs are dark red, rigid, oxygen-depleted, and weighty. Microscopic examination reveals capillary congestion, necrosis of alveolar epithelial cells, edema and hemorrhage in the alveolar interstitium and alveoli, and accumulation of neutrophils in the capillaries (notably in sepsis). A most common feature is the presence of a hyaline membrane, particularly in the dilated alveolar ducts. This membrane is composed of fibrin-rich edematous fluid and necrotic epithelial cell remnants. In the organizing phase, type II alveolar cells proliferate vigorously in an effort to regenerate the alveolar epithelium. Fibrin exudates from the tissue more frequently induce fibrosis in the alveoli than does resolution. Interstitial cell proliferation and collagen deposition substantially thicken the alveolar septa.

8.5.2 Pathogenesis

The alveolar capillary membrane consists of two distinct barriers: the microvascular endothelium and the alveolar epithelium. Injury to endothelial or epithelial cells, or more commonly both, compromises the integrity of this barrier in ARDS. Acute effects of alveolar capillary membrane damage include increased vascular permeability, alveolar flooding, loss of gas exchange, and extensive surfactant abnormalities resulting from type II alveolar cell injury. Recent studies have shown that in ARDS, lung injury is caused by an imbalance between pro-inflammatory and anti-inflammatory mediators, although the cellular and molecular basis of acute lung injury and ARDS is still an area of active research. Lung macrophages increase their production of IL-8, a potent neutrophil chemotactic factor and activator, 30 min after acute injury. The release of this mediator and similar mediators, such as IL-1 and TNF, results in the activation of endothelial cells and the recruitment and activation of neutrophils in pulmonary capillaries. It is believed that neutrophils play an essential role in the pathogenesis of ARDS. In the early stages of the disease, histological examination of the lungs reveals increased neutrophils in the vascular spaces, interstitium, and alveoli. Activated neutrophils discharge numerous substances (e.g., oxidants, proteases, platelet activators, leukotrienes) that injure the alveolar epithelium and endothelium. The combined assault on endothelial and epithelial cells perpetuates vascular leakage and surfactant loss, thereby preventing the proper function of alveolar units. Notably, various endogenous antiproteases, antioxidants, and anti-inflammatory cytokines such as IL-10, which are upregulated by pro-inflammatory cytokines, can counteract the destructive force emitted by neutrophils. The degree of tissue injury and clinical severity of ARDS are ultimately determined by the equilibrium between destructive and protective factors [1, 2].

8.6 Respiratory Distress Syndrome (RDS)

There are numerous causes of respiratory distress in newborns, such as excessive anesthesia of the mother, fetal head injury during delivery, aspiration of blood or amniotic fluid, and intrauterine hypoxia caused by compression of the umbilical cord around the neck. The most prevalent cause, however, is respiratory distress syndrome (RDS), also known as hyaline membrane disease due to the formation of "membranes" in the peripheral air spaces of neonates with this condition [1, 2].

8.6.1 Morphology

Infants with RDS have normal-sized lungs, but they are dense and airless. Atelectasis is characterized by purple mottling, solid microscopic tissue, dysplasia, and alveolar collapse. In the terminal bronchioles and alveolar ducts, only deceased cell fragments will be present if the infant expires within the first few hours of life. In the later phases of the disease, transparent eosinophilic membranes are observed in the respiratory bronchioles, alveolar ducts, and random alveoli. These membranes contain necrotizing epithelial cells and exuded plasma proteins.

8.6.2 Pathogenesis

RDS is predominantly a preterm disease linked to male gender, maternal diabetes, and cesarean delivery. RDS is caused by the inability of immature lungs to produce enough surfactants. Surfactants are complexes of surface-active phospholipids, primarily dipalmitoyllecithin (lecithin), and at least two categories of proteins related to surfactants. The development of acute respiratory failure in infants with congenital surfactant deficiency due to mutations in corresponding genes highlights the significance of surfactant-associated proteins to normal lung function. Surfactant, which is synthesized by type II lung cells, swiftly coats the surface of the alveoli upon the first inhalation of a healthy neonate, thereby reducing the surface tension and, consequently, the pressure required to keep the alveoli open. In lungs deficient in surfactant, the alveoli have a tendency to collapse, necessitating a greater inhalation effort to release them with each breath. The infant experiences difficulty breathing and systemic atelectasis. In the absence of oxygen, epithelial and endothelial cells are damaged, leading to the formation of clear membranes.

8.7 Carcinomas of the Lung

In industrialized nations, lung cancer is unquestionably the leading cause of cancer-related mortality. Since 1987, the incidence has progressively decreased in men but continued to rise in women, with more women being affected. Each year, lung cancer kills more people than breast cancer. These numbers clearly demonstrate the causal connection between smoking and lung cancer. Lung cancer is most prevalent in individuals aged 50–60. More than 50% of patients have distal metastatic disease at the time of diagnosis, while 25% have regional lymph node disease. The 5-year survival rate for all stages of lung cancer combined is approximately 16%, a figure that has not changed significantly over the past three decades. Even when lung cancer is confined to the lungs, the 5-year survival rate is only 45%.

The four major histological forms of lung cancer are adenocarcinoma, squamous cell carcinoma, small cell carcinoma, and large cell carcinoma. In some instances, a combination of histological forms (such as small cell carcinoma and adenocarcinoma) is present. The strongest association between smoking and lung cancer is observed with squamous cell carcinoma and small cell carcinoma.

Lung cancer was previously divided into two main categories: small cell lung cancer and non-small cell lung cancer, which includes adenocarcinoma, squamous cell carcinoma, and large cell carcinoma [1, 2].

Adenocarcinoma* is the histologic classification of malignant epithelial lung tumors.

- Acinar
- Papillary
- Micropapillary
- Solid
- Predominately lepidic
- Mucinous

Squamous cell carcinoma
Large cell carcinoma
Large cell neuroendocrine carcinoma
Small cell carcinoma
Combined small cell carcinoma
Adenosquamous carcinoma
Carcinomas exhibiting pleomorphic, sarcomatoid, or sarcomatous characteristics

- Spindle cell carcinoma
- Giant cell carcinoma

Carcinoid neoplasm

- Standard
- Atypical

Unclassified carcinomas of the salivary glands
*Adenocarcinoma, squamous cell, and large cell carcinomas are collectively referred to as non-small cell lung cancer.

8.7.1 Morphology

Lung cancer originates with microscopic, typically grayish-white, and rigid mucosal lesions. They may form lumen masses that infiltrate the bronchial mucosa or large masses that invade the contiguous lung parenchyma. Due to central necrosis or focal hemorrhage, certain large masses develop cavities. Finally, these tumors may extend into the pleura, invade the pleural cavity and chest wall, and disseminate to neighboring intrathoracic structures. Additional spread is possible via lymphatic or hematogenous routes.

Squamous cell carcinoma is more prevalent in men than in women and is strongly associated with a history of smoking. It tends to appear centrally in the great bronchus and spread to local hilar lymph nodes but disseminates beyond the thorax much later than other histological types. Central necrosis and cavitation may be present in larger lesions. Usually, squamous cell carcinoma is preceded by squamous metaplasia or dysplasia of the bronchial epithelium, which is followed by transformation into carcinoma in situ—a stage that can last for several years. At this stage, atypical cells can be found on cytological specimens of sputum, bronchoalveolar lavage, or brush, despite the lesions being asymptomatic and radiographically undetectable. Small tumors progress to the symptomatic stage when a defined tumor mass begins to obstruct the large bronchial lumen, frequently resulting in distal atelectasis and infection. Simultaneously, lesions infiltrate the adjacent lung tissue (Fig. 8.12a). These tumors range from highly differentiated squamous cell tumors with keratin pearls (Fig. 8.12b) and intercellular bridges to poorly differentiated tumors with very few remaining squamous cell characteristics [1, 2].

Adenocarcinomas can manifest as central lesions, similar to squamous cell variants but are typically peripherally located, with numerous cases exhibiting central scarring. Adenocarcinoma is the most prevalent form of lung cancer among non-smokers and females. In general, adenocarcinomas develop more slowly and form smaller masses than other subtypes but have a propensity to metastasize extensively in the early phases. Possible histological findings include acinar carcinoma, papillary carcinoma, mucinous carcinoma (protomucinous bronchioalveolar carcinoma, which is frequently multifocal and may manifest with pneumonia-like

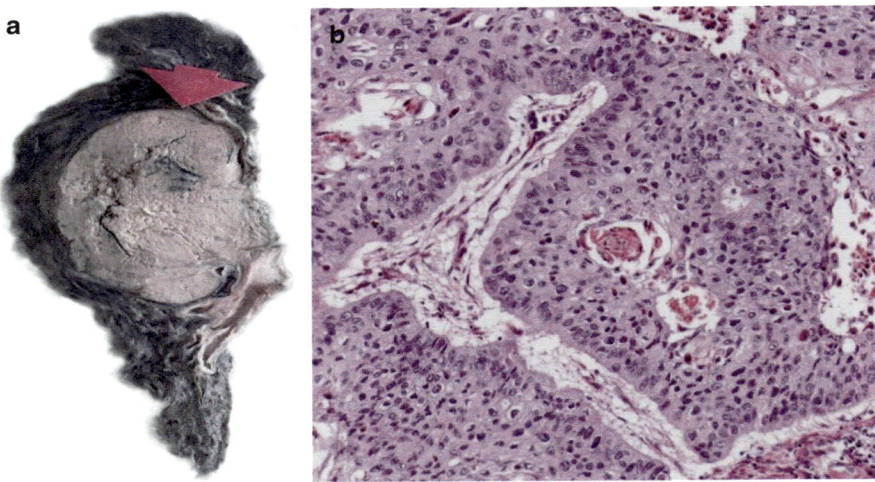

Fig. 8.12 (**a**) As shown, squamous cell carcinoma typically originates as a central (hilar) mass and spreads to the surrounding parenchyma. (**b**) Squamous cell carcinoma with well-differentiated keratinization and pearls

consolidation), and solid carcinoma. To determine adenocarcinoma lineages, demonstrating intracellular mucin production often requires special staining of solid changes (Fig. 8.13).

Adenocarcinoma in situ (AIS), formerly known as bronchioloalveolar carcinoma, frequently affects the lung's periphery and consists of a solitary nodule. AIS is characterized by a diameter of less than or equal to 3 mm, growth along the original structure, and preservation of the alveolar structure. Tumor cells, which may be non-mucus, mucus, or mixed, develop along the alveolar septum epithelium, acting as scaffolding (this is known as a "squamous" growth pattern, alluding to the resemblance of tumor cells to butterflies perched on a fence). By definition, AIS does not demonstrate deterioration of alveolar structure or interstitial infiltration with adherent hyperplasia, which would aid in the diagnosis of adenocarcinoma. By analogy with the colonic adenocarcinoma sequence, it has been hypothesized that some lung invasive adenocarcinomas may be caused by atypical adenomatous hyperplasia in the ortho-invasive adenocarcinoma sequence [1, 2].

Small cell lung cancer manifests as a centrally located, pale gray mass that extends into the mediastinal lymph nodes. These tumors are composed of round to fusiform cells with sparse cytoplasm and finely grained chromatin. Mitotic figures are common (Fig. 12.48). Despite their name, tumor cells are typically twice the size of normal lymphocytes. Necrosis is inevitable and may be widespread. Small biopsies frequently reveal fragments and "crush artifacts" resulting from the

Fig. 8.13 Adenocarcinomas are more common in women than men and are almost always gland-forming

fragility of tumor cells. In cytological specimens of small cell carcinoma, the nucleation resulting from the close packing of tumor cells devoid of cytoplasm is another distinguishing feature. In addition to expressing various neuroendocrine markers, these tumors typically secrete large quantities of polypeptide hormones that can cause paraneoplastic syndromes [1, 2].

Combined patterns need no further explanation. Notably, a small number of lung malignancies display multiple lines of cell differentiation, indicating that all cells originated from a single multipotent progenitor cell. For all of these tumors, it is feasible to locate lymph node chains around the carina, in the mediastinum, in the neck (angular lymph nodes), and in the clavicular region, with eventual distal metastasis. Involvement of the left supraclavicular lymph node (Virchow lymph node) is particularly common, and occult primary tumors occasionally require attention. When these malignancies progress, they frequently spread into the pleura or pericardial cavity, resulting in inflammation and fluid accumulation. They have the potential to compress or infiltrate the superior vena cava, resulting in venous congestion or vena cava syndrome. Apical tumors may invade the humeral or cervical sympathetic plexus, causing severe pain in the ulnar nerve distribution or generating Horner's syndrome (ipsilateral eye invagination, ptosis, pupil reduction, and lack of perspiration). This apical tumor is occasionally referred to as a Pancoast tumor, and the constellation of clinical symptoms is known as Pancoast syndrome. Frequently, Pancoast tumors are accompanied by the elimination of the first and second ribs, and occasionally the thoracic vertebra. As with other malignancies, tumor lymph node metastasis classifications have been established to denote the primary tumor's size and spread.

8.7.2 Pathogenesis

As a consequence of the progressive accumulation of a large number of genetic abnormalities (the number of small cell malignancies is estimated to be in the thousands), benign progenitor cells in the lung transform into tumor cells in smokers with lung cancer.

Regarding carcinogenic effects, there is substantial evidence that smoking and, to a lesser extent, other environmental factors are responsible for the genetic alterations that lead to lung cancer. About 90% of lung malignancies are diagnosed in current or former smokers. The frequency of lung cancer is nearly proportional to the number of years of smoking. Habitual heavy smokers (two cartons per day for 20 years) are 60 times more at risk than nonsmokers. However, since only 11% of regular smokers develop lung cancer, other factors must contribute to the pathogenesis of this fatal disease. For unknown reasons, women are more susceptible than men to the carcinogens found in tobacco. Despite the fact that ceasing smoking can reduce the risk of lung cancer over time, the risk may never revert to its initial level [1, 2].

Other contributing factors may work in conjunction with smoking or may be the cause of certain lung cancers. There is an increasing incidence of this type of tumor among miners of radioactive ore, asbestos workers, and those exposed to dust containing arsenic, chromium, uranium, nickel, vinyl chloride, and mustard gas. Exposure to asbestos increases the risk of lung cancer in nonsmokers by a factor of five. In contrast, the risk of lung cancer among regular smokers exposed to asbestos is approximately 55 times that of nonsmokers who are not exposed to asbestos.

Cancer-causing sequence alterations are best documented in squamous cell carcinoma, but they also occur in other histological subtypes. In essence, there is a linear correlation between the intensity of cigarette smoke exposure and the appearance of epithelial changes that progress from relatively harmless basal cell hyperplasia and squamous metaplasia to squamous dysplasia and carcinoma in situ, and finally to aggressive cancer. Among the main histological subtypes of lung cancer, the association between squamous and small cell carcinomas and tobacco use is the strongest.

8.8 Nasopharyngeal Carcinoma

Because (1) there is a strong epidemiological association with the Epstein-Barr virus (EBV) and (2) the high incidence of this form of cancer in the Chinese population raises the possibility of viral carcinogenesis in the context of genetic predisposition, nasopharyngeal carcinoma is a rare tumor worthy of discussion. It is believed that EBV replicates in the nasopharyngeal epithelium before infecting adjacent tonsil B lymphocytes and thus infecting the host. This results in the transformation of epithelial cells in some individuals. In infectious mononucleosis, EBVs infect B lymphocytes directly, followed by a considerable proliferation of reactive T lymphocytes, resulting in atypical lymphocytosis (observed in the peripheral blood) and enlarged lymph nodes. Similarly, nasopharyngeal carcinoma frequently demonstrates an influx of mature lymphocytes. Therefore, these tumors are referred to as "lymphoepithelioma" despite the fact that lymphocytes are not involved in the tumor process and tumors are not benign. The presence of large tumor cells against a background of reactive lymphocytes may produce an appearance similar to non-Hodgkin's lymphoma and may necessitate immunohistochemical staining to demonstrate the epithelial nature of the malignant cells. Locally, nasopharyngeal carcinoma invades, spreads to lymph nodes in the neck, and then metastasizes far away. Even among patients with advanced cancer, survival rates at 5 years have been reported to be as high as 50% [1, 2].

8.9   Carcinoma of the Larynx

Laryngeal cancer accounts for only 2% of all malignancies. It occurs most frequently after the age of 40 and is seven times more prevalent in men than in women. Environmental factors have a significant impact on its genesis. Asbestos exposure and alcohol consumption may also play a role. Almost all cases have been found in smokers. Human papillomavirus sequences have been identified in approximately 15% of tumors, which have a more favorable prognosis than other malignancies.

Clinically, laryngeal carcinoma is characterized by persistent hoarseness. The location of laryngeal malignancies significantly influences the prognosis. At the time of diagnosis, approximately 90% of glottic malignancies are confined to the larynx. First, they develop symptoms early in the disease's progression due to vocal cord dysfunction. Second, the glottic region has a limited lymphatic supply, and diffusion beyond the larynx is uncommon. In contrast, the supraglottic larynx has numerous lymphatic spaces, and nearly one-third of malignancies have spread to regional lymph nodes (cervical). Subglottic tumors are typically clinically stable and manifest as advanced disease. With surgery, radiation, or a combination of treatments, many patients can be cured of the disease, but approximately one-third succumb to it. Common causes of mortality included distal respiratory infections or widespread infections, as well as metastases and cachexia [1, 2].

8.10   Pleural Lesions

With rare exceptions, pathological involvement of the pleura is secondary to the underlying pulmonary disease. At autopsy, secondary infection and pleural adhesion are common findings. Important primary diseases include (1) primary intrapleural bacterial infection and (2) primary malignant pleural tumors.

8.10.1   Pleural Effusion and Pleuritis

The fluid in a pleural effusion (the presence of fluid in the pleural cavity) may be transudate or exudate. When pleural effusions leak, they are referred to as pleural effusions. Exudates with a protein concentration greater than 2.90 g/dL, which typically contain inflammatory cells, are indicative of pleurisy. The four primary causes of pleural exudation are as follows: (1) microbial invasion through direct extension of pulmonary infection or blood-borne seeding (suppurative pleurisy or empyema), (2) cancer (lung cancer, metastatic tumor of lung or pleural surface, mesothelioma), (3) pulmonary infarction, and (4) viral pleurisy. Infrequent causes of exudative pleural effusion include systemic lupus erythematosus, rheumatoid arthritis, uremia, and previous thoracic surgery. Hemorrhagic pleurisy is characterized by a large volume and is frequently bleeding. Examination of cytology reveals malignant and inflammatory cells.

8.10.2   Malignant Mesothelioma

Malignant mesothelioma is a rare mesenchymal cell carcinoma that typically develops in the parietal pleura or visceral pleura; however, it can also develop in the peritoneum and pericardium, albeit less frequently. It is significant due to its association with occupational asbestos exposure. Approximately 50% of cancer patients with malignant mesothelioma have been exposed to asbestos. Those who work directly with asbestos (such as shipyard workers, miners, and insulators) are at the greatest risk, but malignant mesothelioma can also develop in those who come into contact with asbestos in the vicinity of asbestos factories or through relatives of asbestos workers. The extended incubation period for malignant mesothelioma development, typically 25–40 years after initial asbestos exposure, suggests that tumor transformation requires multiple somatic genetic events.

Morphology: On computed tomography, extensive pleural fibrosis and plaque formation are commonly observed alongside malignant mesothelioma. These tumors originate locally and disseminate through continuous growth or diffuse dissemination on the surface of the pleura over time. At autopsy, the afflicted lung is typically surrounded by a yellowish-white, firm, and occasionally gelatinous tumor (Fig. 12.50). Distant metastasis is uncommon. The tumor may invade the thoracic wall or subpleural lung tissue directly. Normal mesothelial cells are biphasic, generating both pleural lining cells and fibrous tissue underneath. Therefore, histologically, mesothelioma conforms to one of three patterns: (1) biphasic, (2) sarcomatous area, and (3) epithelial area [1, 2].

References

1. Kumar V, Cotran RS, Robbins SL. Bobbins basic pathology. 9th ed. Philadelphia: W.B Saunders; 2007.
2. Kumar V, Abbas AK, Fauston N, et al. Bobbins and Cotran pathologic basis of disease. 8th ed. Philadelphia: Saunders Elsevier; 2009. p. 677–737.

Chapter 9
Digestive System Diseases

Yihui Ma and Jingjing Xu

Contents

Y. Ma (✉) · J. Xu
School of Basic Medical Sciences, Zhengzhou University, Zhengzhou, China

© Zhengzhou University Press 2024
K. Chen et al. (eds.), *Textbook of Pathologic Anatomy*,
https://doi.org/10.1007/978-981-99-8445-9_9

Objectives

1. To master the pathological changes and clinicopathological relationships of common inflammatory lesions of the digestive tract, such as peptic ulcer, chronic gastritis, and viral hepatitis
2. To master the common types, pathological features, and clinicopathological relationships of liver cirrhosis
3. To master the pathological characteristics of common carcinomas of the digestive tract, such as gastric carcinoma, hepatocellular carcinoma, colorectal carcinoma, and esophageal carcinoma
4. To comprehend the pathological characteristics of cholecystitis, pancreatitis, inflammatory bowel disease, and cholecystitis

Key Concepts

1. **Barrett esophagus** is defined as the replacement of the normal stratified squamous epithelium of the lower esophagus by metaplastic columnar epithelium containing goblet cells. Ulceration and canceration may occur.
2. **Inflammatory bowel disease (IBD).** Regional enteritis and ulcerative colitis are two main types of IBD. Regional enteritis, also known as Crohn's disease, mostly affects the terminal ileum and/or colon, although any part of the gastrointestinal tract may be involved. The characteristic features include transmural inflammatory cells; non-caseating, sarcoid-like granulomas; patchy ulceration of the mucosa; and widening and fibrosis of all layers in the submucosa. In comparison, ulcerative colitis is characterized by continuous involvement of the rec-

tum and colon without any uninvolved skip areas. Microscopically, superficial mucosal ulcerations can be seen in the intestinal mucosa in the early stage, penetrating into the muscle coat in severe cases, accompanied by nonspecific inflammatory cell infiltration of neutrophils, lymphocytes, plasma cells, and eosinophils in the mucosa and submucosa. In the late stage, fibrous tissue proliferates.

3. **Viral hepatitis** is a common infectious disease caused by a group of hepatitis viruses, characterized by the main pathological changes of liver parenchyma cell degeneration and necrosis. It has been confirmed that viral hepatitis can be caused by hepatitis viruses A, B, C, D, E, and G. Among them, hepatitis B caused by hepatitis B virus infection is more common. All types of hepatitis show similar changes: hepatocellular degeneration, necrosis, varying degrees of inflammatory cell infiltration, hepatocellular regeneration, and interstitial fibrosis.

4. **Bridging necrosis** is characterized by bands of necrosis linking portal tracts to central hepatic veins, one central hepatic vein to another, or a portal tract to another tract, which is commonly seen in moderate and severe chronic hepatitis.

5. **Liver cirrhosis** is a common chronic liver disease caused by diffuse degeneration and necrosis of liver cells, fibrous tissue hyperplasia, and nodular regeneration of hepatocytes. Advanced patients often have different levels of increased portal pressure and liver dysfunction, which are very harmful to the human body.

Introduction

The digestive system consists of two main parts: the digestive tract and the digestive glands. The digestive tract includes the oral cavity, pharynx, esophagus, stomach, small intestine (duodenum, jejunum, ileum), and large intestine (cecum, appendix, colon, rectum, anal canal). Four of the ten most serious malignancies in China are from the digestive system, including esophageal carcinoma, gastric carcinoma, liver carcinoma, and colorectal carcinoma. Inflammation of the digestive system is also common, such as appendicitis, cholecystitis, cholelithiasis, and acute pancreatitis. This chapter will discuss diseases of the gastrointestinal tract, the liver and biliary tract, and the pancreas.

9.1 Esophagitis

9.1.1 Reflux Esophagitis

Reflux esophagitis is defined as chronic inflammation in the lower esophageal mucosa caused by reflux of gastric juice. It is also known as gastroesophageal reflux disease (GERD).

9.1.1.1 Pathological Changes

Local hyperemia is common, and severe damage causes obvious hyperemia. Epithelial hyperplasia and infiltration of neutrophils and eosinophils can be found at an early stage, and sometimes local epithelial necrosis may develop into superficial ulcers. Fibrosis occurs when inflammation spreads to the esophageal wall. Barrett esophagus can develop as a result of chronic inflammation.

9.1.1.2 Clinical Features

The clinical manifestations include nausea, heartburn, pain, and dysphagia, sometimes accompanied by vomiting and melena. However, the severity of the clinical symptoms is not necessarily consistent with the histological changes of esophagitis.

9.1.2 Barrett Esophagus

Barrett esophagus is defined as the replacement of the normal stratified squamous epithelium of the lower esophagus by metaplastic columnar epithelium containing goblet cells. Ulceration and canceration may occur.

9.1.2.1 Pathological Changes

Barrett esophageal mucosa presents visible orange-red, velvet-like irregular shaped lesions, appearing as a patch-like island or ring on the background of normal esophageal mucosa. Esophagus stenosis and hiatus hernia may be secondary to erosive ulcers.

Barrett esophageal mucosa has similar mucosal epithelial cells and glands to those of the gastric intestinal metaplasia. It can be diagnosed when intestinal goblet cells are found in columnar epithelial cells. In Barrett esophagus, columnar epithelial cells have the ultrastructural and cytochemical characteristics of both squamous and columnar epithelial cells.

9.1.2.2 Clinical Features

The major complications of Barrett esophagus include peptic ulcer, stricture, and bleeding, which are similar to reflux esophagitis. Sometimes, atypical hyperplasia and adenocarcinoma may occur.

9.2 Gastritis

Gastritis is defined as an inflammatory disease of the gastric mucosa. It is common and can be divided into acute and chronic gastritis. Acute gastritis often has definite causes, while the causes and pathogenesis of chronic gastritis are complex and unclear at present.

9.2.1 Acute Gastritis

Acute gastritis is often caused by physical and chemical factors and microbial infections. It usually has four types:

1. **Acute irritant gastritis**: It is usually caused by consuming too much hot or irritant food or drinking strong alcohol. Under gastroscopy, red mucosa, congestion and edema, mucus covering the surface, and sometimes erosion can be found.
2. **Acute hemorrhagic gastritis**: It is often caused by improper medication or excessive consumption of alcohol, in addition to stress response to trauma and surgery. Mucosal acute bleeding, mild erosion, and multiple superficial stress ulcers may be seen.
3. **Corrosive gastritis**: It is often caused by corrosive chemical agents. The mucosa becomes necrotic and dissolved. The lesion may involve deep tissue or even lead to perforation.
4. **Acute infective gastritis**: When pyogenic bacteria such as Staphylococcus aureus, Streptococcus, or Escherichia coli directly infect the stomach through the blood or stomach, it can cause acute phlegmonous gastritis.

9.2.2 Chronic Gastritis

Chronic gastritis is a chronic nonspecific inflammation of the gastric mucosa and has a high incidence.

9.2.2.1 Causes and Pathogenesis

The pathogens of chronic gastritis are still unknown. Pathogenic factors can be divided into four types:

1. Helicobacter pylori infection: Helicobacter pylori is a gram-negative bacillus that exists on the gastric epithelial surface or in the mucus layer of the glands in affected patients. It can secrete urease, cytotoxin-related proteins, and vacuolating cytotoxin, which can cause chronic gastritis.

2. Chronic stimuli: This includes long-term drinking and smoking, abuse of salicylic acid, eating hot or strongly acidic and spicy food, and recurrent acute gastritis.
3. The destruction of the gastric mucosa is due to duodenal liquid reflux.
4. Autoimmune injury damages the gastric mucosa.

9.2.2.2 Types and Pathological Changes

According to different pathological changes, it can be divided into four types:

1. **Chronic Superficial Gastritis**

   Also known as chronic simple gastritis, this type commonly occurs in the antrum and may be multifocal or diffuse. It is one of the most common gastric mucosa diseases. Under gastroscopy, congestion, edema, and sometimes slight bleeding or erosion can be seen. Grayish-yellow or white mucus exudates cover the surface. Chronic inflammatory cells such as lymphocytes and plasma cells infiltrate the lamina propria. Most patients can be cured after therapy or a reasonable diet. A few cases can develop into chronic atrophic gastritis.

2. **Chronic Atrophic Gastritis**

   The gastric mucosa becomes atrophic and thins, with mucosal glands decreasing or disappearing and intestinal metaplasia occurring. Lymphocytes and plasma cells infiltrate the lamina propria.

   There are complex causes for the disease. Some may be associated with smoking, drinking, or improper drug use; some may originate from chronic superficial gastritis; and some are autoimmune diseases. Depending on whether it is associated with autoimmunity or accompanied by pernicious anemia, it can be divided into types A and B. Type A is an autoimmune disease, mainly affecting the body and fundus of the stomach, with pernicious anemia. Antibodies against parietal cells and intrinsic factor are positive. Type B lesions are more common in the gastritis antrum and do not involve pernicious anemia. In China, type B is more common. Mucosal lesions are similar in these two types of gastritis (Table 9.1).

   Under gastroscopy, the mucosa appears gray or gray-green; mucosal folds become thin or disappear; the surface is finely granular; submucosal blood vessels are visible, with rare hemorrhage and erosion.

   Microscopically, lymphocytes and plasma cells infiltrate the lamina propria with the formation of lymphatic follicles. The stomach mucosa becomes thinner and glands are atrophic and sparsely distributed with cystic dilatation. Fibrous tissue hyperplasia can be observed in the gastric mucosa. False pyloric gland and intestinal metaplasia can be found. Intestinal metaplasia refers to the replacement of gastric mucosal epithelium with intestinal epithelium. Atypical hyperplasia can be found in intestinal metaplasia. If there are both goblet cells and absorptive epithelial cells, it is called complete metaplasia. If there are only goblet cells, it is called incomplete metaplasia. It has been reported that incomplete

Table 9.1 Comparison of two types of chronic atrophic gastritis

	Type A	Type B
Etiology and pathogenesis	Autoimmunity	Infection with Helicobacter pylori (60–70%)
Location	Gastric body or fundus	Gastric antrum
Anti-parietal cell and anti-intrinsic factor antibody	Positive	Negative
Serum gastrin level	High	Low
Hyperplasia of G cells	Yes	No
Autoantibody	Positive (>90%)	Negative
Secretion of gastric acid	Significant reduction	Moderate reduction or normal
Serum vitamin B12 level	Reduction	Normal
Malignant anemia	Common	None
Peptic ulcer	None	Common

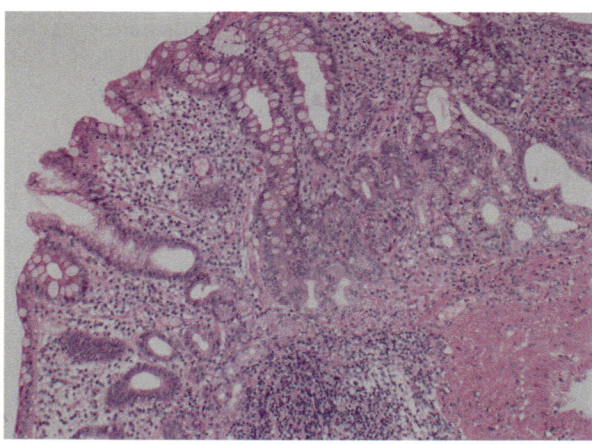

Fig. 9.1 Chronic atrophic gastritis. The gastric mucosa becomes atrophic and thins. Mucosal glands become fewer or disappear, with intestinal metaplasia. Lymphocytes and plasma cells infiltrate the lamina propria

metaplasia in the large intestine is closely related to intestinal carcinoma of the stomach (Fig. 9.1).

False pyloric gland metaplasia refers to the replacement of parietal cells and chief cells in the fundus and body of the stomach with mucin-producing cells.

Clinicopathological Relationship: Secretion of gastric juice is reduced due to the decrease or disappearance of parietal cells and chief cells. Patients suffer from dyspepsia, poor appetite, and upper abdominal discomfort. Type A patients are prone to have malignant anemia because of the significant destruction of parietal cells, lack of intrinsic factor, and impaired absorption of vitamin B12.

Atrophic gastritis is usually accompanied by varying degrees of intestinal metaplasia. During metaplasia, it is inevitably accompanied by the proliferation of local epithelial cells, which may lead to stomach carcinoma.

3. **Chronic Hypertrophic Gastritis**

    Also known as giant hypertrophic gastritis or Menetrier-like disease, the pathogenesis is unknown. It often occurs in the body and fundus of the stomach. Under gastroscopy, mucosal folds become deepened, widened, and coarse, with a gyrus-like appearance. Transverse cracking on mucosal folds is visible with numerous verrucous protrusions; the tops of mucosal protrusions are often accompanied by erosion. Gland hypertrophy and hyperplasia, as well as duct extension, are visible. Sometimes, hyperplastic glands break through the muscularis mucosa. Mucus secretion on the mucosal surface is increased with a higher number of mucin-producing cells. In the lamina propria, inflammatory cell infiltration is inconspicuous.

4. **Gastritis Verrucosa**

    Its etiology is unknown. It is a characteristic lesion that usually occurs in the antrum. Numerous verrucous lesions with depressed centers appear on the gastric mucosa. Under the microscope, gastric epithelial degeneration, shedding, and necrosis are noted, accompanied by acute inflammatory exudate.

9.3 Peptic Ulcer Disease

It is characterized by chronic ulcers in the gastric or duodenal mucosa and is related to the self-digestion by gastric juice. It is more common in adults (20–50 years old and is usually a chronic recurrent disease. Duodenal ulcers are more common than gastric ulcers. The former accounts for 70%, the latter 25%, and mixed ulcers only 5%. Patients are prone to periodic abdominal pain, acid reflux, and belching.

9.3.1 Causes and Pathogenesis

The causes are thought to be related to the following factors:

1. **Infection with Helicobacter pylori**

    A large number of studies have shown that Helicobacter pylori play an important role in the pathogenesis of ulcers in the following ways:

    (a) They can release a bacterial platelet-activating factor, which promotes the formation of thrombosis in capillaries, leading to blood vessel obstruction, mucosal ischemia, and so on. These effects may destroy the gastric and duodenal mucosal defense barrier.

    (b) They secrete urease, which catalyzes the production of free ammonia and proteinase. This contributes to the cleavage of glycoprotein in the gastric mucosa, facilitating direct contact between gastric acid and the epithelium and mucus membrane, promoting the proliferation of G cells in the gastric mucosa and increasing gastric acid secretion.

(c) They also have chemotactic effects on neutrophils. Neutrophils release myeloperoxidase, which then produces hypochlorous acid. If ammonia is present, ammonia chloride is synthesized. Hypochlorous acid and ammonia chloride destroy mucosal epithelial cells and induce peptic ulcers (Fig. 9.2).

2. **Decrease of Mucosal Digestive Ability**

   Under normal conditions, the mucus secreted by the gastric mucosa (mucus barrier) and the lipoprotein of mucosal epithelial cells (mucosal barrier) protect the mucus membrane from being digested by gastric juice. When gastric acid secretion is insufficient or the mucosa is damaged, barrier functions are weakened, reducing resistance to digestion. Hydrogen ions in gastric juice diffuse into the gastric mucosa, damaging capillaries in the mucosa and inducing mast cells to release histamine, leading to local blood circulation disorders and gastric mucosa damage. Secretion of pepsinogen increases, which strengthens the digestive function of gastric juice and leads to ulcer formation. The diffusing capacity of hydrogen ions in the gastric antrum is 15 times higher than in the gastric fundus, while in the duodenum, it is 2–3 times higher than in the gastric antrum. That may explain the higher morbidity in the duodenum and gastric antrum compared to other parts.

   In addition, long-term use of nonsteroidal anti-inflammatory drugs, smoking, and other factors that damage the mucus membrane barrier can induce peptic ulcer disease.

3. **Digestive Effect of Gastric Juice**

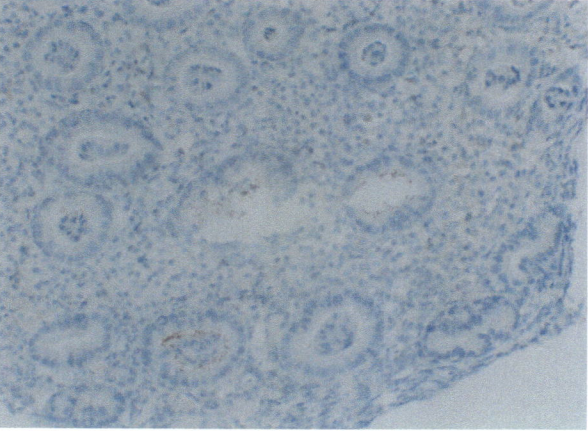

Fig. 9.2 Helicobacter pylori. Pyloric Helicobacter pylori can be seen in chronic active inflammation of the gastric antrum. (Immunohistochemical detection of Helicobacter pylori)

It has been proven that the disease results from the digestion of stomach and duodenal mucosa tissue by gastric acid and pepsin. In duodenal ulcers, the number of parietal cells increases significantly, and the increased gastric acid digests the mucus membrane. The environment of the jejunum and ileum is alkaline, where ulceration rarely occurs. However, after gastrojejunostomy, ulcers may occur at the anastomosis due to the digestion of gastric juice.

4. **Neuroendocrine Dysfunction**

   Mental stimulation causes functional dysfunction of the cerebral cortex, which leads to dysfunction of the autonomic nervous system. When a duodenal ulcer occurs, vagus hyperfunction increases gastric acid secretion. In contrast, in gastric ulcers, vagus nerve excitation decreases, gastric peristalsis weakens, and the secretion of gastric acid is increased by gastrin.

5. **Genetic Factors**

   Ulcers have a high tendency in some families, suggesting that the occurrence of this disease may also be associated with genetic factors.

9.3.2 Pathological Changes

Grossly, gastric ulcers occur at the lesser curvature of the stomach near the pylorus, most commonly in the gastric antrum area; the closer the ulcer is to the stomach pylorus, the more common it is. Usually, the ulcer is single, round or ellipse, about 2 cm in diameter, with neat edges, a clean flat bottom, and often extends deep into the muscular layer, sometimes reaching the serosa. Generally, the ulcer is deep on the side of the cardia and shallow at the edge. The mucosal folds become radial due to the traction of scar formation at the bottom of the ulcer (Table 9.3).

Under the microscope, ulcers can be divided into four layers form inside to outside: inflammatory exudate such as leukocyte and cellulose, necrotic tissue, granulation tissue, and scar tissue (Fig. 9.3).

Table 9.3 Differentiation of gross appearance between benign ulcers and malignant ulcers

	Benign ulcers (Gastric ulcer)	Malignant ulcers (Ulcerative gastric cancer)
Gross appearance	Round or ellipse	Irregular
Size (diameter)	<2 cm	>2 cm
Depth	Deep	Shallow
Marginal	Regular and flat	Irregular and bulging
Bottom	Flat	Irregular, necrotic, and obvious hemorrhage
Peripheral mucous	Mucus folds concentrating to the ulcer	Mucus fold interruption and nodular hypertrophy

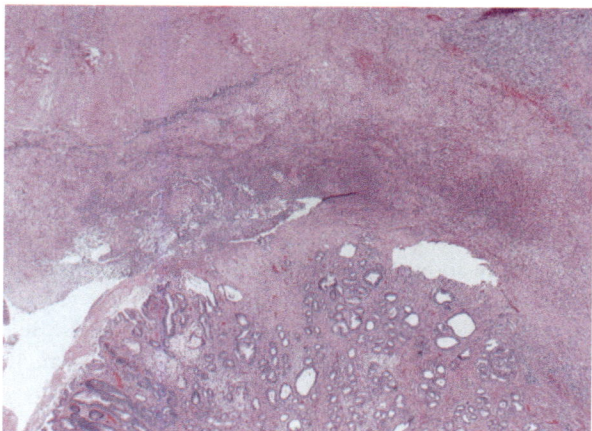

Fig. 9.3 Gastric peptic ulcer. The most superficial ulcer is covered by a small amount of inflammatory exudate such as leukocytes and cellulose; next is necrotic tissue; fresh granulation tissue is seen in the third layer; and in the fourth layer, granulation tissue turns into old scar tissue

Duodenal ulcers are similar to gastric ulcers but primarily occur in the anterior or posterior wall of the duodenum. The ulcers are small, less than 1 cm in diameter, and are easier to heal.

9.3.3 Outcomes and Complications

1. **Healing**

   Exudates and necrotic tissues will be gradually absorbed or expelled if they do not recur. The damaged muscle layer cannot regenerate and will be replaced with scar tissue formed by the hyperplasia of granulation tissue at the bottom. The surrounding mucosal epithelium will regenerate and cover the ulcers.

2. **Complications**

   Hemorrhage (accounts for about 10–35%): Fecal occult blood tests of the patients are positive due to the rupture of capillaries at the bottom of the ulcer. If the large blood vessel ruptures, hematemesis, tarry stools, or even hemorrhagic shock may occur.

   Perforation (accounts for about 5%): Duodenal ulcers are more prone to perforation due to their thinner wall. Gastrointestinal contents leak into the abdominal cavity, causing peritonitis. If perforation occurs in the posterior wall of the stomach, the contents of the stomach leak into the lesser omentum sac.

   Pyloric Stenosis (accounts for about 3%): Prolonged ulcers tend to form a large number of scars. Shrinking of these scars cause pyloric stenosis, making it difficult for gastric contents to pass through and leading to secondary gastric dilatation. Patients may experience frequent vomiting and even alkalosis.

Cancerization (accounts for less than 1%): Cancerization is commonly seen in patients with long-term gastric ulcers and is rarely seen in duodenal ulcers.

9.3.4 Clinicopathological Relationship

Gastric acid stimulates local nerve endings in gastric ulcers, resulting in periodic upper abdominal pain. In addition, it is related to the spasm of smooth muscle in the gastric wall. In duodenal ulcers, increased excitation of the vagus nerve leads to increased secretion of gastric acid, so pain often occurs at night. Belching and acid regurgitation are related to factors such as pyloric sphincter spasm, gastric antiperistalsis, and early pyloric stenosis.

9.4 Appendicitis

Appendicitis is a common disease. Its main clinical manifestations include persistent right lower abdominal pain, vomiting accompanied by elevated body temperature, and elevated peripheral blooding neutrophils. According to the course of the disease, it can be divided into acute and chronic appendicitis.

9.4.1 Causes and Pathogenesis

The appendix is a slender blind tube with a narrow cavity. Feces and bacteria from the intestinal cavity are easily retained. The wall of the appendix is rich in nerve tissue, such as the myenteric plexus. The end of the appendix has a structure similar to a sphincter, so it can shrink and the cavity can become narrower when stimulated.

Appendicitis is caused by bacterial infection, but there is no specific pathogenic bacterium. In 50–80% of cases, appendicitis is accompanied by obstruction of the appendix cavity. Feces and parasites can cause mechanical obstruction. Various stimuli induce contraction of the appendix, leading to blood circulation disorders of the appendix wall and loss of mucosa. These changes promote bacterial infection.

9.4.2 Pathological Changes

1. **Acute Appendicitis**

    It mainly has three subtypes:

    Acute simple appendicitis: This occurs in the early stage. Mild swelling of the appendix and serosal hyperemia are noted. Neutrophil infiltration, fibrin exudates, and inflammatory edema can be seen, especially severe in the mucosa and submucosa.

Acute phlegmonous appendicitis: This develops from simple appendicitis. There is significant swelling and serosal hyperemia. The surface of the appendix is covered with purulent exudation. Microscopically, neutrophils, inflammatory edema, and fibrous exudation extend from the surface layer to the deep layers, directly affecting the muscular and serous layers. Patients may have periappendicitis or localized peritonitis (Fig. 9.4).

Acute gangrenous appendicitis: It is a serious form of appendicitis. Obstruction, abscess, increased pressure in the cavity, and thrombophlebitis induced by appendicular vein inflammation contribute to the disturbance of blood circulation and appendix necrosis. The appendix appears dark red or black, and the lesion may perforate the wall, leading to periappendicitis or localized peritonitis.

2. **Chronic Appendicitis**

   The main lesions are varying degrees of fibrosis and infiltration by chronic inflammatory cells. Patients may experience right lower abdominal pain, which can occasionally present as an acute attack.

9.4.3 Outcomes and Complications

Acute appendicitis generally has a good prognosis with surgical treatment. The main complications include acute diffuse peritonitis and periappendiceal abscess, and occasionally thrombophlebitis or liver abscess. Liver abscess occurs when bacteria or bacterial emboli flow into the liver. If the proximal end of the appendix is blocked, the distal end may become highly expanded, leading to appendiceal abscess or appendiceal mucocele. If the mucocele ruptures, mucus can spill into the peritoneum, forming pseudomyxoma.

Fig. 9.4 Suppurative appendicitis. The inflammatory lesions extend from the surface layer to the deep layer. Each layer is infiltrated by a large number of neutrophils, and inflammatory edema and fibrous exudation are noted

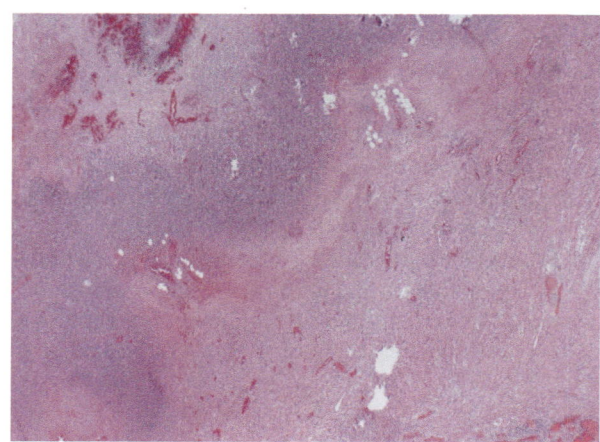

9.5   Inflammatory Bowel Disease

Inflammation bowel disease (IBD) can occur at any age. Regional enteritis and ulcerative colitis are the two main types of IBD.

9.5.1   Regional Enteritis

Regional enteritis is also known as Crohn's disease. It is a systemic disease that mainly involves the terminal ileum. Patients may experience abdominal pain, diarrhea, abdominal mass, perforated ulcer, intestinal fistula, and intestinal obstruction. Immune diseases outside the intestine, such as migrating polyarthritis and ankylosing spondylitis, may also be present.

9.5.1.1   Causes and Pathogenesis

Anticolon antibodies can be detected in the blood of patients. Immune complex deposits can be detected in the site of the lesion by immunofluorescence and enzyme-linked immunosorbent assay (ELISA). Although the causes are unclear, the condition may be related to immune disorders.

9.5.1.2   Pathological Changes

1. **Gross appearance**. The lesions mostly occur in the segment of the terminal ileum and/or colon, though any part of the gastrointestinal tract may be involved. Well-demarcated segments of bowel involvement are interspersed with uninvolved "skip areas." Due to fibrosis and stenosis, the wall of the affected bowel segment becomes thick and hard. The lumen of the affected segment is markedly narrowed.
2. **Microscopic changes**. The lesions are complex and diverse. The characteristic features are as follows:
   (a) Transmural inflammatory cells include chronic inflammatory cells (lymphocytes, plasma cells, and macrophages).
   (b) Non-caseating, sarcoid-like granulomas are present in all layers of the affected bowel wall in 50% of cases.
   (c) Patchy ulceration of the mucosa may form deep fissures, accompanied by lymphocytes and plasma cells infiltrating.
   (d) Due to edema and aggregation of lymphoid foci, the submucosa is widened.
   (e) In chronic cases, fibrosis becomes increasingly prominent in all layers.

9.5.2 Ulcerative Colitis

Ulcerative colitis is a chronic inflammation affecting mainly the mucosa and submucosa of the rectum and descending colon, though it may sometimes involve the entire length of the large bowel and occasionally the ileum. The disease is often accompanied by immune diseases outside the intestine, such as roving polyarthritis, uveitis, and primary sclerosing cholangitis. Patients may experience abdominal pain, diarrhea, and bloody mucus in their feces.

9.5.2.1 Causes and Pathogenesis

The exact etiology of ulcerative colitis remains unknown. However, it is primarily considered an autoimmune disease. Antibodies to colonic cells can be detected in less than half of patients' serum.

9.5.2.2 Pathological Changes

Gross changes: Compared to Crohn's disease, the characteristic feature is the continuous involvement of the rectum and colon without any uninvolved skip areas. The appearance of the colon varies depending on the stage and intensity of the disease due to remissions and exacerbations. Initially, hyperemia, punctate hemorrhage, and small abscesses are seen in the colonic mucosa. Abscesses turn into shallow ulcers due to necrosis and abscission of the mucosa. Shallow ulcers merge to form deep ulcers and may even cause abscess beside the colon, peritonitis, or adhesion with adjacent organs. Due to hyperemia, edema, and hyperplasia, inflammatory "pseudopolyps" may form in the intervening intact mucosa. Microscopic examination shows superficial mucosal ulcerations in the intestinal mucosa in the early stage, which can penetrate into the muscle coat in severe cases, accompanied by nonspecific inflammatory cell infiltration of neutrophils, lymphocytes, plasma cells, and eosinophils in the mucosa and submucosa. In the late stage, fibrous tissue proliferates. In long-standing cases, atypic epithelial pseudopolyps may develop into carcinoma in situ and adenocarcinoma.

9.5.2.3 Complications

Complications of ulcerative colitis include malabsorption, formation of fistulas and strictures, toxic megacolon, and the development of malignancy in late cases. If the lesion is limited to the left colon, the cancer risk is low. If the lesion extends to the whole colon within 20 years, the cancer rate increases to 17%, and after 30 years, it increases to 50%.

9.5.3 *Acute Hemorrhagic Enteritis*

Acute hemorrhagic enteritis (AHE) is a pediatric emergency characterized by acute necrotizing hemorrhagic inflammation. The main clinical manifestations include abdominal pain, hematochezia, fever, vomiting, diarrhea, and even shock leading to death.

9.5.3.1 Causes and Pathogenesis

The exact etiology of AHE is not known. However, it has been reported that the disease is a severe allergic reaction (Schwartzman) caused by nonspecific infections such as bacteria, viruses, or their decomposed products.

9.5.3.2 Pathological Changes

It is characterized by obvious hemorrhage and necrosis, often segmental, most commonly and seriously affecting the jejunum and ileum. The bowel wall thickens, and edema is distinct from normal, with the surface covered by pseudomembrane. Secondary ulcers can cause intestinal perforation. Inflammatory cells infiltrate the submucosa. Smooth muscle fibers rupture in the muscle layer.

9.5.4 *Dysplastic Enteritis*

It is also known as antibiotic enteritis and is often caused by long-term use of broad-spectrum antibiotics.

9.6 Viral Hepatitis

Viral hepatitis is a common infectious disease caused by a group of hepatitis viruses, which lead to the degeneration and necrosis of liver parenchyma cells. It has been confirmed that viral hepatitis can be caused by hepatitis viruses A, B, C, D, E, and G. Among these, hepatitis B caused by hepatitis B virus infection is more common. Studies have shown that it is primarily caused by a cellular immune response, and the humoral immune response may not be significant. Viral hepatitis has a high incidence and is showing a rising trend. Epidemic areas are widespread, and individuals of all ages and genders can be affected, which seriously endangers human health (Table 9.2).

Table 9.2 Characteristics of each type of hepatitis virus and the corresponding hepatitis

Types of Hepatitis virus	Size and character	Incubation period (week)	Channel of infection	Chance of turning into chronic hepatitis	Fulminant hepatitis
HAV	27 nm, single strand	2–6	Intestinal tract	None	0.1–0.4%
HBV	43 nm, DNA	4–26	Intimate contact, blood transfusion, injection	5–10%	<1%
HCV	30–60 nm, single strand	2–26	The same as above	>70%	Seldom
HDV	Defective RNA	4–7	The same as above	<5% when coinfection 80% when overlapping infection	3–4% when coinfection 7–10% when overlapping infection
HEV	32–34 nm, single strand	2–8	Intestinal tract	None	20% concurring pregnancy
HGV	Single strand	Unknown	Blood transfusion, injection	None	Unknown

9.6.1 Etiology and Pathogenesis

The pathogenesis of viral hepatitis is complex and has not yet been fully elucidated, depending on a variety of factors, especially the immune state of the body.

9.6.1.1 Hepatitis A Virus (HAV)

HAV is a benign, self-limited disease with an incubation period of 2–6 weeks. It is characterized by infection of the digestive tract, which can be scattered or endemic. HAV reaches the liver through the portal vein system via the intestinal epithelium, where the virus replicates in liver cells and is secreted into the bile. Therefore, the virus can be detected in the feces. HAV does not directly damage cells but may cause liver cell damage through cellular immune mechanisms. HAV does not lead to chronic hepatitis or a carrier state and only uncommonly causes acute hepatic failure.

9.6.1.2 Hepatitis B Virus (HBV)

HBV was first linked to hepatitis in the 1960s when the Australia antigen (later known as HBV surface antigen) was identified. The mature HBV virion is a spherical, double-layered "Dane particle" with an outer surface envelope of protein, lipid, and carbohydrate. The genome of HBV is a partially double-stranded circular DNA molecule with 3200 nucleotides and four open reading frames: S, C, P, and X genes. The S gene codes for the surface envelope protein, hepatitis B surface antigen (HBsAg); this product is a major protein. HBsAg consists of three related proteins: large, middle, and small HBsAg. Infected hepatocytes can synthesize and secrete massive quantities of noninfective surface protein (mainly small HBsAg). The C gene codes for two nucleocapsid proteins, HBeAg and a core protein termed HBcAg. The P gene is the largest and codes for DNA polymerase. The X gene codes for HBxAg, which plays an important role in the occurrence of hepatocellular carcinoma.

There is strong evidence linking immune pathogenesis with hepatocellular damage. HBV generally does not cause direct hepatocyte injury. Instead, viral antigens (in particular, nucleocapsid proteins HBcAg and HBeAg) are attacked by host cytotoxic CD8+ T lymphocytes. HBV is a major cause of chronic hepatitis in China, which eventually leads to liver cirrhosis. It can also cause acute hepatitis B, acute severe hepatitis, and asymptomatic carriers. HBV is mainly transmitted through blood, blood-contaminated items, drug use, or close contact. Maternal-neonatal transmission is also evident in high-incidence areas.

Ground-glass-like hepatocytes: HBsAg carriers and chronic hepatitis patients show hepatocyte changes such as the presence of finely granular, ground-glass, eosinophilic cytoplasm, which are called ground-glass-like hepatocytes and are positive for HBsAg by immunohistochemistry and immunofluorescence tests. Under electron microscopy, smooth endoplasmic reticulum hyperplasia and an increased number of HBsAg particles in the endoplasmic reticulum are observed.

9.6.1.3   Hepatitis C Virus (HCV)

The main routes of transmission are injection or transfusion. HCV is a single-stranded RNA virus with six main genotypes. The most common genotypes are 1a, 1b, 2a, and 2b, with genotype 1b being closely related to hepatocellular carcinoma. Alcohol consumption can promote HCV replication and activation and liver fibrosis. The HCV virus can directly destroy liver cells. Numerous experiments show that immune factors are also an important cause of liver cell injury. Chronic disease occurs in the majority of HCV-infected individuals (80–90%), and cirrhosis eventually occurs in up to 20% of individuals with chronic HCV infection. Hepatocellular carcinoma may occur in some cases.

9.6.1.4   Hepatitis D Virus

Hepatitis D virus (HDV) is a replication-defective RNA virus that must rely on a complex infection with HBV to replicate. HDV infection and hepatitis B may be simultaneous (co-infection), or HDV may infect a chronic HBsAg carrier (superinfection). It is self -limited and is usually followed by clearance of both viruses.

However, there is a higher rate of acute hepatic failure, especially in intravenous drug users. HBV/HDV complex chronic hepatitis may be responsible for about 80% of cases, and acute severe hepatitis may occur.

9.6.1.5   Hepatitis E Virus

HEV is a single-stranded RNA virus. Hepatitis E is mainly transmitted through the digestive tract and is prevalent during the rainy season and after floods. It often appears in autumn and winter (10–11 months). Sporadic cases occur throughout the year in areas with poor environmental and water hygiene conditions. HEV sometimes infects middle-aged and older people over 35 years old. The proportion of severe HEV hepatitis in pregnancy is high. HEV is mainly prevalent in developing countries such as those in Asia and Africa, especially in India and other countries. HEV infection has a particularly high mortality rate in pregnant women (about 20%) but is otherwise a self-limited disease and has not been associated with chronic liver disease.

9.6.1.6   Hepatitis G Virus

HGV infection primarily occurs in dialysis patients, mainly transmitted through contaminated blood or blood products, and may also be transmitted through sexual contact. Some patients may develop chronic diseases. Whether HGV is a true hepatitis virus is still controversial. It is believed that HGV can replicate in mononuclear cells, so it may not necessarily be classified as a hepatitis virus.

9.6.2   Basic Pathological Changes

All types of hepatitis have similar changes: hepatocellular degeneration, necrosis, varying degrees of inflammatory cell infiltration, hepatocellular regeneration, and interstitial fibrosis.

9.6.2.1   Hepatocyte Degeneration

1. **Cellular Swelling**: This is the most common lesion. Under the light microscope, the hepatocyte appears obviously enlarged, and the cytoplasm is reticular and translucent, a condition referred to as "cytoplasm loosening." With further development, the volume of hepatocytes expands from polygonal to round. The cytoplasm becomes almost completely transparent, a condition known as ballooning degeneration. Under the electron microscope, the endoplasmic reticulum shows varying degrees of expansion, mitochondria swelling, and increased lysosomes.

2. **Acidophilic Degeneration**: This type of degeneration generally affects only a single or a few hepatocytes and is scattered within the hepatic lobules. Under the light microscope, the cytoplasm becomes intensely eosinophilic, and the nucleus becomes small and pyknotic.

9.6.2.2 Hepatocyte Necrosis and Apoptosis

1. **Lytic Necrosis**: This develops from severe cell edema. The range and distribution of this type of necrosis in different types of viral hepatitis can be divided into the following:

    Spotty Necrosis: This refers to the necrosis of a single or a few hepatocytes and is common in acute common hepatitis (Fig. 9.5).

    Piecemeal Necrosis: This involves focal necrosis and disintegration of liver cells at the periphery of the hepatic lobule and is common in chronic hepatitis.

    Bridging Necrosis: This is characterized by bands of necrosis linking portal tracts to central hepatic veins, one central hepatic vein to another, or one portal tract to another. It is commonly seen in moderate and severe chronic hepatitis.

    Massive Necrosis: This refers to a large area of hepatocyte necrosis that involves the entire hepatic lobule and is common in severe hepatitis (Fig. 9.6).

2. **Apoptosis**: It has been considered eosinophilic necrosis in the past. With acidophilic degeneration, the cytoplasm becomes intensely eosinophilic, and the nucleus is eventually extruded from the cell, leaving behind a necrotic, acidophilic mass called an acidophil body as part of the apoptosis process.

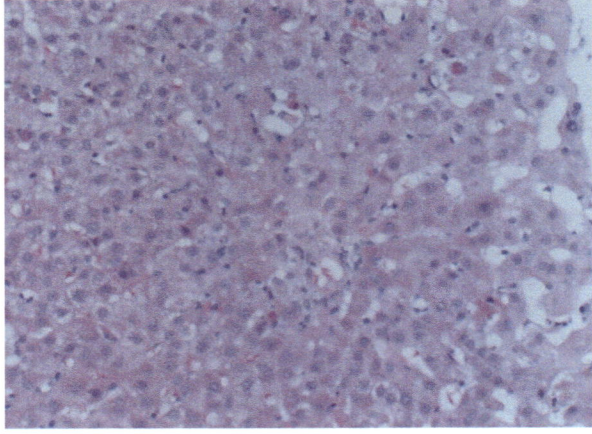

Fig. 9.5 Focal necrosis in the liver. It refers to the necrosis of a single or several hepatocytes, which is common in acute hepatitis

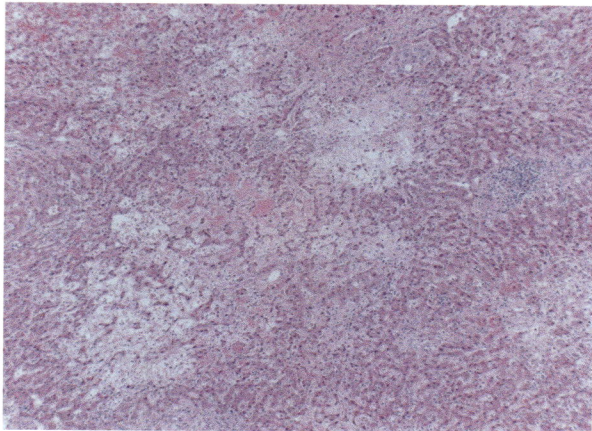

Fig. 9.6 Lytic necrosis in the liver. It refers to a large area of hepatocyte necrosis that involves the entire hepatic lobule and is common in severe hepatitis

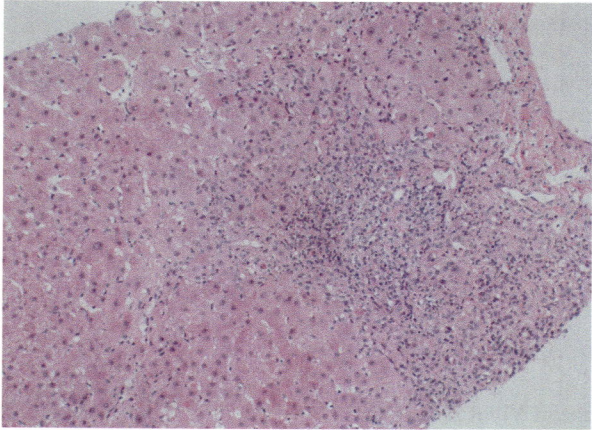

Fig. 9.7 Interfacial inflammation in the liver. Significant plasma cell infiltration of chronic inflammatory cells is found at the junction of hepatic lobules and the portal area

9.6.2.3 Inflammatory Cell Infiltration

Lymphocytes and mononuclear cells are primarily infiltrated in the hepatic lobule or the portal tracts (Fig. 9.7).

9.6.2.4 Regeneration

1. **Hepatocyte regeneration**: Necrotic hepatocytes are repaired by direct or indirect division of the surrounding liver cells. The regenerated hepatocytes are large in size, with basophilic cytoplasm, large and deeply stained nucleus, and sometimes binucleation. The regenerated hepatocytes can be arranged along the original framework. However, if necrosis is severe, the reticular scaffold in the primary lobule collapses, and the regenerated hepatocytes are organized into a bulky mass, called nodular regeneration.
2. **Interstitial reactive hyperplasia and small bile duct hyperplasia**: Kupffer cells proliferate and can migrate into the sinus cavity to become phagocytic cells that participate in inflammatory cell infiltration. Mesenchymal stem cells and fibroblasts also proliferate and are involved in the injury repair. In cases of severe hepatocellular necrosis, hepatic fibrosis and hepatic cirrhosis may develop due to the proliferation of a large number of fibroblasts, and small bile duct hyperplasia can be seen in the portal area or large necrotic foci.

9.6.2.5 Fibrosis

In general, fibrosis is mostly irreversible, but some believe that liver fibrosis can be absorbed in certain cases, making it reversible. The deposition of collagen in fibrosis significantly affects liver blood flow and hepatocyte perfusion. Early fibrosis can be distributed around the portal area or the central vein, or collagen can be directly deposited within the Disse cavity. As fibrosis progresses, the liver becomes divided into fibrous nodules, eventually developing into liver cirrhosis.

9.6.3 Clinicopathological Types

9.6.3.1 Common Viral Hepatitis

Divided into Two Types: Acute and Chronic

1. **Acute Common Hepatitis**: This can be further classified into jaundice and non-jaundice types. In our country, there are more non-jaundice cases, with hepatitis B virus being the predominant type. Jaundice-type hepatitis tends to be severe and has a shorter course. It is more common in hepatitis A, D, and E. The pathological changes in jaundice-type and non-jaundice-type hepatitis are essentially the same.

    Macroscopically, the liver is swollen, soft, and smooth.

    Microscopically, hepatocytes exhibit extensive degeneration, primarily characterized by cellular swelling and ballooning changes. Hepatocytes increase in size and become disordered and crowded, and the hepatic sinusoids are compressed and narrowed. Cholestasis can be observed in liver cells. Slight necrosis and eosinophilic bodies are found in the hepatic lobules. Mild inflammatory cell infiltration is noted in both the hepatic lobules and the portal area. Jaundice-type necrosis is often more severe, with cholestasis and embolus formation in the capillary bile ducts.

Clinicopathological relationship Diffuse hepatocyte swelling causes the liver volume to increase and the capsule to tighten, resulting in liver pain. The degeneration of liver cells leads to the release of liver enzymes into the blood and an increase in serum alanine aminotransferase, which can cause abnormal liver function and severe jaundice.

Conclusion Most hepatitis patients can be cured within 6 months, with spotty necrotic hepatocytes often completely regenerated and repaired. However, hepatitis B and C often recover slowly, with about 5–10% of hepatitis B and approximately 70% of hepatitis C cases progressing to chronic hepatitis.

2. **Chronic (Common) Hepatitis**

Chronic hepatitis is defined as continuing or relapsing hepatic disease for more than 6 months. Factors leading to chronic hepatitis include the type of virus infection, improper treatment, malnutrition, other infectious diseases, alcohol consumption, drug-induced liver damage, and immune factors, all of which clinicians should pay attention to. Historically, chronic hepatitis was classified into chronic persistent hepatitis and chronic active hepatitis. Currently, it has been noted that a high percentage of HCV patients progressed from chronic hepatitis to cirrhosis, regardless of the degree of the initial liver disease. Therefore, understanding the pathogenesis of chronic hepatitis is crucial. Based on the degree of inflammation, necrosis, and fibrosis, chronic hepatitis can be divided into three types:

(a) **Mild Chronic Hepatitis**: Spotty necrosis, few chronic inflammatory cell infiltrations in the portal area, and a small amount of fibrous tissue proliferation around. The hepatic plate of the lobule is integrated and the structure of the lobule is clear.

(b) **Moderate Chronic Hepatitis**: Liver cell degeneration and necrosis are evident, with moderate fragment necrosis and characteristic bridging necrosis. There is a fibrous septum in the lobule, but the structure of the lobule is mostly preserved.

(c) **Severe Chronic Hepatitis**: There is severely fragmented necrosis and extensive bridging necrosis. In the necrotic area, hepatocytes exhibit irregular regeneration, and the fibrous septum separated the hepatic lobule structure.

In the late stage, the condition gradually transforms into liver cirrhosis. If a large fresh area of necrosis occurs on the basis of chronic hepatitis, it will progress to severe hepatitis.

9.6.3.2 Severe Viral Hepatitis

The most serious type of viral hepatitis is rare. It is categorized into two types based on the disease course and severity: severe acute and subacute severe.

1. **Acute Severe Hepatitis**: This type is rare, with an abrupt onset and a short duration, typically lasting about 10 days. It is characterized by severe lesions and high mortality. Clinically, it may be referred to as outbreak type, electric shock type, or malignant hepatitis.

   Macroscopically, the liver volume is significantly reduced, especially in the left lobe. The capsule is wrinkled with a soft texture, and the liver appears yellow or reddish-brown with some areas showing red and yellow stripes. This appearance is also referred to as acute yellow atrophy or acute atrophy of the red liver.

   Microscopically, necrosis of hepatocytes is extensive and severe, with liver cell cords dissociated and liver cells dissolved. Necrosis begins at the center of the hepatic lobule and spreads rapidly to the periphery. Only a few degenerated hepatocytes remain around the lobule. The necrotic hepatocytes were quickly removed, leaving only reticulated scaffolds. The hepatic sinusoids were dilated, showing hyperemia and even bleeding. Kupffer cells proliferated and hypertrophied, with active phagocytosis. A large number of lymphocytes and macrophages infiltrated the hepatic lobule. Several days later, the reticular scaffold collapsed, and the residual hepatocytes showed no obvious regeneration. The extensive necrosis of liver cells can lead to several conditions: a large amount of bilirubin entering the blood, causing severe hepatocellular jaundice; impaired synthesis of coagulation factors, resulting in a significant bleeding tendency; liver failure, with disrupted detoxification of various metabolites leading to hepatic encephalopathy. In addition to bilirubin metabolic disorders and blood circulation issues, it can also induce renal failure (hepatorenal syndrome).

   Outcome: Most patients died in the short term. The main cause of death was liver failure (hepatic encephalopathy), followed by massive hemorrhage of the digestive tract, renal failure and disseminated intravascular coagulation. A small number of patients with delayed healing progressed to subacute severe hepatitis.

2. **Subacute Severe Hepatitis**: The onset is slower than in acute severe hepatitis, and the duration is longer (weeks to months). Most cases progress from severe acute hepatitis, while a few are caused by the worsening of acute common hepatitis.

   Macroscopically, the liver size is reduced, the surface of the capsule is uneven, the texture quality varies, and parts of the liver have nodular regions of different sizes. The necrotic areas are red-brown or yellow, and the regenerated nodules are yellow-green due to cholestasis.

   Microscopically, it is characterized by extensive necrosis of hepatocytes and regeneration of nodular hepatocytes. In the necrotic areas, reticular fiber scaffolds collapse and become collagenated (sclerotic). Therefore, the remaining hepatocytes cannot be arranged along the original scaffolds and form nodules. There is significant infiltration of inflammatory cells inside and outside the hepatic lobules, mainly lymphocytes and monocytes. Small bile ducts are present in the periphery of the lobules, and connective tissue hyperplasia is observed in older lesions.

Outcome: If treated reasonably and in a timely manner, the disease can stop progressing and may be cured. Most cases, however, often develop into necrotic cirrhosis.

9.6.4   Characteristics of Various Types of Viral Hepatitis

9.6.4.1   Pathological Changes of Hepatitis A

It mainly causes acute hepatitis and, less commonly, cholestatic and severe hepatitis. The main pathological changes are as follows:

1. **Degeneration and necrosis of liver cells**: The most common are early hepatocyte swelling and ballooning degeneration, accompanied by the formation of eosinophilic bodies in hepatocytes and the disappearance of hepatic sinusoids, leading to disordered liver cells in lobules. Liver cells in the central vein of the hepatic lobules may be dissolved and necrotic.
2. **Infiltration of inflammatory cells** in the portal area, mainly large mononuclear cells and lymphocyte cells.
3. **Proliferation of Kupffer cells** in the hepatic sinusoids.

The above lesions are reversible.

9.6.4.2   Pathological Changes of Hepatitis B

Glassy-like cells are a special morphological feature of hepatitis B.

9.6.4.3   Pathological Changes of Hepatitis C

Microscopically, in addition to the typical characteristics of chronic hepatitis, there are some unique changes in chronic hepatitis C: fatty degeneration of liver cells, caused by changes in lipid metabolism of infected hepatocytes or insulin resistance, also known as metabolic syndrome; periportal lymphocyte infiltration, sometimes with formation of lymph follicles; and bile duct injury, which may be associated with direct viral infection of bile duct epithelial cells.

9.6.4.4 Pathological Changes of Hepatitis D

The liver cells exhibit eosinophilic and vesicular fatty degeneration, accompanied by infiltration of inflammatory cells and inflammatory reaction in the confluence area. Patients with chronic HBV infection have increased liver tissue lesions after overlapping infection with HDV.

9.6.4.5 Pathological Changes of Hepatitis E

Inflammation changes in the portal area, showing a large number of Kupffer cells and polymorphonuclear leukocytes but rare lymphocytes. Hepatocyte cytoplasm and bile capillaries exhibit cholestasis. Hepatocellular necrosis shows focal or patchy to subarea or large area necrosis, especially around the portal vein.

9.6.4.6 Pathological Changes of Hepatitis G

Hepatitis G pathologies characterized by a single HGV infection are generally less damaging. Acute hepatitis mainly involves hepatocyte swelling and portal inflammatory cell infiltration. Chronic hepatitis presents with hepatocyte swelling, lobular or focal necrosis, portal area inflammatory cell infiltration, and fibrosis, mainly hyperplasia.

9.6.5 Other Types of Hepatitis

9.6.5.1 Hepatitis Caused by Other Viral Infections

1. Epstein-Barr virus infection can cause mild hepatitis in the acute stage.
2. Cytomegalovirus infection, especially in almost all liver cells, including hepatocytes, bile duct epithelial cells, and endothelial cells, can cause virus-related giant cell changes.
3. Yellow fever virus infection is a major and serious cause of hepatitis in tropical countries. It can cause massive hepatocyte apoptosis, with the apoptosis of liver cells showing strong eosinophilic. This phenomenon is known as Councilman bodies.
4. Herpes simplex virus infection of the liver cells in newborns or immunosuppressed individuals leads to characteristic pathological changes of cells and necrosis of liver cells.

9.6.5.2 Drug-/Toxin-Mediated Injury Mimicking Hepatitis

Many drug effects can mimic the characteristics of acute or chronic viral or autoimmune hepatitis.

9.6.5.3 Autoimmune Hepatitis

Autoimmune hepatitis is a chronic, progressive hepatitis. Its histological features are difficult to distinguish from chronic viral hepatitis. Although the damage pattern of autoimmune hepatitis is similar to that of acute and chronic viral hepatitis, its histologic progress is different. In the early stage, there is severe cell damage, inflammation, and scar formation. The mortality of patients with severe untreated autoimmune hepatitis is approximately 40% within 6 months of diagnosis, and cirrhosis develops in at least 40% of survivors.

9.7 Alcoholic Liver Disease

Alcoholic liver disease is one of the main manifestations of chronic alcoholism. According to statistics from foreign countries, patients with alcoholic liver disease due to alcoholism account for 25–30%. Although there is no definite statistical data on the incidence of alcoholic liver disease in our country, there has been a noticeable increase in recent years.

9.7.1 Pathological Changes

Chronic alcoholism mainly causes three types of liver damage: hepatocellular steatosis or fatty change, alcoholic hepatitis, and alcoholic cirrhosis. These conditions can appear individually, simultaneously, or successively.

9.7.1.1 Fatty Liver

The most common liver disease associated with alcoholism is steatosis. Macroscopically, fatty liver in individuals with chronic alcoholism appears as a large, soft, yellow organ. Hepatocytes are swollen and round, and when they contain large lipid droplets, the nucleus can be pushed to the side of the cells. The central lobules are significantly affected, sometimes accompanied by various degrees of hydropic degeneration of hepatic cells. Simple fatty liver is often asymptomatic. If the lesion does not progress to fibrosis, abstinence can restore the fatty liver.

9.7.1.2 Alcoholic Hepatitis

There are three kinds of pathological changes in patients with clinical symptoms of liver disease: steatosis, alcoholic hyaline, and focal hepatocyte necrosis with neutrophil infiltration.

9.7.1.3 Alcoholic Cirrhosis

It is believed that this cirrhosis is caused by the progression of fatty liver and alcoholic hepatitis. If alcohol consumption continues, fatty liver develops into alcoholic hepatitis, followed by cirrhosis. Alcoholic hepatitis is often accompanied by prominent activation of sinusoidal stellate cells and portal fibroblasts, leading to fibrosis. Liver fibrosis results in the division and destruction of the normal structure of the hepatic lobules and the formation of alcoholic cirrhosis.

9.7.2 Pathogenesis

The liver is the main site for alcohol metabolism and degradation. Alcohol directly damages the liver through the following mechanisms:

Alcohol is transformed into acetaldehyde in the liver by the action of ethanol dehydrogenase and microsomal ethanol oxidase and then into acetic acid. This reaction converts NAD to NADH, resulting in an increased NADH-to-NAD ratio, which inhibits the mitochondrial tricarboxylic acid cycle. This leads to reduced oxidation of fatty acids in liver cells and the accumulation of fat, causing fatty liver. Increased NADH can also raise lactic acid levels, affecting liver metabolism. Alcohol causes free radical damage to the membrane system through the microsomal oxidation system in hepatocytes. Acetaldehyde has strong lipid peroxidation and toxicity, which can destroy liver cell structures and induce an immune response. In addition, alcoholism often leads to malnutrition, particularly protein and vitamin deficiencies.

9.8 Liver Cirrhosis

Liver cirrhosis is a common chronic liver disease characterized by diffuse degeneration and necrosis of liver cells, fibrous tissue hyperplasia, and nodular regeneration of hepatocytes. Advanced cases often present with varying levels of increased portal pressure and liver dysfunction, which are highly detrimental to the body. The onset typically occurs between the ages of 20 and 50, with no significant difference in incidence between males and females. Due to the diverse causes and complex pathogenesis of liver cirrhosis, no unified classification exists. It is generally classified based on the size of the nodules or the underlying cause. Internaticnally, liver cirrhosis is categorized into four types: large nodule type, small nodule type, mixed large and small nodule type, and incomplete segmentation type. The following are three common types of liver cirrhosis in our country:

9.8.1 Portal Cirrhosis

Portal cirrhosis is the most common type of liver cirrhosis worldwide. It is equivalent to small nodular cirrhosis in international morphological classification.

9.8.1.1 Etiology and Pathogenesis

The etiology and pathogenesis are not yet fully understood. Most studies suggest that various factors can cause liver cell damage, which may then progress to liver cirrhosis. The common factors include the following:

1. **Viral Hepatitis**
   This is the main cause of liver cirrhosis in China, particularly hepatitis B and C. There is substantial epidemiological, clinical, and pathological evidence supporting this association. Viral hepatitis remains a significant concern.
2. **Chronic Alcoholism**
   Chronic alcoholism is another major factor contributing to liver cirrhosis, especially in some European and American countries. The acetaldehyde produced during alcohol metabolism has a direct toxic effect on liver cells, leading to fatty degeneration and progression to liver cirrhosis.
3. **Damage from Toxic Substances**
   Many chemicals can damage liver cells, including carbon tetrachloride and cinchophen. Prolonged exposure to these substances can result in liver damage and cirrhosis.
4. **Malnutrition**
   Long-term deficiencies in methionine or choline can impair the liver's ability to synthesize phosphatidylcholine, potentially leading to fatty liver and, ultimately, liver cirrhosis.
   All of the above factors can cause diffuse damage to liver cells. If the damage persists over time and is repeatedly aggravated, it can result in extensive proliferation of collagen fibers in the liver. There are two sources of this increased collagen fiber: After hepatocyte necrosis, the original mesh scaffold in the hepatic lobule collapses and accumulates collagen (without cell sclerosis), or hepatic stellate cells transform into myofibroblast-like cells that produce collagen fibrosis. Fibroblasts in the portal area proliferate and secreted collagen fibers. After the collapse of the hepatic lobule, the regenerated hepatocytes cannot be arranged along the original scaffold, leading to the formation of irregular

regenerated hepatocyte nodules. On one hand, extensive proliferation of collagen fibers extends into the lobules, dividing the liver lobules. On the other hand, these fibers connect with collagenous fibers in the hepatic lobules, forming fibrous septum that surround the original or regenerative liver cell clusters and create pseudolobules. These lesions, along with the continuous necrosis and regeneration of liver cells, eventually lead to pseudolobules pervading the entire liver, resulting in hepatic blood circulation remodeling, hepatic dysfunction, and liver cirrhosis.

9.8.1.2 Pathological Changes

Macroscopically, in the early stages, the liver volume is normal or slightly increased, the weight is elevated, and the texture is normal or slightly hard. In advanced stages, the liver volume is significantly reduced, the weight is decreased, and the hardness is increased. The surface and tangent surfaces exhibit small nodules that are dispersed throughout the liver. These nodules are similar in size, ranging from 0.15 to 0.5 cm in diameter, and generally do not exceed 1 cm. The liver is thickened by the membrane, and round or circular island-like structures of similar size to the nodules are present on the surface, surrounded by gray-white fibrous tissue cords or spaced around (Fig. 9.8).

Microscopically, the structure of the normal lobule is destroyed and replaced by false lobules. The extensively proliferated fibrous tissue divides the original hepatic lobules, encircling round or rounded hepatocyte masses of varying sizes to form pseudolobules. The liver cells in the pseudolobules are arranged in disorder, with degeneration, necrosis, and regeneration present. The central vein is often absent or partially present, or there may be more than two central veins. Regenerated hepatocyte nodules are also observed, characterized by disordered arrangement of

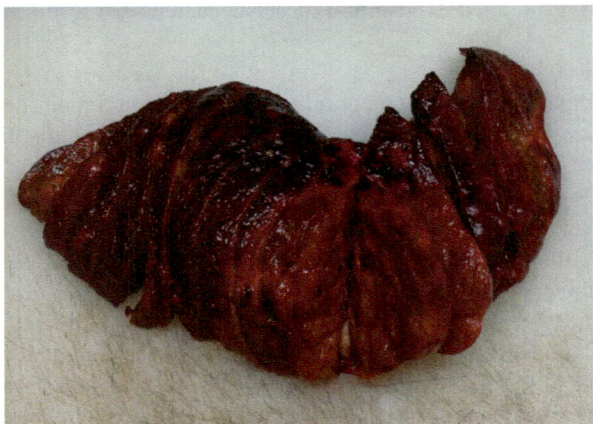

Fig. 9.8 Nodular cirrhosis

hepatocytes, large volumes of regenerated hepatocytes, large nuclei with deep staining, and visible binucleation. A small number of fibrous spacers wrap around the pseudolobules, with a small number of infiltrated lymphocytes and mononuclear cells, and hyperplasia of small bile ducts is also noted (Fig. 9.9).

9.8.1.3 Clinicopathological Relationship

1. **Portal Hypertension**

The causes of increased portal pressure are as follows:

(a) A wide range of connective tissue hyperplasia in the liver causes hepatic sinusoid occlusion and impedes portal circulation.

(b) The pseudolobule compresses the sublobular veins and affects the portal vein blood flowing into the hepatic sinusoids.

(c) Intrahepatic small branches of the hepatic artery and small branches of the portal vein form abnormal anastomoses before entering the hepatic sinusoids, allowing high-pressure arterial blood to flow into the portal vein.

After the portal vein pressure is elevated, a series of symptoms and signs are often present in patients. The main manifestations are as follows:

Splenomegaly: Long-standing congestion may cause splenomegaly. The degree of splenic enlargement varies widely and may reach as much as 1000 g. About 70–85% of patients with cirrhosis have splenomegaly. Splenomegaly may cause hematologic abnormalities, such as hypersplenism, thrombocytopenia, or even pancytopenia. The dilatation of the splenic sinus, proliferation and enlargement of the sinusoidal endothelial cells, atrophy of the splenic corpuscle,

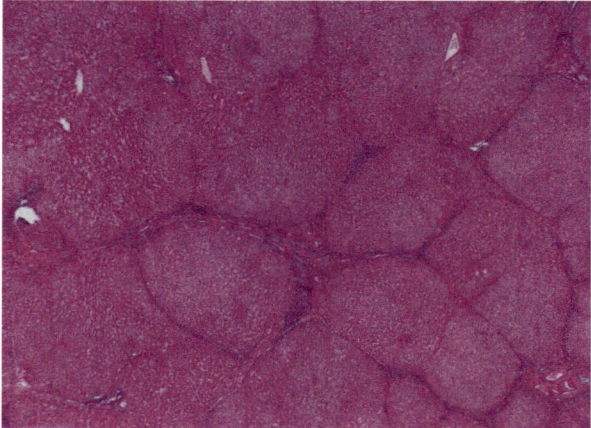

Fig. 9.9 Nodular cirrhosis of the liver. The structure of the normal lobule is destroyed and replaced by pseudolobules. The liver cells in the pseudolobules are arranged in disorder. The central vein is often absent, partial, or more than two

proliferation of the fibrous tissue in the red pulp, and some iron nodules are seen. Splenomegaly can cause hypersplenism.

Ascites: The accumulation of excess fluid in the peritoneal cavity is called ascites. Ascites is the translucent yellowish transudate, which can lead to bulging of the abdomen. In 85% of cases, ascites is caused by cirrhosis. The pathogenesis of ascites is complex, involving the following mechanisms: The elevation of portal vein pressure increases the hydrostatic pressure of the capillary fluid in the portal system, the permeability of the vessel wall increases, and the fluid leaks into the abdominal cavity. Liver dysfunction reduces the inactivation of aldosterone and antidiuretic hormone, increases the blood volume, and causes the formation of ascites through water and sodium retention. The presence of hypoproteinemia in the patient reduces plasma colloid osmotic pressure, which is also related to the formation of ascites.

Portosystemic Shunts: With the rise in portal system pressure, flow is reversed from portal to systemic circulation due to the dilation of collateral vessels and the development of new vessels. Venous bypasses develop because systemic and portal circulations share common capillary beds. Principal sites include veins around and within the rectum (manifesting as hemorrhoids), the esophagogastric junction (producing varices), the retroperitoneum, and the falciform ligament of the liver (involving periumbilical and abdominal wall collaterals). Although hemorrhoidal bleeding may occur, it is rarely massive or life-threatening. More importantly, esophagogastric varices appear in about 40% of individuals with advanced cirrhosis of the liver and cause massive hematemesis and death in about half of them. Each episode of bleeding is associated with a 30% mortality rate. Abdominal wall collaterals appear as dilated subcutaneous veins extending from the umbilicus toward the rib margins (caput medusae) and constitute an important clinical hallmark of portal hypertension.

Gastrointestinal Congestion and Edema: The increased portal vein pressure and the blockage of reflux from the gastrointestinal venous blood result in congestion and edema of the gastrointestinal wall, affecting the digestion and absorption functions of the stomach. Patients may experience symptoms such as abdominal distension and loss of appetite.

2. **Hepatosis**

The liver parenchyma (liver cells) is primarily damaged by repeated long-term injury. When liver cells cannot fully regenerate and compensate for the function of the injured liver cells, symptoms and signs of liver dysfunction may appear:

Protein Synthesis Disorder: When liver cells are damaged, their ability to synthesize protein is reduced, leading to decreased plasma protein levels. Additionally, due to absorption from the gastrointestinal tract, some antigenic substances may bypass the liver cells through collateral circulation and enter the systemic circulation, stimulating the immune system to increase globulin synthesis. Serological examination may reveal decreased albumin levels, a decreased or inverted albumin-globulin ratio.

Bilirubin Metabolic Disorder: This is mainly related to hepatocyte necrosis and capillary cholestasis. Patients often exhibit hepatocyte jaundice in the clinic.

Hemorrhagic Tendency: Patients with cirrhosis may experience skin, mucus membrane, or subcutaneous bleeding, mainly due to reduced liver synthesis of coagulation factors. In addition, this is related to splenomegaly, hypersplenism, and excessive destruction of platelets.

Impaired Hormone Inactivation: Patients with chronic liver failure may develop palmar erythema (a reflection of local vasodilatation) or spider angiomas on the skin. Spider angiomas result from the impaired inactivation of estrogen in the liver, leading to increased estrogen levels in the body and dilation of peripheral arterioles. They often appear on the neck, chest, and face. In men, hyperestrogenemia can also cause hypogonadism and gynecomastia. In women, hypogonadism may lead to oligomenorrhea, amenorrhea, and infertility.

Hepatic encephalopathy (hepatic coma): This is the most serious consequence and manifestation of extreme liver function failure. It is also another important cause of death in patients with liver cirrhosis.

9.8.2   Postnecrotic Cirrhosis

Postnecrotic cirrhosis is equivalent to the large nodular type and large nodules mixed cirrhosis in the international classification. It is formed on the basis of massive necrosis of hepatocytes.

9.8.2.1   Etiology and Pathogenesis

1. **Viral Hepatitis**

   Most cases are delayed by subacute severe hepatitis. In the course of repeated episodes of chronic hepatitis, if the necrosis is severe, it can also develop into this type of liver cirrhosis.
2. **Drug and Chemical Poisoning**

   Some drugs or chemicals can cause diffuse toxic necrosis in liver cells, which then leads to nodular regeneration, eventually developing into postnecrotic cirrhosis.

9.8.2.2   Pathological Changes

Macroscopically, the liver is narrowed and hardened, especially in the left lobe. The difference from portal cirrhosis is that the liver is more obviously deformed, and the nodules are much larger. The diameter of the largest nodule can reach up to 5–6 cm, and the connective tissue space within the section is wide and uneven in thickness.

Microscopically, the extent of liver cell necrosis and their irregular shape result in pseudolobules of various morphologies, such as semilunar, map-like, or nearly circular shapes. Large pseudolobules may sometimes contain a few complete liver lobules, and some areas may still show remnants of the portal area. Within the pseudolobules, liver cells exhibit varying degrees of degeneration and necrosis. If viral hepatitis is present, hepatocyte edema, eosinophilia, or the formation of eosinophilic bodies is often observed. There is wide fiber spacing, with significant infiltration by inflammatory cells and hyperplasia of small bile ducts.

Conclusion

Hepatic necrosis in postnecrotic cirrhosis is more severe, and the course of the disease is shorter. Therefore, liver dysfunction appears more obvious and earlier compared to portal cirrhosis. Portal hypertension is mild and occurs late. The canceration rate of this type of liver cirrhosis is also higher than that of portal cirrhosis.

9.8.3 Biliary Cirrhosis

Biliary cirrhosis is a rare disease resulting from biliary obstruction and cholestasis leading to liver cirrhosis. According to the different causes, it is divided into two types: primary and secondary.

Primary biliary cirrhosis is rare in China. The cause is unknown but may be related to an autoimmune reaction, as autoantibodies can be detected in patients' blood. It can be caused by chronic nonsuppurative cholangitis of the small bile ducts in the liver.

The causes of secondary biliary cirrhosis are associated with two factors: long-term extrahepatic bile duct obstruction and upper biliary tract infection. Long-term obstruction of the bile duct leads to cholestasis, liver cell degeneration and necrosis, and secondary connective tissue proliferation, ultimately resulting in liver cirrhosis.

9.8.3.1 Pathological Changes

Macroscopically, liver shrinkage is not as obvious as in the first two types of cirrhosis. The texture is medium hard, and the surface is smooth with small nodules or no obvious nodules. The color is dark green or green-brown.

Microscopically, primary lobar biliary cirrhosis shows early edema and necrosis of interlobular bile duct epithelial cells, surrounded by lymphocytic infiltration. The destruction of the small bile duct eventually results in connective tissue hyperplasia, which then extends into the hepatic lobules. The pseudolobules are incompletely divided. Microscopic examination of secondary biliary cirrhosis shows obvious hepatocyte degeneration and necrosis, with necrotic hepatocytes appearing swollen, having loose cytoplasm, reticular formation, nuclear disappearance, reticular or feathery necrosis, and incomplete segmentation of connective tissue around the pseudolobule.

9.9 Metabolic Liver Disease and Circulatory Disorders

9.9.1 Metabolic Liver Disease

9.9.1.1 Hepatolenticular Degeneration

Hepatolenticular degeneration, also called Wilson disease, is a hereditary disease transmitted by a recessive gene on chromosome 13, with many familial characteristics. Most patients are children and adolescents. The hallmark of this disease is copper metabolism disorder, where copper cannot be discharged normally and accumulates in various organs. The liver is the first organ affected. Once the liver is saturated, copper redeposits in the central nervous system, leading to neurologic symptoms. Copper can also accumulate in the cornea, forming a green-brown ring around the cornea (Kayser-Fleischer ring). In the liver, lipofuscin and copper-binding proteins, along with iron, are visible in the liver cells. Copper or copper-binding proteins (such as rhodamine, rubeanic acid, etc.) can be detected by histochemical staining. In the early stages, large particles or crystal sediments can be seen in the mitochondrial matrix of hepatocytes. The liver is often accompanied by acute or chronic hepatitis and liver cirrhosis. The central nervous system is most prominently affected in the striatum, thalamus, and globus pallidus.

9.9.1.2 Hemosiderosis

Hemosiderosis refers to the accumulation of stainable iron (hemosiderin) in liver tissue. The etiology of hemosiderin deposition is mainly due to massive erythrocyte destruction and hemoglobin decomposition, such as chronic hemolytic anemia caused by hemolysis and intrahepatic hemorrhage. Hemosiderin is mainly deposited in the liver cells and is often seen in Kupffer cells. The pigmentation of Kupffer cells due to blood transfusion is more pronounced.

Hemochromatosis is a systemic disease with congenital abnormalities in iron metabolism. The pathogenesis is unknown. Liver disease is part of the systemic condition, characterized by severe hemosiderin deposition in the liver, giving the entire liver a rusty appearance. In later stages, it is accompanied by liver fibrosis or cirrhosis.

9.9.1.3 Glycogenosis

Glycogenosis is the deposition of abnormal qualities and increased amounts of glycogen in tissues, caused by congenital autosomal recessive inheritance. It mainly involves the liver, heart, kidneys, and muscle tissue, with symptoms including hypoglycemia, developmental delay, and ketonuria. Depending on the stage of abnormal glucose metabolism and the specific enzymes involved, the disease is now classified into types O–XI, including several subtypes.

Visible hepatomegaly, with some livers being up to three times larger than normal and lighter in color, is observed. Microscopically, the liver cells are obviously swollen, with pale cytoplasm, loose granules, and bright areas. In frozen section, periodic acid-Schiff staining reveals red glycogen granules in the liver cells. The amylase digestion reaction is stable. In later stages, many types of glycogenosis can be accompanied by liver fibrosis or cirrhosis. It should be noted that the diagnosis and classification of glycogenosis cannot be based solely on histopathological changes and must be combined with clinical findings and enzyme analysis from liver biopsy samples.

9.9.1.4 Lipoidosis

Lipoidosis is the accumulation and deposition of lipids within tissues, caused by congenital lipid metabolism disorders. It mainly involves the deposition of glycolipids, phospholipids, and cholesterol. The mechanism is mostly due to a genetic deletion of enzymes involved in certain aspects of lipid metabolism, leading to the accumulation of the corresponding substrate (lipid) in tissues due to impaired catabolism.

1. **Phosphatide Lipoidosis**

    The increase and accumulation of phosphatidylcholine without glycerin is also known as Niemann-Pick disease, or neuro-phosphatidylcholine. The deficiency of sphingomyelinase, caused by autosomal recessive inheritance, prevents the hydrolysis of phospholipids, leading to their deposition in tissues. This condition can also be accompanied by other lipid storage disorders. The disease primarily affects the liver, spleen, bone marrow, and lymph nodes. In children, the nervous system is also involved. The main lesion is hepatomegaly. Microscopically, a large number of Kupffer cells and macrophages accumulate in the hepatic sinusoids and portal areas. The cells are enlarged, with foamy cytoplasm and small, centrally located nuclei, referred to as "Pick cells." There is also fat in the liver cells, mainly neutral fat and cholesterol. Under the electron microscope, Pick cells are filled with numerous spherical inclusion bodies arranged in a ring-like pattern. This disease often occurs in young children, and the prognosis is poor.

2. **Glycogenosis Disease**

Glycolipid refers to lipids such as cerebrosides and gangliosides, which do not contain phosphoric acid. Their metabolic disorders can lead to the deposition of cerebrosides (such as in Gaucher disease) and gangliosides.

Gaucher disease, also known as brain glycoside deposition disease, is a disorder of cerebroside catabolism due to a deficiency of β-glucosidase, caused by autosomal recessive inheritance. It mainly affects the mononuclear phagocyte system, including the liver, spleen, lymph nodes, and bone marrow. This disease often occurs in infants and is fatal. The main lesions are liver and spleen enlargement, with splenomegaly being particularly pronounced, up to 20 times the normal weight of the spleen. Microscopically, a large number of highly swollen lipid-laden macrophages accumulate in the liver. Some of these cells have foamy cytoplasm, while others have cytoplasm that appears with red stripes arranged in a wrinkled pattern. The nucleus is small, round or oval, and centrally located, known as Gaucher's cells. These cells are mainly distributed in the hepatic sinusoids and portal areas near the central vein of the lobule. Occasionally, liver fibrosis and cirrhosis mat occur (Fig. 9.10).

9.9.2 Circulatory Disorders

9.9.2.1 Portal Vein Obstruction

Portal vein obstruction is rare. It is caused by conditions such as liver cirrhosis, liver carcinoma, pancreatic carcinoma, and other liver or pancreatic diseases invading the intrahepatic portal vein, as well as by purulent peritonitis or neonatal umbilical cord

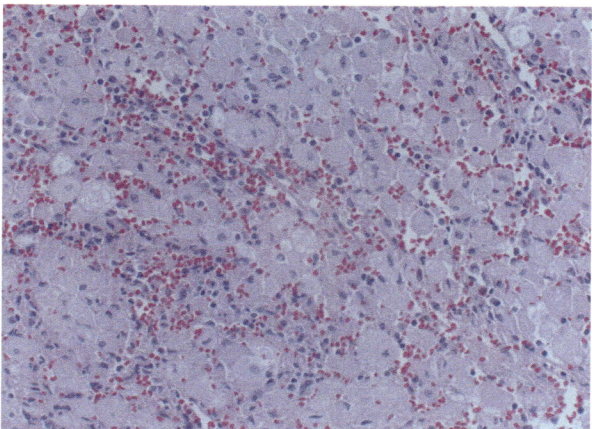

Fig. 9.10 Gaucher disease of the liver. A large number of highly swollen lipid-laden macrophages are found in the liver. Some have foamy cytoplasm, while others show red stripes arranged in a wrinkled pattern. The nucleus is small, round or oval, and centrally located within the cells

suppurative infections, leading to portal vein thrombosis or embolism. The thrombosis does not cause ischemic infarction but instead results in a sharply demarcated area of red-blue discoloration known as Zahn infarct. There is no necrosis, only severe hepatocellular atrophy and marked stasis in distended sinusoids. The lesion appears round or rectangular, dark red, and clearly defined. The local hepatocytes are atrophic, necrotic, or disappearing. A new anastomosis forms around the occlusion of the portal vein during the recovery period. This condition has little impact on the body, but it can be a source of intraperitoneal bleeding.

9.9.2.2 Hepatic Venous Obstruction

Hepatic venous obstruction is generally divided into two types: obstruction of the hepatic vein to the inferior vena cava, known as Budd-Chiari syndrome, and obstruction of the intrahepatic small hepatic veins, known as veno-occlusive disease.

Obstruction of two or more major hepatic veins leads to liver enlargement, pain, and ascites, a condition known as Budd-Chiari syndrome. Hepatic vein thrombosis is associated with myeloproliferative disorders such as polycythemia vera, inherited coagulation disorders, antiphospholipid antibody syndrome, paroxysmal nocturnal hemoglobinuria, and intraabdominal cancers, particularly hepatocellular carcinoma. During pregnancy or with oral contraceptive use, it may occur in conjunction with an underlying thrombogenic disorder. The main pathological changes include atrophy, degeneration, and necrosis of liver cells. In addition, hepatic hemorrhage occurs due to the deposition of red blood cells in the hepatic sinusoids, entering the Disse space with lower pressure outside the sinus, and the atrophy of the hepatic plate. Chronic cases can develop into congestive liver cirrhosis.

9.10   Cholecystitis and Cholelithiasis

9.10.1   Cholecystitis

Cholecystitis is often caused by bacteria and is usually associated with cholestasis. The main bacteria involved are E. coli, staphylococcus, and others. If the inflammation primarily affects the gallbladder, it is referred to as cholecystitis. If it mainly affects the bile ducts, it is called cholangitis.

9.10.1.1   Pathological Changes

1. **Acute Cholecystitis and Cholangitis**
   This is characterized by mucosal hyperemia and edema, epithelial cell denaturation, neutrophil infiltration, necrosis, and exfoliation. If it occurs in the

gallbladder, it is referred to as catarrhal cholecystitis, which can develop into honeycombed cholecystitis. Gangrenous cholecystitis occurs when blood circulation in the bile ducts or gallbladder wall is disturbed by spasm, edema, obstruction, or congestion. If the wall is perforated, bile peritonitis may occur.

2. **Chronic Cholangitis and Cholecystitis**

   This condition often develops from recurrent acute cholangitis and cholecystitis. The mucosa of the bile ducts and gallbladder wall become atrophic. Lymphocyte and monocyte infiltration, along with obvious fibrosis, can be observed in each layer.

9.10.2 Cholelithiasis

In the biliary system, some components of bile (bile pigment, cholesterol, mucosal substance, calcium, etc.) can precipitate and form calculus. Calculi formed in the bile duct are called bile duct calculus, while those formed in the gallbladder are called cholelithiasis (Fig. 9.11).

9.10.2.1 Causes and Pathogenesis

1. **Changes in Physical and Chemical Properties of Bile**

   Bilirubin is mostly combined with glucuronic acid to form esters in normal bile. The glucuronidase in intestinal bacteria, such as E. coli, decomposes these esters to release free bilirubin. Excess free bilirubin combines with calcium in the bile to form insoluble bilirubin calcium. Excess cholesterol can lead to the formation of cholesterol calculus. Loss of bile salts in certain intestinal diseases promotes cholesterol formation.

2. **Stagnation of Bile**

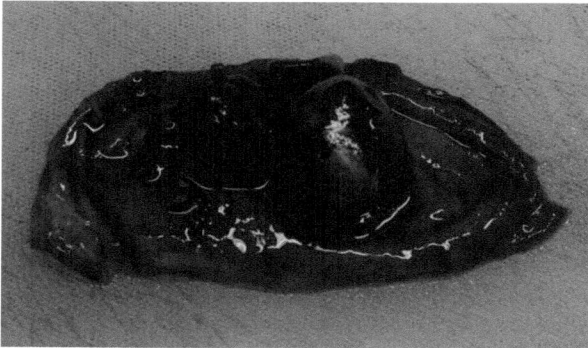

Fig. 9.11 Cholecystolithiasis

If too much water is absorbed from bile, bile pigment or cholesterol will become concentrated and form gallstones.

3. **Infection**

Inflammatory edema and chronic fibrosis can thicken the biliary wall, resulting in cholestasis. Cells transudate during inflammation, along with deciduous epithelium, worm residues, and insect eggs, can act as the nucleus of stones, promoting the formation of cholelithiasis.

9.10.2.2 Types and Characteristics of Gallstones

1. **Pigmentary Gallstone**

It is muddy or arenaceous, often multiple, and usually found in the bile ducts.
2. **Cholesterol Gallstone**
3. It is often single, large, and round and typically found in the gallbladder.**Mixed Gallstone**

It consists of more than two main components. In China, the most common type is the mixed gallstone, with bilirubin as the main component. It is usually not single, found in the gallbladder or bile ducts, and mostly polyhedral and colorful. The outer layer is often very hard and shows multiple layers upon sectioning.

9.11 Pancreatitis

Pancreatitis is the pancreatic self-digestion caused by abnormal activation of pancreatic enzymes.

9.11.1 Acute Pancreatitis

It often occurs in middle-aged men after binge eating or due to biliary disease.

9.11.1.1 Pathological Types and their Features

1. **Acute Edematous (Interstitial) Pancreatitis**

Acute edematous pancreatitis is more common and has better prognosis. It is mostly confined to the tail of the pancreas. The disease is characterized by swelling of the pancreas, hyperemia, interstitial edema, and infiltration of neutrophils and mononuclear cells. Sometimes localized fat necrosis may occur. A small amount of exudate can be found in the abdominal cavity.
2. **Acute Hemorrhagic Pancreatitis**

The onset is rapid, and the condition is critical. It is characterized by extensive bleeding and necrosis.

Grossly, the pancreas appears soft, swollen, and dark red, with the original lobulated structure disappeared. Scattered muddy yellow-white spots (where fat has been hydrolyzed into glycerol and fatty acids and combined with calcium ions into insoluble calcium soap) or small focal fatty necrosis (where adipose tissue is necrotic and catabolized by pancreatic juice) can be seen in the pancreas, omentum, and mesentery.

Microscopically, it is characterized by massive coagulation necrosis and unclear cell structure. Necrosis of the interstitial small vessel wall leads to massive hemorrhage. Mild inflammatory cells infiltration is observed around the necrotic pancreatic tissue. If the patient survives the crisis, inflammatory exudation and bleeding can be absorbed, be cured by fibrosis, or progress to chronic pancreatitis.

9.11.1.2 Clinicopathological Relationship

1. **Shock**: The main causes are as follows: severe abdominal pain caused by stimulation of the peritoneum from extravasated pancreatic juice, massive body fluid loss and electrolyte disorders caused by extensive bleeding and vomiting, and poisoning caused by tissue necrosis and protein decomposition.
2. **Peritonitis**: It is usually caused by the stimulation of extravasated pancreatic juice, resulting in severe pain that radiates to the back.
3. **Enzyme Changes**: The extravasated pancreatic juice contains high levels of amylase and lipase, which can be absorbed into the blood and expelled through urine. These enzyme levels can be detected to help in diagnosis.
4. **Serum Ion Changes**: The levels of calcium, potassium, and sodium ions in the serum of patients are decreased. When pancreatic islet A cells are stimulated, glucagon secretion causes the thyroid to release calcitonin, which inhibits calcium release from the bone. Persistent vomiting also reduces potassium and sodium levels in the blood.

9.11.2 Chronic Pancreatitis

It develops from recurrent acute pancreatitis and is often accompanied by biliary tract diseases and sometimes diabetes. Chronic alcoholism is also a contributing cause.

Grossly, the pancreas appears nodular, atrophic, and hard. On sectioning, diffuse fibrotic interstitial hyperplasia, pancreatic duct dilatation, and stone formation are visible. There may also be focal necrosis or pseudocysts wrapped in fibrous tissue.

Microscopically, chronic pancreatitis is characterized by fibrosis, atrophy, and infiltration of lymphocytes and plasma cells.

9.12   Common Tumors in the Digestive System

9.12.1   Carcinoma of the Esophagus

Carcinoma of the esophagus is defined as a malignant tumor of the esophageal mucosal epithelium or glands. Worldwide, about 300,000 people die from it. It is more common in men and in patients over 40. Clinically, it is characterized by varying degrees of dysphagia.

9.12.1.1   Causes

The etiology is not yet clear. The related factors are as follows:

1. **Lifestyle**: Long-term drinking, consumption of overheated and harsh diets, smoking, and nitrate intake are related to carcinoma of the esophagus.
2. **Chronic Inflammation**: Unhealed chronic esophagitis may be a precancerous lesion for carcinoma of the esophagus.
3. **Genetic Factors**: In high-incidence areas, familial aggregation of the disease is more obvious. The relationship between the gene diversity of metabolic enzymes and susceptibility to the disease has attracted scholarly attention.

9.12.1.2   Pathological Changes

It often occurs in three physiological structures, with the most common site being the middle section, followed by the distal section, and the least common in the upper part.

1. **Early Cancer**: There are no obvious clinical symptoms in the early stage. The lesion is limited, often in situ or intramucosal, without muscle layer invasion or lymph node metastasis.

    Grossly, the cancerous mucosa appears as mildly erosive, grainy, or tiny papillary. X-ray barium meal examination shows a normal or mildly rigid wall.

    Microscopically, most cases are squamous cell carcinoma
2. **Middle and Advanced Cancer**: Typical clinical symptoms such as dysphagia appear at this stage. Based on gross changes, it can be divided into four types:

    Fungating Type: This is the most common type. The invasive growth of the carcinoma tissue involves the whole or most of the esophagus. The wall of the tube thickens and the lumen becomes smaller. The cancer section is soft, like brain tissue, with a gray color; sometimes ulcers cover the surface.

    Ulcerated Type: The cancer presents as a flat, round lump, convex to the esophagus, with superficial ulcers and an ectropion at the edge. The cancer tissue invades part or most of the esophageal wall.

Infiltrating Type: The cancer has deep ulcers covering the surface, extending even to the muscle layer, with an uneven bottom.

Narrow Type: The cancer tissue is hard, with significant connective tissue proliferation and infiltration into the esophagus. The local esophageal wall narrows annularly, with obvious expansion above.

Microscopically, more than 95% of Chinese patients have squamous cell carcinoma, with adenocarcinoma being the second most common. Most adenocarcinomas develop from the cardia, with few originating from the esophageal mucosal glands.

The incidence of adenocarcinoma of the esophagus, which develops from esophageal canceration, has increased in recent years among White people.

9.12.1.3 Spread

1. **Direct Spread**
   The cancer tissue continuously infiltrates the surrounding tissues and organs after penetrating the wall of the esophagus.
2. **Metastasis**
   Lymphatic Metastasis: It metastasizes following the lymphatic drainage. In the upper segment, it metastasizes to the upper mediastinal lymph nodes and neck; in the middle segment, to the paraesophageal or hilar lymph nodes; and in the lower segment, to the adjacent esophageal or cardia lymph nodes and upper abdominal lymph nodes.
   Hematogenous Metastasis: It often metastasizes to the lungs or liver in the later stages.

9.12.1.4 Clinicopathological Relationship

There is no obvious infiltration or mass in the early stage, so symptom are not prominent. Patients may experience mild retrosternal pain, burning sensation, or choking feeling, which may be due to esophageal spasm or mucosal infiltration. In the moderate or advanced stages, the esophageal wall becomes narrow. Patients may have dysphagia or even lose the ability to ingest food, leading to cachexia and eventual death.

9.12.2 Carcinoma of the Stomach

Carcinoma of the stomach is a malignant cancer occurring in the epithelial cells of the gastric mucosa and glandular epithelium, often in the lesser curvature of the gastric antrum. It typically affects patients aged 40–60 years old and is more common in men.

9.12.2.1 Causes

The causes are not entirely clear and may be related to the following factors:

1. **Dietary and Environmental Factors**

   The incidence is somewhat related to geographical distribution. For example, in Japan, Chile, Columbia, Costa Rica, Hungary, and some regions in China, the incidence of carcinoma of the stomach is four to six times higher than that in the United States and Western Europe. Epidemiological surveys of immigration show that the incidence in the next generation decreases when people move from high-incidence areas to low-incidence areas. However, the incidence in the next generation increases when moving from low-incidence areas to high-incidence areas.

2. **Nitroso Compounds**

   Animal experiments have shown that feeding nitroguanidine to rats, mice, and dogs can successfully induce carcinoma of the stomach. Secondary amines and nitrites can be transformed into nitroso compounds under the action of gastric acid.

3. **Helicobacter pylori**

   Epidemiological studies have revealed that Helicobacter pylori infection may be associated with the occurrence of gastric cancer. It has been shown that Helicobacter pylori infection can lead to CpG island methylation of tumor-related genes in gastric epithelial cells and apoptosis.

4. **Some long-term unhealed chronic gastric diseases**, such as chronic atrophic gastritis, gastric ulcer disease with abnormal hyperplasia, and intestinal metaplasia of the gastric mucosa, are the pathological basis of gastric cancer.

9.12.2.2 Pathological Changes

1. **Early Carcinoma of the Stomach**

   The cancer tissue is limited to the mucosa or submucosa, without lymph node metastasis. Microcarcinoma refers to cancers less than 0.5 cm in diameter. Small carcinoma of the stomach refers to cancers 0.6–1.0 cm in diameter.

   Grossly, early carcinoma of the stomach can be divided into three types as follows:

   Protruded carcinoma of the stomach: Carcinoma protrudes from the surface of the mucosa like a polyp. It is rare.

   Superficial carcinoma of the stomach: The tumor is flat, slightly swollen on the mucosal surface.

   Concave carcinoma of the stomach: It is an early carcinoma in the surrounding mucosa of an ulcer, also known as cancerous erosion around ulcers. It is the most common type.

Microscopically, cancer in situ and highly differentiated tubular adenocarcinoma are more common, followed by papillary adenocarcinoma, with undifferentiated carcinoma being the rarest.

The 5-year survival rate for early gastric cancer is over 90%, the 10-year survival rate is 75%, and the 5-year survival rate for small gastric cancer and microcarcinoma is 100%. Early diagnose can improve the 5-year survival rate after operation and enhance the prognosis of gastric cancer.

2. **Advanced Carcinoma of the Stomach**

It refers to carcinoma infiltrating through the submucosa to the entire layer of the gastric wall. The deeper it invades, the worse the prognosis. Grossly, it can be divided into three types:

Polypoid or fungoid type: It is also known as the tubercular fungoid type. The carcinoma grows outward from the surface of the mucus membrane, resembling polyps or mushrooms, protruding into the stomach cavity.

Ulcerated type: The necrotic carcinoma exfoliates and forms ulcers The ulcer is generally large, like a dish with an unclear boundary. It can also rise like a volcano, with clear edges and an uneven bottom (Fig. 9.12; Table 9.3).

Infiltrating type: Carcinoma infiltrates locally or diffusely into the gastric wall, with unclear edges compared to surrounding normal tissue. Most of the surface folds of the gastric mucosa have disappeared. When diffusely infiltrated, the gastric wall thickens and hardens, and the gastric cavity becomes narrow, resembling leather, which is referred to as linitis plastica.

Colloid carcinoma: When cells secrete a lot of mucus, the carcinoma appears as translucent jelly grossly. It can present as any of the three forms above.

Microscopically, the main type is adenocarcinoma, commonly tubular adenocarcinoma and mucus carcinoma. A few cases may also be adenoacanthoma or squamous cell carcinoma, which is common in the cardia.

More than two types can exist in the same specimen.

9.12.2.3 Spread

1. **Direct spread**

The cancer tissue can infiltrate all layers of the stomach wall. It continues to spread to the surrounding tissue and organs, such as the liver and omentum majus, when penetrating the wall.

2. **Metastasis**

Lymphatic metastasis: This is the main metastasis. Initially, it metastasizes to local lymph nodes, with the most common being the lymph nodes of the lesser curvature under the pylorus. It then spreads to the paraaortic lymph nodes, portal or mesenteric lymph nodes, and later to the left supraclavicular lymph nodes (Virchow) via the thoracic duct.

Hematogenous metastasis: It often metastasizes to organs such as the liver, lungs, brain, or bones via the portal vein, typically in later stages.

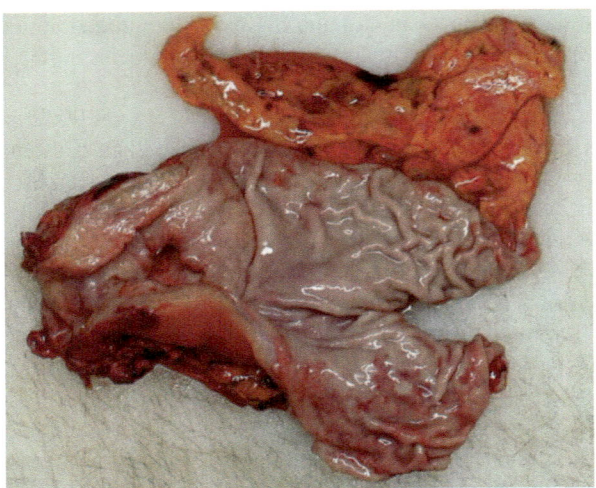

Fig. 9.12 Ulcerative gastric cancer

Transcoelomic metastasis: In carcinoma of the stomach, especially in mucinous carcinoma, cells infiltrate the surface of the serous membrane, fall into the abdominal cavity, and grow on the serous membrane of the abdominal and pelvic organs. Krukenberg tumor refers to metastatic mucinous carcinoma in the bilateral ovaries.

9.12.2.4 Histogenesis of Gastric Cancer

1. **The Cell Origin of Gastric Cancer**
   Carcinoma of the stomach mainly arises from tissue stem cells in the cervix of the gastric gland and the bottom of the gastric pits, where regeneration and repair are active, and canceration begins. These cells can differentiate into either the epithelium of the stomach or the intestinal epithelium.
2. **Intestinal Metaplasia and Carcinogenesis**
   The detection rate of intestinal metaplasia in gastric carcinoma is 88.2%. It is observed that the activity of aminopeptidase, lactate dehydrogenase, and its isozyme increases in the cytoplasm of intestinal metaplastic cells and cancer cells, while these enzymes are inactive in normal gastric mucosal cells.
3. **Atypical Hyperplasia and Carcinogenesis**
   Severe atypical hyperplasia is common in paracancerous mucosa and is sometimes associated with carcinogenesis.

9.12.3   Carcinoma of the Large Intestine

Carcinoma of the large intestine is a malignant tumor that occurs in the epithelium or glands of the intestine, including cancer of the colon and rectum. It is the third most common malignant tumor worldwide and is often seen in Europe, North America, and regions with British ancestry. Globally, China has a low incidence of large intestine carcinoma, but it is the fifth most common malignant tumor in China. The incidence of the disease is rising, particularly for colon carcinoma, which is increasing rapidly, with a faster growth rate is faster in large cities. In China, the incidence is higher in cities compared to rural areas, higher in large cities compared to small cities, and higher in men than in women. This may be closely related to improvements in lifestyle and dietary changes.

The main clinical manifestations include anemia, weight loss, increased stool frequency, mucus stools, abdominal pain, abdominal mass, and intestinal obstruction.

9.12.3.1   Causes and Pathogenesis

1. **Genetic Factors**

   Genetic carcinoma of the large intestine mainly has two types: one is familial adenomatous polyposis, which is caused by mutations in the adenomatous polyposis coli (APC) gene. The other is hereditary nonpolyposis colorectal cancer, which is caused by mutations in mismatch repair genes, such as hMSH2, hMLH1, and others.
2. **Eating Habits**

   A diet high in nutrition and low in fiber is related to the occurrence of this disease. A highly nutritious diet may hinder regular defecation and prolong the contact time between the intestinal mucosa and potential carcinogens in food.
3. **Chronic Intestinal Diseases**

   Some chronic intestinal diseases accompanied by intestinal mucosal hyperplasia, such as intestinal polypoid adenoma, proliferative polyposis, juvenile polyposis, villous adenoma, chronic schistosomiasis, and chronic ulcerative colitis, can progress to cancer due to hyperplasia of the mucosal epithelium.
4. **Molecular Biological Basis of Gradual Canceration in Large Intestinal Mucosa Epithelium**

   In addition to a small number of hereditary tumors, the development of the disease involves many genetic changes and interactions, such as those involving APC, C-myc, RAS, p53, p16, DCC, MCC, DPC4, or mismatch repair genes. Overexpression of the C-myc is seen in 90% of large intestine carcinomas. Most cases have mutations in the p53 genes and defects in the von Hippel-Lindau gene. Recent studies have found that abnormal expression of some proteins may be related to the occurrence of large intestine carcinoma.

   The pathogenesis of large intestine carcinoma mainly involves four types as follows:

Adenoma Canceration: The majority of colorectal carcinomas arise from preexisting adenomas, a process known as adenoma-carcinoma sequence. The occurrence of sporadic large intestine carcinoma is considered to be related to abnormalities in the APC-β-catenin-Tcf pathway, methylation of specific genes, dysfunction in mitosis regulation, and other factors.

Ulcerative Colitis-Associated Cancer Pathway: Ulcerative colitis-associated carcinoma is different from sporadic large intestine carcinoma. It has an earlier age of onset, similar incidence rates across different segments of the intestine, and a different molecular mechanism. For example, the abnormality of the p53 gene in sporadic colorectal carcinoma usually occurs in the adenocarcinoma stage or later, whereas in ulcerative colitis-associated carcinoma, it occurs in the early stage of epithelia proliferation. Morphologically, there are multiple lesions with flat infiltrating foci. Low-differentiated adenocarcinoma and mucinous adenocarcinoma are common.

Serrated Route to Carcinoma: Hyperplastic polyposis and malignant transformation of a serrated adenoma are caused by the inhibition of gene expression due to methylation in the promoter region of mismatch repair gene and functional incapacitation.

Juvenile Polyposis-Carcinoma Pathway: The occurrence of some juvenile polyposis is due to mutations in the Smad4 gene.

9.12.3.2 Pathological Changes

The most common site for the disease occurs is the rectum (50%), followed by the sigmoid colon (20%), cecum and colon (16%), and descending colon (6%).

Grossly, the general forms are divided into the following four types:

1. **Protruded Type**: The tumor is polypoid or discoid and protrudes into the lumen of the intestine and may be accompanied by superficial ulcers. This type is mostly well-differentiated adenocarcinoma.
2. **Ulcerative Type**: There is a deep ulcer or crater on the surface of the tumor. This type is more common.
3. **Infiltrative Type**: Carcinoma tissue infiltrates into the deep wall of the intestine, often involving the entire wall, resulting in thickening and stiffening.
4. **Gelatinous Type**: The surface and section of the tumor are translucent and jelly-like. The prognosis is worse.

The gross form of large intestine carcinoma varies slightly between the left and right colon. In the left colon, colorectal carcinoma is more commonly of the infiltrative type, which can easily cause narrowing of the intestinal wall and early symptoms of obstruction. In the right colon, the protruding type is more common.

Microscopically, the histologic types are as follows:

1. Papillary adenocarcinoma: The papillary tissues are thin and rarely have mesenchyme.

2. Tubular adenocarcinoma: It can be divided into three grades according to differentiation.
3. Mucinous adenocarcinoma: It is characterized by the formation of large mucus lakes.
4. Undifferentiated adenocarcinoma.
5. Adenosquamous carcinoma.
6. Squamous cell carcinoma. .

Carcinoma of the large intestine is mainly seen as highly differentiated tubular adenocarcinoma and papillary adenocarcinoma. Undifferentiated adenocarcinoma and squamous adenocarcinoma are less common. Squamous adenocarcinoma often occurs near the rectum and anus.

9.12.3.3 Stage and Prognosis

The stages significant implications for prognosis. The staging system currently in use was proposed by Astler-Coller in 1954 and has been revised many times since by Dukes. It is based on the spread of colorectal carcinoma and whether there is regional lymph node metastasis or distant organ metastasis.

The definition of large bowel carcinoma is clearly defined by the World Health Organization stages. Carcinoma of the large intestine is classified as carcinoma only when the invasion of the mucosal layer reaches the submucosa. It is termed intraepithelial neoplasia when the invasion does not exceed the muscularis mucosae. Severe atypical hyperplasia and carcinoma in situ are classified as high-grade intraepithelial neoplasia, and intramural carcinoma is called intramucosal neoplasia. Studies have shown that the 5-year survival rate of intramural carcinoma (not extending beyond the muscularis mucosae) is as high as 100%. However, once invasion extends to the submucosa, the 5-year survival rate decreases significantly.

9.12.3.4 Spread

1. **Direct Spread**
   When the invasion reaches the serous layer through the muscle layer, the disease can spread to nearby organs such as the prostate, bladder, and peritoneum.
2. **Metastasis**
   Lymphatic Metastasis: When cancerous tissue does not penetrate the intestinal muscle layer, lymphatic metastasis rarely occurs. Once penetration is evident, the rate of lymphatic metastasis increases. It generally spreads to local lymph nodes at the cancer site and then to distant lymph nodes along the lymphatic drainage route. Rarely, it can reach the supraclavicular lymph nodes via the thoracic duct.

Hematogenous Metastasis: In advanced carcinoma, cells may spread to the liver, or even to more distant organs such as the lungs or brain through hematogenous metastasis.

Implantation Metastasis: After the carcinoma tissue breaks through the serous membrane of the intestinal wall, it can spread to the surface of the abdominal cavity. The carcinoma cells may detach and spread into the abdominal cavity, forming implantation metastasis.

9.12.4 Primary Carcinoma of the Liver

Primary carcinoma of the liver is a malignant tumor of hepatocytes or intrahepatic bile duct epithelial cells. The incidence of this carcinoma is high in our country, making it one of the common tumors here. It predominantly affects middle-aged individuals. There are no clinical symptoms in the early stages of liver carcinoma, so it is often discovered late, leading to a high mortality rate. In recent years, the widespread use of alpha-fetoprotein and imaging examinations has significantly increased the detection rate of early liver carcinoma. Some early hepatomas with a diameter below 1 cm have been identified, leading to satisfactory results.

9.12.4.1 Etiology

The relevant factors are not entirely clear but are as follows:

1. **Hepatitis Virus**: Epidemiological and pathological data show that HBV is closely related to hepatocellular carcinoma, followed by HCV. HBV infections are present in 60–90% of patients with primary liver carcinoma. Scholars have found that common HBV genes are integrated into the genome of hepatoma cells. The HBV genome encoding HBx protein can inhibit the function of p53 protein and activate mitogen-activated protein kinase and Janus family tyrosine kinase signal transduction pathways. Repeated regeneration of liver cells may accumulate cell gene mutations, leading to malignant transformation.
2. **Cirrhosis**: Liver carcinoma is often associated with liver cirrhosis in China, most commonly with necrotic cirrhosis. Statistics show that liver carcinoma can develop approximately 7 years after cirrhosis.
3. **Fungus and Their Toxins**: Aspergillus flavus, Penicillium, and other fungi can cause experimental liver carcinoma. The close relationship between aflatoxin B1 and hepatocellular carcinoma is highly valued.
4. **Alcohol**: Alcohol is a carcinogenic factor for liver carcinoma, primarily acting indirectly through liver cirrhosis, which later contributes to liver carcinoma during the repair process.

9.12.4.2   Pathological Changes

1. **Early Liver Carcinoma (Small Liver Carcinoma)**: This refers to primary liver carcinoma where the maximum diameter of a single carcinoma nodule is less than 3 cm, or the total maximum diameter of two cancer nodules is less than 3 cm. Morphological characteristics include being mostly spherical with a clear boundary, a uniform section, and no bleeding or necrosis.
2. **Advanced Liver Carcinoma**: The liver volume increases significantly, often reaching 2000–3000 g, and the general morphology is divided into the following three types:

    Massive Type: The tumor is huge, sometimes as large as a child's head. It is round and more common in the right lobe. The central part of the section is often bleeding and necrotic, with many different satellite cancer nodules around the tumor. This is often associated with no or only mild cirrhosis (Fig. 9.13).

    Multiple Nodule Type: This is the most common type, usually accompanied by cirrhosis of the liver. Cancer nodules are scattered, round or oval, with different sizes. When nodules merge, they form larger masses.

    Diffuse Type: Cancer tissue is diffused throughout the liver, with no obvious nodules. It often occurs on the basis of liver cirrhosis and can be easily confused with cirrhosis. This type is rare.

    Microscopically, there are three types of tissue:

1. **Hepatocellular Carcinoma**: This occurs in hepatocytes and is the most commonly seen type. The degree of differentiation varies. Cancer cells with higher differentiation resemble hepatocytes and secrete bile. These cells are arranged in nests with many blood vessels (similar to hepatic sinusoids) and fewer interstitial substances. In low differentiation, the cancer cells show significant heteromorphism, with variations in size and shape.
2. **Cholangiocarcinoma**: This is a malignant tumor arising from the intrahepatic bile duct epithelium. The neoplastic cells are arranged in a tubular glandular

Fig. 9.13 Massive liver carcinoma

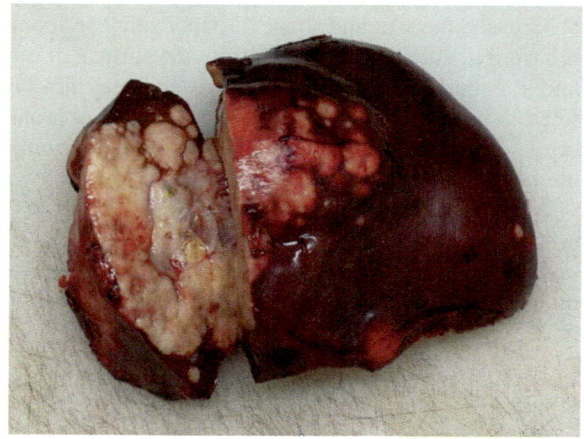

manner, can secrete mucus, and have more interstitial cancerous tissue. Liver cirrhosis is usually not associated with this type.

3. **Mixed-Cell-Type Liver Carcinoma**: This rare type contains two components: hepatocellular carcinoma and cholangiocarcinoma.

9.12.4.3 Spread

The cancer tissue first spreads directly within the liver and metastasizes along the portal vein, causing multiple metastatic nodules in the liver. Through the lymphatic system, extrahepatic metastasis can occur in the hilar lymph nodes, upper abdominal lymph nodes, and retroperitoneal lymph nodes. In the late stage, metastasis can extend to the lungs, adrenal glands, brain, and kidneys. After cancerous cells on the surface of the liver detach, implantation metastases can form.

9.12.5 Carcinoma of the Pancreas

Carcinoma of the pancreas is a rare digestive tract tumor, accounting for 1% of all body cancers in China. However, according to statistics from some countries, the incidence of pancreatic carcinoma has been increasing in recent years. The age of patients is typically between 60 and 80 years. The main environmental risk factor is smoking, which can double the risk. About 90% of patients have a KRAS mutation. In addition, there is overexpression of C-myc and mutation of the p53 gene.

9.12.5.1 Pathological Changes

Carcinoma of the pancreas can occur in the head (60%), body (15%), or tail (5%) or involve the entire pancreas, with the head being the most common site.

Grossly, the size and shape of pancreatic carcinoma vary. Sometimes, the tumor is a rigid nodule that protrudes from the surface of the pancreas. At other times, the tumor is buried within the pancreas and may not be visible externally. Common sclerosis of the peritumoral tissue can be so pronounced that the entire gland becomes hard, making it difficult to distinguish from chronic pancreatitis even during laparotomy.

Microscopically, the common histological types are ductal adenocarcinoma, cystadenocarcinoma, mucinous carcinoma, and solid carcinoma. There are also undifferentiated carcinoma or pleomorphic carcinoma and rare types such as squamous cell carcinoma or adenosquamous carcinoma.

9.12.5.2 Diffusion and Transfer

Carcinoma of the head of the pancreas can directly spread to adjacent tissues and organs, such as the bile duct and the duodenum. It then transfers to the lymph nodes around the head of the pancreas and the common bile duct. Intrahepatic metastasis via the portal vein is common, especially for cancers of the body or tail, can lead to invasion of the celiac plexus perilymphatic space and metastasis to the lungs and bones. Cancers of the pancreatic body and tail are often accompanied by multiple venous thrombosis.

9.12.5.3 Clinicopathological Relationship

The main symptom of pancreatic head carcinoma is painless jaundice. Tumors in the body and tail of the pancreas typically cause deep pain due to invasion of the celiac plexus, ascites from invasion of the portal vein, and splenomegaly from compression of the splenic vein. In addition, symptoms such as anemia, hematemesis, and constipation may occur without jaundice, and extensive thrombosis can develop. Without early diagnosis, the prognosis is poor, with most patients dying within a year.

9.12.6 Tumor of Biliary Tract

9.12.6.1 Carcinoma of the Gallbladder

Pathological Features: Carcinoma of the gallbladder often occurs at the bottom and neck of the gallbladder.

Grossly, the gallbladder wall is thickened, hardened, and gray, typically showing diffuse infiltrative growth. It can also present as polyps with a wide basal part.

Microscopically, most cases are adenocarcinomas, with some being adenosquamous carcinomas or squamous cell carcinomas.

Clinical Manifestation: The condition is more common in women and older individuals. Because it is difficult to detect early, the prognosis is poor. Its occurrence is associated with cholelithiasis and chronic cholecystitis.

9.12.6.2 Extrahepatic Cholangiocarcinoma

Pathological Features: This carcinoma often occurs at the confluence of the common bile duct, the hepatic duct, and the cystic duct.

Grossly, it can be polypoid, nodular, or deeply infiltrated into the wall of the bile duct.

Microscopically, most cases are adenocarcinomas, including papillary adenocarcinoma, mucinous adenocarcinoma, and sclerosing cholangiocarcinoma with abundant fibrous interstitium. A few cases are adenosquamous cell carcinoma or squamous cell carcinoma.

Clinical Manifestation: It is more common in the elderly, with symptoms including obstructive jaundice, abdominal pain, and abdominal mass.

9.12.7 Gastrointestinal Stromal Tumors

Gastrointestinal stromal tumors are a class of tumors that originate from gastrointestinal mesenchymal tissue and are mainly found in the elderly.

Pathological Features: These tumors are most commonly found in the stomach, followed by the small intestine, and are less commonly seen in the large intestine and the esophagus. They occasionally occur in the omentum and mesentery. Most tumors do not have a complete capsule and can be accompanied by cystic degeneration, necrosis, and focal hemorrhage. The degree of malignancy is related to the tumor size, mitotic activity, and the site of occurrence. Tumors larger than 5 cm in diameter are considered malignant, and the risk of gastrointestinal stromal tumor is higher in the small intestine is higher than in the stomach.

Microscopical features: About 70% of gastrointestinal stromal tumors are composed of spindle cells, while 20% are epithelioid cells. Immunohistochemically, gastrointestinal stromal tumors are positive for the cell surface antigen CD117. Approximately 60–70% of gastrointestinal stromal tumors are CD34 positive.

Further Reading

Li Y. Pathology, 8th ed. People's Health Publishing House.
Odze RD, Goldblum JR. Odze and Goldblum surgical pathology of the GI tract, liver, biliary tract, and pancreas, 3rd ed. Elsevier Saunders.
Underwood JCE General and systematic pathology, 2nd ed. Harcourt Asia.

Chapter 10
The Diseases of Hematopoietic and Lymphoid Systems

Yun Pan

Contents

Objectives
1. To master the subtypes and pathological characteristics of Hodgkin's lymphoma
2. To master the commonalities and differences between Hodgkin's lymphoma and non-Hodgkin's lymphoma

Y. Pan (✉)
Department of Pathology, The First Affiliated Hospital of Dali University, Dali, China

© Zhengzhou University Press 2024
K. Chen et al. (eds.), *Textbook of Pathologic Anatomy*,
https://doi.org/10.1007/978-981-99-8445-9_10

317

3. To comprehend the pathological features of common types of non-Hodgkin's lymphoma
4. To understand the classification and clinicopathologic characteristics of myeloid neoplasms

Key Concepts
1. Reed-Sternberg cell
2. Classic Hodgkin lymphoma
3. Follicular hyperplasia
4. Follicular lymphoma

Introduction
Disorders of the hematopoietic and lymphoid systems include a variety of different diseases that are classified as disorders primarily affecting red cells, white cells, and the hemostatic system. Diseases of red cells, white cells, and the hemostatic system include anemia, leukopenia, and thrombocytopenia. On the contrary, proliferation may be reactive, such as in reactive lymphadenitis, leukocytosis, and thrombocytosis, or neoplastic, such as in leukemias and malignant lymphomas. The hematopoietic and lymphoid systems, unlike other organ systems, are not confined to a single anatomic site. Therefore, when considering diseases of the hematopoietic and lymphoid systems, it is important to keep in mind that hematopoietic and lymphoid cells are spread throughout the body. Therefore, a patient with lymphoma diagnosed by lymph node biopsy can also have neoplastic lymphocytes in the bone marrow and blood. Neoplastic lymphoid cells in the bone marrow can inhibit hematopoiesis, leading to cytopenias, and the further spread of malignant cells to the spleen and liver may cause splenomegaly and hepatomegaly. Thus, in benign and malignant hematolymphoid disorders, a single underlying abnormality can result in various systemic manifestations. In this chapter, we first briefly introduce the structure and function of hematopoietic and lymphoid systems, describe some non-neoplastic conditions, and then mainly focus on white cell disorders.

10.1   Structure and Function of the Hematopoietic System

The hematopoietic and lymphoid systems are composed of lymphoid tissue (thymus, spleen, lymph nodes, and extranodal lymphoid tissues) and myeloid tissue (bone marrow). The thymus and bone marrow are often termed "central lymphoid tissues" because they are central to the prenatal development of the immune system, but they do not participate in the immune response in adults. The remaining lymphoid organs are actively involved in the immune response and constitute the peripheral lymphoid tissue.

The lymph nodes are surrounded by a capsule of connective tissue capsule, with trabeculae that extend into the substance of the node and provide a framework for the contained cellular elements. Beneath the capsule is a slit-like space, the subcapsular sinus. There are three distinct regions in normal lymph nodes: (1) the cortex, which contains nodules of B lymphocytes, either primary or germinal centers; (2) the paracortex or deep cortex, which is the T-cell-dependent region of the lymph node; and (3) the medullar, containing the medullary cords and sinuses, which drain into the hilum.

The appearance of the follicles varies according to their activity status. The primary follicles appear as small, round nodules with small lymphocytes. Secondary follicles appear following antigenic stimulation and are characterized by the presence of germinal centers. The cells present in these formations are B lymphocytes, known as follicle center cells, including centroblasts, centrocytes, macrophages, and follicular dendritic cells. The germinal center is surrounded by a mantle zone of small B lymphocytes.

There are many kinds of immune cells in the lymphatic system (lymph nodes, spleen, tonsils, etc.), where lymphocytes, macrophages, and other immune cells are arranged in various ways conducive to the formation of immune responses. Lymphatic tissues mainly gather at the entrance of the entrance of the antigen: tonsillar (mouth and nose), respiratory and gastrointestinal submucosa (inhaling and ingested antigen), lymph nodes (lymphatic drainage for the skin and organs), and the spleen (such as blood filters). The histologic appearances of lymphoid tissue mainly depend on the degree of antigenic stimulation. Reactive follicles (foci of B-cell proliferation) appear only under exposure to antigen. Similarly, immunoblasts are present only under antigenic stimulation.

In embryos, the formation of blood may occur in various parts of the body, such as the liver and spleen. In addition to the fact that lymphocytes and large mononuclear cells continue to mature in lymphoid tissue, the main center of hematopoietic activity is located in the bone marrow. After birth, the bone marrow is the only site of production of erythrocytes, granulocytes, and platelets. It also produces blood monocytes, which are part of the macrophage system. At birth, hematopoietic marrow is present in the medullary cavity of bones. In adults, the hematopoietic marrow is replaced by adipose tissue in the bones of the extremities, and hematopoietic marrow is found only in the axial skeleton.

10.2   Infection and Reactive Proliferation

10.2.1   Reactive Lymphadenitis

Lymph nodes undergo reactive changes in response to a wide variety of stimuli, including microbial infections, drugs, environmental pollutants, tissue injury, immune complexes, and malignant neoplasms. Reactive lymphadenitis is the most common benign lesion. Immune response against foreign antigens can lead to lymph node enlargement (lymphadenopathy). The histologic appearance of reactive lymphadenitis is nonspecific, with a few variant forms. Microscopically, the features of three patterns of reactive lymphoid hyperplasia are as follows:

10.2.1.1 Follicular Hyperplasia

This is the most common pattern. There is marked enlargement and prominence of the germinal centers of the lymphoid follicles (Fig. 10.1), consisting of numerous activated B cells, scattered T cells, phagocytic macrophages (containing nuclear debris), and a meshwork of antigen-presenting follicular dendritic cells. The reactive follicles develop from primary follicles in the cortex of the lymph node. The first phase of follicular formation appears to be antigen capture by dendritic reticulum cells, which then play a role in stimulating B-cell proliferation, leading to the development of a group of active B lymphocytes (the secondary or reactive follicles). Follicular hyperplasia can be found in rheumatoid arthritis, toxoplasmosis, and early human immunodeficiency virus (HIV) infection.

10.2.1.2 Paracortical Hyperplasia

This is due to hyperplasia of the T-cell-dependent area of the lymph node. When activated, parafollicular T cells transform into a large number of proliferating immunoblasts that can affect the follicles of B cells. In some instances, the T-cell response may be dominant; follicles may be inconspicuous. Paracortical hyperplasia commonly occurs in viral infections (such as Epstein-Barr virus), vaccinations (e.g., smallpox), and drug-induced immune responses (especially phenytoin).

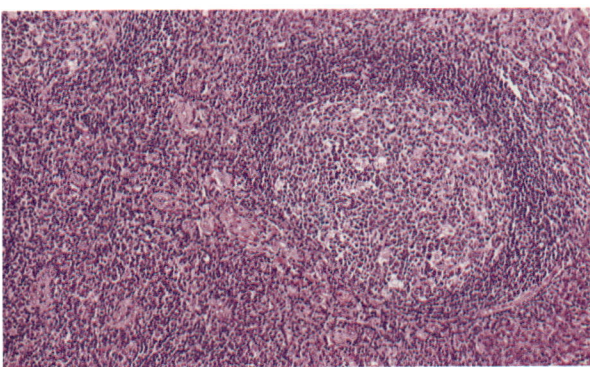

Fig. 10.1 Reactive hyperplasia in a lymph node shows features of a predominantly B-cell response, characterized by enlarged follicles with prominent reactive centers

10.2.1.3   Sinus Histiocytosis

It is characterized by distention and prominence of the lymphatic sinuses, which are filled with macrophages (histiocytes) and hypertrophy of the lining endothelial cells. It often occurs in lymph nodes draining cancers and may represent an immune response to the tumor or its products.

10.2.2   Infectious Mononucleosis

It is caused by the B-lymphocytotropic Epstein-Barr virus (EBV). Infected B cells express viral antigens on their surface, resulting in strong immune responses from T cells. Infectious mononucleosis is characterized by florid T-cell hyperplasia, which is so extensive that follicles are obscured. The most striking feature is the expansion of paracortical areas with a large number of immunoblasts and trans-formed large T cells. B-cell areas (follicles) may also be hyperplastic, but usually mild. Occasionally, EBV-infected B cells resembling Reed-Sternberg cells may be found, leading to possible misdiagnosis as malignant lymphoma. An increased number of large transformed lymphocytes can also be found in the peripheral blood (so-called Downey cells). Infectious mononucleosis is more common in adolescents and young adults and is transmitted via the upper respiratory tract. Patients are characterized by high fever, sore throat, lymphadenitis, and hepatosplenomegaly. Lymph nodes are enlarged throughout the body, principally in the posterior cervical, axillary, and groin regions. The spleen is enlarged in most cases and is usually soft and fleshy, with a hyperemic cut surface. The histologic changes of the spleen show an expansion of white pulp follicles and red pulp sinusoids due to the presence of numerous activated T cells. It may be diagnosed by the peripheral blood appearance (lymphocytosis with Downey cells). It should be noted that cytomegalovirus infection induces a similar syndrome, which can be differentiated only by serologic methods.

10.2.3   Specific Infections of Lymph Nodes

Characteristics of various special infections within the lymph node are listed as follows: (1) Specific microbial antigens, special pathological features of inflammation, special staining in the diseased tissue, and secretions or body fluids can be found associated with microbial pathogens. (2) Special drug therapy may be required in the clinic. (3) The presence of specific reagents in lymph nodes is often diagnosed through the culture of lymph nodes, which should be required in the biopsy.

10.2.3.1 Tuberculosis Lymphadenitis

It usually occurs in the cervical nodes (scrofula) and is frequently associated with extrapulmonary tuberculosis. It tends to be unifocal and localized. A group of enlarged lymph nodes can merge together to form a large lump. The cut surface reveals cheesy material (caseation). The basic histopathological lesion is caseous necrosis with granulomatous inflammation. The centers of granulomas undergo caseous necrosis, resulting in featureless areas of eosinophilic material. Langhans cells may be observed. .

10.2.3.2 Cat Scratch Disease

It is a self-limited lymphadenitis caused by the bacterium Bartonella henselae. Ninety percent of the patients are younger than 18 years of age. It presents as regional lymphadenopathy, most frequently in the axilla and the neck. The nodal enlargement appears approximately 2 weeks after a feline scratch or, less commonly, after a splinter or thorn injury. An inflammatory nodule, vesicle, or eschar is sometimes visible at the site of the skin injury. In most patients, the lymph node enlargement regresses over a period of 2–4 months.

The pathological changes in the lymph node in cat scratch disease are quite characteristic. Initially, sarcoid-like granulomas form, but these then undergo central necrosis associated with an infiltrate of neutrophils. These irregular stellate necrotizing granulomas are similar to those seen in a limited number of other infections, such as lymphogranuloma venereum. The microbe is extracellular and can be visualized with silver stains. The diagnosis is based on a history of exposure to cats, the characteristic clinical findings, a positive result on serologic testing for antibodies to Bartonella, and the distinctive morphologic changes in the lymph nodes.

10.3 Lymphoid Neoplasms

Lymphoid neoplasms, both malignant lymphomas and lymphocytic leukemias, are a group of tumors with clinical manifestations and behaviors that vary widely. In fact, they are the same disease in different clinical stages, resulting in diverse manifestations. Both lymphomas and lymphocytic leukemias may initially arise from lymphoid tissue or hematopoietic tissue and then evolve into each other. Lymphocytic leukemia is used for neoplasms that present with widespread involvement of the bone marrow and (usually, but not always) the peripheral blood. Lymphoma is used for proliferations that arise as discrete tissue masses. Originally, these terms were attached to what were considered distinct entities, but now these divisions have blurred. Many entities called "lymphoma" occasionally have leukemic presentations.

In daily medical practice, lymph nodes and bone marrow biopsy, bone marrow aspiration cytology, and blood cytology are the most important methods in the diagnosis of lymphoid and hematopoietic system diseases. Many tumors also involve gene-phenotype changes. Therefore, molecular biology, immunohistochemistry, and flow cytometry have become indispensable tools in the diagnosis of blood diseases.

10.3.1 Classification of Lymphoid Neoplasms

Lymphoid neoplasms show enormous variation in clinical behavior and response to therapy. The classification aims to identify homogeneous subgroups that behave in a predictable way. Lymphoid neoplasms are derived from cells that recapitulate stages of normal B-, T-, and NK-cell differentiation and function, so to some extent, they can be classified according to the corresponding normal counterpart. Several morphologic stages can be identified in the process of lymphocyte differentiation from stem cells to mature cells, according to the differentiation pattern of lymphocytes proposed by Lennert et al. in 1975. However, the classification method varies greatly because lymphocytic morphology changes can occur at any stage of lymphocyte differentiation.

The World Health Organization (WHO) classification of lymphoid neoplasms is based on morphology, cell origin (determined by immunophenotyping), clinical features, and genotype (e.g., karyotype, the presence of viral genomes) of each entity. All lymphoid neoplasms can be classified on the basis of cell origin: (1) precursor lymphoid neoplasms, (2) mature B-cell neoplasms, (3) mature T-and NK-cell neoplasms, and (4) Hodgkin lymphomas (Table 10.1).

10.3.2 Precursor B- and T-Cell Lymphoblastic Leukemia/ Lymphoma

These are high-grade non-Hodgkin lymphomas (NHLs) composed of diffuse sheets of medium-sized immature lymphocytes (lymphoblasts). They may be of B- or T-cell lineage, which are morphologically similar, presenting similar signs and symptoms, and treated similarly. Thus, precursor B- and T-cell lymphoblastic leukemia/lymphoma are classified together.

Acute lymphoblastic leukemia/lymphoblastic lymphoma (ALL) occurs predominantly in children and young adults. ALL accounts for 80% of childhood leukemia, with the peak age being 4 years old, and most cases are pre-B-cell origin. Pre-T-cell tumors are most common among males between 15 and 20 years old. Just as B-cell precursors normally develop within the bone marrow, pre-B-cell tumors usually present in the bone marrow and peripheral blood as leukemia. Similarly, pre-T-cell

Table 10.1

1.	Tumor-like lesions with B-cell predominance
2.	Precursor B-cell neoplasms
3.	Mature B-cell neoplasms
	Pre-neoplastic and neoplastic small lymphocytic proliferations
	Splenic B-cell lymphomas and leukemias
	Lymphoplasmacytic lymphoma
	Marginal zone lymphoma
	Follicular lymphoma
	Cutaneous follicle center lymphoma
	Mantle cell lymphoma
	Transformations of indolent B-cell lymphomas
	Large B-cell lymphomas
	Burkitt lymphoma
	KSHV/HHV8-associated B-cell lymphoid proliferations and lymphomas
	Lymphoid proliferations and lymphomas associated with immune deficiency and dysregulation
	Hodgkin lymphoma
4.	Plasma cell neoplasms and other diseases with paraproteins

tumors commonly present as masses involving the thymus, the normal site of early T-cell differentiation. However, pre-T-cell lymphomas often progress rapidly to the leukemic stage, while other pre-T-cell tumors may seem to involve only the bone marrow. Hence, both pre-B- and pre-T-cell tumors usually present clinical manifestations at specific times in their process.

10.3.2.1 Morphology

Microscopically, lymph nodes are composed of small to medium-sized blast cells with scant cytoplasm and inconspicuous nucleoli. Most pre-T-cell lymphomas present as mediastinal masses and progress rapidly to leukemia stage, but other cases present with marrow involvement only. Pre-B-cell lymphoma usually presents with bone marrow involvement. Meningeal infiltration is an important feature. In the blood smear slide, the nuclei of lymphoblasts with Wright-Giemsa staining show somewhat coarse and clumped chromatin and one or two nucleoli; myeloblasts tend to have fine chromatin and more cytoplasm, which may contain granules. It is essential to distinguish ALL from acute myeloid leukemia (AML), as these two diseases have different therapies.

10.3.2.2 Immunophenotypic and Genetic Features

Immunophenotyping is very useful in distinguishing ALL from AML. Terminal deoxynucleotidyl transferases (TdT), a DNA polymerase and a utility marker for these diseases, is present in more than 95% of ALL cases. It is necessary to perform specific immunologic staining to further subtype into pre-B- and pre-T-cell types, such as CD19 (B cell) and CD3 (T cell).

Approximately 90% of ALLs have nonrandom karyotypic abnormalities. Most common in childhood, pre-B-cell tumors are hyperdiploidy (more than 50 chromosomes per cell) and the presence of a cryptic (12;21) translocation involving the *ETV6* and *RUNX1* genes, while about 25% of adult pre-B-cell tumors harbor the (9;22) translocation involving the ABL and BCR genes. Pre-T-cell tumors are associated with diverse chromosomal aberrations, including frequent translocations involving the T-cell receptor loci and transcription factor genes such as TAL1.

10.3.2.3 Clinical Features

The clinical characteristics of ALL are similar to those of AML. The manifestations include anemia, bleeding (petechiae, ecchymoses, epistaxis, gum bleeding), and infection, as well as related symptoms, characterized by an abrupt clinical onset. Intensive combination chemotherapy, using several anticancer agents simultaneously in various combinations, has dramatically improved the prognosis of ALL patients. Recently, acute leukemias have been treated more aggressively with the intention of destroying all the hematopoietic cells in the bone marrow, including leukemic cells, followed by bone marrow transplantation.

10.3.3 Mature B-Cell Neoplasms

Many mature B cells originate from the follicular growth pattern of normal B cells. Thus, in some B-cell tumors, tumor cells are clustered into identifiable nodules similar to normal follicles. These tumors are called follicular lymphomas. Other B-cell tumors do not produce nodules but diffuse into lymph nodes. This structure is referred to as diffuse lymphoma. The normal structure of the lymph node disappears.

10.3.3.1 Chronic Lymphocytic Leukemia/Small Lymphocytic Lymphoma

It is a neoplasm composed of monomorphic small, round to slightly irregular B lymphocytes in the peripheral blood, bone marrow, spleen, and lymph nodes, admixed with prolymphocytes and paraimmunoblasts forming proliferation centers in tissue infiltrates. Chronic lymphocytic leukemia (CLL) and small lymphocytic lymphoma (SLL) are considered the same underlying disease, just with different appearances. If the peripheral blood lymphocyte count exceeds 5000 cells/µL, the patient is diagnosed with CLL. Tumors mainly involving lymph nodes, spleen, or extranodal locations are diagnosed as SLL. CLL/SLL is a disease of adults, and most (>75%) people newly diagnosed with CLL are over the age of 50, with the majority being men.

1. Morphology

   Histologically, enlarged lymph nodes in patients with CLL/SLL show effacement of the architecture, with a pseudofollicular pattern of regularly distributed pale areas corresponding to proliferation centers containing larger cells in a dark background of small cells. The predominant cells are small, resting lymphocytes with dark, round nuclei and scanty cytoplasm. Proliferation centers contain a continuum of small, medium, and large cells. Prolymphocytes are small to medium-sized cells with relatively clumped chromatin and small nucleoli; para-immunoblasts are larger cells with round to oval nuclei, dispersed chromatin, central eosinophilic nucleoli, and slightly basophilic cytoplasm. In addition to the lymph nodes, the bone marrow, spleen, and liver are involved in almost all cases. In most patients, there is an absolute lymphocytosis featuring small, mature-looking lymphocytes. The circulating tumor cells are fragile and, during the preparation of smears, are frequently disrupted, producing characteristic smudge cells. Variable numbers of larger activated lymphocytes are also usually found in the blood smear.

2. Immunophenotypic and Genetic Features

   CLL/SLL is a neoplasm of mature B cells expressing the pan-B-cell markers CD19, CD20, and CD23 and surface immunoglobulin heavy and light chains. The tumor cells also express CD5, which is a helpful diagnostic clue since, among B-cell lymphomas, only CLL/SLL and mantle cell lymphoma commonly express CD5. Approximately 50% of tumors have karyotypic abnormalities, the most common of which are trisomy 12 and deletions of chromosomes 11, 13, and 17. Deep sequencing of CLL/SLL cell genomes has identified activating mutations in the Notch1 receptor in a subset of cases that predict a worse outcome. Unlike in other B-cell neoplasms, chromosomal translocations are rare.

3. Clinical Features

   CLL/SLL is often asymptomatic. Most cases are diagnosed as a result of routine blood tests or clinical examination for other reasons. The most common clinical signs and symptoms are nonspecific, including fatigue, weight loss, and anorexia. Fifty to sixty percent of patients have lymphadenopathy and hepatosplenomegaly. White blood cell count may only increase slightly in SLL or may exceed 200,000 cells/µL. Clinical hypogammaglobulinemia develops in more than 50% of patients, usually in later stages, and leads to an increased susceptibility to bacterial infections. Rare cases of autoimmune hemolytic anemia and thrombocytopenia are seen. DNA analysis has distinguished two major types of CLL/SLL with different survival times. CLL/SLL that is positive for the marker ZAP-70 has an average survival of 8 years, while CLL/SLL that is negative for

ZAP-70 has an average survival of more than 25 years. Many patients, especially older ones, with slowly progressing disease can be reassured and may not need any treatment in their lifetimes. Approximately 2–8% of CLL/SLL patients develop diffuse large B-cell lymphoma (DLBCL), and less than 1% develop classic Hodgkin lymphoma (CHL).

10.3.3.2   Follicular Lymphoma

Follicular lymphoma is a tumor derived from germinal center B cells, characterized by a follicular or nodular architecture. It accounts for about 20% of all lymphomas, with the highest incidence in the USA and Western Europe. In China, it accounts for only about 10% of NHL cases.

1. Morphology

   Microscopically, the closely packed neoplastic follicles replace the normal lymph node architecture. Neoplastic follicles are mainly composed of centrocyte (CC) and centroblast (CB) cells. CC cells are slightly larger than resting lymphocytes, with angular nuclei that have prominent indentations and linear infoldings. The nuclear chromatin is coarse and condensed, and nucleoli are indistinct. CB cells are larger cells with vesicular chromatin, several nucleoli, and modest amounts of cytoplasm. In most tumors, CB cells are a minor component of the overall cellularity, mitoses are infrequent, and single necrotic cells (cells undergoing apoptosis) are not found (Fig. 10.2). These features help to distinguish follicular lymphoma from reactive follicular hyperplasia, in which mitoses and apoptosis are prominent. As the disease progresses, the number of CB cells gradually increases. The growth pattern changes from follicular to diffuse, resulting in more aggressive clinical behavior.

2. Immunophenotypic and Genetic Features

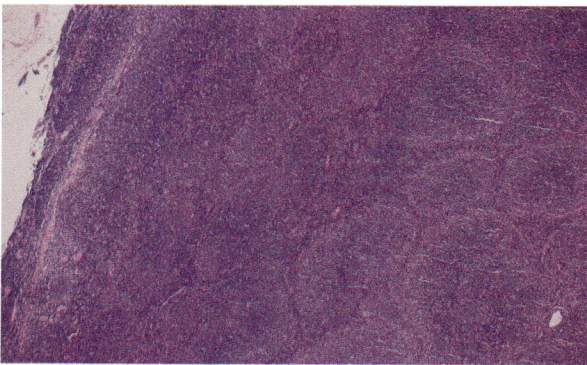

Fig. 10.2 Follicular lymphoma. The neoplastic follicles are closely packed and show an almost back-to-back pattern in focal areas. Nodular aggregates of lymphoma cells are present throughout the lymph node

These neoplastic cells express pan-B-cell markers (CD19 and CD20), CD10, BCL6, and BCL2 (Fig. 10.3). Most cases have a specific chromosome translocation involving the immunoglobulin heavy chain promoter region on chromosome 14 and the anti-apoptotic gene BCL2 on chromosome 18, (t14;18) (q32;q21), which results in high expression of the BCL2 gene and protein.

3. Clinical Features

Follicular lymphoma is more common in adults over the age of 50. Most patients present with painless lymphadenopathy. Bone marrow is commonly involved in the diagnosis, while visceral disease is rare. It is an indolent disease and the median survival time is 7–9 years. Approximately 40% of follicular lymphoma patients develop DLBCL, resulting in more aggressive clinical behavior.

10.3.3.3 Mantle Cell Lymphoma

Mantle cell lymphoma is composed of monomorphic small to medium-sized cells resembling naive B cells of mantle zones. It accounts for approximately 4% of all NHLs and occurs mainly in men more than 50 years old.

1. Morphology

Mantle cell lymphoma may involve lymph nodes in a diffuse or nodular pattern. Tumor cells are usually slightly larger than normal, with irregular nuclei, inconspicuous nucleoli, and scant cytoplasm. Bone marrow involvement occurs in most cases, but peripheral blood involvement is rare. Sometimes, it originates from extranodal lymphoid tissue, such as the gastrointestinal tract, presenting as multifocal submucosal nodules that grossly resemble polyps (lymphomatous polyposis).

2. Immunophenotypic and Genetic Features

Tumor cells usually express pan-B-cell antigens (CD19 and CD20) and CD5.

Mantle cell lymphoma is genetically characterized by the translocation (11;14) and cyclin D1 gene rearrangement, which results in high expression of cyclin D1 gene and protein.

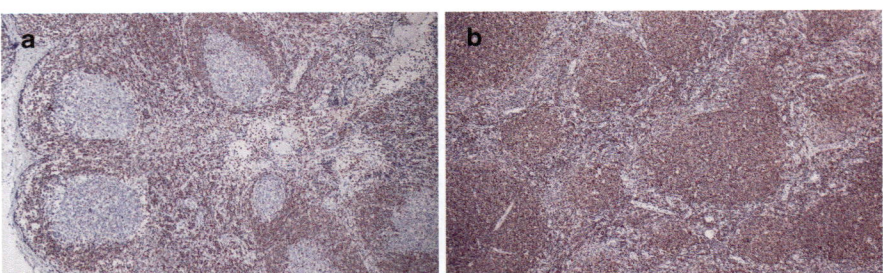

Fig. 10.3 BCL2 expression in reactive and neoplastic follicles. In reactive follicles (**a**), BCL2 is present in mantle zone cells but not in follicular center B cells. In follicular lymphoma (**b**), BCL2 staining is strong in follicular lymphoma cells

Cyclin D1, a cell cycle regulator, is believed to be an important mediator of uncontrolled tumor cell growth.

3. Clinical Features

Most patients present with fatigue and lymphadenopathy and are found to have generalized disease involving the bone marrow, spleen, liver, and (often) the gastrointestinal tract. The tumor is moderately aggressive, with a median survival time of 3–5 years.

10.3.3.4 Diffuse Large B-Cell Lymphomas

Diffuse large B-cell lymphomas (DLBCL) is the most common type of high-grade lymphoma, accounting for approximately 50% of adult NHLs.

1. Morphology

DLBCL is characterized by a diffuse outgrowth of large B cells, which may display centroblastic or immunoblastic cytology. Nuclear size is equal to or exceeds normal macrophage nuclei or is more than twice the size of a normal lymphocyte.

Three common and additional minor morphological variants have been recognized:

(a) **Centroblastic variant**: This is the most common variant. CBs are medium-sized to large lymphoid cells with oval to round, vesicular nuclei containing fine chromatin. There are two to four nuclear membrane-bound nucleoli. The cytoplasm is usually scanty and amphophilic to basophilic.

(b) **Immunoblastic variant**: More than 90% of the cells in this variant are immunoblasts with a single centrally located nucleolus and an appreciable amount of basophilic cytoplasm. Immunoblasts with plasmacytoid differentiation may also be present.

(c) **Anaplastic variant**: This variant is characterized by large to very large round, oval, or polygonal cells with bizarre pleomorphic nuclei that may resemble, at least in part, Hodgkin and Reed-Sternberg (HRS) cells.

2. Immunophenotypic and Genetic Features

Tumor cells express pan-B-cell antigens, such as CD20, CD19, and CD79a. Some cases also show expression of surface IgM and/or IgG. Other antigens (e.g., CD10, CD5) are variably expressed. A variety of cytogenic abnormalities can be seen in DLBCL. The (14;18) translocation is the most common deregulation, and about one-third of cases show rearrangement of the BCL6 gene.

3. Clinical Features

DLBCL occurs primarily in older individuals, with a median age at diagnosis of around 70 years. It is an aggressive tumor that can arise in virtually any part of the body. The first sign of this illness is typically the observation of a rapidly growing mass at single or multiple nodal or extranodal sites, such as the gastrointestinal tract, skin, bone, or brain. Most patients are asymptomatic, but when symptoms are present, they are highly dependent on the sites of involvement. Without treatment, the prognosis of DLBCL is poor. With intensive combination chemotherapy and anti-CD20 immunotherapy, complete remission is achieved in 60–80% of patients.

10.3.3.5   Burkitt Lymphoma

It is a B-cell lymphoma with an extremely short doubling time that often presents in extranodal sites or as an acute leukemia. It is named after Denis Parsons Burkitt, a surgeon who first described the disease in 1958 while working in equatorial Africa. It is endemic in parts of Africa and occurs sporadically in other areas. The disease is associated with EBV infection.

1. Morphology
   The tumor consists of sheets of a monotonous (i.e., similar in size and morphology) population of medium-sized lymphoid cells with high proliferative activity and apoptotic activity. A "starry sky" pattern is usually present, imparted by numerous benign macrophages that have ingested apoptotic tumor cells (Fig. 10.4). Tumor cells have round or oval nuclei, two to five distinct nucleoli, and a moderate amount of basophilic or amphophilic cytoplasm.
2. Immunophenotypic and Genetic Features
   Tumor cells express CD20, CD19, CD10, and BCL6, a phenotype consistent with germinal center B cells. Tumor cells do not express BCL2. The high mitotic activity of Burkitt lymphoma is confirmed by nearly100% of the cells staining positive for Ki-67. Most cases have MYC translocation at band 8q24 to the immunoglobulin heavy chain region,14q32, or, less commonly, at the lambda, 22q11, or kappa, 2p12, light chain loci. These translocations lead to MYC protein overexpression.

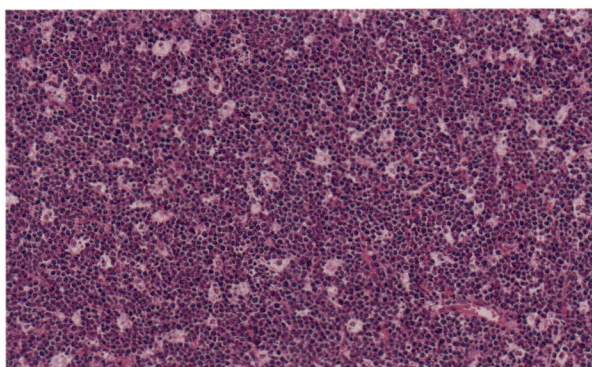

Fig. 10.4 Burkitt lymphoma. Numerous pale tingible body macrophages are evident, producing a macrophage appearance

3. Clinical Features

   Both the endemic and nonendemic sporadic forms affect mainly children and young adults, accounting for approximately 30% of childhood NHLs. It often occurs in the mandible, skull, bones, abdominal organs, and the central nervous system, forming a rapidly growing mass. Leukemic presentations are rare and should be distinguished from ALL. Clinically, it is a highly invasive tumor. Intensive combination chemotherapy regimens result in cure rates of up to 90% in patients with early-stage disease and 60–80% in patients with advanced-stage disease. The results are better in children than in adults.

10.3.4 Mature T- and NK-Cell Neoplasms

These categories represent a heterogeneous group of neoplasms resembling mature T cells or NK cells. Peripheral T-cell tumors make up about 5–10% NHLs in the USA and Europe but are more common in Asia. NK-cell tumors are rare in Western countries but more common in the Far East.

10.3.4.1 Peripheral T-Cell Lymphomas, Not Otherwise Specified

T-cell lymphomas are relatively common in Asia, accounting for 20–30% NHL in the region. Although the WHO classification includes several distinct peripheral T-cell lymphomas, many cases are still difficult to categorize. Furthermore, morphology is not a good indicator of clinical behavior. Thus, a majority of these tumors have been classified under the "wastebasket diagnosis" of peripheral T-cell lymphoma, not otherwise specified. Histologically, tumor cells diffusely infiltrate the lymph node paracortex with angiogenesis. The cytological spectrum of tumor cells is extremely broad, ranging from highly polymorphous to monomorphous, and is accompanied by numerous non-neoplastic cells, such as eosinophils, plasma cells, macrophage, and epithelioid histiocytes. The tumor cells are positive for CD2, CD3, CD5, and other mature T-cell markers (i.e., $\alpha\beta$ or $\gamma\delta$ T-cell receptors). Some also express CD4 or CD8, which are considered markers of helper or cytotoxic T-cell origin, respectively. However, the phenotype of many tumors does not resemble any known normal T cells. T-cell receptor gene rearrangement analysis can confirm the presence of a monoclonal rearrangement, helping to distinguish lymphoma from lymphoid hyperplasia. Most patients present with lymphadenopathy, sometimes accompanied by eosinophilia, pruritus, fever, and weight loss. These are highly aggressive lymphomas, with a poor response to therapy, frequent relapses, and a low 5-year overall survival rate (20–30%).

10.3.4.2 Mycosis Fungoides

Mycosis fungoides (MF) is the most common form of cutaneous T-cell lymphoma. Most cases are found in people over 20 years of age, and it is more common in men than women. Histologically, the epidermis and upper dermis are infiltrated by neoplastic T cells, which often have a cerebriform appearance due to marked folding of the nuclear membrane. The neoplastic T cells form a band-like upper dermal infiltrate with a moderate degree of epidermal infiltration, often forming small aggregates of cells within the epidermis (termed Pautrier microabscesses). It generally affects the skin but may progress internally over time.

Clinically, the cutaneous lesions of mycosis fungoides usually undergo three different stages: the patch stage, characterized by erythematous macules usually occurring in areas not exposed to sunlight; the plaque stage, with elevated scaly plaques that may be pink or red/brown and are often intensely pruritic; and the tumor stage, with dome-shaped firm tumors that may ulcerate. With the development of the disease, the density of lymphoid infiltrate increases from the patch to the tumor stage. Although MF is initially confined to the skin, involvement of lymph nodes and viscera is particularly common in the late stage of the disease. Patients with limited cutaneous disease have a good prognosis, with a mean survival time of 8–9 years. When visceral organs are involved, the median survival time is 2.5 years.

10.3.4.3 Extranodal NK-/T-Cell Lymphomas

The tumors are derived from cytotoxic T cells or NK cells and are highly aggressive. They are closely associated with EBV, with 80–100% of cases being positive for EBV DNA or its encoded proteins. The disease primarily occurs in the nasal region, followed by the jaw and throat, often involving the nasopharynx and nasal sinuses. About 10–20% of nasal NK-/T-cell lymphomas may also have skin involvement simultaneously. In advanced stages, tumors may disseminate rapidly to various sites, including the skin, digestive tract, testis, brain, and spleen. The histological features of extranodal NK-/T-cell lymphoma are similar regardless of the site of involvement. In a background of coagulative necrosis, neoplastic lymphoid cells are scattered or diffusely distributed, mixed with inflammatory cells. The tumor cells vary in size and shape, with irregular, hyperchromatic nuclei and one to two nucleoli. Tumor cells infiltrate into the vessel walls, a phenomenon known as angiocentric infiltration, leading to lumen stenosis, atresia, and rupture of the elastic membrane. Fibrinoid changes can be seen in the blood vessels even in the absence of angioinvasion. Tumor cells often express T-cell antigens (CD2, cytoplasmic CD3), cytotoxic proteins (such as granzyme B), and CD56, which is a useful NK-cell marker. The prognosis of nasal NK-/T-cell lymphoma is variable, with some patients responding well to therapy and others succumbing of disseminated disease despite aggressive treatment.

10.4 Hodgkin Lymphoma

Hodgkin lymphoma is a group of primary malignant tumors of lymphoid tissues, characterized by the presence of giant tumor cells known as Reed-Sternberg cells.

Hodgkin lymphomas have the following characteristics: (1) They usually arise in lymph nodes, preferentially in cervical lymph node. (2) The majority of the patients are young adults. (3) Large mononucleated and multinucleated tumor cells account for only a minority of the total number of cells, dispersed in an abundant heterogeneous admixture of non-neoplastic inflammatory and accessory cells. (4) The tumor cells are often ringed by T cells in a rosette-like manner.

1. Morphology

    Macroscopically, lymph nodes are enlarged with a homogeneously pale white cut surface or a nodular or fibrotic appearance. Microscopically, the lymph node structure is destroyed and composed of a mixed infiltrate containing lymphocytes, histiocytes, plasma cells, eosinophils, as well as Reed-Sternberg cells. These large malignant cells are abundant, with slightly eosinophilic cytoplasm and a diameter of 15–45 μm. A typical Reed-Sternberg cell is binuclear, in a face-to-face arrangement, symmetric to each other, forming the so-called mirror cells (mirror image cells). HRS cells are generally considered precursor cells of classical Reed-Sternberg cells. There are also several special types of tumor cells (Fig. 10.5):

    (a) **Reed-Sternberg cell, mononuclear variant**: The cells are large and irregular in shape. Nuclei have coarse chromatin, with an obvious nucleolus. Mitotic figures are common, usually multipolar mitosis.
    (b) **Lacunar cells**: Their cytoplasm retracts when fixed in formalin, so the nuclei give the appearance of cells lying in empty spaces (called lacunae) between them.
    (c) **Lymphocyte predominant cells (previously lymphocytic variant Reed-Sternberg cells)**: The cells are large and usually have one large nucleus and scant cytoplasm. The nuclei are folded or multilobulated and have also been termed histiocytes. The nucleoli are usually multiple, basophilic, and smaller than those seen in classical Reed-Sternberg cells.

    Although the diagnosis of Hodgkin lymphoma depends upon the discovery of classic Reed-Sternberg cells, there is little evidence of nucleic acid synthesis or proliferative activity in these cells. Large mononuclear Reed-Sternberg cells (called Hodgkin cells) are proliferating cells in Hodgkin lymphoma. The histologic feature of Hodgkin lymphoma is particularly notable in that the neoplastic Reed-Sternberg cells are few and are admixed with variable numbers of lymphocytes, plasma cells, histiocytes, eosinophils, neutrophils, and fibroblasts, all of which are considered reactive. However, malignant cells are dominant in other tumors. The lymph node can be totally destroyed, involving the spleen, liver, bone marrow, and extralymphatic tissues.

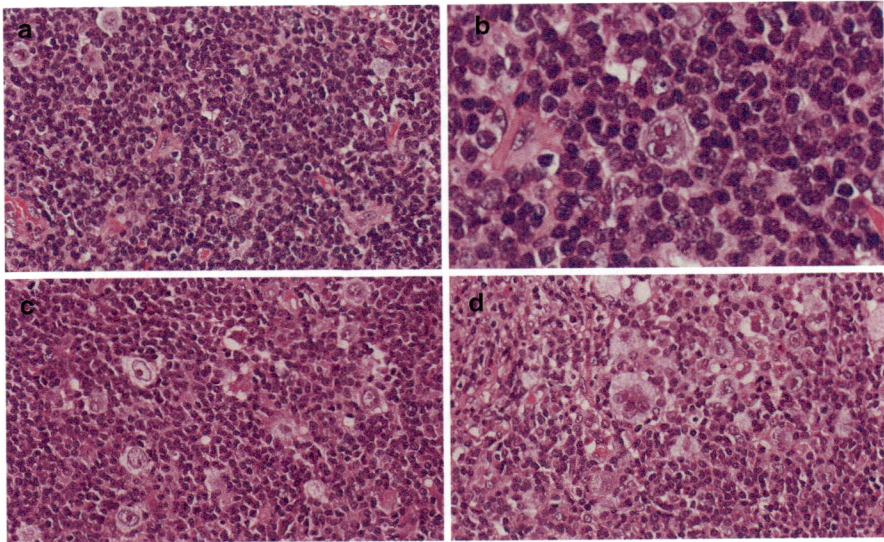

Fig. 10.5 Reed-Sternberg cells and variants. (**a, b**) Diagnostic Reed-Sternberg cell, with two nuclear lobes, large inclusion-like nucleoli, and abundant cytoplasm, surrounded by lymphocytes, macrophages, and an eosinophil. (**c**) Reed-Sternberg cell, mononuclear variant. (**d**) Reed-Sternberg cell, lacunar variant. This variant has a folded or multilobulated nucleus and lies within an open space, which is an artifact created by disruption of the cytoplasm during tissue sectioning

2. Classification of Hodgkin Lymphoma

According to 2017 WHO classification, Hodgkin lymphoma is classified into two subtypes: nodular lymphocyte-predominant Hodgkin lymphoma and CHL. Based on Reed-Sternberg cell morphology and the composition of the reactive cell infiltrate seen in the lymph node biopsy specimen, CHL can be subclassified into four pathologic subtypes: nodular sclerosis classical Hodgkin lymphoma (NSCHL), mixed cellularity classical Hodgkin lymphoma (MCCHL), lymphocyte-rich classical Hodgkin lymphoma (LRCHL), and lymphocyte-depleted classical Hodgkin lymphoma (LDCHL) (Table 10.2).

The subtypes of Hodgkin lymphoma are recognized as follows:

(a) Nodular lymphocyte-predominant Hodgkin lymphoma.
(b) Classic Hodgkin lymphoma: (1) nodular sclerosis, (2) mixed cellularity, (3) lymphocyte rich, and (4) lymphocyte depletion.

10.4.1 Nodular Lymphocyte-Predominant Hodgkin Lymphoma

It is characterized by the presence of lymphohistiocytic (L&H) variant Reed-Sternberg cells that have a delicate multilobed, puffy nucleus resembling popped corn ("popcorn cell"). L&H variants are usually found within large nodules

Table 10.2 WHO classification of Hodgkin's lymphoma

Histologic subtype	Incidence	Main pathology	Reed-Sternberg cells	Prognosis
(a) Nodular lymphocyte-predominant Hodgkin lymphoma				
	5%	Proliferation of small lymphocytes, nodular pattern of growth	Sparse number of RS cells, CD20+, CD45+, EMA+, CD30−, CD15−	Chronic relapsing, may transform into large B-cell NHL
(b) Classic Hodgkin lymphoma				
Lymphocyte predominance	5%	Proliferating lymphocytes, a few histiocytes	Few, classic and polyploid type, CD30−, CD15−, CD20+	Excellent
Nodular sclerosis	30–60%	Lymphoid nodules, collagen bands	Frequent, lacunar type CD30+, CD15+	Very good
Mixed cellularity	20–40%	Mixed infiltrate	Numerous, classic type CD30+, CD15+	Good
Lymphocyte depletion (diffuse fibrotic and 2% reticular variants)	<2%	Scanty lymphocytes, atypical histiocytes, fibrosis	Numerous pleomorphic type CD30+, CD15+	Poor

containing numerous mature-looking small B cells admixed with a variable number of macrophages. Other types of reactive cells, such as eosinophils, neutrophils, and plasma cells, are scanty or absent, and typical Reed-Sternberg cells are rare. Unlike the Reed-Sternberg variants in "classical" forms of Hodgkin lymphoma, L&H variants express B-cell markers, such as CD20 and CD79a, but lack CD15 and CD30. Most patients present with cervical or axillary lymphadenopathy and have a good prognosis.

10.4.2   Classical Hodgkin Lymphoma

Classical Hodgkin lymphoma (CHL) is a monoclonal lymphoid neoplasm composed of mononuclear Hodgkin cells and multinucleated Reed-Sternberg cells residing in an infiltrate containing a variable mixture of non-neoplastic small lymphocytes, eosinophils, neutrophils, histiocytes, plasma cells, fibroblasts, and collagen fibers. CHL represents approximately 95% of all Hodgkin lymphomas. A bimodal age distribution is seen with peaks in the 10–35 years old age group and in those over 50 years old. Patients are most frequently in the 10–35 age group and older adults. Seventy-five percent of cases involve cervical lymph nodes.

10.4.2.1 Nodular Sclerosis Classical Hodgkin Lymphoma (NSCHL)

It is the most common subtype. The incidence of NSCHL is similar in males and females and peaks at ages 15–34 years. It has a striking propensity to involve the lower cervical, supraclavicular, and mediastinal lymph nodes. Mediastinal involvement occurs in 80% of cases, bulky disease in 54%, splenic involvement in 10%, and bone marrow involvement in 3%. Morphologically, it is characterized by the presence of nodular sclerosis, broad bands of collagen circumscribing nodules of involved tissue, and large Reed-Sternberg cell variants. This large cell, termed the lacunar cell, has a single multilobed nucleus, multiple small nucleoli, and abundant, pale-staining cytoplasm. In sections of formalin-fixed tissue, the cytoplasm is often torn away, leaving the nucleus lying in an empty space. The immunophenotype of lacunar variants is identical to that of other Reed-Sternberg cells found in classical subtypes.

10.4.2.2 Mixed Cellularity Classical Hodgkin Lymphoma (MCCHL)

It is more frequent in patients with HIV infection and in developing countries. A bimodal age distribution is not seen. The median age is 38 years and approximately 70% are males. Histologically, lymph node architecture is usually obliterated, although an inner-follicular growth pattern may be seen. Interstitial fibrosis may be present, but the lymph node capsule is usually not thickened, and there are no broad bands of fibrosis as seen in NSCHL. The Reed-Sternberg cells are typical in appearance. The background consists of numerous inflammatory cells, which are rich in T lymphocytes, histiocytes, eosinophils, and plasma cells. This type is most often associated with EBV infection.

10.4.2.3 Lymphocyte-Rich Classical Hodgkin Lymphoma (LRCHL)

It is a subtype of CHL with scattered Reed-Sternberg cells and a nodular or, less commonly, diffuse cellular background consisting of small lymphocytes and an absence of neutrophils and eosinophils.

10.4.2.4 Lymphocyte-Depleted Classical Hodgkin Lymphoma (LDCHL)

It is the rarest subtype of CHL, composed of large numbers of pleomorphic Reed-Sternberg cells with only a few reactive lymphocytes, which may easily be confused with diffuse large cell lymphoma. Prior to modern therapy, the course of LDCHL was aggressive. Histologically, compared with the lymphocytes in the background, Reed-Sternberg cells are dominant. There are two patterns: diffuse fibrotic and reticular cell. In the diffuse fibrotic pattern, lymph node cells are significantly reduced, with an irregular arrangement of reticular fibers and amorphous protein substances. There are few diagnostic Reed-Sternberg cells, and lymphocytes are rare. Necrosis is usually. In the reticular cell pattern, it is rich in cells, comprising many pleomorphic Reed-Sternberg cells, a small number of diagnostic Reed-Sternberg cells, and spindle tumor cells. Mature lymphocytes, eosinophils, plasma cells, neutrophils, and histiocytes are rare. Compared to other types of Hodgkin lymphoma, necrosis is more prominent. Reed-Sternberg cells are positive for CD15 and CD30. Most cases express EBV-encoded LMP-1 and EBER.

10.4.3   Staging of Hodgkin Lymphoma

Treatment of Hodgkin lymphoma is based on clinical, and occasionally pathological, staging of the disease. The modified Ann Arbor staging system is used (Table 10.3).

Younger patients with the more favorable subtypes tend to present with stage I or II disease, without systemic manifestations. Patients with advanced disease (stages III and IV) are more likely to have systemic complaints such as fever, weight loss, pruritus, and anemia.

10.4.4   Clinical Features of Hodgkin Lymphoma

Hodgkin Lymphoma is more common in males and shows a peak incidence in early adulthood. Like NHLs, it usually manifests as painless lymphadenopathy, most often in the upper half of the body, with involvement of cervical and/or axillary lymph nodes, and spreads to anatomically contiguous nodes. Radiological evidence of mediastinal involvement is present in over 40% of patients and, on occasion, may be massive. With the advancement of the disease, the involvement of the spleen, liver, bone marrow, and other organs and tissues may appear. Some patients with the disease have systemic symptoms, such as weight loss, unexplained pyrexia

Table 10.3 Clinical staging of Hodgkin and non-Hodgkin lymphomas

Stage		Distribution of disease
I	I	Involvement of a single lymph node region
	I_E	Involvement of a single extralymphatic organ or site
II	II	Involvement of two or more lymph node regions on the same side of the diaphragm
	II_E	(or) with limited contiguous involvement of an extranodal organ or site
III	III	Involvement of lymph node regions on both sides of the diaphragm
	III_E	(or) with localized contiguous involvement of an extranodal organ or site
	III_S	(or) with involvement of the spleen
	III_{ES}	(or) both features of IIIE and IIIS
IV		Multiple or disseminated involvement of one or more extralymphatic organs or tissues with or without lymphatic involvement

Table 10.4 Clinical differences between Hodgkin and non-Hodgkin lymphomas

HL	NHL
More often localized to a single axial group of nodes (cervical, mediastinal, paraaortic)	More frequent involvement of multiple peripheral nodes
Orderly spread by contiguity	Noncontiguous spread
Mesenteric nodes and Waldeyer ring rarely involved	Mesenteric nodes and Waldeyer ring commonly involved
Extranodal involvement uncommon	Extranodal involvement common

exceeding 39 °C, anemia, and drenching night sweats. Despite intensive research and a wealth of immunologic data, the diagnosis of Hodgkin lymphoma is still based entirely upon histologic examination—the finding of the classic Reed-Sternberg cell in pathologic tissue is considered essential for diagnosis.

Localized forms of Hodgkin lymphoma may be treated with either radiation or chemotherapy. Chemotherapy is highly effective when multiple agents are used and may lead to cures even in patients with disseminated disease. The 5-year survival rate of patients with stage I or II disease is close to 100%. Fifty percent of patients with advanced disease (stages III and IV) can achieve a 5-year disease-free survival.

Although a definitive distinction from NHL can be made only by examination of a lymph node biopsy, several clinical features favor the diagnosis of Hodgkin lymphoma (Table 10.4).

10.5 Myeloid Neoplasms

Myeloid neoplasms arise from hematopoietic progenitors and typically give rise to clonal proliferations that replace normal bone marrow cells. There are three broad categories of myeloid neoplasia:

1. Acute myeloblastic leukemias (AMLs): The neoplastic cells are blocked at an early stage of myeloid cell development.
2. Myelodysplastic syndromes: Terminal differentiation occurs but in a disordered and ineffective fashion, leading to the appearance of dysplastic marrow precursors and peripheral blood cytopenias.
3. Myeloproliferative neoplasms: The neoplastic clone continues to undergo terminal differentiation but exhibits increased or dysregulated growth. Commonly, these are associated with an increase in one or more of the formed elements (red cells, platelets, and/or granulocytes) in the peripheral blood.

Although these three categories provide a useful starting point, the divisions between the myeloid neoplasms sometimes blur. Both myelodysplastic syndromes and myeloproliferative neoplasms often transform to AML, and some neoplasms have features of both myelodysplasia and myeloproliferative neoplasms. Because all myeloid neoplasms arise from early multipotent progenitors, the close relationship among these disorders is not surprising.

10.5.1   Acute Myeloblastic Leukemia (AML)

AML is a very heterogeneous neoplasm. It primarily affects older adults; the median age is 50 years. The clinical signs and symptoms closely resemble those produced by ALL and usually are related to the replacement of normal marrow elements by leukemic blasts. Fatigue, pallor, abnormal bleeding, and infections are common in newly diagnosed patients, who typically present within a few weeks of the onset of symptoms. Splenomegaly and lymphadenopathy generally are less prominent than in ALL, but on rare occasions, AML mimics a lymphoma by manifesting as a discrete tissue mass (a so-called granulocytic sarcoma). The diagnosis and classification of AML are based on morphologic, histochemical, immunophenotypic, and karyotypic findings. Of these tests, the karyotype is most predictive of outcome.

1. Morphology

    By definition, AML myeloid blasts or promyelocytes make up more than 20% of the bone marrow cellular component. Myeloid blasts have delicate nuclear chromatin, which has three to five nucleoli, and fine azurophilic cytoplasmic granules. In some cases, there are red-staining rod-like structures (Auer rods) which are more often present in the promyelocytic variant. Therefore, Auer rods are specific for neoplastic myeloblasts and a helpful diagnostic clue when present.

2. Immunophenotypic Features

    The expression of immunologic markers is varied in AML. Most tumors express some combination of myeloid-associated antigens, such as CD13, CD14, CD15, CD64, or CD117 (c-KIT). Because multipotent stem cells express CD33, myeloid progenitor cells are positive for CD33.

3. Clinical Features

    AML is a devastating disease. The onset is often very rapid and progresses to death because of anemia, hemorrhage, or infection occurring within weeks without treatment. Severe anemia causes pallor and hypoxic symptoms. Thrombocytopenia may lead to abnormal bleeding or purpura. Neutropenia results in infections. AMLs with t(8;21) or inv(16) have a better prognosis with conventional chemotherapy, particularly in the absence of c-KIT mutations, which have a 50% chance of long-term disease-free survival. However, the overall long-term disease-free survival is only 15–30% with conventional chemotherapy. Currently, allergenic marrow transplantation seems to be the only method to cure the disease.

10.5.2   Chronic Myelogenous Leukemia (CML)

CML is a myeloproliferative neoplasm characterized by the presence of a chimeric BCR-ABL gene, which is the product of a (9;22) translocation that moves the ABL gene on chromosome 9 to a position on chromosome 22 adjacent to the BCR gene. It accounts for 15–20% of all leukemias and peaks at ages 40–50 years.

1. Morphology

    CML is characterized by the presence of very high peripheral blood cell counts. The leukocyte count often exceeds 300×10^9/L. The circulating cells are predominantly neutrophils, metamyelocytes, and myelocytes, but basophils and eosinophils are also prominent. A small proportion of myeloblasts, usually less than 5%, can be seen in the peripheral blood. The bone marrow is hypercellular due to the proliferation of granulocytic and megakaryocytic precursors. Similar to the bone marrow, the red pulp of the spleen presents extensive extramedullary hematopoiesis. This burgeoning mass often compromises the local blood supply, leading to splenic infarcts.

2. Clinical Features

    The onset of CML is usually slow, and the initial symptoms are often nonspecific (e.g., easy fatigability, weakness, weight loss). Massive splenomegaly is common in CML patients. Sometimes the first symptom is a dragging sensation in the abdomen, caused by extreme splenomegaly. More than 90% of cases have the (9;22)(q34;q11) translocation (the so-called Philadelphia chromosome [Ph]),which can be tested for the presence of the BCR-ABL fusion gene. In other cases, the BCR-ABL fusion gene is formed by cytogenetic complexity or recessive rearrangement. CML patients are in the accelerated or blast crisis stage when the bone marrow contains more than 5% myeloblasts. The natural history of CML is initially slow, with a median survival of 3 years without treatment. After a variable (and unpredictable) period, approximately half of CML cases enter an accelerated phase, characterized by increasing anemia and new thrombocytopenia, with additional cytogenetic abnormalities, and eventually transformation into a picture resembling acute leukemia (i.e., blast crisis).

10.6   Histiocytic Neoplasms

Histiocytic neoplasms originate from mononuclear phagocytes (macrophages and dendritic cells) or histiocytes. Some are malignant tumors, such as the very rare histiocytic lymphoma. Others are benign, such as reactive histiocytic proliferations in lymph nodes. Langerhans cell histiocytic disease is relatively rare and is characterized by clonal proliferation of Langerhans cells. Clonal proliferation of others are benign, such as the reactive histiocytic proliferations in lymph nodes.

    In the past, these disorders were regarded as histiocytosis X and subdivided into three   categories: Letterer-Siwe   syndrome   (generalized   histiocytosis),

Hand-Schüller-Christian disease, and eosinophilic granuloma. Based on recent studies, these three categories represent three different stages of the same disease. The proliferating Langerhans cells are human leukocyte antigen DR (HLA-DR) and CD1 antigen positive. HX bodies (Birbeck granules) in the cytoplasm of these cells are characteristic. Under light microscopy, proliferating Langerhans cells have abundant, vacuolated cytoplasm and vesicular nuclei, which is different from their normal dendritic counterparts.

Clinically, Langerhans cell histiocytosis can be divided into three types: acute disseminated Langerhans cell histiocytosis, unifocal eosinophilic granuloma, or multifocal eosinophilic granuloma.

10.6.1   Acute Disseminated Langerhans Cell Histiocytosis

Acute disseminated Langerhans cell histiocytosis (also called Letterer-Siwe syndrome) appear to represent the aggressive end of the spectrum, with widespread lesions of bone and lymphoid tissue, and occurs usually before 2 years of age. The dominant clinical signs are skin lesions due to Langerhans cell infiltration. Most patients have hepatosplenomegaly, lymphadenopathy, pulmonary lesions, and destructive osteolytic bone lesions. Extensive infiltration of the marrow leads to anemia, thrombocytopenia, and recurrent infections. Thus, the clinical signs and symptoms may resemble those of acute leukemia. Without treatment, the prognosis is poor. With intensive chemotherapy, 50% of the patients survive 5 years.

10.6.2   Eosinophilic Granulomas

Both unifocal and multifocal eosinophilic granulomas are characterized by expanding, erosive accumulations of Langerhans cells, usually within the medullary cavities of bones. The proliferative Langerhans cells are variably admixed with eosinophils, lymphocytes, plasma cells, and neutrophils. The eosinophilic component ranges from scattered mature cells to sheet-like masses of cells. The calvarium, ribs, and femur are commonly involved.

Unifocal lesions are relatively benign and typically involve bone, particularly the skull and ribs of children and young adults, although long bones are sometimes involved. They may be asymptomatic or may cause pain, tenderness, and pathologic fractures. Radiologically, they present as well-demarcated lytic lesions. This disorder may heal spontaneously or be cured by local excision or irradiation.

Multifocal lesions have a less favorable prognosis. They usually affect children and present with fever; diffuse eruptions, particularly on the scalp and in the ear canals; and frequent bouts of otitis media. Lymphadenopathy, hepatomegaly, and splenomegaly are associated with the infiltration of Langerhans cells. The base of the skull is characteristically involved, producing the triad of proptosis, lytic bone lesions in the skull, and diabetes insipidus (the latter due to the destruction of the posterior pituitary). The combination of calvarial bone defects, diabetes insipidus, and exophthalmos is referred to as the Hand-Schüller-Christian triad. Most patients with this disease experience spontaneous regression, and others can be treated with chemotherapy.

CD1a and S-100 protein are positive in Langerhans cell with immunohistochemistry techniques. Birbeck bodies can be found with electron microscopy. These indicators are useful for the diagnosis of Langerhans cell histiocytosis.

Chapter 11
Diseases of the Immune System

Zhiping Hou

Contents

Objectives

1. To master the main pathologic morphological changes, both on microscopy and gross specimen.
2. To comprehend the epidemic characteristics of each major diseases.
3. To comprehend the major pathogenesis.

Immune response is an important process of "recognizing oneself and eliminating dissidents." It refers to protection against infections by defending against outside pathogens. Immune deficiencies increase the body's susceptibility to infections. However, abnormal immune reactions, including too little or too much immune activity, ultimately lead to tissue damage. This chapter briefly introduces some common types of immune diseases [1, 2].

Z. Hou (✉)
School of Basic Medical Sciences, Chengde Medical University, Chengde, China

© Zhengzhou University Press 2024
K. Chen et al. (eds.), *Textbook of Pathologic Anatomy*,
https://doi.org/10.1007/978-981-99-8445-9_11

11.1   Autoimmune Diseases

Autoimmune disease is the underlying condition that results from autoimmune reactions. Compared with other diseases, autoimmune diseases have their own characteristics: (1) In many of these diseases, autoimmune antibodies and/or autoreactive T lymphocytes can be detected. (2) In some cases, autoimmune antibodies and/or autoreactive T lymphocytes are known to cause pathological abnormalities and functional disorders. (3) These diseases often exhibit recurrent flare-ups and chronic progression, with symptoms that may be delayed or deferred over time.

11.1.1   Factors and Mechanisms

Autoimmune disease results from a breakdown of self-tolerance to self-antigens and mutations of related genes. The exact factors causing autoimmune disease have not been fully elucidated, but they may be related the following causes: immunologic tolerance breakdown, genetic factors in autoimmunity, and microorganisms.

11.1.1.1   Immunologic Tolerance Breakdown

The termination and destruction of autoimmune self-tolerance may be the fundamental mechanism of autoimmune diseases. Self-tolerance refers to the lack of immune response to self-tissue antigens. According to the maturity of T and B cells and the amount of autoantigens exposed to one's own, it can work through the following mechanisms to select against autoreactivity and prevent immune response against autoantigens:

Acquired Central Immune Tolerance

The principal mechanism is the antigen-induced death/apoptosis of self-reactive T and B lymphocytes during their mutation in generative lymphoid organs, such as the bone marrow for B cells and the thymus for T cells. Many autologous protein antigens in the thymus are processed and presented by thymic APCs (antigen-presenting cell) in association with the self-major histocompatibility complex. Any immature T cells that recognize self-antigens in the central lymphoid organs undergo deletion, also called negative selection. Similarly, some immature B cells may be deleted by apoptosis, while other self-reactive B cells may not die but undergo receptor editing.

Acquired Peripheral Immune Tolerance

Mature lymphocytes that recognize self-antigens in peripheral tissues lose function and become anergic, may be suppressed by T cells, or may die by apoptosis.

11.1.1.2 Genetic Factors in Autoimmunity

Many susceptibility genes play an important role in the occurrence of autoimmune diseases: (1) Autoimmune diseases tend to occur within families, and the incidence of the same autoimmune disease is greater in monozygotic twins than in dizygotic twins. (2) There is evidence that human leukocyte antigen (HLA) loci, such as HLA-DR and HLA-DQ, are linked with several autoimmune diseases and are associated with an increased odds ratio or relative risk. (3) Genetic polymorphism studies associated with families have revealed a connection to autoimmune diseases. Some loci appear to be linked with general mechanisms of self-tolerance and immune regulation, while other loci are reported to influence specific self-antigens or affect organ sensitivity.

11.1.1.3 Microorganisms

A variety of bacteria, mycoplasmas, and viruses have been implicated as triggers for autoimmunity. The underlying mechanisms of microbe-induced autoimmune reactions may include three factors: (1) Viruses and other bacteria, such as streptococci and klebsiella, may have cross-reacting epitopes with self-antigens, a process known as molecular mimicry. (2) Microbial infections with inflammation and resultant tissue necrosis may lead to the upregulation of costimulatory molecules on APCs in the tissue, thereby contributing to the breakdown of T-cell anergy and subsequent activation of T cells. (3) Infected tissue antigens may act as triggers for autoimmune responses.

11.1.2 Systemic Lupus Erythematosus

Systemic lupus erythematosus (SLE) is a relatively common disease with variable clinical behavior and labile manifestations. SLE is also a multisystem autoimmune disease, with a prevalence of about 1 in 2500 people in certain populations. Similar to other immune diseases, there is a strong (approximately 9:1) female preponderance, affecting 1 in 700 women of childbearing age. The onset of SLE is usually between 10 and 30 years old, but it may manifest at any age, even in early childhood (Fig. 11.1).

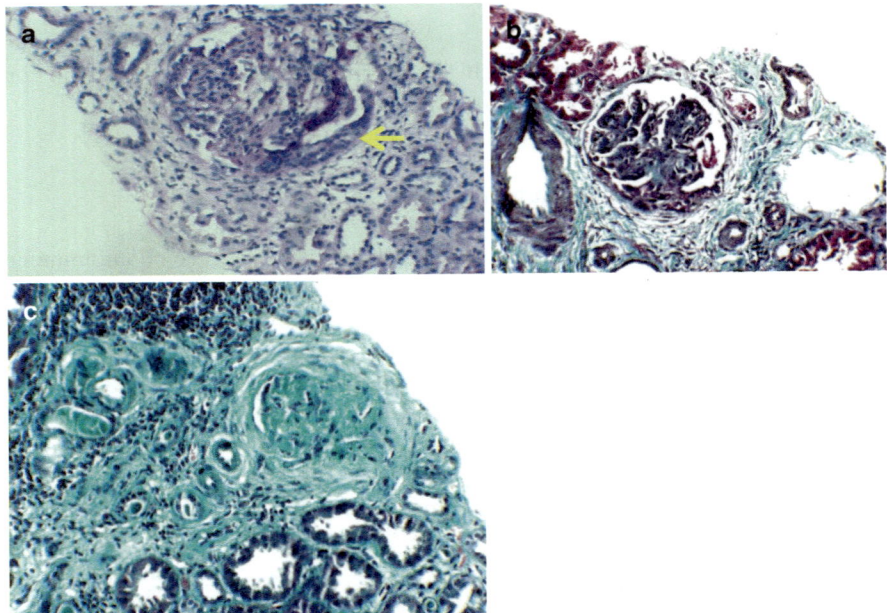

Fig. 11.1 Lupus nephritis. (**a**) Cellular crescent in renal capsule (H&E stain). (**b**) Cellular crescent in renal capsule (Masson stain). (**c**) Interstitial tissue hyperplasia, renal tubular atrophy (Masson stain)

11.1.2.1 Etiology and Pathogenesis

The major defect in SLE is the termination and breakdown of self-tolerance, leading to the production of multiple autoantibodies, which can injure tissues either directly or through the formation of immune complex deposits. Among them, antinuclear antibody is the most important in the autoantibody spectrum, including anti-histone antibody, anti-DNA antibody, anti-nucleolar antigen antibody, and antibodies to nonhistone proteins bound to RNA. Recent reports have demonstrated several factors influencing the pathogenesis of SLE:

1. Genetic Factors: Genome-wide association studies show that family members have up to a 20% increased risk of developing SLE.
2. Environmental Factors: Examples include ultraviolet radiation, cigarette smoking, and sex hormones.
3. Immune Factors: Failure of B-cell tolerance, which may be influenced by CD4$^+$ T_H cells, plays a role in its pathogenesis.

11.1.2.2 Morphology

The morphologic changes in SLE are variable, and the most characteristic changes depend on the deposition of natural autoantibodies and/or immune complexes. The fundamental morphologic changes include acute necrotizing arteritis and arteriolitis, which are present in almost all patients and involve all organs. In the active phase, fibrinoid necrosis is the major change; in the chronic phase, the major changes include necrosis, fibrinoid deposits involving antibodies, fibrinogen, and complement fragments, along with leukocytic infiltrate in the transmural and perivascular wall, as well as fibrous thickening of the vessel walls and narrowing of the lumen.

Kidney

Kidney failure is the most common cause of death in SLE. The classical pathological feature is glomerular pathology, although pathologic lesions are also observed in the interstitial and tubular areas. When the kidney is involved in SLE patients, almost all histological types of primary glomerulonephritis can be shown. As presented in the WHO (World Health Organization) morphologic classification, there are six nonspecific classifications of SLE glomerular disease, as shown in Table 11.1.

Skin

Erythematous or maculopapular lesions appear on the malar bones and the bridge of the nose, forming a "butterfly pattern," which is the most characteristic lesion in about half of SLE patients. A similar red rash appears on sun-exposed areas, hence the term photosensitivity. Histopathologically, the basal layer of the epidermis shows liquefactive degeneration, dermoepidermal junction edema, as well mononuclear infiltration. Immune complex deposition at the dermoepidermal junction and in other apparently uninvolved skin can be observed on immunofluorescence microscopy.

Joints

Ninety-five percent of SLE patients experience varying degrees of joint lesions, presenting with joint deformity associated with significant pathological changes. Pathological changes involve mononuclear cell and lymphocyte infiltration, swelling of the synovial membranes, and focal fibroid necrosis in the connective tissue adjacent to the synovial cells.

Table 11.1 Classification of glomerular disease in systemic lupus erythematosus

Class	Name	Light microscopic change	Immune complexes	Clinical symptoms	Prevalence (%)
Class I	Minimal mesangial lupus nephritis	No concomitant structural alterations	Present in the mesangium	Rarely	Rarely detected in renal biopsies
Class II	Mesangial proliferative lupus nephritis		Deposited in the cellularity and mesangial matrix	Mild clinical symptoms	10–25%
Class III	Focal lupus nephritis	Segmentally or globally displayed within less than half of each glomerulus			20–35%
Class IV	Diffuse lupus nephritis	Endothelial and mesangial proliferation, producing epithelial crescents that fill Bowman's space	Immune complexes are shown in subendothelial, mesangium, and other parts of the capillary wall, causing a granular fluorescent staining form	Hematuria, moderate to severe proteinuria, hypertension, and renal insufficiency	35–60%
Class V	Membranous lupus nephritis	By widespread thickening of the capillary wall	Immune complexes are shown in subepithelial immune complexes and basement membrane-like material	Severe proteinuria with overt nephrotic syndrome	10–15%
Class VI	Advanced sclerotic lupus nephritis	Complete sclerosis of more than 90% of glomeruli		End-stage renal disease	

Central Nervous System

Central nervous system (CNS) lesions can also be involved and are typically accompanied by focal neurologic deficits and neuropsychiatric symptoms. Most CNS disease in SLE is due to vascular changes causing ischemia or multifocal cerebral microinfarcts. The most common pathological lesion is small vessel angiopathy with endothelial proliferation; frank vasculitis is uncommon.

Cardiovascular System

The heart is involved in about half of SLE cases, primarily manifesting as nonbacterial verrucous endocarditis, which displays warty deposits on either surface of the bicuspid and tricuspid valves.

Other Organs

Many other organs, such as the spleen, may be involved. The pathological lesions are caused by acute vasculitis of the small vessels. The affected spleen may be enlarged, with capsular fibrous thickening, thickening of central penicilliary arteries, and perivascular fibrosis showing an onion skin lesion.

11.1.2.3   Clinical Manifestations

SLE is a multisystem disease with variable clinical presentations. The typical manifestations include nephritis, skin lesions, arthritis, and hematologic and/or neurologic abnormalities.

11.1.3   Rheumatoid Arthritis

Rheumatoid arthritis (RA) is a systemic chronic autoimmune disease, primarily characterized by nonsuppurative proliferative synovitis. RA is a very common disease. The onset of RA can occur at any age, but the peak incidence is between 30 and 50 years old. The incidence rate of RA in the population is about 1%, with a male-to-female prevalence ratio of approximately 1:3–5. With frequent exacerbations and remissions of inflammation, RA can cause the destruction of articular cartilage and underlying bone, eventually leading to disabling ankylosis.

11.1.3.1   Pathogenesis

RA is an immune-mediated inflammatory disease attributed to a breakdown of tolerance to self-antigens, influenced by genetic predisposition and environmental factors. The underlying mechanism involves the activation of CD4$^+$ T cells, which respond to an arthritogenic agent, such as microbial or self-antigens. Cytokines activated by T cells first activate macrophages and other cells in the joint space, which release specific enzymes or other factors that perpetuate inflammation. Secondly, cytokines can activate B cells to target self-antigens in the joint.

According to a variety of clinical and experimental observations, approximately 80% of RA patients have serum IgM antibodies that bind to the Fc fragment of self-IgG. These antibodies are therefore named rheumatoid factor (RF). The deposition of RF immune complexes in joints and other tissues is the primary cause of inflammation and tissue damage. The increased frequency of RA in first-degree relatives suggests that genetic factors play an important role in the pathogenesis. Moreover, specific agents activated by infectious agents, such as Epstein-Barr virus, Mycoplasma species, mycobacteria, and parvoviruses, may also activate T or B cells.

11.1.3.2 Morphology

Joint Lesions

Typical RA presents as multiple and symmetrical arthritis, especially affecting small joints of the hands, feet, wrists, knees, elbows, and shoulder. Histopathologically, the affected joints display chronic synovitis, with the classical features being (1) proliferation and hyperplasia of synovial cells, forming multiple layers, with villous processes sometimes present; (2) formation of lymphoid follicles surrounding perivascular areas in the synovium, infiltrated by CD4$^+$ T cells, macrophages, and plasma cells; (3) increased vascularity with endothelial cells showing high levels of adhesion molecules due to angiogenesis; (4) aggregation of neutrophils and organizing fibrin on the synovial surface and in the joint space; and (5) increased osteoblast activity in the related bone, leading to synovial penetration and bone erosion. The most characteristic appearance is the pannus, formed by proliferating synovial lining cells mixed with inflammatory cells, granulation tissues, and fibrous connective tissue; the smooth synovial membrane is transformed into lush, edematous, frondlike projections.

Extraarticular Lesions

Extraarticular lesions, which develop in one-fourth of RA patients, can affect multiple organs and tissues and form rheumatoid subcutaneous nodules. Rheumatoid subcutaneous nodules occur in subcutaneous tissue, especially along the extensor surfaces of the forearm or other tissues subject to mechanical pressure. Rarely, they form in solid viscera, such as the lungs, heart, and spleen. Typical features of rheumatoid subcutaneous nodules include an oval or rounded shape, normally more than 2 cm in diameter and a firm texture, and they are non-tender. Microscopically, the central part shows fibroid necrosis, surrounded by epithelioid cells arranged radially, with granulation tissue at the periphery.

11.2 Rejection of Transplants

Immunologic rejection of organ transplants is the major barrier in allografts (transplants from one individual to another of the same species). Immunologic transplant rejection is a complex process mediated by both cell and antibody reactions that damage the graft. The key to successful transplantation is minimizing rejection and developing effective therapies. How grafts are recognized as foreign or rejected is discussed as follows:

11.2.1 Immune Recognition of Allografts

Initially, MHC (major histocompatibility complex) antigens (on the graft) were discovered as transplantation antigens, which are the main antigen system causing transplantation rejection and play a major role in the process of allograft immune recognition. The polymorphism of MHC molecules enables the population to produce an appropriate immune response to various pathogens and cope with changing environmental conditions to maintain population stability. There are two main mechanisms by which the host immune system recognizes and responds to the MHC on the graft: (1) direct recognition, where foreign allogeneic MHC is recognized by host T cells, and (2) indirect recognition, where host-activated CD4$^+$ T cells initiate delayed-type hypersensitivity (DTH) by recognizing MHC, accompanied by microvascular damage, tissue ischemia, and macrophage damage.

11.2.2 Effector Mechanisms of Graft Rejection

11.2.2.1 T-Cell-Mediated Rejection

T lymphocytes from donors, especially CTLs (cytotoxic T lymphocytes), can cause necrosis of parenchyma and endothelial cells once they are recognized by the host's lymphocytes. Activated CD4$^+$ T cells induce inflammation-related DTH within grafted tissue and blood vessels. Meanwhile, CD8$^+$ cells differentiate into mature CD8$^+$ CTLs, which dissolve and destroy the grafted tissue.

11.2.2.2 Antibody-Mediated Rejection

Although T-cell-mediated rejection plays an important role, antibody-mediated rejection also plays a role. If the recipient has received a blood transfusion, been infected with bacteria or viruses, or undergone multiple pregnancies, antibodies against the donor's HLA due to cross-reactivity with some surface antigens may be present in the blood circulation. In this case, rejection can occur immediately after transplantation. This is because circulating antibodies can bind to the vascular endothelium of the graft, activating the complement system and causing vascular endothelial damage, leading to vascular wall inflammation, thrombosis, and tissue necrosis. Even in individuals who are not sensitized, with the development of T-cell-mediated rejection, anti-HLA antibodies can form in the vascular endothelium of the graft simultaneously, resulting in graft damage.

11.2.3   Morphology

Based on the morphologic changes and mechanisms of rejection reactions, there are three classic patterns: hyperacute, acute, and chronic rejection. This section introduces the pathological changes of rejection in kidney transplantation as an example. The morphological changes are similar in other vascularized organ transplants.

11.2.3.1   Hyperacute Rejection

This pattern normally occurs within minutes to a few hours after organ transplantation due to vascular anastomosis. Hyperacute rejection is caused by pre-sensitization of the host, such as the presence of donor-specific HLA antibodies in the recipient's blood circulation, or a mismatch in ABO blood types between the recipient and donor. Compared to a non-rejected kidney graft, a kidney experiencing hyperacute rejection becomes cyanotic with red spots, and urine output is reduced to only a few drops of bloody fluid. Histologically, the graft endothelium shows widespread acute arteritis and arteriolitis accompanied by vessel thrombosis and ischemic necrosis. Affected arterioles and arteries are characterized by precipitated fibroid necrosis of the vessel walls, narrowing or occlusion of their lumens, and cellular debris.

11.2.3.2   Acute Rejection

Acute rejection is a more common pattern and often occurs within several days after transplantation, though it can also occur months or years later. The underlying mechanisms may involve both cellular and humoral immune reactions, and in each patient, one mechanism may predominate, or both may be involved simultaneously. On histopathology, cellular rejection is indicated by interstitial mononuclear cell infiltrate accompanied by edema and cellular injury, while humoral rejection is associated with vasculitis.

Acute Cellular Rejection

Generally, acute cellular rejection is detected within the first few months after transplantation and is normally associated with manifestations of renal failure. Under the microscope, CD4$^+$ and CD8$^+$ T cells, predominantly monocytes, infiltrate the interstitial kidney, accompanied by hemorrhage and edema. Monocytes can be found in the capillaries of the glomeruli and peritubular areas, causing focal tubular necrosis. They can also invade the endothelium, leading to endotheliitis.

Acute Humoral Rejection (Rejection Vasculitis)

Acute humoral rejection is mainly caused by anti-donor antibodies. The deposition of these antibodies and activated complement may lead to necrotizing vasculitis, accompanied by infiltrated neutrophils, endothelial necrosis, fibrin deposition, and thrombosis. Histopathologically, lesions may show renal parenchyma ischemic necrosis; intima thickening due to the proliferation of fibroblasts, myocytes, and foamy macrophages; and narrowed arterioles.

11.2.3.3 Chronic Rejection

Chronic rejection, characterized by a slowly rising serum creatinine level, occurs months to 1 year after transplantation. Typical features include vascular changes, interstitial fibrosis, and atrophy of renal parenchyma without ongoing cellular parenchymal infiltrates. The vascular changes mainly occur in the arteries and arterioles, leading to the proliferation of intimal smooth muscle cells and synthesis of extracellular matrix. These lesions ultimately result in renal ischemia or hyalinization of glomeruli, interstitial fibrosis, and tubular atrophy. The vascular lesions may be caused by cytokines released from activated T cells that affect the cells of the vascular wall, and this may represent the end stage of proliferative arteritis. Studies have shown that excessive accumulation of renal extracellular matrix is a significant event in chronic renal rejection, renal fibrosis, and renal tubular atrophy. In the process of renal fibrosis, TGF-β1 can increase the formation of extracellular matrix and promote renal fibrosis through various mechanisms, and it may be the primary way TGF-β1 induces renal tubular epithelial-mesenchymal transition (EMT).

11.3 Immunodeficiency Diseases

Immunodeficiency diseases are a group of conditions leading to immune function deficiency caused by an underdeveloped or damaged immune system. Clinically, individuals with immune deficiency show increased susceptibility to uncontrolled opportunistic infections, autoimmune diseases, and certain types of cancer.

Immunodeficiency diseases can be divided into two categories: (1) primary immune deficiencies, also known as congenital immune deficiencies, which are related to genetics and mostly affect infants and young children, and (2) secondary immune deficiencies, also called acquired immune deficiencies, which can occur at any age. Acquired immunodeficiency syndrome (AIDS) will be discussed as a prominent example of secondary immune deficiency.

11.3.1 Acquired Immunodeficiency Syndrome (AIDS)

AIDS is caused by human immunodeficiency virus (HIV) infection and is characterized by the depletion of $CD4^+$ T lymphocytes, secondary tumors, and immune deficiency accompanying opportunistic infection. The symptoms of HIV infection vary depending on the stage of infection. People living with HIV tend to be the most infectious in the first few months after infection, but many people do not realize their status until later. In the initial weeks of infection, individuals may be asymptomatic or experience flu-like symptoms such as fever, headache, rash, or sore throat. As the viral infection weakens the immune system, other symptoms and signs may develop, including lymphadenopathy, weight loss, fever, diarrhea, and cough. Without treatment, serious diseases such as tuberculosis, cryptococcal meningitis, severe bacterial infections, and cancer (such as lymphoma and Kaposi's sarcoma) may occur. The WHO observes December 1 each year as World AIDS Day because the first case was diagnosed on this day in 1984. HIV remains a major global public health problem, causing nearly 33 million deaths to date. By the end of 2019, an estimated 38 million people were living with HIV. In 2019, 690,000 people died from HIV-related diseases, and 1.7 million people became newly infected. As of the end of 2019, 25.4 million people were receiving antiretroviral treatment. Between 2000 and 2019, new HIV infections dropped by 39%, and HIV-related deaths dropped by 51%. Antiretroviral drug treatment has saved approximately 15.3 million lives [3].

11.3.1.1 Etiology and Pathogenesis

HIV, a retrovirus with a single strand of RNA, leads to AIDS. The main routes of HIV infection include sexual contact, parenteral inoculation, blood transmission, and vertical mother-to-child transmission. HIV is divided into two subtypes, HIV-1 and HIV-2, which were discovered in 1983 and 1985, respectively. The three major mechanisms are as follows: (1) a large number of $CD4^+$ T cells are destroyed by HIV, damaging their functions and leading to cellular immune deficiency; (2) monocytes-macrophages in HIV infection act as gatekeepers, providing a portal for initial transmission; and (3) DCs (dendritic cells) in HIV infection, particularly follicular DCs in the germinal centers of lymph nodes, can also be infected by HIV and become a reservoir for the virus.

11.3.1.2 Morphology

The pathological features can be summarized as systemic lymphoid tissue changes, opportunistic infections, and malignant tumors.

1. Changes in lymphatic tissue: Lymph nodes are enlarged in the early stage, with marked follicular hyperplasia. Under the microscope, there are abundant plasma cells in the medulla. As the disease progresses, the lymphocytes in the outer layer of follicles decrease or disappear, small blood vessels proliferate, and the germinal centers become fragmented. In the late stage, the lymph nodes appear "burn-out," atrophic, and small. Sometimes, a large number of pathogens, such as mycobacteria and fungi, can be detected by special staining, but cellular immune response diseases such as granuloma formation are rarely seen.
2. Secondary infections: Multiple opportunistic infections are another feature of this disease. The infections can affect various organs, with the CNS, lungs, and digestive tract being the most common.
3. Kaposi's sarcoma and lymphoma are the common tumors: As the disease progresses, there is severe follicular involution and generalized lymphocyte depletion due to B-cell proliferation. The organized network of follicular DCs is disrupted, and the follicles may become atrophic. These "burnt-out" lymph nodes are small and may harbor numerous opportunistic pathogens. Non-Hodgkin lymphomas, often involving extranodal sites such as the brain, are mainly aggressive B-cell neoplasms.

References

1. 翟启辉, 周庚寅. 病理学. 北京大学医学出版社; 2009.
2. Kumar V, Abbas AK, Aster JC. Robbins Basic Pathology. 9th ed. 2013. Elsevier saunders. Printed in Canada.
3. https://www.who.int/zh/news-room/fact-sheets/detail/hiv-aids

Chapter 12
Diseases of the Urinary System

Jing Cui, Jiateng Zhong, and Haijun Wang

Contents

Objectives

1. To master the basic pathological changes of glomerulonephritis and the pathological changes and clinicopathological correlation of acute diffuse proliferative glomerulonephritis, rapidly progressive glomerulonephritis, chronic glomerulonephritis, and chronic pyelonephritis.
2. To become familiar with the clinical manifestations, pathological types, and characteristics of glomerulonephritis, the pathologic features and clinicopathological correlation of renal cell carcinoma and bladder cancer, and the pathological features of other common diseases of the urinary system.
3. To understand the cause and pathogenesis of glomerulonephritis, pyelonephritis, renal cell carcinoma, and bladder cancer.

J. Cui · J. Zhong · H. Wang (✉)
School of Basic Medical Sciences, Xinxiang Medical University, Xinxiang, China

© Zhengzhou University Press 2024
K. Chen et al. (eds.), *Textbook of Pathologic Anatomy*,
https://doi.org/10.1007/978-981-99-8445-9_12

Key Concepts
Glomerulonephritis, crescent, nephrotic syndrome, granular atrophy of the kidneys, pyelonephritis.

12.1 Introduction

The urinary system consists of the kidneys, ureters, bladder, and urethra. The main function of the urinary system is to eliminate waste, excess water, and inorganic salts. The kidney is the most important organ in the urinary system; its main functions are to generate and drain urine metabolites and regulate water, electrolyte, and acid-base balance, and it also has endocrine function, secreting renin, erythropoietin, prostaglandins, and so on.

The kidney unit is the basic structure and functional unit of the kidney, consisting of the glomerulus and tubule. There are about two million kidney units on both sides of the human body, with powerful compensatory functions. The glomeruli mainly perform filtration functions, while the renal tubule reabsorbs the original urine composition and excretes waste. The glomeruli are composed of a central vascular ball and a capsule wrapped around the kidney capsule. The blood vessel is composed of capillary loops. The capillary wall of the glomeruli is the filtration membrane, which is composed of capillary endothelial cells, basement membrane, and visceral epithelial cells (also known as foot cells). Under normal circumstances, glomerular filtration membranes are highly permeable to water and small molecular solutes, while macromolecules such as proteins are almost completely unable to pass (Fig. 12.1).

The glomerular mesangium is composed of mesangial cells and mesangial matrix. The mesangium is located between the capillaries, forming the central axis of the glomerulus and providing support and protection for the capillaries. Mesangial cells have functions such as contraction, phagocytosis, proliferation, and synthetizing membrane matrix and collagen fibers. They can secrete a variety of vasoactive substances and cytokines to participate in the inflammatory response.

The renal capsule, also known as the Bowman capsule, has an inner layer of visceral epithelial cells and an outer layer of parietal epithelium cells. Between these two layers is the balloon lumen, and its urinary pole is connected with the proximal convoluted tubule.

The renal tubules, including the proximal convoluted tubules, thin segment, and distal convoluted tubules, are composed of a single layer of epithelium.

Diseases of the urinary system include renal and urinary tract lesions, such as inflammation, tumors, urinary obstruction, vascular diseases, metabolic diseases, and congenital malformations. Based on the main part of the lesion, renal diseases are divided into glomerular diseases, renal tubular diseases, renal interstitial diseases, and vascular diseases. This chapter mainly introduces primary glomerular disease, renal tubulointerstitial nephritis, and common neoplasms of the kidney and bladder.

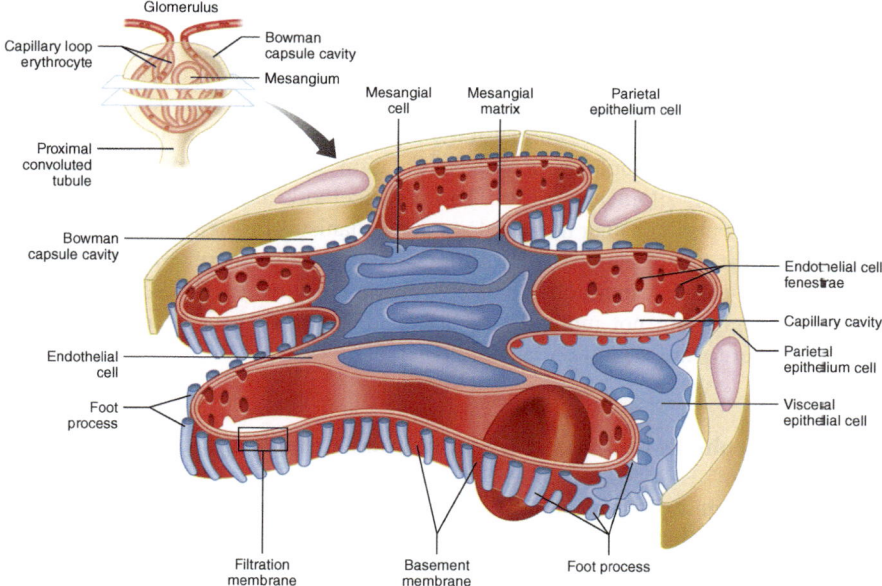

Fig. 12.1 Schematic diagram of glomerular structure

12.2   Glomerular Disease

Glomerular disease, also called glomerulonephritis (GN), is a group of diseases characterized by glomerular damage and changes. Glomerular disease is divided into primary glomerulonephritis, secondary glomerular disease, and hereditary diseases. Primary glomerulonephritis is an independent kidney disease, primarily affecting the kidneys. Secondary glomerular disease is a manifestation of other diseases or systemic conditions, such as lupus nephritis, diabetic nephropathy, and allergic purpura nephritis. This section mainly discusses primary glomerulonephritis.

12.2.1   Etiology and Pathogenesis

The etiology and pathogenesis of primary glomerulonephritis have not been fully elucidated, but it has been confirmed that most types are caused by immune mechanisms.

There are many known antigens that can cause glomerulonephritis, which can be divided into two categories: endogenous and exogenous. Endogenous antigens include glomerular antigens (glomerular basement membrane antigen, foot cells, endothelial cells, mesangial cell membrane antigen, etc.) and non-glomerular antigens (nuclear antigens, DNA, immunoglobulin, tumor antigens, thyroglobulin, etc.). Exogenous antigens include bacteria, fungi, viruses, parasites, helminths, drugs, exogenous lectin, heterogeneous serum, etc.

The antigen-antibody response is the main cause of glomerular injury. The damage related to antibodies occurs mainly through two mechanisms: one is the reaction of the antibody with glomerular antigens in situ; the other is the formation of antigen-antibody complexes in blood circulation, which are then deposited in the glomeruli, causing glomerulopathy. These two methods are the basic mechanisms of glomerulonephritis. Other immunological mechanisms involved in the occurrence of nephritis include cytotoxic reactions induced by anti-glomerular cell antibodies, activation of cellular immunity, and complement pathways. These pathways of immune damage are not mutually exclusive; different damage mechanisms may work together to cause glomerular lesions.

12.2.1.1 Nephritis Caused by In Situ Immune Complexes

Antibodies interact directly with the antigen components of the glomerulus itself or with antigens implanted from blood circulation into the glomerulus, forming in situ immune complexes in the glomeruli, causing glomerulopathy. Different kinds of antigens can cause different types of glomerulonephritis. There are three main types:

1. Nephritis caused by anti-glomerular basement membrane antibodies: This type of nephritis is caused by the interaction between the antibody and the glomerular basement membrane itself (Fig. 12.2).

Fig. 12.2 Schematic diagram of nephritis caused by anti-glomerular basement membrane antibodies

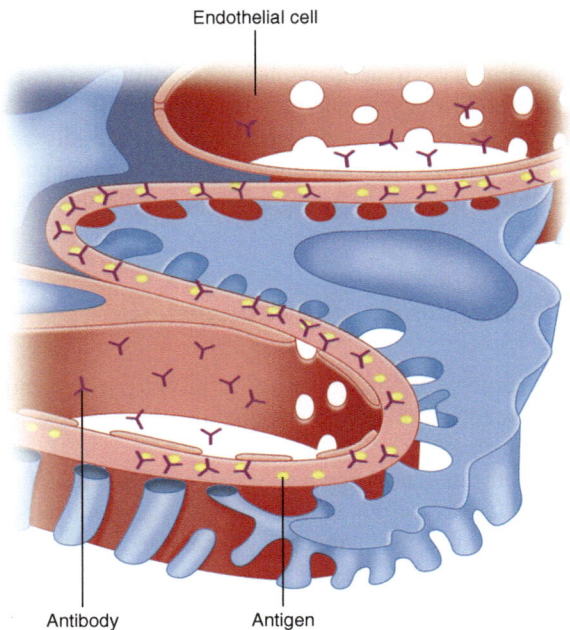

Endothelial cell

Antibody   Antigen

Fig. 12.3 Schematic diagram of Heymann nephritis

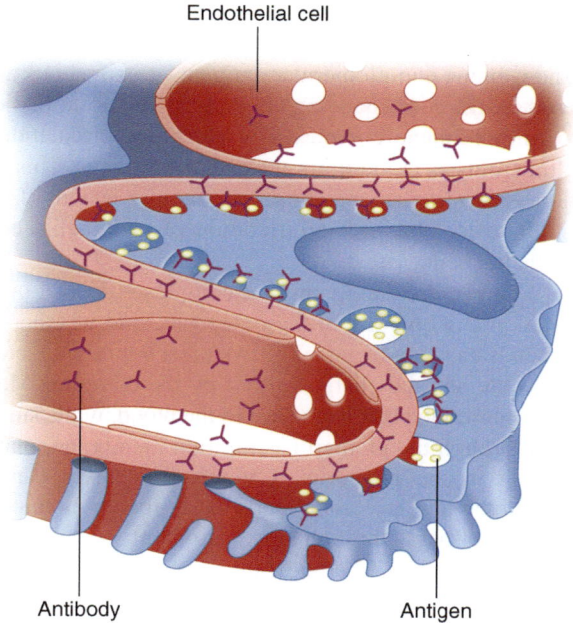

Endothelial cell

Antibody   Antigen

The formation of glomerular basement membrane (GBM) antigens may result from infection or other factors that alter the basement membrane structure, rendering it antigenic and stimulating the body to produce autoantibodies. It may also involve a cross-reaction due to shared antigenicity between pathogenic microorganisms and GBM components. The antibodies are deposited along the GBM, and immunofluorescence examination shows characteristic continuous linear fluorescence.

2. Heymann nephritis: Heymann nephritis is a classic animal model for studying human primary membranous glomerulopathy. The brush border of the proximal convoluted tubule shares antigenicity with foot cells. The former stimulates the production of antibodies, leading to a cross-immunoreaction with podocytes, forming immune complexes that deposit in the subepithelial region and cause glomerulonephritis (Fig. 12.3). Immunofluorescence shows diffuse granular immunoglobulin or complement deposition.

3. The reaction of antibodies with implantable antigens: Some non-glomerular antigens can bind with components in the glomerulus, forming implantable antigens that lead to the formation of in situ immune complexes, causing glomerulonephritis. Immunofluorescence examination shows scattered granular fluorescence.

12.2.1.2 Nephritis Caused by Circulating Immune Complexes

Endogenous non-glomerular antigens or exogenous antigens stimulate the body to produce the corresponding antibodies. These antibodies combine with the antigens to form immune complexes in the blood circulation, which then deposit in the glomerulus as blood flows through the kidneys. This deposition, often in combination with complement, causes glomerular lesions (Fig. 12.4). Neutrophils usually infiltrate locally, with proliferation of endothelial cells, mesenchymal cells, and epithelial cells.

Whether circulating immune complexes deposit in the glomerulus, as well as the location and quantity of these deposits, and whether they cause glomerular injury depends on many factors. The two most important factors are the size of the complex molecule and the charge carried by the complex. Macromolecular complexes are often scavenged by phagocytes in the blood, while small molecule complexes easily pass through the glomerular filtration membrane, making both types less likely to deposit in the glomeruli. Cationic complexes can pass through the basement membrane and are often deposited in the epithelium. Anionic complexes are less likely to pass through the basement membrane and are often deposited subendothelially. Charge-neutral complexes are more easily deposited in the

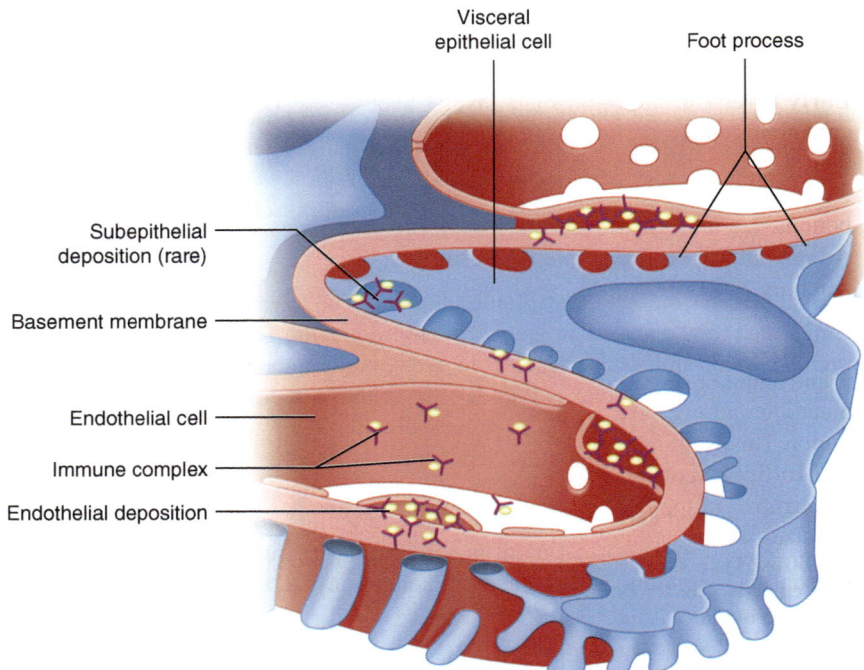

Fig. 12.4 Schematic diagram of nephritis caused by circulating immune complexes

Fig. 12.5 Immunofluorescence staining showed
discontinuous granular
fluorescence

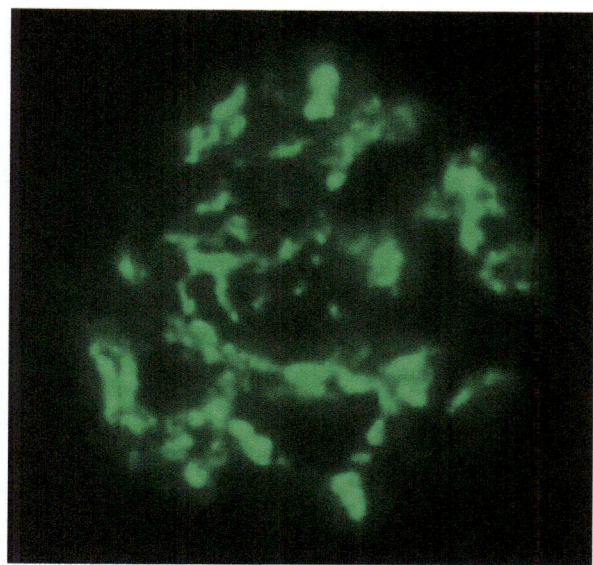

mesangial region. Immunofluorescence examination shows discontinuous granular fluorescence in glomerular lesions (Fig. 12.5).

The presence of immune complexes or sensitized T lymphocytes in glomeruli requires various mediators to participate in glomerular injury. These mediators include cells and macromolecules bioactive substances, such as inflammatory mediators released by neutrophils, macrophages, and platelets, as well as intrinsic glomerular cells.

The formation and deposition of immune complexes are characteristic of most types of glomerulonephritis, and the clinical detection of immune complexes is an important aspect of glomerulonephritis pathological diagnosis and research. Electron microscopy and immunofluorescence examination are common and indispensable methods for this purpose and are key pathological features of glomerulonephritis.

Different types of glomerulonephritis have different locations for immune complex deposition and formation. Immune complexes can deposit under the endothelium (in the basement membrane and endothelial cells), in the basement membrane, under the epithelium (between the basement membrane and podocyte), or in the mesangial area. Under the electron microscope, the immune complex appears as electron-dense material. Immunofluorescence confirms that the immune complex contains immunoglobulin and complement, showing continuous linear fluorescence or discontinuous granular fluorescence in different positions of the glomeruli, depending on the type.

In summary, the mechanisms of glomerular injury are as follows: (1) Antibody-mediated immune injury is an important mechanism of glomerular injury, primarily involving complement and leukocyte-mediated functions. (2) Most antibody-mediated nephritis is caused by circulating immune complex deposition, with immunofluorescence examination showing a granular distribution of immune complexes. (3) Anti-GBM antibodies can induce anti-GBM nephritis, with immunofluorescence showing a linear distribution of antibodies. (4) Antibodies can react with implanted glomerular antigens to form immune complexes in situ, with immunofluorescence examination showing granular fluorescence.

12.2.2 Morphology

Pathological examination of renal tissue plays an irreplaceable role in diagnosing glomerular disease. Renal puncture tissue can be routinely examined by light microscopy, immunofluorescence, and transmission electron microscopy. In addition to hematoxylin and eosin (HE) staining, tissue sections are routinely stained with periodic acid-Schiff (PAS) staining, periodic acid-silver methenamine (PASM) staining, and Masson trichrome staining. Renal biopsy tissue is routinely examined using immunofluorescence to detect immunoglobulin (IgG, IgM, and IgA) and complement component (C3, C1q, and C4) deposits. An electron microscope is used to observe ultrastructural changes and the condition and location of immune complex deposition.

The basic pathological changes of glomerulonephritis include the following:

1. Hypercellularity: The number of glomerular cells increases, with proliferation of mesangial cells, endothelial cells, and epithelial cells, accompanied by infiltration of neutrophils, macrophages, and lymphocytes. Epithelial cell hyperplasia can lead to the formation of an inner crescent in the kidney capsule.
2. Basement Membrane Thickening: The thickening of the basement membrane can result from the thickening of the membrane itself or from the deposition of immune complexes in the subendothelial, subepithelial, or basement membrane areas. PAS and PASM staining show thickening of the basement membrane.
3. Inflammatory Exudation and Necrosis: In acute inflammation, the glomeruli may exhibit infiltration by neutrophils and other inflammatory cells, with exudation of fibrin. The capillary walls may show fibrinoid necrosis and thrombosis.
4. Hyalinization and Sclerosis: Hyalinization of the glomeruli is characterized by homogeneous eosinophilic deposition in HE staining. In severe cases, there is a decrease or disappearance of glomerular intrinsic cells, stenosis and occlusion of the capillary lumen, increased collagen fibers, and eventual glomerulosclerosis. Glomerular hyalinization and sclerosis are the final results of various glomerular lesions.
5. Renal Tubules and Interstitial Changes: The epithelial cells of the renal tubules often undergo degeneration, with protein, cell debris, or cellular casts forming in the lumen. Hyperemia, edema, and infiltration of inflammatory cells can occur

in the renal interstitium. When the glomeruli show hyalinization and sclerosis, the corresponding renal tubules may atrophy or disappear, and interstitial fibrosis may develop.

According to the extent of the lesions, glomerulonephritis is classified into two categories: focal and diffuse. Focal glomerulonephritis involves less than half of the glomeruli, while diffuse glomerulonephritis involves more than half or all of the glomeruli. If a glomerulus shows lesions involving all or most of the glomerular capillary loops, it is termed a global lesion; if the lesions involve only some of the capillary loops (no more than 50% of the glomerular section), it is termed a segmental lesion.

12.2.3 Clinical Course

The clinical course of glomerulonephritis mainly includes changes in urine volume (oliguria, anuria, polyuria, or nocturia), changes in urine characteristics (hematuria, proteinuria, and tubular urine), edema, and hypertension.

The clinical course of glomerulonephritis is closely related to the pathological type, but there is not always a direct correspondence. The same pathological type can produce different symptoms and signs, while different pathological changes can also cause similar clinical manifestations. In addition, clinical manifestations are related to the degree and stage of the disease. The main clinical manifestations of glomerulonephritis are divided into the following types:

1. Acute Nephritic Syndrome: This condition has an acute onset, with prominent hematuria, mild to moderate proteinuria, edema, and hypertension as the main clinical manifestations. Severe cases may present with azotemia. Acute glomerulonephritis syndrome is often observed in acute diffuse proliferative glomerulonephritis.
2. Rapidly Progressive Nephritic Syndrome: This condition has an acute onset and rapid progression, manifested by edema, hematuria, proteinuria, rapid oliguria or anuria, azotemia, and acute renal failure. Rapidly progressive nephritic syndrome primarily occurs in rapidly progressive glomerulonephritis.
3. Nephritic Syndrome: The main clinical features include proteinuria (urinary protein greater than 3.5 g/d), severe edema, hypoproteinemia, hyperlipemia, and lipids in urine, commonly referred to as "three highs and one low." Various pathological types can present with nephrotic syndrome, including minimal change glomerulopathy, membranous glomerulopathy, focal segmental glomerulosclerosis, membranous proliferative glomerulonephritis, and mesangial proliferative glomerulonephritis.
4. Asymptomatic Hematuria or Proteinuria: The main clinical feature is persistent or recurrent macroscopic or microscopic hematuria, with mild proteinuria,

which can occur simultaneously. The corresponding pathological type is IgA nephropathy.

5. Chronic Nephritic Syndrome: This can occur in the end stage of various glomerulonephritis, mainly presenting as polyuria, nocturia, hypobaric urine, hypertension, anemia, azotemia, and uremia and gradually developing into renal failure.

Glomerular lesions can lead to a decrease in the glomerular filtration rate and an increase in blood urea nitrogen and plasma creatinine levels, resulting in azotemia. Uremia occurs late in both acute and chronic renal failure, presenting with symptoms and signs of self-poisoning, including gastrointestinal tract changes, nerve and muscle symptoms, and cardiovascular issues, such as gastroenteritis, uremic peripheral neuropathy, and fibrinous pericarditis.

12.2.4 Pathological Type of Glomerulonephritis

The pathological types of glomerulonephritis are numerous, with complex clinical courses, varying therapeutic effects, and different prognoses. In recent years, the application of renal biopsy techniques has made pathomorphological classification significant for the treatment and prognosis of glomerulonephritis. This section mainly introduces common primary glomerulonephritis.

12.2.4.1 Acute Diffuse Proliferative Glomerulonephritis

The characteristics of acute diffuse proliferative glomerulonephritis include diffuse hyperplasia of glomerular capillary endothelial cells and mesangial cells, with infiltration by neutrophils and macrophages. Clinically, it is abbreviated as acute nephritis and is primarily manifested as acute nephritic syndrome. Acute nephritis is mainly caused by infection and is also known as post-infectious glomerulonephritis or endocapillary proliferative glomerulonephritis.

Group A β-hemolytic streptococcus is the most common pathogen, with some cases associated with other bacterial or viral infections. Most patients experience angina, scarlet fever, or skin streptococcal infection 1–4 weeks before onset. Moreover, there is an increase in the serum antibody titer of anti-streptolysin O and other anti-streptococcus antigens and a decrease in serum complement levels. The pathogenesis of this disease involves nephritis mediated by circulating immune complexes.

Morphology

Grossly, the kidneys are bilaterally enlarged, with increased membrane tension, hyperemia on the surface, and a red color, sometimes referred to as "red kidney." Occasionally, the surface and section of the kidneys show scattered hemorrhagic

Fig. 12.6 Acute diffuse
proliferative
glomerulonephritis

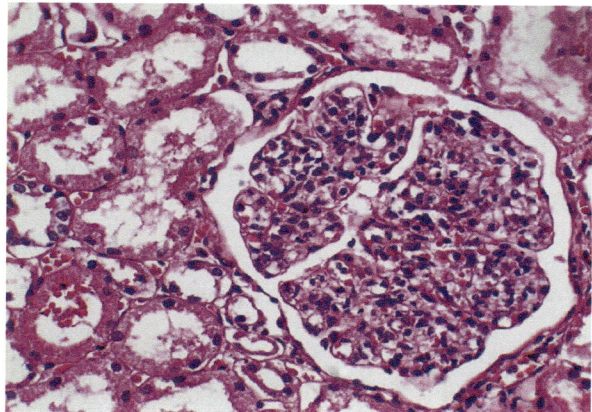

points, resembling flea bites, known as "flea-bitten kidney." The cortex is thicker, and the demarcation between the cortex and medulla is clear.

Histological changes involve most of the glomeruli. The volume of the glomerulus increases, with proliferation of endothelial cells and mesangial cells, swelling of endothelial cells, and infiltration by neutrophils and macrophages. A significant increase in the number of cells in the glomerulus can exert pressure on the capillaries, causing stenosis or occlusion of the lumen (Fig. 12.6). In severe lesions, fibrous necrosis and thrombus formation in the glomerular capillaries can be accompanied by significant bleeding. The epithelial cells of the renal proximal convoluted tubule may show denaturation, with the presence of red cell casts, white cell casts, and granular casts. The renal interstitial blood vessels exhibit hyperemia, edema, and infiltration by inflammatory cells. Immunofluorescence examination shows deposition of IgG, IgM, and C3 in the glomeruli with granular fluorescence. Electron microscopic examination reveals hump-like deposits between the visceral epithelial cells and the GBM.

Clinicopathological Correlation

Acute nephritis is more commonly seen in children, with clinical courses presenting as acute nephritic syndrome. Patients often exhibit symptoms of fever, oliguria, and hematuria approximately 10 days after infection in the pharynx or other areas. Hematuria and mild proteinuria result from increased permeability of glomerular capillaries and blood vessels. Hematuria is a common symptom; most patients have microscopic hematuria, with a few showing macroscopic hematuria. Urinalysis may reveal protein casts, cell casts, and granular casts. Due to the proliferation of glomerular cells, compression of capillaries leads to stenosis and occlusion, reducing the glomerular filtration rate and causing oliguria and azotemia in severe cases. Patients often experience edema and mild to moderate hypertension. Edema is primarily caused by decreased glomerular filtration rate, leading to water and sodium retention or hypersensitivity increasing capillary permeability. Water and sodium retention increase blood volume and elevate blood pressure.

The prognosis for children is generally good; most cases gradually subside, with symptoms relieved or disappearing. However, less than 1% of children may develop acute glomerulonephritis, and some may progress to chronic glomerulonephritis. The prognosis for adults is poorer, with potential progression to acute nephritis or chronic glomerulonephritis.

12.2.4.2 Rapidly Progressive Glomerulonephritis

The pathological changes of rapidly progressive glomerulonephritis (RPGN) are characterized by the formation of crescents in the epithelial cells of the glomerular wall layer, so it is also known as crescentic glomerulonephritis. This disease is characterized by rapid onset, rapid progression, and poor prognosis. Its clinical course is rapidly progressive glomerulonephritis syndrome.

Rapidly progressive glomerulonephritis can be primary, secondary, or associated with other glomerular diseases. According to immunological and pathological tests, RPGN can be divided into three subtypes:

1. Type I: Glomerulonephritis caused by anti-GBM antibodies. There is a cross-reaction between the anti-GBM antibody and the alveolar basement membrane, causing pulmonary hemorrhage, accompanied by hematuria, proteinuria, hypertension, and other nephritis symptoms, often progressing to renal failure. This lesion is called Goodpasture syndrome.
2. Type II: Immune complex nephritis, which is more common in China. It can develop from immune complex nephritis caused by streptococcal glomerulonephritis, systemic lupus erythematosus, IgA nephropathy, anaphylactoid purpura, and other causes.
3. Type III: Immunoreactive deficiency-type nephritis. No anti-GBM antibodies or immune complex deposits are found in the glomeruli, and it is considered to be caused by glomerular vasculitis.

Morphology

Gross appearance: The bilateral kidney volume is increased, the color is pale, and the surface of the kidney may show scattered points of hemorrhage. The cortical thickness is also increased.

Histological changes: Most glomeruli exhibit a characteristic crescent formation. Crescents are mainly composed of proliferating epithelial cells and mononuclear cells, with infiltration of neutrophils and lymphocytes. These crescents or ring structures are attached to the outer wall layer of the capillaries (Fig. 12.7). Early crescents are primarily made up of cellular components, known as cellular crescents. Over time, collagen fibers gradually increase, transforming into fibrocellular

Fig. 12.7 Rapidly progressive glomerulonephritis

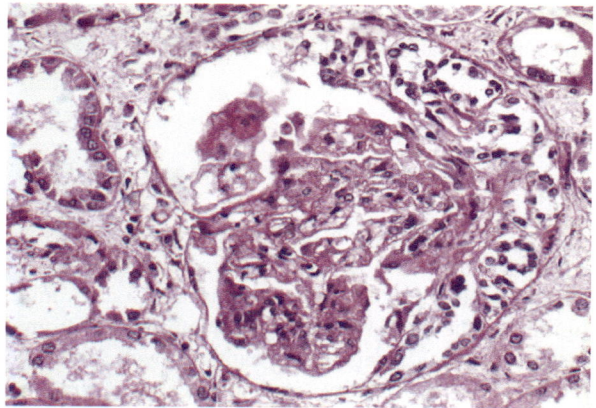

crescents, and eventually, the crescent becomes completely fibrotic, forming a fibrous crescent. The crescent often contains more cellulose, and cellulose exudation is a key reason for crescent formation. After the crescent or annular body forms, the glomerular capillaries may be compressed, leading to narrowing or blockage of the glomeruli. This results in atrophy, fibrosis, and hyalinization of the glomerular capillaries. Renal tubule epithelial cells show degeneration due to protein absorption and intracellular hyalinization, with partial atrophy or disappearance of renal tubular epithelial cells. Renal interstitial edema and inflammatory cell infiltration are present, with late fibrosis.

Immunofluorescence results: The results vary with the type of acute nephritis. Type I shows linear fluorescence, type II shows granular fluorescence, and type III is negative.

Electron microscopy: Electron microscopy reveals crescentic formations and defects or breakages in the GBM. Electron-dense deposits are found in type II cases.

Clinicopathological Correlation

The rapid progression of rapidly progressive glomerulonephritis is characterized by rapid progressive nephritis syndrome. Due to glomerular capillary necrosis, basal membrane defects, and hemorrhage, patients exhibit obvious hematuria and moderate proteinuria. A large number of crescent bodies form, which can rapidly reduce urine output and lead to anuria. The retention of metabolic waste in the body causes nitrogenemia, eventually leading to uremia and kidney failure. Extensive fibrosis and hyalinization in the glomeruli can cause renal glomerular ischemia and high blood pressure through the renin-angiotensin system and water and sodium retention. Patients with pulmonary hemorrhagic nephritis syndrome may experience recurrent hemoptysis and severe outcomes.

Acute glomerulonephritis is a serious disease with rapid development and poor prognosis. If not treated promptly, most patients may die from acute renal failure within weeks to months. The prognosis is related to the number and proportion of crescent formations; a higher number of crescents is associated with a worse prognosis.

12.2.4.3   Nephritic Syndrome and Associated Nephritis

The main manifestation of nephrotic syndrome is a large amount of proteinuria, with protein content in urine reaching or exceeding 3.5 g/d. It is also characterized by hypoalbuminemia (plasma albumin <30 g/L), high edema, hyperlipidemia, and lipiduria. The main symptoms of nephrotic syndrome are interrelated. The critical lesion involves damage to the capillary wall of the glomerulus, increased filtration membrane permeability, and increased plasma protein filtration. When the membrane damage is relatively light, albumin and transferrin, which are primarily low molecular weight proteins, are selectively filtered, resulting in selective proteinuria. When the damage is more severe, large molecular weight proteins can also be filtered, resulting in non-selective proteinuria.

Multiple primary glomerulonephritis and systemic diseases can cause nephrotic syndrome. In children, nephrotic syndrome is mainly caused by primary glomerulonephritis, whereas in adults it may be associated with systemic diseases. The common types of nephrotic syndrome are as follows:

12.2.4.4   Membranous Glomerular Disease

Membranous glomerular disease is the most common cause of adult nephrotic syndrome. The lesion is characterized by diffuse thickening of glomerular capillaries. In the early stage of the lesion, glomerular inflammatory changes are not obvious and the condition is also known as membranous nephropathy.

Membranous glomerulonephritis is a chronic immune complex-mediated disease. Primary membranous glomerular disease is an autoimmune disease similar to Heymann nephritis. Autoantibodies react with glomerular epithelial membrane antigens, leading to the formation of immune complex deposits between the glomerular epithelial cells and the basement membrane, causing damage to the capillary wall and protein leakage through complement.

Morphology

Gross appearance shows bilateral kidney volume increase, with pale color and the appearance of a "white kidney," accompanied by obvious thickening of the cortex. Under the microscope, early glomeruli appear basically normal. As the lesion progresses, the glomerular capillary walls exhibit diffuse thickening, and the tubular lumen gradually narrows or even becomes blocked, eventually leading to glomeruli

Fig. 12.8 Membranous glomerular disease (PASM staining)

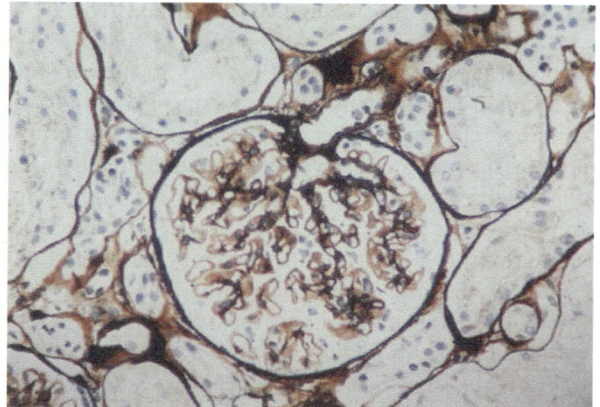

sclerosis. The epithelial cells of proximal convoluted tubules often contain small droplets of absorbed protein and are infiltrated by inflammatory cells such as lymphocytes and macrophages.

Immunofluorescence examination showed that IgG and C3 were deposited along the lateral margin of the glomeruli, displaying a discontinuous high-intensity fine granular fluorescence. Electron microscopy observed that the epithelial cells were swollen and the foot processes had disappeared. There were a large number of electron-dense deposits between the basement membrane and epithelial cells, and the basement membrane hyperplasia formed numerous spikes inserted between the deposits. The basement membrane, dyed black with PASM staining, displayed a thickened membrane with vertical spikes resembling a comb (Fig. 12.8). In the early stage of the lesion, there were fewer deposits, and the spikes were small, with depositions gradually increasing. As the spikes extended to the surface of the deposits and covered them, the basement membrane thickened significantly. The deposits were gradually dissolved and absorbed in the thickened basement membrane, forming gaps resembling a worm-eaten appearance.

Clinicopathological Correlation

Membranous glomerular disease is common in adults. The clinical course is nephrotic syndrome. Due to severe injury to the GBM, the permeability of the filtration membrane is significantly increased, resulting in non-selective proteinuria. Some patients may have hematuria or mild hypertension.

The clinical manifestations of membranous glomerular disease are chronic, with a long course, and adrenocortical hormones are often not effective. Some patients can be relieved or controlled, while others may progress to renal failure due to extensive glomerular fibrosis.

12.2.4.5 Minimal-Change Glomerulopathy

Minimal-change glomerulopathy, also known as minimal-change glomerulonephritis or minimal-change nephrosis, is the most common cause of nephrotic syndrome in children. The lesion was characterized by the diffuse disappearance of the foot processes of glomerular visceral epithelium cells. Under the light microscope, the glomerulus appeared basically normal, but lipid deposition was observed in the epithelial cells of the renal tubules, leading to the term lipid nephrosis.

The occurrence of this disease may be related to abnormal immune function, especially T-lymphocyte dysfunction and mutations in glomerular protein genes. Immune dysfunction causes the release of cytokines and damage to visceral epithelial cells, leading to proteinuria.

Morphology

Gross appearance: The bilateral kidney volume increased, showing yellow and white streaks due to lipid deposition in renal tubular epithelial cells. Under the microscope, the glomerular structure appeared basically normal, with a large number of lipid droplets and small proteins in the epithelial cells of the proximal convoluted tubule. Immunofluorescence examination showed no immunoglobulin or complement deposition. Under the electron microscope, the main changes included the diffuse disappearance of foot processes in the visceral epithelium cells, cytoplasmic swelling, the formation of vacuoles in the cytoplasm, and the proliferation of microvilli on the surface of the cells.

Clinicopathological Correlation

This disease occurs in children and is characterized by nephrotic syndrome, particularly with highly selective proteinuria. Edema is often the earliest symptom; proteinuria is selective, usually without hematuria and hypertension. Corticosteroid therapy is effective for more than 90% of children. The effect of corticosteroids in adult patients is less pronounced.

12.2.4.6 Focal Segmental Glomerulosclerosis

Focal segmental glomerulosclerosis is characterized by partial sclerosis of some glomeruli. The main clinical manifestation is nephrotic syndrome.

The pathogenesis of this disease has not yet been fully elucidated but is mainly attributed to injury and alteration of visceral epithelial cells. Due to the significant increase in local permeability, plasma proteins and lipids deposit in the extracellular matrix, activating mesangial cells and causing segmental glomerulosclerosis.

Morphology

Under the microscope, the lesion showed a focal distribution, initially affecting only the glomeruli at the junction of the skin and medulla and gradually involving the entire cortex. In the affected glomerular capillary loops, the mesenchymal matrix is increased, the basement membrane collapses, and the lumen is blocked, resulting in glomerular sclerosis, renal tubular atrophy, and interstitial fibrosis.

Immunofluorescence examination revealed IgM and C3 deposition at the lesion site. Electron microscopy showed that the epithelial cells in the diffuse layer had disappeared, and some epithelial cells were detached from the glomeruli.

Clinicopathological Correlation

The main clinical course of this disease is nephrotic syndrome, with some cases presenting only proteinuria. The disease responds poorly to corticosteroids and is progressive, and most cases develop into chronic glomerulonephritis. The prognosis is better for children.

12.2.4.7 Membranoproliferative Glomerulonephritis

The pathological features of membranoproliferative glomerulonephritis include thickening of glomerular capillaries and proliferation of glomerular cells and mesangial matrix. Due to the significant proliferation of mesangial cells, it is also known as mesangiocapillary glomerulonephritis. The disease is characterized by nephrotic syndrome and can also cause hematuria and proteinuria.

Primary membranoproliferative glomerulonephritis is divided into two types according to the characteristics of immunofluorescence and ultrastructure (Fig. 12.9): Type I is caused by cyclic immune complex deposition and complement activation. Type II is associated with an autoantibody called C3 in the patient's serum, and the occurrence of this type of nephritis is related to the activation of alternative complement pathways, with significantly lower serum C3 levels in patients.

Morphology

Gross appearance: The kidneys did not change significantly. Under the microscope, the glomeruli were enlarged, with increased mesangial cells and endothelial cells and leukocyte infiltration observed. There was increased proliferation of mesangial cell and mesangial matrix along the capillary endothelial cells, which were widely inserted into capillary basement membrane, leading to diffuse thickening of the blood capillary basement membrane and broadening and lobulation of glomerular

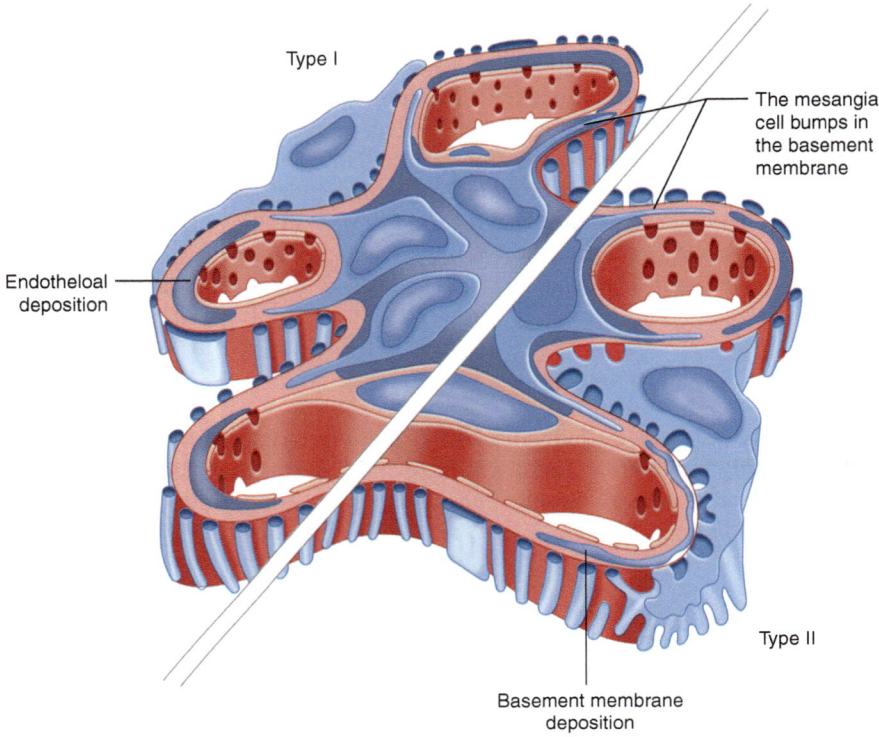

Type I

The mesangial cell bumps in the basement membrane

Endotheloal deposition

Type II

Basement membrane deposition

Fig. 12.9 Membranoproliferative glomerulonephritis

lobules. PASM staining showed that the thickened substrate had a double-track appearance. Type I is common; electron microscopy shows the main electron-dense deposits under the mesangial area and endothelial cells. Immunofluorescence examination showed C3 granular deposits, and early complement components, such as IgG, C1q, and C4. Type II shows a large number of dense, high electron density deposits along the basement membrane layer with zonal distribution, and immunofluorescence examination showed C3 deposition.

Clinicopathological Correlation

This disease usually occurs in children. The clinical is chronic and the prognosis is poor. There may be only mild proteinuria or hematuria in the early stage, with the appearance of nephrotic syndrome as the disease progresses. In the late stage, hypertension and renal failure are caused by the narrowing and even occlusion of the glomerular capillaries, sclerosis of the mesangium, and glomerular fibrosis. About 50% of patients will develop chronic renal failure within 10 years. The effectiveness of hormone and immunosuppressive therapy is often limited. The disease frequently recurs after kidney transplantation.

12.2.4.8 Mesangial Proliferative Glomerulonephritis

Mesangial proliferative glomerulonephritis is characterized by diffuse mesangial cell proliferation and increased mesangial matrix.

There may be multiple pathogenetic pathways, such as circulating immune complex deposition or in situ immune complex formation. Immune responses stimulate mesangial cells through mediators, resulting in the proliferation of mesangial cells and increased mesangial matrix.

Morphology

Under the microscope, the mesangial area is broadened with diffuse mesangial cell proliferation and increased mesangial matrix. The hyperplastic mesangial tissue can compress capillary loops, leading to narrowing of the lumen, with infiltration of a few macrophages and neutrophils in the mesangial membrane. Immunofluorescence often shows varying results. In China, IgG and C3 deposits are most common, while IgM and C3 deposits are more frequent in other countries. In some cases, only C3 deposition is observed, or immunofluorescence examination may be negative. Electron-dense deposits are observed in the mesangial area.

Clinicopathological Correlation

This disease is very common in young people. It can be characterized by asymptomatic proteinuria or hematuria and nephrotic syndrome. This disease can be treated with hormones and cytotoxic drugs, and the prognosis is generally good. If the lesion is severe, it can progress to chronic sclerosing glomerulonephritis.

12.2.4.9 IgA Nephropathy

The pathological feature of IgA nephropathy is IgA deposition in the mesangial area. It is characterized by recurrent microscopic or gross hematuria. This disease has a regional occurrence and may be the most common type of glomerulonephritis worldwide. The disease was first described by Berger, so it is also known as Berger disease.

The occurrence of IgA nephropathy is related to congenital or acquired immunomodulation abnormalities. Bacterial, viral, and food proteins stimulate an increase in IgA synthesis in the respiratory or digestive mucosa. The immune complex of IgA or IgA deposits in the mesangial region activates complement and causes glomerular damage. Serum levels of IgA are higher in patients.

Morphology

Under the microscope, there are a variety of lesions, which can show mild pathological changes, focal segmental hyperplasia or sclerosis, diffuse capillary hyperplasia, mesangial proliferation, membrane proliferation, crescent formation, and even glomerulosclerosis, with mesangial proliferative lesions being the most common. Immunofluorescence showed IgA deposition in the mesangial area, often accompanied by C3 deposition, or with a small amount of IgG and IgM deposition, showing high-intensity granular fluorescence. Electron microscopy showed dense deposits in the mesangial area.

Clinicopathological Correlation

IgA nephropathy often occurs in children and young people. Before occurrence, there is usually an upper respiratory tract infection. A few cases occur after gastrointestinal or urinary infections. The clinical manifestations of IgA nephropathy are chronic, mainly manifested as recurrent hematuria and mild proteinuria. A few patients are characterized by nephrotic syndrome or acute nephritic syndrome. The prognosis is related to the type of disease. Patients with older age, significant albuminuria, high blood pressure, glomerulosclerosis on renal biopsy, severe hyperplasia, or crescent formation have a poorer prognosis.

12.2.4.10 Chronic Glomerulonephritis

Chronic glomerulonephritis is a common result of the progression of different types of glomerulonephritis. It is characterized by extensive glomerular hyalinization and sclerosis and is also called chronic sclerosing glomerulonephritis. Clinically, chronic nephritis syndrome is a typical manifestation.

Chronic glomerulonephritis develops from different types of glomerulonephritis, each with different pathogenesis. Most patients have a history of nephritis, but some patients may have no clear history of nephritis, with lesions already in the chronic stage when discovered. The glomerular damage caused by various factors results in glomerular fibrosis, hyalinization, sclerosis, and corresponding renal tubular epithelial cell atrophy and renal interstitial fibrosis.

Morphology

Gross appearance: Bilateral kidneys shrink symmetrically, weight decreases, texture becomes hardened, and the surface has diffuse fine granularity. The renal cortex thins significantly, with an unclear boundary between the cortex and medulla. The walls of small arteries become thickened and hardened, and there is increased fat around the renal pelvis. The general lesion of chronic nephritis is referred to as

Fig. 12.10 Chronic
glomerulonephritis

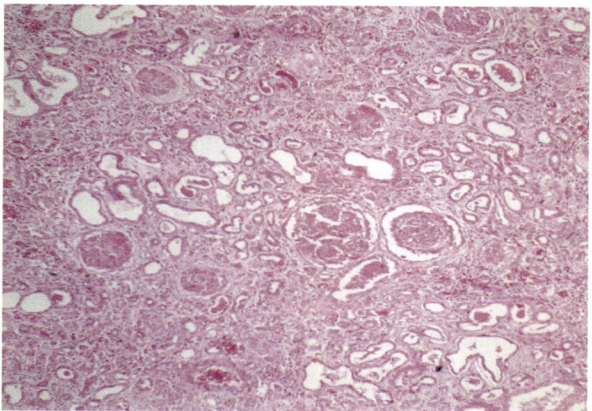

secondary granular contracted kidney. Under the microscope, early-stage pathological types of primary nephritis can be observed. In the later stages, there is diffuse hyalinization and sclerosis of the glomeruli, and renal tubules atrophy and disappear due to ischemia. In light lesions, glomeruli show compensatory hypertrophy, and renal tubules are dilated with various tubular types visible in the lumen. Renal interstitial fibrous tissue undergoes hyperplasia with infiltration of lymphocytes and plasma cells. Interstitial fibrosis causes glomeruli to become closely packed, forming glomerular concentration. Due to high blood pressure caused by nephritis, fine and small arteries in the kidney may exhibit hyalinization and sclerosis, resulting in thickening of the vessel walls and stenosis (Fig. 12.10). Because of extensive glomerular sclerosis, immunofluorescence and electron microscopy are often negative. In relatively mild cases, immunoglobulin and complement deposition can sometimes be seen, and electron microscopy may reveal electron-dense deposits.

Clinicopathological Correlation

This disease is more common in adults. Some patients have a history of other types of nephritis, while others may have an insidious onset of the disease. In the early stage, symptoms may include poor appetite, vomiting, fatigue, and anemia. Some patients may also show proteinuria, edema, hypertension, or azotemia. Advanced stages of chronic nephritis syndrome often lead to chronic renal failure.

1. **Polyuria, nocturia, and low urine specific gravity:** Due to significant damage and loss of function in many renal units, blood flow through the remaining nephrons increases, resulting in a higher glomerular filtration rate. However, renal tubular reabsorption function is limited, reducing the ability to concentrate urine. This results in polyuria, nocturia, and low urine specific gravity.
2. **Hypertension:** Extensive glomerular sclerosis leads to severe renal tissue ischemia and increased renin secretion. High blood pressure exacerbates arterial

sclerosis, worsening renal ischemia and further increasing blood pressure. Chronic hypertension can result in left ventricular hypertrophy and severe heart failure and may also lead to cerebral hemorrhage.

3. **Anemia:** This is caused by the destruction of kidney tissue, reduced erythropoietin secretion, and the accumulation of metabolites that inhibit hematopoietic function.

4. **Increased nitrogenous waste and uremia:** Extensive damage to renal units lead to significant accumulation of metabolites in the body, causing disturbances in water, electrolyte, and acid-base balance. Elevated blood urea and creatinine levels result in nitrogenous waste accumulation and uremia.

The rate of progression of chronic glomerulonephritis varies widely, and the duration of the disease also differs, but the prognosis is generally poor. Early, rational treatment can help control the progression of the disease. If dialysis or kidney transplantation is not performed in time, patients often die from heart failure, cerebral hemorrhage, or secondary infections resulting from chronic renal failure or hypertension.

It is important to note that the diagnosis and differential diagnosis of glomerulonephritis in clinical practice must be based on a comprehensive analysis of the patient's history, clinical manifestations, laboratory examinations, and pathological findings. The pathological types of primary glomerulonephritis, along with the characteristics observed in light microscopy and the main clinical manifestations, are summarized as follows (Table 12.1).

Table 12.1 The classification and clinicopathologic features of glomerulonephritis

Pathological types	Characteristics in light microscopy	Main clinical manifestations
Acute diffuse proliferative glomerulonephritis	Diffuse glomerular capillary endothelial cells and mesangial cell hyperplasia proliferation	Acute nephritic syndrome
Rapidly progressive glomerulonephritis	Crescent formation	Rapidly progressive syndrome
Membranous glomerular disease	Diffuse GBM thickening, spike formation	Nephritic syndrome
Minimal-change glomerulopathy	Normal glomeruli, renal tubule lipid deposit	Nephritic syndrome
Focal segmental glomerulosclerosis	Focal segmental hyalinization and sclerosis	Nephritic syndrome, proteinuria
Membranoproliferative glomerulonephritis	Mesangial cell proliferation, insertion, basement membrane thickening, double-track	Nephritic syndrome, hematuria, proteinuria, chronic renal failure
Mesangial proliferative glomerulonephritis	Mesangial cell proliferation and increased mesangial matrix	Albuminuria, hematuria, nephritic syndrome
IgA nephropathy	Focal segmental hypertrophy or diffuse mesangial broadening	Repeated episodes of hematuria or proteinuria
Chronic glomerulonephritis	Glomerular hyalinization and sclerosis	Chronic nephritic syndrome

12.3 Tubulointerstitial Nephritis

Tubulointerstitial nephritis (TIN) refers to a group of inflammatory diseases that involve the renal tubules and interstitium of the kidney and can be divided into acute and chronic types. Acute TIN is mainly characterized by interstitial edema and tubular and interstitial neutrophil infiltration and typically combined with focal tubular necrosis. Chronic TIN manifests as lymphocytic and mononuclear cell infiltration, renal interstitial fibrosis, and tubular atrophy. This section mainly discusses pyelonephritis and tubulointerstitial nephritis caused by drugs and poisoning.

12.3.1 Pyelonephritis

Pyelonephritis is an inflammatory disease of the renal pelvis, interstitium, and renal tubules. It is a common kidney disease caused by bacterial infection and is more likely to affect females. According to the clinical course and pathological features, pyelonephritis can be divided into two types: acute and chronic. Acute pyelonephritis is mainly manifested by fever, chills, low back pain, hematuria, pyuria, and signs of bladder irritation. Chronic pyelonephritis may present with symptoms of hypertension and renal insufficiency in addition to changes in urine.

12.3.1.1 Etiology and Pathogenesis

Acute pyelonephritis, a common suppurative inflammation of the kidney and renal pelvis, is caused by bacterial infection. Escherichia coli is by far the most common causative pathogens (about 60–80%). Other important pathogens include Proteus, Aerobacter aerogenes, Staphylococcus, and fungi. Acute pyelonephritis is usually caused by a single bacterial infection, while chronic pyelonephritis often involves a mixed infection of two or more bacteria.

Two Main Routes of Infection in Pyelonephritis

1. Hematogenous Infection. Most of the causative bacteria are Staphylococcus aureus. These bacteria invade the blood vessels from an infection site elsewhere in the body, enter the kidneys via the bloodstream, and settle in the glomeruli or the capillary vessels around the renal tubules, causing inflammation. As the bacteria spread through the blood, they can reach the renal medulla, calyces, and renal pelvis, which is why hematogenous infection is also known as descending infection. Pyelonephritis caused by hematogenous infection is rare and can occur in cases of sepsis or infective endocarditis. The lesions often involve both kidneys.

2. Ascending Infection. This is the primary route for causing pyelonephritis. Most pathogens are Escherichia coli. Ascending infection often originates from the lower urinary tract, such as urethritis, prostatitis, or cystitis. The bacteria can travel up the ureter or periureteral lymphatic vessels to the renal pelvis, calyces, and renal interstitium, also known as retrograde infection. The lesions can involve one or both kidneys.

The Occurrence of Pyelonephritis

The urinary system has a defensive mechanism under physiological conditions, and simple bacterial invasion may not result in pyelonephritis. When the body's resistance is reduced or the local defense function of the urinary system is weakened, pathogenic bacteria can exploit the opportunity, enter and grow in the urinary tract, and eventually cause pyelonephritis.

Common causes of pyelonephritis include the following:

1. Urethral Mucosa Injury: Procedures such as cystoscopy, urethral catheterization, and retrograde pyelography can lead to urinary tract mucosal injury, allowing pathogenic bacteria to enter and cause infection. Long-term indwelling catheters are a significant factor in inducing this disease.
2. Urinary Tract Obstruction: Conditions such as urinary calculi, prostatic hyperplasia, pregnant uterus, tumor compression, cicatricial stenosis, and congenital urinary tract malformations can cause urinary tract stenosis. This leads to reduced local defense, resulting in urine retention where bacteria can easily invade and proliferate, causing disease.
3. Vesicoureteral Reflux (VUR): Congenital abnormalities, such as abnormal opening of the ureters, spinal cord injury, and other factors can cause relaxation of the bladder, leading to vesicoureteral reflux. VUR increases residual urine volume, which supports bacterial proliferation. The bacteria can then invade the renal pelvis and calyx via the reflux.
4. Intrarenal Reflux: Due to the renal papillae, located at the upper or lower poles of the kidney, being extremely flat and concave, and the central papillae being convex, intrarenal reflux is more likely to occur in the upper and lower poles of the kidney. Urine can pass through the open ducts at the tips of the papillae and move further into the renal parenchyma.
5. Decreased Body Resistance: Factors such as chronic wasting disease, paraplegia, long-term use of hormones and immunosuppressive drugs, and other factors can decrease body resistance, making the development of pyelonephritis more likely.

12.3.1.2   Acute Pyelonephritis

Acute pyelonephritis, an acute suppurative inflammation of the renal pelvis, renal interstitium, and renal tubules, is mainly caused by bacterial infection. It is common in children, pregnant women, and elderly patients with prostatic hyperplasia.

Morphology

Gross specimen: The kidney size is enlarged, and the renal surface is congested and studded with discrete, yellowish, raised abscesses surrounded by purple-red hyperemia. These abscesses may be widely scattered or limited to one region of the kidney, or they may coalesce to form a single large area of suppuration. The mucosa of the renal pelvis shows hyperemia and edema, and the mucosal surface has purulent exudates, which can accumulate in the renal calyces in severe cases.

Histological section: The characteristic histologic feature of acute pyelonephritis is liquefactive necrosis with abscess formation within the renal parenchyma. The renal pelvis is initially involved in the inflammation caused by ascending infection, resulting in local mucosal hyperemia, edema, and a large number of neutrophils infiltrating the area. In the early stage, suppuration is limited to the interstitial tissue, but later abscesses rupture into the tubules. Large masses of intratubular neutrophils frequently extend within the involved nephrons into the collecting ducts, giving rise to the characteristic white cell casts found in the urine. The glomeruli of the renal cortex and the surrounding interstitium are initially involved in inflammation caused by hematogenous infection and then spread to adjacent tissues and the renal pelvis.

Complications

1. Renal Papillary Necrosis: Common in patients with urinary obstruction or diabetes, renal papillary necrosis occurs due to ischemia and suppuration. The pathognomonic gross feature of papillary necrosis is sharply defined gray-white or yellow necrosis of the papillae, with the papillary tips showing characteristic coagulative necrosis.
2. Pyelonephritis: In cases of severe urinary obstruction, especially when the upper urinary tract is blocked, purulent exudate cannot be discharged and accumulates in the renal pelvis. In severe cases, renal tissue may atrophy due to prolonged compression, and the entire kidney may become filled with pus.
3. Perinephric Abscess: In severe cases of renal perinephric abscess, the capsule may be breached by the suppurative inflammation of the kidney, forming an abscess in the perinephric tissue.

Clinical Course

1. Systemic Symptoms: These are usually sudden, with common symptoms including fever, chills, and peripheral blood leukocytosis.
2. Local Manifestations: Patients often experience waist and kidney pain, urinary frequency, urinary urgency, and symptoms of bladder irritation. Additionally, symptoms may include pus in the urine, bacteriuria, proteinuria, tubular casts, and hematuria. The presence of leukocyte casts in the renal tubules is clinically significant for pyelonephritis.

The prognosis of acute pyelonephritis is generally good, with most patients recovering with timely treatment. However, if treatment is incomplete or predisposing factors persist, the disease may become recurrent or chronic. Acute renal failure can occur if accompanied by renal papillary necrosis.

12.3.1.3   Chronic Pyelonephritis

Chronic pyelonephritis is inflammation of the renal tubules and interstitium that can develop from acute pyelonephritis. The lesions are characterized by predominantly interstitial inflammation and scarring of the renal parenchyma, which is associated with grossly visible scarring and deformity of the pelvicalyceal system. Chronic pyelonephritis is a significant cause of chronic renal failure and can be divided into two forms: chronic obstructive pyelonephritis and chronic reflux-associated pyelonephritis.

Morphology

Gross specimen: One side or both kidneys may be involved. Even when bilateral, the kidneys may not be equally damaged and therefore not equally contracted. The volume of the affected kidney is reduced and the texture is hard. The surface shows uneven scarring (Fig. 12.11). The medullary margin of the renal cortex is not distinct, the renal papillae are atrophied, and the renal pelvis and calyces are deformed due to scarring. The mucosa of the renal pelvis is thickened and coarse.

Histological section: Chronic pyelonephritis is a nonspecific inflammation of the renal tubules and interstitium. Histologically, it is characterized by local infiltration of lymphocytes and plasma cells and interstitial fibrosis. Renal tubules may show atrophy or compensatory dilation, with many dilated tubules containing homogenous

Fig. 12.11 Chronic pyelonephritis

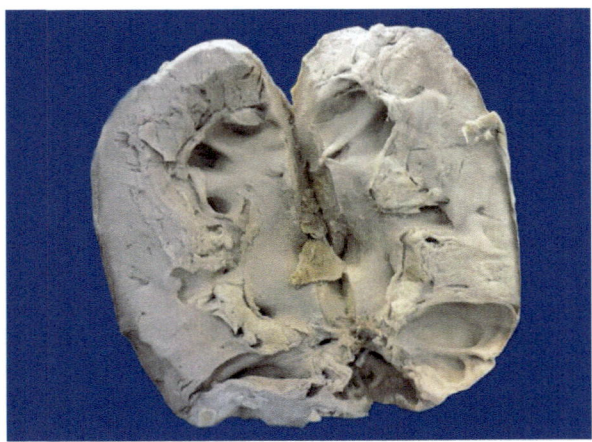

red-dyed protein, resembling thyroid tissue. Renal interstitial fibrosis, along with infiltration by lymphocytes and plasma cells, is accompanied by intrarenal arterial hypertension, leading to hyaline degeneration and sclerosis. In the early stage, the glomeruli are rarely involved, with fibrosis occurring around the renal capsule. In the late stage, some glomeruli show hyaline degeneration and fibrosis (Fig. 12.12).

Clinical Course

Many individuals with chronic pyelonephritis seek medical attention relatively late in the disease course due to the gradual onset of renal insufficiency or incidental findings on routine laboratory tests. Clinical manifestations include polyuria, nocturia, and electrolyte imbalances such as hyponatremia, hypokalemia, and metabolic acidosis due to excessive urine loss of sodium, potassium, and bicarbonate. Renal tissue fibrosis and small vascular sclerosis lead to ischemia, increased renin secretion, and subsequent hypertension. In the late stage, severe renal tissue damage results in azotemia and uremia.

Chronic pyelonephritis can present acutely, with symptoms similar to those of acute pyelonephritis, including fever, low back pain, pyuria, and bacteriuria. The characteristic X-ray image shows the affected kidney as asymmetrically contracted, with blunting and deformity of the calyceal system (caliectasis).

Chronic pyelonephritis has a longer duration and can often be recurrent. Removal of predisposing factors in a timely manner can still control the development of lesions, while renal function remains in the compensatory period. Severe lesions can be life-threatening due to uremia or hypertension-induced heart failure.

12.3.2 Drug-Induced Interstitial Nephritis

In this era of widespread antibiotic and analgesic use, drugs have emerged as an important cause of renal injury. Drugs and intoxication can induce an interstitial immune response, which can cause acute hypersensitivity interstitial nephritis and also result in chronic tubular injury and chronic renal insufficiency.

12.3.2.1 Acute Drug-Induced Interstitial Nephritis

Acute drug-induced interstitial nephritis can be caused by antibiotics, nonsteroidal anti-inflammatory drugs (NSAIDs), diuretics, and other drugs. The drug acts as a hapten and binds to the cytoplasm or extracellular components of the tubular epithelial cells, producing antigenicity, causing an immune response, and leading to immunological damage and inflammatory reactions of the renal tubular epithelial cells and the basement membrane. The main pathological changes include severe

Fig. 12.12 Chronic
pyelonephritis

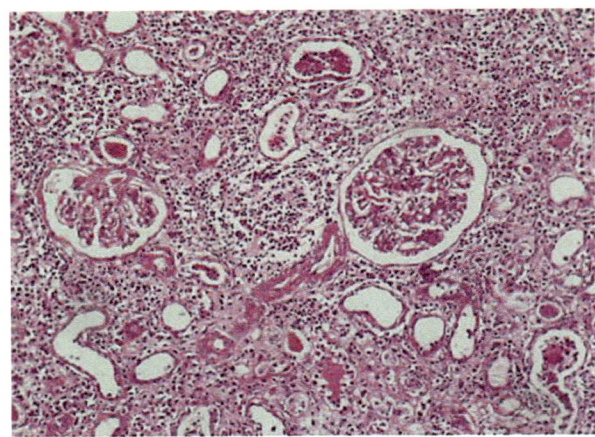

edema of renal interstitium, infiltration of lymphocytes, macrophages, a large number of eosinophils, neutrophils, and degeneration and necrosis of renal tubules.

The disease begins about 15 days (range, 2–40 days) after exposure to the drug and is characterized by fever, transient eosinophilia, a rash, and renal abnormalities. Urinary findings include hematuria, minimal or no proteinuria, and leukocyturia. Clinical recognition of drug-induced kidney injury is imperative, as withdrawal of the offending drug is followed by recovery, although it may take several months for renal function to return to normal.

12.3.2.2 Analgesic Nephropathy

Analgesic nephropathy is a chronic kidney disease caused by the mixed administration of analgesics. The lesions are characterized by chronic tubular-interstitial inflammation with renal papillary necrosis. Most people who develop this nephropathy consume at least two analgesics, such as aspirin and phenacetin. The toxic effects of these drugs and ischemia result in kidney damage. The thickness of the renal cortex varies, and the surface of the necrotic papilla may show subsidence. Necrosis, calcification, and shedding can occur in the renal papilla. Microscopically, necrosis of the renal papilla appears in the early stage. In severe cases, entire renal papillary necrosis is observed, with the local structure destruction but preservation of tubular outlines. The renal cortex and tubules are atrophic, with interstitial fibrosis and lymphocyte infiltration.

Patients often present with chronic renal failure, hypertension, and anemia. Renal papillary necrosis can cause hematuria and renal colic. Imaging examination reveals renal papillary necrosis and calcification. Withdrawal of analgesics can stabilize the disease and may restore renal function. A small number of patients may develop transitional cell carcinoma of the renal pelvis.

12.3.2.3 Aristolochic Acid Nephropathy

Aristolochic acid nephropathy (AAN) is a chronic kidney interstitial disease. The incidence of AAN is closely related to the use of aristolochic acid-containing Chinese herbs, including Aristolochia, Tian Xian Teng, Qing Mu Xiang, Guang Fang Ji, Xun Gu Feng, and Guan Mu Tong.

Acute AAN manifests as acute renal failure with pathological features of acute tubular necrosis. AAN can also cause renal tubular dysfunction, manifested as acidosis. Most cases present as chronic AAN, with a slow and latent onset; a few cases may rapidly progress to uremia. There is no established treatment plan for this disease. Drugs should be withdrawn and symptomatic treatment should be initiated. Corticosteroids may alleviate the disease in patients during the early and midterm stages.

12.4 Common Tumors of the Kidney and Bladder

Urinary system tumors can occur anywhere, with kidney and bladder tumors being common, and most are malignant. This section focuses on renal cell carcinoma, nephroblastoma (Wilms tumor), and urinary tract and bladder epithelial tumors.

12.4.1 Renal Cell Carcinoma

Renal cell carcinoma is a malignant tumor that originates from the renal tubular epithelium, also known as renal adenocarcinoma. Most tumors are yellow, and the morphology of the cells under the microscope often resembles adrenal cortical cells, which is why it is also called an adrenal-like tumor. Renal cell carcinoma is the most common malignant tumor of the kidney, representing 80–85% of all primary malignant tumors of the kidney in adults, with men being affected about twice as often as women.

12.4.1.1 Etiology and Pathogenesis

The occurrence of renal cell carcinoma is related to smoking, chemical carcinogens, and genetic factors. Smoking is the most important risk factor for renal cell carcinoma. Obesity, hypertension, long-term exposure to asbestos, petroleum products, and heavy metals are also risk factors for renal cell carcinoma.

Renal cell carcinoma has two types: hereditary and sporadic. The vast majority of renal cell carcinomas are sporadic, which typically have a later age of onset and occur mostly on one side of the kidney. Hereditary renal cell carcinoma accounts for only 4% and follows an autosomal dominant inheritance pattern. It is characterized by an earlier age of onset, and the tumors are often bilaterally multifocal. Hereditary renal cell carcinoma can be divided into three types:

1. von Hippel-Lindau (VHL) syndrome: This is a familial tumor syndrome characterized by a predisposition to a variety of neoplasms, particularly hemangioblastomas of the cerebellum and retina. Hundreds of bilateral renal cysts and bilateral, often multiple, clear-cell carcinomas develop in 40–60% of affected individuals. Those with VHL syndrome inherit a germline mutation of the VHL gene on chromosomal band 3p25 and lose the second allele by somatic mutation.
2. Hereditary (familial) clear-cell carcinoma: This type of renal clear-cell carcinoma may involve changes in VHL and related genes but does not present with VHL syndrome or other associated conditions.
3. Hereditary papillary carcinoma: These tumors are frequently multifocal and bilateral and often present as early-stage tumors. The tumor cells are arranged in a papillary pattern. This type of tumor does not have VHL gene mutations but may have other cytogenetic abnormalities and mutations in the oncogene MET.

12.4.1.2 Morphology

1. Gross Specimen. Renal cell carcinomas are mostly located at the poles of the kidneys, especially the upper poles. They are usually solitary and large when symptomatic (spherical masses 3–15 cm in diameter). The cut surface of clear-cell renal-cell carcinomas is yellow to orange to gray-white, with prominent areas of cystic softening or of hemorrhage, either fresh or old (Fig. 12.13).

   As the tumor enlarges, it is often accompanied by hemorrhage and cystic degeneration. Tumors often have a pseudocapsule well-defined from the surrounding tissue. Advanced tumors can invade the renal pelvis, calyx, and ureter

Fig. 12.13 Renal cell carcinomas

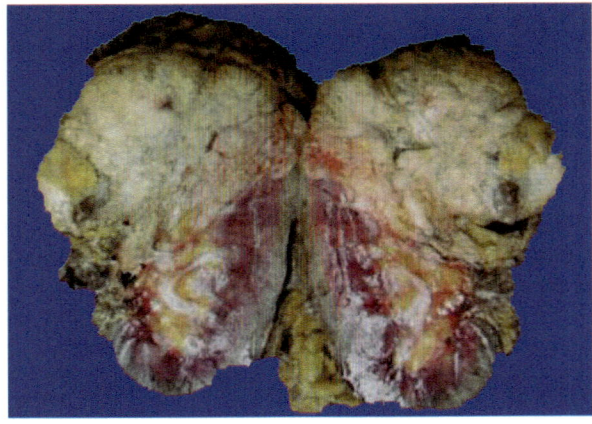

and often invade the renal vein, growing as a solid column within this vessel. Papillary carcinoma can be multifocal and bilateral.

2. Histological Section. The main histological types of renal cell carcinoma are as follows:

   (a) Clear-cell carcinoma: This is the most common type of renal cell carcinoma, accounting for 70–80%. Microscopically, tumor cells are larger, round or polygonal, with clear outlines. Depending on the amounts of lipid and glycogen present, the tumor cells may appear almost vacuolated or may be solid. The nucleus is small and round, and the tumor stroma is rich in capillaries and sinusoids (Fig. 12.14).

   Most cases are sporadic and associated with changes in the VHL gene. Immunohistochemically, tumor cells express low molecular weight cytokeratin, CK8, CK18, vimentin, CD10, and EMA.

   (b) Papillary renal cell carcinomas: These account for 10–15% of renal cell carcinomas. Tumor cells are cubic or short columnar, arranged as papillary with fibrovascular cores. Psammoma bodies and foam cells are often found in the mesenchyme of the papilla, and edema can occur.

   This type includes both familial and sporadic forms. Immunohistochemical staining shows that the tumor cells are CK7 positive.

   (c) Chromophobe-type renal cell carcinomas: These account for approximately 5% of renal cell carcinomas. Microscopically, the tumor cells are arranged in nests or alveoli, and the cells usually have clear, flocculent cytoplasm with very prominent, distinct cell membranes. The nuclei are surrounded by halos of clear cytoplasm.

Cytogenetic examinations show multiple chromosome deletions and subdiploidy. These patients generally have a good prognosis. Immunohistochemically, tumor cells show positive staining for EMA and CK, and negative or weak staining for vimentin.

Other types of renal cell carcinoma include renal collecting duct carcinoma, unclassified renal carcinoma, multi-cystic renal cell carcinoma, renal medullary carcinoma, mucinous small tubular carcinoma, and spindle cell carcinoma.

Fig. 12.14 Clear-cell carcinoma

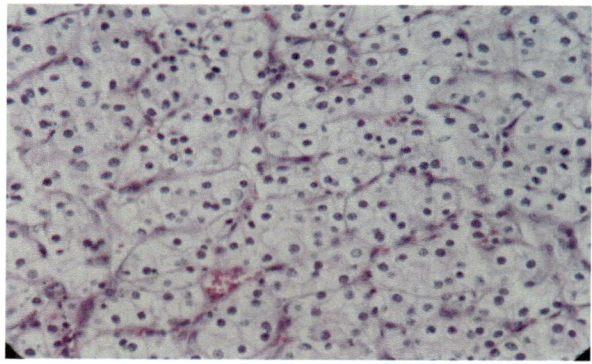

12.4.1.3 Tumor Spread

Renal cell carcinoma is prone to metastasize, most commonly to the lungs and bones but also to the regional lymph nodes, liver, adrenal glands, and brain. Common spread pathways are as follows:

1. Direct Spread. Cancer tissue can invade the renal calyx, pelvis, and even the ureter, causing obstruction and resulting in hydronephrosis. It can also penetrate the renal capsule, invading the adrenal gland and surrounding soft tissue of the kidney.
2. Hematogenous Metastasis. Due to the rich blood supply of renal cancer, hematogenous metastasis can occur early. The most common sites for metastases are the lungs and bones, but the liver, adrenal glands, and brain can also be affected.
3. Lymphatic Metastasis. Typically, tumor cells metastasize to the hilar and paraaortic lymph nodes.

12.4.1.4 Clinical Course

The early symptoms of renal cell carcinoma are not obvious. Three typical clinically significant symptoms are hematuria, lumbago, and renal masses. Painless hematuria is the main symptom of kidney cancer and is mostly caused by cancer tissue eroding blood vessels or invading the renal calyx and renal pelvis. Sometimes, blood clots can cause renal colic when discharged through the ureter. When the tumor is larger or invades the kidney capsule, it can cause pain in the kidney area and a palpable mass may be detected.

Renal cell carcinoma can produce a variety of heterotopic hormones and hormone-like substances that cause paraneoplastic syndromes, such as increased production of erythropoietin causing polycythemia, increased renin production causing hypertension, increased parathyroid hormone causing hypercalcemia, and increased production of adrenal glucocorticoids causing Cushing syndrome.

The prognosis for patients with renal cell carcinoma is poor, with a 5-year survival rate of about 45%. If the tumor invades the renal vein and surrounding tissue of the kidney, the 5-year survival rate drops to only 15–20%. However, patients without metastasis can have a survival rate of up to 70%.

12.4.2 Nephroblastoma

Nephroblastoma, also known as Wilms' tumor, originates from the kidney's residual embryonic tissue and is the most common malignancy of childhood kidneys. The disease is more common in children and occasionally occurs in adults.

Most nephroblastomas are sporadic and are rarely inherited in an autosomal dominant manner. Its occurrence is mostly related to the deletion or mutation of the

WT-1 gene (Wilms tumor-associated gene 1) located at 11p13. Some patients have various congenital malformations. The occurrence of nephroblastoma may be due to a differentiation disorder of mesenchymal basal cells in the posterior renal tissue and their continuous proliferation.

12.4.2.1 Morphology

1. Gross Specimen. Nephroblastoma usually occurs in a unilateral kidney, with a few cases being bilateral and multifocal. Most tumors present as a single solid mass of large size. The tumor has a clear boundary, and the surrounding kidney tissue may form a pseudocapsule. Tumors are soft, with a cut surface that is fish-like and gray or gray-red and may show bleeding, necrosis, or cystic degeneration.
2. Histological Section. Nephroblastoma has different histological structures at various developmental stages, and its composition and structure are complex. Histological features include naive glomerular or tubular-like structures at different stages of development. The cellular components include epithelioid cells, embryonic naive cells, and cells of mesenchymal tissue. Epithelial-like cells are small and round or polygonal and can form small ball-like and tubular-like structures. Embryonic naive cells are small, round or ovoid primitive cells with scant cytoplasm. Mesenchymal-derived cells are mostly fibrous or mucinous and may be small, spindle-shaped or star-shaped. There may also be differentiation into striated muscle, bone, cartilage, or fat.

Immunohistochemical staining showed that tumor cells can express NSE, CK, desmin, vimentin, WT-1, and other markers.

12.4.2.2 Tumor Spread

Nephroblastoma can invade the perirenal or renal veins, metastasize to the renal hilum and paraaortic lymph nodes via lymphatics, and can also metastasize to the lungs or liver.

12.4.2.3 Clinical Course

Abdominal mass is the most common symptom of nephroblastoma, which can compress adjacent organs and cause abdominal pain or intestinal obstruction. Some children may have high blood pressure, potentially associated with tumor compression of the renal artery and increased renin production. Hematuria occurs when the tumor invades the renal pelvis.

The current treatment for renal tumor is mainly a comprehensive approach combining surgical resection with chemotherapy and radiotherapy. The long-term survival rate of non-metastatic cases is up to 90%, and metastatic cases can also achieve satisfactory results with treatment.

12.4.3   Urinary Tract and Bladder Epithelial Tumors

Urothelial tumors can occur in the renal pelvis, ureters, bladder, and urethra, with the bladder being the most common site. About 95% of bladder tumors originate from epithelial tissue. The majority of these tumors are urothelial or transitional epithelial tumors. The bladder may also develop squamous cell carcinoma, adenocarcinoma, and mesenchymal tumors, though these are rare. Bladder cancer occurs mostly in men, with a ratio of about 3:1 between males and females. Most patients are over the age of 50 at onset.

12.4.3.1   Pathogenesis

The occurrence of bladder cancer is associated with smoking, long-term exposure to aromatic amines, radiation, chronic inflammation of the bladder mucosa, and schistosomiasis infection in Egypt, with smoking being the most important risk factor for bladder cancer.

Cytogenetic and molecular changes in urothelial carcinoma are heterogeneous. In some cases, there are single chromosome 9 deletions or deletions of 9p, 9q, 17p, 13q, 11p, and 14q. The molecular model of bladder cancer includes two pathways. One pathway involves the loss of tumor suppressor genes located on 9p and 9q, leading to superficial papillary tumors. Another pathway involves mutations in the p53 gene that cause cancer in situ, followed by deletions on chromosome 9 that result in invasive cancer.

12.4.3.2   Morphology

According to the World Health Organization (WHO) and International Society of Urinary Pathology (ISUP) classification, urothelial tumors are classified into urothelial papilloma, papillary urothelial neoplasm of low malignant potential, low-grade papillary urothelial carcinoma, and high-grade papillary urothelial carcinoma.

1. Gross Specimen. Urothelial carcinoma often occurs in the bladder's lateral wall, trigone area, and near the ureteral opening. Tumors can be single or multiple, varying in size from a few millimeters to several centimeters or more. They are mostly papillary, polypoid, or cauliflower-like protrusions from the mucosal surface or present as local thickening of the mucous membrane as flat plaques. Tumors may be invasive or noninvasive.

2. Histological Section. Microscopically, urothelial papilloma accounts for about 1% of bladder tumors and is more common in youth. The tumor is papillary, with the center of the papilla consisting of fibrous connective tissue and capillaries. The surface is well-differentiated urothelium with no atypia or mitotic activity.

The histological features of papillary urothelial neoplasm of low malignant potential are similar to those of papilloma, except that the epithelial cell layer is thicker, the papillae are more pronounced, and the nuclei are generally enlarged.

Low-grade papillary urothelial carcinoma has more regular cellular and histological structures. The tumor cells are often papillary with increased layers and normal polarity. The nuclei are heterogeneous with dense staining, and a small number of tumor cells on the basement membrane show nuclear mitosis and mild nuclear polymorphism. Low-grade urothelial papillary carcinoma can recur after removal but is rarely invasive.

High-grade papillary urothelial carcinoma cells are disordered in arrangement, with increased layers and loss of polarity. The nuclei are densely stained, increased in size, with some cells showing obvious atypicality, more mitoses, and possible pathological mitoses. High-grade papillary urothelial carcinomas are mostly invasive and prone to metastasis (Fig. 12.15).

Immunohistochemical staining showed that tumor cells could express CK7, CK8, CK18, CK20, EMA, survivin, and other markers.

12.4.3.3 Tumor Spread

Invasive urothelial carcinoma can affect adjacent structures such as the prostate, seminal vesicle, and ureter and can also form a fistula communicating with the vagina or rectum. About 40% of invasive urothelial carcinomas can metastasize to regional lymph nodes. Hematogenous metastasis can occur in the late stage and affect organs such as the liver, lungs, bones, kidneys, and adrenal glands.

Fig. 12.15 High-grade papillary urothelial carcinoma

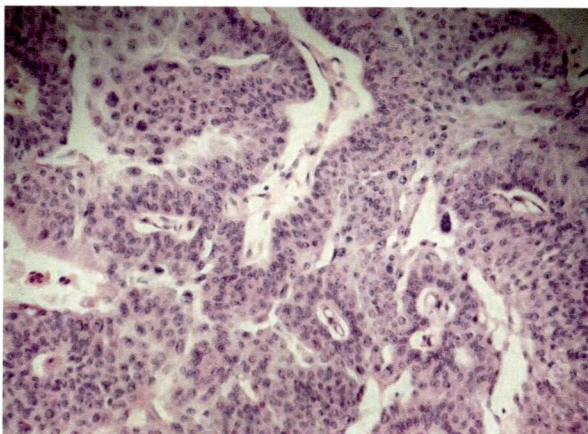

12.4.3.4 Clinical Course

The dominant clinical presentation of bladder cancer is painless hematuria. Rupture of tumor papillae, tumor surface necrosis, and ulceration can cause hematuria. Tumor invasion of the bladder wall, stimulation of the bladder mucosa, or concurrent infection can cause bladder irritation, including urinary frequency, urgency, and pain. Obstruction of the ureter by the tumor can lead to pyelonephritis, hydronephrosis, and even sepsis.

Bladder tumors of transitional cell origin tend to recur after surgery, and some relapsed tumors may show poor differentiation. The prognosis of patients with urothelial tumors is closely related to the grade and infiltration of the tumors. The 10-year survival rate for papilloma, papillary urothelial neoplasm of low malignant potential, and low-grade papillary urothelial carcinomas is 90%. The 10-year survival rate for patients with high-grade papillary urothelial carcinoma is about 40%. Cystoscopy and biopsy are the main methods for diagnosing bladder cancer.

Chapter 13
Female Genital System and Breast

Wenjuan Yang, Huijie Jia, and Xiaojiang Qin

Contents

W. Yang (✉)
School of Basic Medicine, Dali University, Dali, China

H. Jia
School of Basic Medical Sciences, Xinxiang Medical University, Xinxiang, China

X. Qin
School of Public Health, Shanxi Medical University, Taiyuan, China

© Zhengzhou University Press 2024
K. Chen et al. (eds.), *Textbook of Pathologic Anatomy*,
https://doi.org/10.1007/978-981-99-8445-9_13

Objectives
1. To master the morphology of cervical intraepithelial neoplasia.
2. To master the morphology and diffusion of invasive carcinoma of the cervix and clinicopathologic relationships.
3. To master the morphology of breast cancer and clinicopathologic relationships.

Key Concepts
Cervicitis, cervical intraepithelial neoplasia, INVASIVE carcinoma of the cervix, breast cancer.

13.1   Introduction

This chapter includes common diseases of the reproductive system of women and the breast, which are affected by endocrine and diseases with special pathological changes. In addition to inflammation and tumors, there are some pregnancy-related diseases caused by endocrine disorders. Reproductive system inflammatory diseases are quite common, and the pathological changes are relatively simple. Therefore, the tumors of the female genital system and breast are the focus of this chapter.

13.1.1   Cervix Disease

13.1.1.1   Cervicitis

Inflammations of the cervix are extremely common and are associated with a purulent vaginal discharge. Cytologic examination of the discharge reveals white cells and inflammatory atypia of shed epithelial cells, as well as possible microorganisms. These inflammations can be subclassified as noninfectious and infectious cervicitis [1–3].

Etiology

The major causes of this disease are streptococcus, enterococcus, staphylococcus, and special pathogenic microorganisms such as Chlamydia trachomatis, Neisseria gonorrhoeae, human papillomavirus (HPV), and herpes simplex virus, which are encountered in sexually transmitted disease (STD) clinics. In addition, childbirth and mechanical damage are also causes of the cervicitis [1–3].

Morphology

Nonspecific cervicitis may be either acute or chronic. The relatively uncommon acute form is limited to women in the postpartum period and usually is caused by staphylococci or streptococci. Chronic cervicitis consists of inflammation and epithelial regeneration, some degree of which is common in all women of reproductive age. The cervical epithelium may show hyperplasia and reactive changes in both squamous and columnar mucosae. Eventually, the columnar epithelium undergoes squamous metaplasia. Chronic cervicitis can be divided into three types based on gross appearance: erosion, nabothian cyst, and cervical polyps. Erosion can be true or pseudo. True erosion shows cervix injury with squamous cell erosion, while pseudo-erosion shows columnar ectopy. Nabothian cysts filled with mucus can be seen in the cervix. A cervical polyp is a smooth mass of tissue with stalks protruding outward from the surface of the cervical mucosa [1–3].

Clinical Features

Cervicitis are commonly noticed during routine examination or due to symptoms such as increased leucorrhea, vaginal bleeding, and vulvar itching. Culture of the discharge must be interpreted cautiously because, as mentioned previously, commensal organisms are almost always present. Only the identification of known pathogens is helpful. When the lesion is severe, it can make it difficult to differentiate between carcinoma and inflammation from cytologic preparations and even with colposcopy. Differentiating inflammatory changes from premalignant dysplasia may also be challenging on cervical biopsy specimen [1–3].

13.1.1.2 Cervical Intraepithelial Neoplasia (CIN) and Invasive Carcinoma of the Cervix

Most tumors of the cervix are of epithelial origin. During development, the columnar, mucus-secreting epithelium of the endocervix is joined to the squamous epithelial covering of the exocervix. The exposed columnar cells may undergo metaplasia to squamous cells, forming a region called the transformation zone, where tumors most commonly arise.

Cervical carcinoma was once the most frequent form of cancer in women worldwide. Since the introduction of the Papanicolaou (Pap) smear technique 50 years ago, the incidence of cervical cancer has plummeted. Today, cytology remains the most successful cancer screening test ever developed. In populations that are screened regularly, many precancerous lesions and early cancers are detected early, significantly reducing the incidence of invasive cancers compared to the past. The annual survival rate and cure rate have significantly improved. Moreover, the rate of detection of cervical intraepithelial lesions has increased significantly. Cervical cancer mostly occurs in women aged 30–60 years, with a trend toward younger age in recent years [1–3].

Etiology and Pathogenesis

Important risk factors directly related to the development of CIN and invasive carcinoma include early age at first intercourse, multiple sexual partners, smoking, cervical trauma, and so on. Clinical epidemiological investigations indicate that premature sexual activity and sexual disorder are the main causes of cervical cancer. Sexually transmitted HPV infection may be the primary pathogenic factor of cervical cancer, especially HPV-16, HPV-18, HPV-31, HPV-33, HPV-58, etc. According to the risk of HPV related to the occurrence of cancer, it is classified into low-risk and high-risk subtypes. For example, HPV-16 and HPV-18 are closely related to cervical cancer and are considered high-risk viruses, while HPV-6 and HPV-11 are related to the occurrence of genital wart diseases and are low-risk viruses. High-risk HPV infection is the most important risk factor for the development of squamous intraepithelial lesion (SIL), which can progress to carcinoma. HPV is detectable by molecular methods, and persistent infection by high-risk types HPV-16 and HPV-18 accounts for approximately 70% of cases of SIL and cervical carcinoma. Although HPV infection occurs in the most immature squamous cells of the basal layer, replication of HPV DNA takes place in more differentiated overlying squamous cells. Squamous cells at this stage of maturation do not normally replicate DNA, but HPV-infected squamous cells do, as a consequence of the expression of two potent oncoproteins encoded in the HPV genome called E6 and E7. The E6 and E7 proteins bind and inactivate two critical tumor suppressors, the p53 gene and the Rb gene, respectively. This leads to a promoted growth rate and increased susceptibility to additional mutations, which may eventually result in carcinogenesis [1–3].

The p16 gene is a novel tumor suppressor gene whose encoded product, the p16 protein, can inhibit the phosphorylation of retinoblastoma protein (pRb) and prevent cells from progressing from the G1 phase into the S phase. However, the product of the high-risk HPV E7 gene can bind to pRb, inactivating the p16 gene, which leads to the overexpression of p16 protein as part of a disrupted feedback mechanism. This disordered epithelial cell cycle results in cellular immortalization and initiates the process of carcinogenesis. Therefore, the p16 protein is an indicator of the gene expression and activity of high-risk HPV types and is also an important auxiliary marker for the early detection of cervical lesions [1–3].

At present, the preventive vaccine against HPV has been marketed globally and has long-term effectiveness in preventing cervical cancer and precancerous lesions in women who have not yet been infected with HPV.

Cervical Intraepithelial Neoplasia

Morphology

Cervical intraepithelial neoplasia (CIN) is a precancerous lesion in which the cervical epithelium is replaced by cells with varying degrees of atypia. This condition is characterized by the presence of hollow cells or squamous epithelial cells of different sizes, nuclear enlargement, an altered nuclear-to-cytoplasm ratio, nuclear division, and disordered cell polarity.

In the past, CIN precancerous changes are graded as follows: cervical intraepithelial neoplasia I (CIN I), cervical intraepithelial neoplasia II (CIN II), and cervical intraepithelial neoplasia III (CIN III):

1. **CIN I:** In CIN I, 2/3 of the cells on the surface of the squamous epithelium are mature. The superficial cells are slightly heterogeneous with very slight nuclear abnormalities. However, the lower 1/3 of the epithelium shows more obvious atypia, though nuclear division and pathological mitotic figures are rare.
2. **CIN II:** In CIN II, the upper 1/3 layer of the squamous epithelial cells is relatively mature, but the entire epithelial layer exhibits obvious cell dysplasia, with nuclear division limited to the lower 2/3 layer.
3. **CIN III:** CIN III includes what was previously classified as severe dysplasia and carcinoma in situ. It refers to a lack of differentiation and maturity throughout the entire epithelial layer or in only the superficial layers. This stage is characterized by significant cell hyperplasia, marked nuclear atypia throughout most or all layers, and the presence of nuclear division and pathological mitotic figures in all layers of the epithelium. Heterotypic cells may spread from the surface along the basement membrane through the cervical gland mouth to the cervical gland, replacing part or all of the glandular epithelium, but without breaking through the gland's basement membrane. This condition, where CIN involves the gland (Fig. 13.1), was formerly known as in situ carcinoma involving the gland [1–3].

The current terminology uses a two-tiered system that not only reflects the biology of the disease but also aids in patient management. It is more reproducible than the previous three-tiered system. According to the degree of lesions, it is classified into low-grade squamous intraepithelial lesions (LSIL) and high-grade squamous intraepithelial lesions (HSIL). LSIL is still often referred to as CIN I, while HSIL encompasses CIN II and CIN III of the previous three-tiered system [1–3].

Fig. 13.1 Partial replacement of endocervical glandular epithelium by CIN III

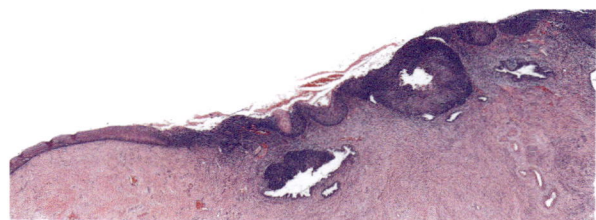

The occurrence of most SIL is closely related to high-risk HPV infection. The integration of viral genes with epithelial genes can be detected by molecular biology techniques. Immunohistochemical staining of p16 and Ki-67 helps to identify LSIL and HSIL. Diffuse strong nuclei and/or cytoplasmic positivity with p16 and diffuse strong nuclear positivity with Ki-67 immunohistochemical staining are useful for the auxiliary diagnosis of HSIL. HSIL is a SIL that carries a significant risk of invasive cancer development if not treated. There is a proliferation of squamous cells most frequently in the zone of metaplasia and near the current squamocolumnar junction. The cells exhibit abnormal nuclear features, including increased nuclear size, irregular nuclear membranes, and increased nuclear-to-cytoplasmic ratios, accompanied by mitotic figures. There is less cytoplasmic differentiation than in LSIL as the proliferating cell compartment extends into the CIN II and CIN III of the epithelium [1–3].

Clinical Features and Prognosis

Patients with CIN typically have no noticeable symptoms, and there are no special changes visible to the naked eye. Not all CIN I lesions develop into CIN II and CIN III or even invasive carcinoma. Most CIN I lesions are reversed or cured with appropriate treatment. The likelihood of progression to CIN III and invasive carcinoma is related to the degree of intraepithelial neoplasia. The higher the lesion grade, the greater the chance of progression, and the shorter the time required. About half of CIN I lesions can undergo spontaneous regression. Approximately 10% of CIN I lesions progress to CIN II and CIN III over 10 years. Less than 2% of CIN I lesions eventually progress to invasive carcinoma. The incidence of developing invasive cancer within 10 years in cases of CIN III is up to 20% [1–3].

The junction between squamous epithelium and columnar epithelium of the cervix is a high-risk site for disease. Suspicious areas can be identified using the iodine liquid test. Normal cervical squamous epithelium is rich in glycogen, so it stains with iodine, while areas affected by lesions do not stain, indicating possible abnormalities. Moreover, acetic acid can be applied to the cervix to highlight CIN changes. Diagnosis still requires further cytological or histopathological examination. LSIL can be monitored through follow-up. HSIL requires cervical conical resection to prevent the progression of cervical squamous epithelial lesions. If cytology suggests cancer, a cervical histology examination is needed to confirm the diagnosis and determine further treatment. Cervical cytology screening has become an effective method for the prevention and treatment of cervical cancer [1–3].

Invasive Carcinoma of the Cervix

Morphology
There are four types based on gross appearance: erosion, exogenous cauliflower, deeply infiltrative, and ulcer:

1. Erosion type: The mucous membrane is red, granular, brittle, and prone to bleeding. Histologically, it is mostly carcinoma in situ and early invasive carcinoma.
2. Exogenous cauliflower type: The cancer tissues grow mainly on the surface of the cervix, forming papillary or cauliflower-patterned masses, with necrotic and superficial ulceration on the surface.
3. Deeply infiltrative type (Fig. 13.2): The cancer tissues mainly infiltrate into the cervix, causing thickening of the cervical lip, which often appears smooth.
4. Ulcer type: The carcer tissues infiltrate deep into the cervix, resulting in a large necrotic lesion on the surface that forms an ulcer resembling a crater [1–3].

Microscopically, the most common cervical carcinomas are squamous cell carcinomas (Fig. 13.3), accounting for about 80%. Adenocarcinomas account for 15%, and the remaining 5% are adenosquamous cell carcinomas and neuroendocrine tumors. In recent years, adenocarcinoma has been on the rise, accounting for about 20% of cervical cancer cases.

Fig. 13.2 The cancer tissue infiltrated into the cervical canal and bilateral muscle wall

Fig. 13.3 Squamous cell carcinomas, cancer nests infiltrate into the cervical interstitium

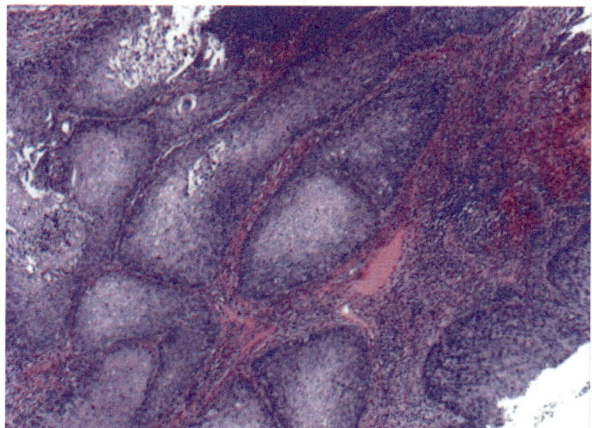

Based on progression, squamous cell carcinomas are classified as early microinvasive and invasive carcinoma. Early invasive carcinoma, or microinvasive squamous cell carcinoma, occurs when tumor cells break through the basement membrane and infiltrate into the stromal membrane, forming irregular tumor cell nests or cables. However, the infiltration depth should not exceed 5 mm below the basement membrane, and the infiltration width should not exceed 7 mm. Early invasive cancer generally cannot be judged by the naked eye and can only be diagnosed under the microscope. Early microinvasive carcinoma typically has a good prognosis. Invasive carcinoma is characterized by tumor tissue growing into the stroma, with a depth exceeding 5 mm below the basement membrane or a width exceeding 7 mm. Based on differentiation, squamous cell carcinomas are classified as keratinized (well-differentiated) and non-keratinized (poorly differentiated) types [1–3].

There is no obvious macroscopic difference between cervical adenocarcinoma and squamous carcinoma. Adenocarcinoma can be classified based on tissue structure and cell differentiation into high-grade, moderate-grade, and poor-grade cervical adenocarcinoma. The tissue structure of highly differentiated adenocarcinoma is similar to normal cervical ductal glands, with cells showing tall columns, cytoplasm rich in mucus, and nucleus located at the base. Moderately differentiated adenocarcinoma is the most common type of cervical adenocarcinoma, with an obvious duct-like structure, glands scattered in the stroma, irregular lumen shapes, varying sizes, cell hierarchies, and variable mucus content in the cytoplasm. Poorly differentiated adenocarcinoma often lacks glandular structure or has only a few glands, with cancer cells arranged into solid nests, showing significant heterogeneity and forming visible mucinous lakes. Cervical adenocarcinoma is not sensitive to radiotherapy and chemotherapy, and the prognosis is poor [1–3].

Spreading

Direct Spreading

The upward infiltration of cancer tissue destroys the entire cervix but rarely invades the uterine body. It can involve the vaginal dome and vaginal wall downward and can invade the uterine and pelvic wall tissues to the sides. If the tumor invades or compresses the ureter, it can cause hydronephrosis and advanced renal failure. Forward, it can invade the bladder; backward, it can involve the rectum [1–3].

Lymphatic Metastasis

This is the most common and significant metastatic route for cervical carcinoma. Carcinoma cells first metastasize to the parauterine lymph nodes, then to the obturator nodes, lymphoglandulae hypogastricae, external iliac lymph nodes, common iliac lymph nodes, inguinal lymph nodes, and presacral lymph nodes. Advanced tumor cells may also metastasize to supraclavicular lymph nodes [1–3].

Hematogenous Metastasis

This is rare, but in advanced stages, it can be spread to the lungs, bones, and liver.

Clinical Features

Invasive cervical cancer is most often seen in women who have never had a Pap smear or who have not been screened for many years. In such cases, cervical cancer often is symptomatic, with patients presenting with unexpected vaginal bleeding, leukorrhea, painful coitus (dyspareunia), or dysuria. Treatment typically involves surgical options such as hysterectomy and lymph node dissection; small microinvasive carcinomas may be treated with cone biopsy. Mortality is most strongly related to tumor stage and, in the case of neuroendocrine carcinomas, to cell type. Most patients with advanced disease die as a result of local invasion rather than distant metastasis. In particular, renal failure due to obstruction of the urinary bladder and ureters is a common cause of death [1–3].

13.2 Body of Uterus

13.2.1 Endometriosis

Endometriosis is defined by the presence of endometrial glands and stroma in a location outside the uterus. The sites most frequently involved are the ovaries (80%), with other tissues and organs, in descending order of frequency, including uterine ligaments, the rectovaginal septum, pelvic peritoneum, abdominal scar, navel, vagina, vulva, and appendix. In contrast, when endometrial glands and stroma are abnormally present within the myometrium (2 mm beyond the basal layer of the endometrium), this condition is called adenomyosis (Fig. 13.4).

Fig. 13.4 Adenomyosis (100×, hematoxylin and eosin stain). Endometrial glands and stroma are present within the myometrium

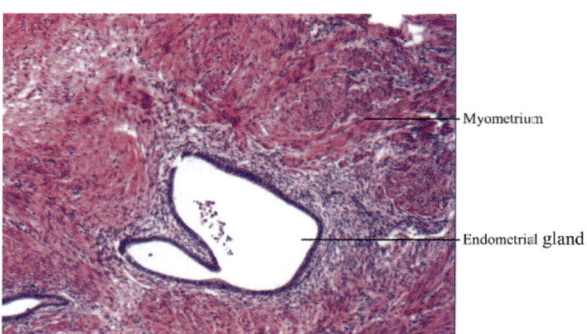
Myometrium

Endometrial gland

13.2.1.1 Pathophysiology

The pathogenesis of endometriosis is uncertain. Several theories are proposed to explain the origin of these dispersed lesions. The regurgitation theory proposes that endometrial tissue refluxes through the fallopian tube to the abdominal organs during menstruation. The benign metastases theory suggests that endometrial tissue can implant in surgical incisions due to operations or spread to distant organs via blood vessels. The metaplastic theory holds that heterotopic endometrium arises directly from metaplasia of coelomic epithelium.

13.2.1.2 Morphology

Endometriotic lesions bleed periodically in response to intrinsic hormonal stimulation. The earliest lesions of endometriosis usually appear grossly as red-blue or yellow-brown nodules with a soft texture, similar to mulberries. When lesions are extensive, organizing hemorrhage causes fibrous adhesions with surrounding organs. If the ovaries are involved, repeated hemorrhage enlarges the lesions and may form large, blood-filled cysts containing inspissated, chocolate-colored material ("chocolate cysts") [1].

The histologic diagnosis of endometriosis at all sites depends on finding two of the following three features within the lesions: endometrial glands, endometrial stroma, and hemosiderin pigment. Occasionally, healed foci may contain only fibrous tissue and hemosiderin-laden macrophages, which by themselves are not diagnostic.

13.2.1.3 Clinical Features

Symptoms of endometriosis depend on the distribution of the lesions. It commonly results in dysmenorrhea or menoxenia.

13.2.2 Endometrial Hyperplasia

Endometrial hyperplasia is the hyperplasia of endometrial glands or stroma in response to an abnormal hormonal state of excess endogenous or exogenous estrogen. It occurs mostly in women of childbearing age and those in climacteric. Endometrial hyperplasia, atypical hyperplasia, and endometrial cancer present as a continuous process of evolution both in morphology and biology, with very similar pathogenesis.

13.2.2.1 Morphology

Endometrial hyperplasia is divided into three types based on cell morphology and the degree of proliferation and differentiation of the gland structure [2]:

1. Simple Hyperplasia: Simple hyperplasia is also called "mild hyperplasia" or "cystic hyperplasia." The morphological features (Fig. 13.5) include thickening of the endometrium, an increase in the number of glands, and punctuated cysts. The lining epithelium is usually a single layer or a pseudo-complex layer. The cells are columnar and show no heteromorphosis. The morphology and arrangement of the cells are similar to those of the endometrium in the proliferative stage. These lesions rarely progress to adenocarcinoma (approximately 1%).

2. Complex Hyperplasia: Complex hyperplasia is also known as adenomatous hyperplasia. The glandular hyperplasia is obvious and crowded, with a complex and irregular gland structure. The endometrial stroma is notably reduced and there is no cell heterotopia. This type may evolve into atypical hyperplasia (Fig. 13.6). Adenocarcinoma develops in 3%.

3. Atypical Hyperplasia: Atypical hyperplasia is characterized by crowded aggregates of cytologically altered tubular or slightly branching glands. The proliferated glands are commonly back-to-back and often have complex outlines due to branching structures. Individual cells are rounded and lose their normal perpendicular orientation to the basement membrane. In addition, the glands display nuclear atypia, with nuclei showing open chromatin and conspicuous nucleoli (Fig. 13.7). The features of atypical hyperplasia have considerable overlap with those of well-differentiated endometrioid adenocarcinoma, and accurate distinction from cancer may not be possible without hysterectomy. Approximately 1/3 of patients with atypical hyperplasia will develop adenocarcinoma.

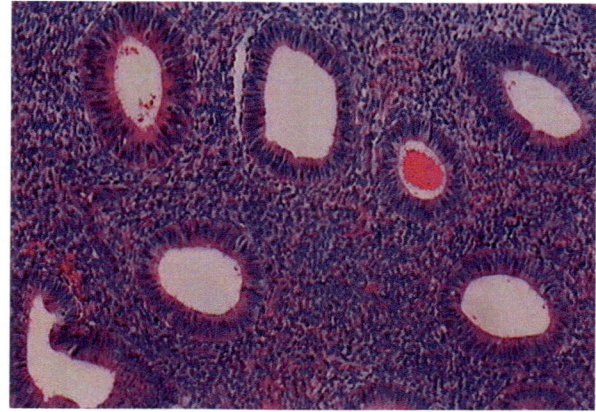

Fig. 13.5 Simple hyperplasia (200×, hematoxylin and eosin stain). It shows anovulatory or "disordered" endometrium containing dilated glands with multiple layers of epithelial cells

Fig. 13.6 Complex
hyperplasia (200×,
hematoxylin and eosin
stain). It is characterized
by nests of closely packed
glands

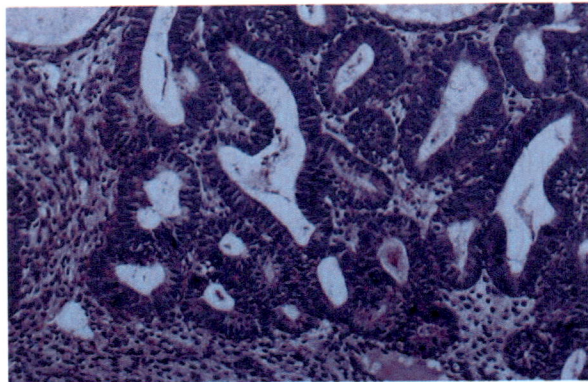

Fig. 13.7 Atypical
hyperplasia (400×,
hematoxylin and eosin
stain). It shows glandular
crowding and cellular
atypia

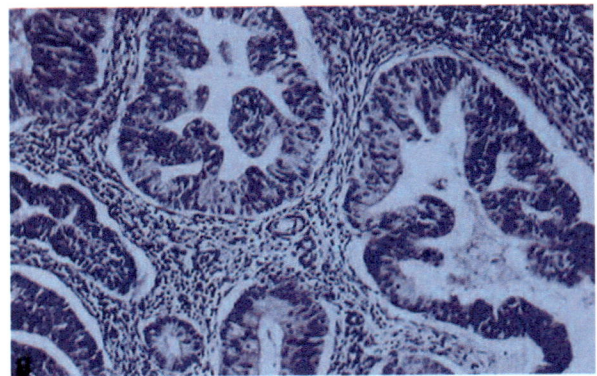

13.2.2.2 Clinical Features

The main clinical manifestations are irregular vaginal bleeding and menorrhagia, also called functional uterine bleeding.

13.2.3 Tumors of the Uterus

13.2.3.1 Endometrial Adenocarcinoma

Endometrial adenocarcinoma is a malignant tumor derived from the epithelial cells of the endometrium. It generally appears in menopausal and postmenopausal women, between the ages of 55 and 65 years.

Pathophysiology

Endometrioid adenocarcinoma accounts for 80% of cases of endometrial adenocarcinoma, which arise in association with estrogen excess. Endometrioid adenocarcinoma is designated "endometrioid" due to its histologic similarity to normal endometrial glands. Risk factors for this adenocarcinoma include obesity, diabetes, infertility, hypertension, and smoking. Many of these risk factors result in increased estrogenic stimulation of the endometrium and are associated with endometrial hyperplasia. Microsatellite instability and mutation of the tumor suppressor gene PTEN are important events in the stepwise development of endometrioid adenocarcinoma.

In addition, some endometrial adenocarcinomas appear to be unrelated to increased estrogen and intimal hyperplasia but arise in the setting of an inactive or atrophic endometrium. This type of endometrial adenocarcinoma, which arises in the context of endometrial atrophy in older postmenopausal women, is much less common, accounting for roughly 15% of tumors. Some of these tumor tissues are similar to ovarian serous cystadenocarcinoma and are referred to as endometrial serous carcinoma. Immunohistochemistry often reveals high levels of p53 in this type, a finding that correlates with the presence of TP53 mutations. The prognosis of this type is also worse than that of estrogen-related endometrial carcinoma.

Morphology

Grossly, endometrial adenocarcinoma is divided into diffuse and localized types. The diffuse type is characterized by diffuse thickening of the endometrium, a rough surface, gray-white color, crisp texture, hemorrhage, necrosis or ulceration, and varying degrees of infiltration into the myometrium (Fig. 13.8). The localized type is mostly located at the bottom or the corner of the uterus, often presenting as polyps or papillae within the uterine cavity. If the cancer tissue is small and superficial, it can be scraped out during diagnostic curettage, and cancer tissue may not be found in the excised uterus.

On microscopic examination, there are three histologic grades of endometrial adenocarcinoma [3]: well-differentiated, moderately differentiated, and poorly differentiated, with well-differentiated accounting for the vast majority. In well-differentiated endometrial adenocarcinoma, the glandular tubes are crowded and disorganized, the stroma is reduced, and a "back-to-back" phenomenon is often observed. The cells have mild atypical structures that resemble hyperplasia of endometrial glands. Moderately differentiated endometrial adenocarcinoma shows irregular and disorder glands. The cancer tissue grows into the glands, forming papillary or sieve-like structures, and cancer foci can be seen. The cancer cells exhibit obvious atypia, and pathological mitosis is often observed (Fig. 13.9). In poorly

Fig. 13.8 Endometrial
adenocarcinoma (diffuse
type). It presents as a
fungating mass in the
fundus of the uterus

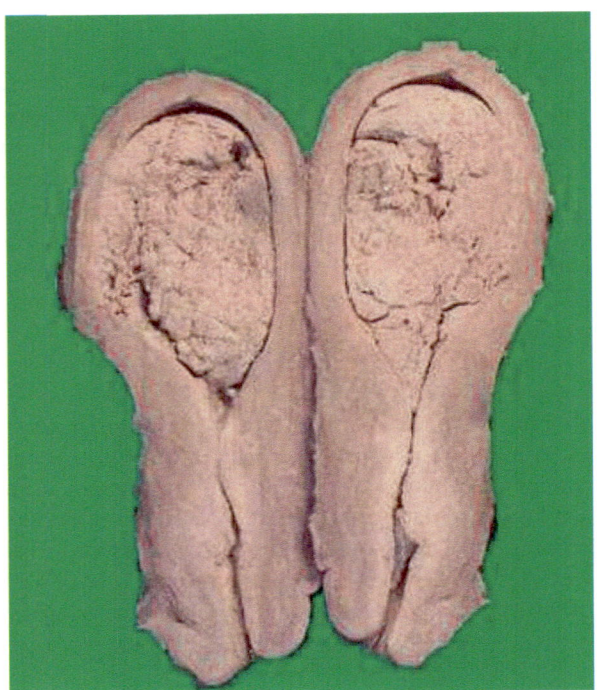

Fig. 13.9 Endometrioid
adenocarcinoma (200×,
hematoxylin and eosin
stain). It shows infiltrating
myometrium and growing
in a cribriform pattern

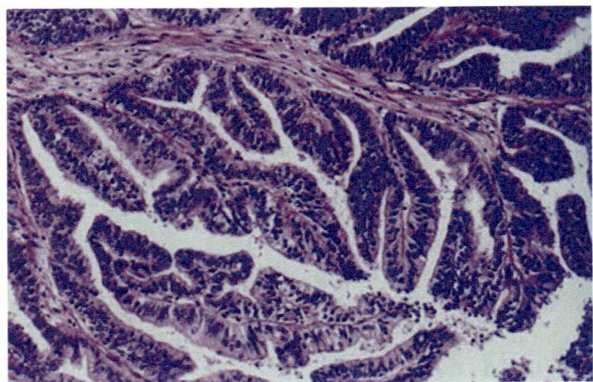

differentiated adenocarcinoma, the cancer cells exhibit poor differentiation and rarely form adenoid structures. The invasive lesions consist of cells with marked cytologic atypia, including a high nuclear-to-cytoplasmic ratio, atypical mitotic figures, hyperchromasia, and prominent nucleoli.

About 1/3 of endometrioid adenocarcinomas contain foci of squamous differentiation. Squamous elements may appear histologically benign when associated with well-differentiated adenocarcinomas. Less commonly, moderately or poorly differentiated endometrioid adenocarcinomas contain squamous elements that appear frankly malignant. Sometimes endometrioid adenocarcinoma is accompanied by metaplasia of benign squamous cell foci, referred to as endometrioid adenocarcinoma with squamous cell metaplasia. If the squamous epithelium is malignant, it is called adenosquamous carcinoma.

Metastasis

Endometrial adenocarcinoma generally grows slowly and can remain confined to the uterine cavity for many years. It is often detected early due to irregular uterine bleeding, which contributes to a better prognosis compared to other gynecological tumors. Metastasis typically occurs relatively late, with local infiltration and lymph node metastasis being more common. Hematogenous spread is rare.

Endometrial adenocarcinoma frequently metastasizes to the ovaries and fallopian tubes when the cancer is located in the upper part of the uterus and to the cervix when the cancer is in the lower part of the uterus. Usually, this cancer first spreads into the myometrium and the serosa and then into other reproductive and pelvic structures. When the lymphatic system is involved, the pelvic and paraaortic nodes are usually the first to become affected. More distant metastases are spread through the blood and often occur in the lungs, liver, and bones.

Clinical Features

Although it may be asymptomatic for a period of time, endometrial adenocarcinoma usually manifests with excessive leucorrhea and irregular bleeding, often in postmenopausal women. With progression, the uterus enlarges and may become affixed to surrounding structures as the cancer infiltrates adjacent tissues. These tumors usually are slow to metastasize but, if left untreated, eventually disseminate to regional nodes and more distant sites. Fortunately, postmenopausal bleeding often leads to early detection, and cures are possible in most patients. The diagnosis of endometrial adenocarcinoma must be established by histologic examination of tissue obtained by biopsy or curettage. With therapy, the 5-year survival rate for early-stage tumors is 90%, but survival drops precipitously in higher-stage carcinomas. As anticipated, the prognosis depends heavily on the clinical stage at diagnosis, as well as histologic grade and subtype.

13.2.3.2 Leiomyomas

Leiomyomas are the most common benign tumors in females. Including minute tumors, leiomyomas occur in 75% of women over age 30. They are rare before age 20 and most regress after menopause. Estrogens and possibly oral contraceptives stimulate their growth.

Morphology

Grossly, leiomyomas are sharply circumscribed, discrete, round, firm, gray-white tumors varying in size from small, barely visible nodules to massive tumors that fill the pelvis. Their cut surface bulges, and borders are smooth and distinct from the neighboring myometrium. Most leiomyomas are intramural, but some are submucosal, subserosal, or pedunculated. Larger neoplasms may develop foci of ischemic necrosis with areas of hemorrhage and cystic softening, and after menopause, they may become densely collagenous and even calcified. Whatever their size, the characteristic whorled pattern of smooth muscle bundles on cut section usually makes these lesions readily identifiable [4] (Fig. 13.10).

On microscopic examination, leiomyomas are typically composed of bundles of smooth muscle cells that resemble the uninvolved myometrium. The cytoplasm is abundant, eosinophilic, and fibrillar. Foci of fibrosis, calcification, ischemic necrosis, cystic degeneration, and hemorrhage may be present (Fig. 13.11).

Leiomyomas are usually benign. Benign leiomyomas rarely transform into sarcomas, and the presence of multiple lesions does not increase the risk of harboring a malignancy.

Clinical Features

Leiomyomas of the uterus may be entirely asymptomatic and discovered only on routine pelvic or postmortem examination. Many intramural leiomyomas are symptomatic because of their sheer bulk, and large ones may interfere with bowel or bladder function or cause dystocia in labor. Moreover, leiomyomas can lead to spontaneous abortion, abnormal fetal exposure, and postmenopausal bleeding.

Recurrence after removal is common with leiomyosarcomas, and many metastasize, typically to the lungs, yielding a 5-year survival rate of about 40%.

Fig. 13.10 Leiomyomas. The uterus is opened to reveal multiple tumors in submucosal (bulging into the endometrial cavity), intramural, and subserosal locations that display a firm white appearance on sectioning

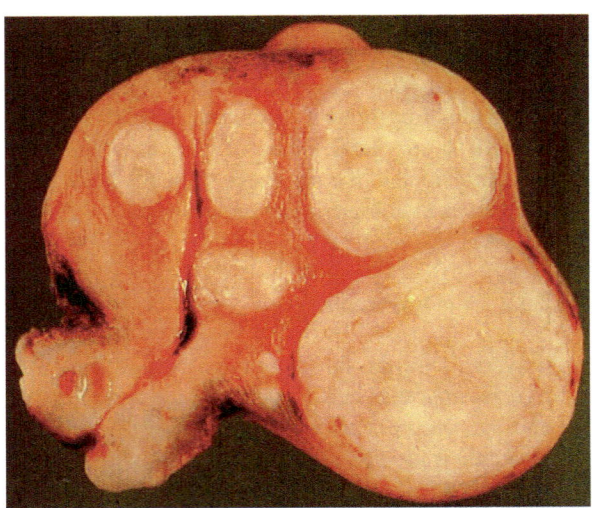

Fig. 13.11 Leiomyomas (40×, hematoxylin and eosin stain). Microscopic appearance of leiomyoma reveals well-differentiated, regular, and spindle-shaped smooth muscle cells

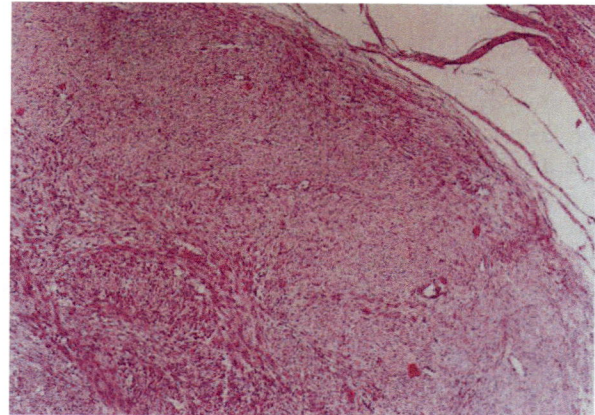

13.3 Gestational Trophoblastic Diseases

The major disorders of gestational trophoblastic diseases (GTD) are hydatidiform mole (complete and partial), invasive mole, choriocarcinoma, and placental site trophoblastic tumor (PSTT). These demonstrate a range of aggressiveness from benign hydatidiform moles to highly malignant choriocarcinomas. The common feature of these diseases is the abnormal proliferation of trophoblastic cells, and most of them are related to pregnancy. All elaborate human chorionic gonadotropin (hCG), which can be detected in the blood and urine at levels considerably higher than those found during normal pregnancy. It can be used as an auxiliary index for clinical diagnosis, follow-up observation, and evaluation of curative effect.

13.3.1 Hydatidiform Mole

Hydatidiform mole is a benign lesion of placental villi, characterized by high edema of villous stroma and varying degrees of trophoblastic cell hyperplasia. This disease occurs at any age during the childbearing period and may be related to ovarian insufficiency or recession. There are two distinctive subtypes of hydatidiform moles: complete and partial.

13.3.1.1 Pathophysiology

The etiology and pathogenesis of hydatidiform mole have not been fully elucidated. In recent years, studies of hydatidiform chromosomes have shown that chromosomal abnormalities may play a leading role.

Complete hydatidiform moles are not compatible with embryogenesis and never contain fetal parts. They result from the fertilization of an egg that has lost its female chromosomes, so the genetic material is completely paternally derived. All of the chorionic villi are abnormal, and the chorionic epithelial cells are diploid (46, XX or, uncommonly 46, XY). Eighty percent have a 46, XX karyotype stemming from the duplication of the genetic material of one sperm (a phenomenon called androgenesis). The remaining 10% result from the fertilization of an empty egg by two sperm; these may have 46, XX or 46, XY karyotype. In complete moles, the embryo dies very early in development and therefore is usually not identified.

Partial hydatidiform moles are compatible with early embryo formation and therefore may contain fetal parts, have some normal chorionic villi, and are almost always triploid (e.g., 69, XXY) or occasionally tetraploid (92, XXXY).

Both types result from abnormal fertilization. In a complete mole, the entire genetic content is supplied by two spermatozoa (or a diploid sperm), yielding diploid cells containing only paternal chromosomes. In a partial mole, a normal egg is fertilized by two spermatozoa (or a diploid sperm), leading to a triploid karyotype with a preponderance of paternal genes.

13.3.1.2 Morphology

The uterus may be of normal size in early moles, but in more advanced cases, the uterine cavity is expanded. The classic appearance of hydatidiform moles is a delicate, friable mass of thin-walled, translucent, cystic, grapelike structures consisting of swollen edematous (hydropic) villi (Fig. 13.12). Fetal parts are rarely seen in complete moles but are common in partial moles [5, 6].

Microscopically, hydatidiform moles have three characteristics:

1. Due to the loose, myxomatous, edematous stroma, the chorionic villi are enlarged and hydropic, with swelling and scalloped shapes and central cavitation.
2. The blood vessels in the villous stroma are either absent or contain a small amount of nonfunctional capillaries with no red blood cells.
3. The chorionic epithelium almost always shows some degree of proliferation of both cytotrophoblasts and syncytiotrophoblasts, which lose their normal order, display a multilayered or patchy aggregation, and exhibit mild atypia.

Proliferation of trophoblasts is the most important characteristic of hydatidiform mole [5, 6] (Fig. 13.13).

13.3.1.3 Clinical Features

Most patients experience symptoms between 11 and 25 weeks of pregnancy. Women with partial and early complete moles often present with spontaneous miscarriage or undergo curettage due to ultrasound findings of abnormal villous enlargement.

Fig. 13.12 Hydatidiform
mole. The uterus consists
of numerous swollen
(hydropic) villi

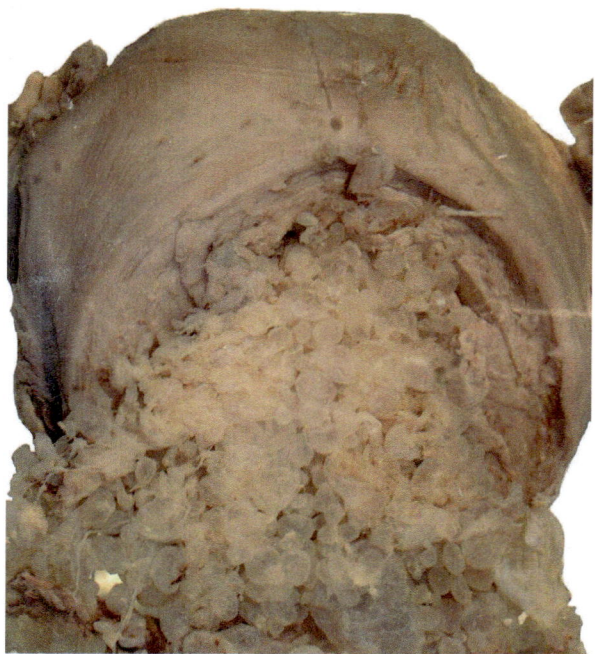

Fig. 13.13 Complete
hydatidiform mole (100×,
hematoxylin and eosin
stain). In this microscopic
image, marked edematous
villous enlargement,
edema, and circumferential
trophoblast proliferation
are evident

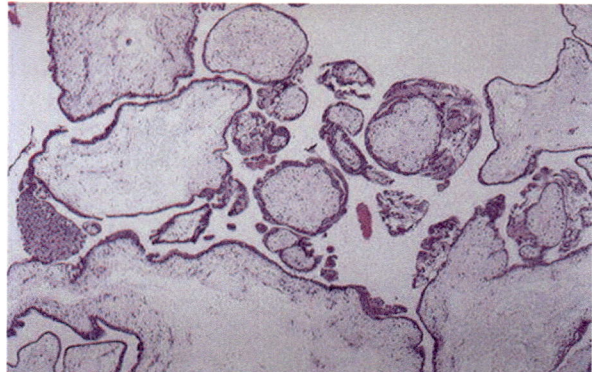

Due to placental villus edema, the volume of the uterus increases significantly and
exceeds that of a normal pregnancy. Because of early embryonic death, there are
neither fetal heart sounds nor fetal movements even after 5 months of pregnancy.

In complete moles, human chorionic gonadotropin (HCG) levels greatly exceed
those of a normal pregnancy of similar gestational age. In addition, the rate at which
HCG levels rise over time in molar pregnancies exceeds that seen with normal single or multiple pregnancies. Most moles can be successfully removed by curettage.
Patients are subsequently monitored for 6 months to a year to ensure that HCG

levels decrease to non-pregnant levels. Because trophoblast cells invade the blood vessels, the uterus may have repeated irregular bleeding, and occasionally, the grapelike substance can be seen in the outflow.

The majority of hydatidiform moles can be cured after curettage. Continuous elevation of HCG may indicate a persistent or invasive mole, which develops in up to 10% of molar pregnancies and is seen more frequently with complete moles. In addition, 2% of complete moles may give rise to subsequent choriocarcinoma. Partial moles have an increased risk of persistent molar disease but are not associated with choriocarcinoma.

13.3.2   Invasive Mole

Invasive mole, also called malignant mole, is defined as a mole that penetrates or even perforates the uterine wall and is a borderline tumor between hydatidiform mole and choriocarcinoma. Invasive moles are complete moles that are more invasive locally but do not have the aggressive metastatic potential of a choriocarcinoma.

13.3.2.1   Morphology

There is invasion of the myometrium by hydropic chorionic villi, which penetrate the uterine wall deeply, possibly causing rupture and sometimes life-threatening hemorrhage. Hydropic villi may embolize to distant sites, such as the lungs and brains, but do not grow in these organs as true metastases; even without chemotherapy, they eventually regress. On microscopic examination, the epithelium of the villi shows atypical changes, accompanied by proliferation of both cytotrophoblasts and syncytiotrophoblasts [7].

13.3.2.2   Clinical Features

The tumor is manifested clinically by vaginal bleeding and irregular uterine enlargement. It is always associated with persistently elevated serum HCG. The tumor responds well to chemotherapy but may result in uterine rupture and necessitate hysterectomy.

13.3.3 Choriocarcinoma

This very aggressive malignant tumor arises either from gestational chorionic epithelium or, less frequently, from totipotent cells within the gonads or elsewhere. Women under age 20 and over age 40 are at high risk. Fifty percent of choriccarcinomas arise in complete hydatidiform moles, 25% follow previous abortions, and 22% follow normal pregnancies, with the remainder occurring in ectopic pregnancies. Those instances may suggest an origin from an abnormal ovum rather than from retained chorionic epithelium.

13.3.3.1 Morphology

Grossly, choriocarcinoma is a soft, fleshy, yellow-white tumor that usually has large pale areas of necrosis and extensive hemorrhage (Fig. 13.14). Sometimes the necrosis is so extensive that little viable tumor remains. Indeed, the primary lesion may "self-destruct" and only the metastases tell the story. Very early, the tumor insinuates itself into the myometrium and into vessels [8].

Histologically, it does not produce chorionic villi and consists entirely of proliferating syncytiotrophoblasts and cytotrophoblasts (Fig. 13.15). In contrast to hydatidiform moles and invasive moles, chorionic villi are not formed; instead, the tumor is purely epithelial, composed of anaplastic cuboidal cytotrophoblast and syncytiotrophoblast [8].

Fig. 13.14 Choriocarcinoma. It presents as a bulky hemorrhagic mass invading the uterine wall

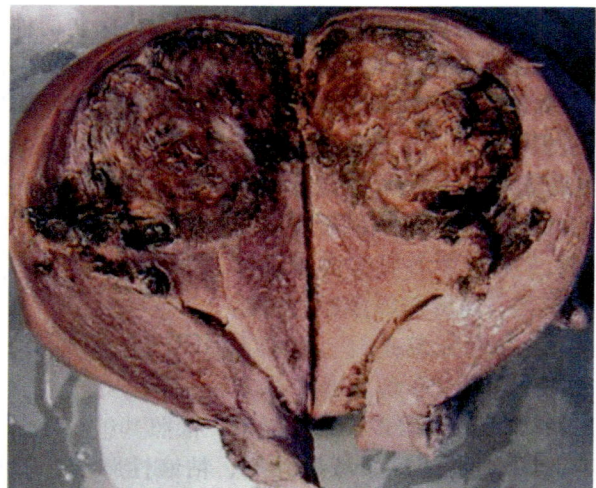

Fig. 13.15 Choriocarcinoma (200×, hematoxylin
and eosin stain). Photomi-
crograph illustrates both
neoplastic cytotrophoblast
and syncytiotrophoblast

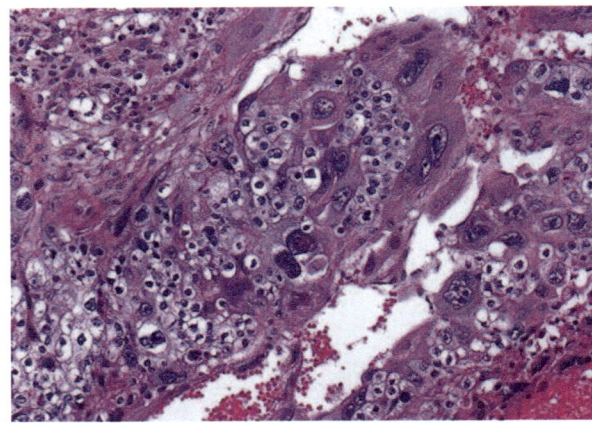

13.3.3.2 Metastasis

Choriocarcinoma easily invades and destroys blood vessels; therefore, in addition to local infiltration, vascular metastasis is the most common route of spread. The most common metastatic sites are the lungs and vagina wall, followed by the brain, liver, spleen, kidneys, and intestines. In a few cases, the metastases can regress after primary resection.

13.3.3.3 Clinical Features

Uterine choriocarcinoma usually manifests as irregular vaginal spotting of a bloody, brown fluid. This discharge may appear during an apparently normal pregnancy, after a miscarriage, or after curettage. By the time most choriocarcinomas are discovered, there is usually widespread dissemination via the blood, most often to the lungs, vagina, brain, liver, and kidneys. Lymphatic invasion is uncommon.

The treatment of gestational choriocarcinoma depends on the stage of the tumor and usually consists of uterine evacuation and chemotherapy. Almost 100% of affected patients can be cured, even those with metastases at distant sites such as the lungs. Many cured patients have had normal subsequent pregnancies and deliveries.

13.3.4 Placental Site Trophoblastic Tumor

Placental site trophoblastic tumors (PSTT) comprise less than 2% of gestational trophoblastic neoplasms. These tumors are derived from the placental site or intermediate trophoblast. These uncommon diploid tumors, often XX in karyotype, typically arise a few months after pregnancy (in half of the cases). Besides normal pregnancy, they may also follow a spontaneous abortion or hydatidiform mole.

13.3.4.1 Morphology

In normal pregnancy, extravillous trophoblasts are found in nonvillous sites such as the implantation site, in islands of cells within the placental parenchyma, and in the placental membranes. Normal extravillous trophoblasts are polygonal mononuclear cells that have abundant cytoplasm and produce human placental lactogen. Histologically, PSTT is composed of malignant trophoblastic cells diffusely infiltrating the endomyometrium [7, 8].

13.3.4.2 Clinical Features

An indolent clinical course is typical, with a generally favorable outcome if the tumor is confined to the endomyometrium. PSTT presents as a uterine mass, accompanies by either abnormal uterine bleeding or amenorrhea. Because intermediate trophoblasts do not produce HCG in large amounts, HCG concentrations are only slightly elevated. Of note, PSTTs are not as sensitive to chemotherapy as other trophoblastic tumors, and the prognosis is poor when spread has occurred beyond the uterus. Patients with localized disease have an excellent prognosis, but about 10–15% of women die of disseminated disease.

13.4 Breast Disease

Lesions of the female breast are much more common than lesions of the male breast and usually take the form of palpable, sometimes painful nodules or masses. Breast cancer is the most common cancer in women (excluding neoplasia of the skin) and is second only to lung cancer as a cause of cancer-related death.

13.4.1 Proliferative Change

13.4.1.1 Fibrocystic Change of the Breast

The designation "fibrocystic" refers to a type of non-neoplastic lesion characterized by the expansion of distal ducts and acini, interstitial fibrous tissue, and epithelial hyperplasia. Fibrocystic changes are the most common breast abnormality seen in premenopausal women. These changes tend to arise between the ages of 25 and 45 and are most likely a consequence of the cyclic breast changes that occur normally in the menstrual cycle. The incidence is related to ovarian endocrine disorders (especially progesterone and estrogen.). However, the exact pathogenesis is still not fully understood [9–12].

Morphology

Nonproliferative Changes

It consists of three major morphologic changes: cysts, fibrosis, and adenosis. These cysts are interlaced with proliferative interstitial fibers, creating a variegated appearance.

Gross appearance: It is often bilateral, with multifocal small nodular distribution, indistinct borders, and variable cyst sizes. The cysts range is from less than 1 cm and up to 5 cm in diameter. Large cysts are called "blue-top cysts" because they are filled with watery, turbid fluid and have a blue outer surface.

Histology: The cysts are covered with columnar epithelium and cuboidal epithelium, but most are covered with flat epithelium, or only the fibrous cyst wall is seen without epithelium covering. Calcification is occasionally seen in the lumen. If the cyst ruptures, the content overflows into the surrounding stroma, causing an inflammatory reaction and interstitial fibrous tissue hyperplasia. The fibrotic stroma may further undergo hyaline degeneration (Fig. 13.16). Frequently, the epithelial cells are large with abundant granular, eosinophilic cytoplasm and small, round, deeply chromatic nuclei. This morphology is called apocrine metaplasia [9–12].

Proliferative Changes

Proliferative disease includes lobular epithelial hyperplasia and ductal epithelial hyperplasia, in addition to cysts and fibrosis. Epithelial hyperplasia can increase the layers and form a papillary pattern protrusion into the lumen, with the tips of the papillae often connecting to form a cribriform structure. It has the potential to evolve into breast cancer and is regarded as a precancerous lesion, especially if cysts are associated with epithelial hyperplasia, particularly atypical epithelial hyperplasia [9–12].

Fig. 13.16 Nonproliferative fibrocystic changes

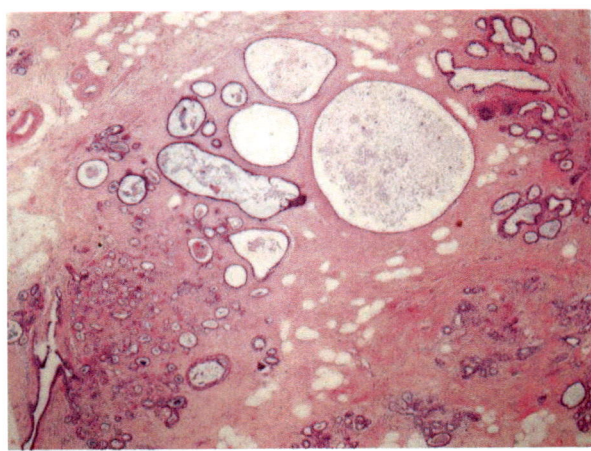

Nonproliferative fibrocystic lesions have no further risk of developing invasive carcinoma. The results of long-term follow-up of patients with usual ductal hyperplasia (UDH) of the breast showed that their risk of developing invasive cancer was 1.5–2 times that of the general population. However, patients with atypical lobular hyperplasia (ALH) and atypical ductal hyperplasia (ADH) have a fivefold increased chance of developing invasive carcinoma, and those with carcinoma in situ of ducts and lobules are ten times more likely to develop invasive carcinoma. This indicates that breast fibrocystic changes—whether clinical, radiological, or pathological—share some similarities with breast cancer and indeed have a certain relationship with its occurrence. However, whether breast cancer develops mainly depends on the degree of ductal and acinar epithelial hyperplasia and the presence of atypical hyperplasia [9–12].

13.4.1.2 Sclerosing Adenosis

Sclerosing adenosis is a rare type of proliferative fibrocystic change, mainly characterized by lobular acini being compressed and distorted due to hyperplasia of central or interlobular fibrosis, generally without cyst formation. It is significant because its clinical and morphologic features may be deceptively similar to those of carcinoma.

Morphology

Grossly, the lesion has a hard, rubbery consistency, and the surrounding mammary gland boundary is not clear. Histologically, the number of acini in each terminal duct increases, with the lobular enlargement and contour remaining intact. The variable hyperplasia of fibrous tissue in the central part of the lesion may compress and distort the lumina of the acini and ducts, while the acini around the lesion are dilated. Sometimes, the acini and ducts are completely compressed, causing the lumen to disappear, so they appear as solid cords of cells. This pattern may be difficult to distinguish histologically from invasive carcinoma. The presence of double layers of epithelium and myoepithelial cells is helpful in suggesting a benign diagnosis, distinguishing it from invasive carcinomas [9–12].

13.4.2 Fibroadenoma

Fibroadenoma is the most common benign tumor of the female breast. Fibroadenomas usually appear at any age after puberty, mostly between 20 and 30 years old. The lesion is usually single or multiple, with unilateral or bilateral occurrence. This benign lesion originates in the terminal duct lobular unit with biphasic proliferation of epithelial and stromal components.

13.4.2.1   Morphology

Grossly, fibroadenomas are solid, ovoid, and well-circumscribed, freely movable nodules, 1–10 cm in diameter. They are gray-white, tough, and lobulated on the cut surface, with slit-like areas often showing a myxoid appearance.

Histologically, tumors are mainly composed of hyperplastic fibrous stroma and glands with a pericanalicular pattern and intracanalicular pattern. The glands are round or oval or squeezed by the surrounding fibrous connective tissue. The stroma is usually loose, rich in mucopolysaccharides, or dense, with vitreous change or calcification (Fig. 13.17).

13.4.2.2   Clinical Features

Clinically, fibroadenomas usually present as solitary, discrete, movable masses. They may enlarge late in the menstrual cycle and during pregnancy. After menopause, they may regress and calcify. Cytogenetic studies reveal that the stromal cells are monoclonal and represent the neoplastic element of these tumors. The basis of ductal proliferation is not clear; perhaps the neoplastic stromal cells secrete growth factors that induce proliferation of epithelial cells [9–12].

13.4.3   Breast Cancer

Breast cancer is a malignant tumor originating from the epithelium of the terminal ductal lobular unit of the breast. The incidence has been increasing over the past 50 years, and it has now become the most common malignant tumor in women.

Fig. 13.17 Fibroadenoma with intracanalicular pattern

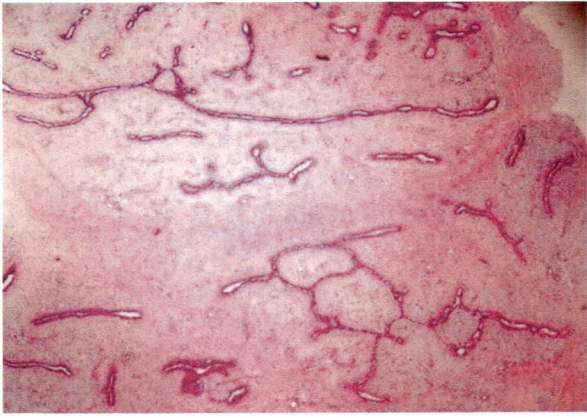

Breast cancer often occurs in women aged 40–60 years and is less common in women under 35 years. Male breast cancer is rare, accounting for about 1% of all breast cancers. Breast cancer most commonly occurs in the upper quadrant of the breast, followed by the central breast and other quadrants [9–12].

13.4.3.1 Etiology and Pathogenesis

The pathogenesis of breast cancer has not been fully elucidated and may be related to the following factors:

1. Hormonal action: The occurrence of breast cancer is related to the level of estrogen, based on the following observations: early menarche, late amenorrhea, late fertility or infertility, and long-term use of estrogen—other factors leading to high estrogen levels are high-risk factors for breast cancer.
2. Genetic factors: About 10% of breast cancer patients have a family genetic predisposition, and the incidence of breast cancer in women with a family history is two to three times higher than that of those without a family history. The incidence of breast cancer occurs at an earlier age and is often accompanied by tumors in other organs. The BRCA1 gene, located on chromosome 17 (17 g21), is a tumor suppressor gene that plays an important role in DNA repair with germ cell or somatic mutations. Women with point mutations and deletions in the BRCA1 gene have an 85% incidence of breast cancer. About half of women with hereditary breast cancer have mutations in the BRCA1 gene. BRCA1 is also a susceptibility gene for ovarian cancer. Moreover, mutations in the BRCA2 gene located on chromosome 13q have also been implicated in hereditary breast cancer onset, in 70% of non-BRCA1 mutation cases associated with BRCA2 mutations. Women with mutations in one copy of BRCA2 have a 30–40% lifetime risk of developing breast cancer [9–12].
3. Environmental factors: There is a distinct geographical distribution of breast cancer, with the highest incidence in North America and Northern Europe and a lower incidence in most Asian and African countries. After migration from a low-incidence area of breast cancer, second- or third-generation descendants gradually show an increased incidence, with the incidence among white women in high-incidence areas converging, while the risk decreases [9–12].
4. Radiation: Long-term and high-dose radiation examination or treatment is considered an inducing factor for breast cancer, with younger age at radiation exposure being associated with a higher future incidence of breast cancer.

Many other less well-established risk factors, such as obesity, alcohol consumption, and a diet high in fat, have been implicated in the development of breast cancer based on population studies. Obesity is a recognized risk factor in postmenopausal women.

13.4.3.2 Morphology

Breast cancer morphology is very complex, with many types that can be roughly divided into noninvasive cancer and invasive cancer:

Noninvasive Carcinoma

Noninvasive carcinoma is classified into ductal carcinoma and lobular carcinoma in situ, both of which originate from terminal ductal lobular unit epithelial cells. Both are confined to the basal membrane and do not infiltrate into the interstitium or lymphatic vessels. Noninvasive carcinoma has a tendency to develop into invasive cancer; the 2012 WHO classification of breast tumors included it in the precancerous lesion category [9–12].

Ductal Carcinoma In Situ (DCIS)

The duct is significantly expanded, with cancer cells confined to the dilated duct and the ductal basal membrane remaining intact. Due to advancements in radiographic examination and general breast cancer surveys, the detection rate has increased significantly, from 5% of all breast cancer cases in the past to 15–30%.

DCIS is a noninvasive carcinoma in situ confined to the ducts of the breast. X-ray examination typically shows clusters of tiny calcifications. DCIS is divided into three grades: low, intermediate, and high, based on nuclear grading, necrosis, and pathological nuclear mitotic figures. High-grade DCIS is often composed of large pleomorphic cells with obvious nucleoli and nuclear mitotic figures, usually proliferating in solid, cribriform, or micropapillary patterns. Some systems also included comedo DCIS as an architectural type. Comedo necrosis is seen as sheets of granular eosinophilic material in the duct lumen with a large amount of necrotic debris (Fig. 13.18). Low-grade DCIS lesions range over 2 mm and consist of small, monomorphic cells with arcade, micropapillary, cribriform, or solid patterns. The nuclei

Fig. 13.18 DCIS with comedo necrosis

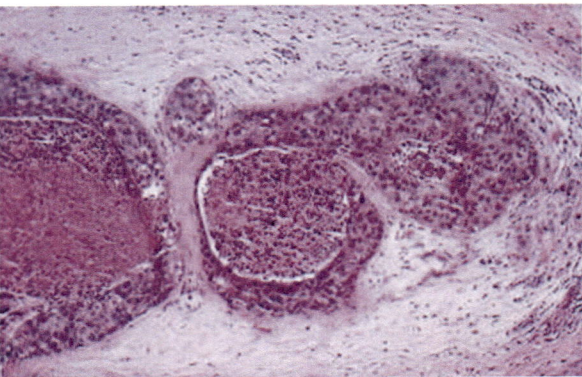

are of uniform size and have a regular chromatin pattern with inconspicuous nucleoli; mitotic figures are rare. Intermediate-grade DCIS is composed of cells with moderate variability in size, shape, and nuclear prominence, falling between high-grade and low-grade DCIS.

Thirty percent of biopsy-proven DCIS without any treatment will develcp into invasive carcinoma after 20 years, indicating that not all DCIS will transform into invasive carcinoma. If it does transform into invasive carcinoma, it usually takes several years or more to do so, and the risk of invasive carcinoma is related to the histological type [9–12].

Lobular Carcinoma In Situ (LCIS)

The terminal ductal lobular unit of the breast is dilated and filled with tumor cells in a solid arrangement, while the lobular structure remains present. The tumor cells are smaller than those of ductal carcinoma in situ, with a uniform shape and round or oval nuclei. Mitotic figures are rare. Because of its small size, LCIS is generally not palpable clinically and is difficult to distinguish from breast lobular hyperplasia. LCIS has a relatively low risk of developing into invasive carcinoma, characterized by a long cancerous interval, bilateral breast involvement, and multiple quadrants. The lifetime probability of developing into invasive carcinoma is 5–32%, with an average cancer rate of 8% [9–12].

Invasive Carcinoma

Invasive carcinoma includes invasive carcinoma of no special type (invasive ductal carcinoma) and invasive carcinoma of special types (inflammatory carcinoma, invasive lobular carcinoma, carcinoma with medullary features, mucinous carcinoma, metaplastic carcinoma, tubular carcinoma, etc.). The microscopic appearance is quite heterogeneous, ranging from tumors with well-developed tubule formation and low-grade nuclei to tumors consisting of sheets of anaplastic cells. The tumor margins are usually irregular but may occasionally be pushing and circumscribed. Invasion of lymphovascular spaces or along nerves may be seen. Advanced cancers may cause dimpling of the skin, retraction of the nipple, or fixation to the chest wall.

Invasive Ductal Carcinoma (Invasive Carcinoma of No Special Type)

Invasive ductal carcinoma develops from ductal carcinoma in situ, where cancer cells break through the basement membrane of the duct and invade the stroma. It is the most common type of breast cancer, accounting for about 70% of breast cancers. Microscopically, the histological morphology is varied. The tumor cells are arranged in nests or cords or with a small amount of glandular structure; some of the original ductal carcinoma in situ structure may be retained, or it may be completely absent. The tumor cells vary in size and shape, with obvious pleomorphism, frequent mitotic figures, and local tumor cell necrosis. Dense fibrous tissue proliferation is

Fig. 13.19 Invasive
carcinoma of no special
type, tumors with strip
shapes infiltrated into the
interstitium

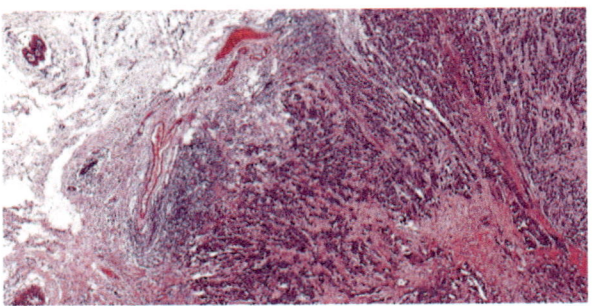

present in the tumor stroma, and the cancer cells grow and infiltrate in the fibrous stroma, with varying proportions (Fig. 13.19).

Macroscopically, the tumor is gray-white and hard in texture, with a gritty cut surface, no capsule, unclear boundary with surrounding tissue, and poor activity. The tumor tissue often invades the adjacent tissues like tree roots, and larger tumors can reach the fascia. If the cancer tissue obstructs the lymphatic vessels in the dermis, it can cause skin edema, with the skin over the sweat glands and hair follicles appearing relatively sunken, showing an orange peel appearance. If cancer tissue breaks through the skin, it can form ulcers. About 2/3 of tumors express estrogen or progesterone receptors, and about 1/3 overexpress HER2/NEU [9–12].

Invasive Carcinoma of Special Type

1. Invasive Lobular Carcinoma

    Invasive lobular carcinoma results from lobular carcinoma in situ invading through the basement membrane and into the stroma, accounting for 5–10% of breast cancers. The tumor cells are small and uniform in size, with rare mitoses, and are morphologically similar to the tumor cells of lobular carcinoma in situ. They are often aligned in strands or chains and occasionally surround cancerous or normal-appearing acini or ducts, creating a so-called bull's-eye pattern. Invasive lobular carcinoma lacks E-cadherin expression due to biallelic deletion of the CDH1 gene, which encodes the cell adhesion protein E-cadherin. The adhesion between tumor cells is poor, which helps differentiate it from ductal carcinoma by immunohistochemistry. Loss of E-cadherin was also observed in lobular carcinoma in situ and atypical lobular hyperplasia, suggesting that loss of E-cadherin expression is an early event in invasive lobular carcinoma. Macroscopically, the cut surface is rubber-like, grayish-white, and flexible, with no clear boundary with the surrounding tissue. Approximately 20% of invasive lobular carcinomas involve bilateral breasts with a diffuse multifocal distribution in the same breast, making them not easily detected by clinical and imaging examinations. The spread and metastasis of invasive lobular carcinoma also have particular characteristics, often metastasizing to the cerebrospinal fluid, serosal surfaces, ovary, uterus, and bone marrow [9–12].

2. Inflammatory Carcinoma

Inflammatory carcinoma is defined by the clinical presentation of an enlarged, swollen, erythematous breast, usually without a palpable mass. The underlying carcinoma is generally poorly differentiated and diffusely invades the breast parenchyma. The blockage of numerous dermal lymphatic spaces by carcinoma results in the clinical appearance. True inflammation is minimal or absent. Most of these tumors have distant metastases, and the prognosis is extremely poor.

3. Carcinoma with Medullary Features

Carcinoma with medullary features consists of sheets of large anaplastic cells with pushing, well-circumscribed borders. Clinically, they can be mistaken for fibroadenomas. There is invariably a pronounced lymphoplasmacytic infiltrate. DCIS is usually absent or minimal. Medullary carcinomas, or medullary-like carcinomas, occur with increased frequency in women with BRCA1 mutations, although most women with medullary carcinoma are not carriers.

4. Mucinous Carcinoma

Mucinous (colloid) carcinoma is also a rare subtype. The tumor cells produce abundant quantities of extracellular mucin that dissects into the surrounding stroma. Like carcinoma with medullary features, they often present as well-circumscribed masses and can be mistaken for fibroadenomas. Grossly, the tumors are usually soft and gelatinous.

5. Tubular Carcinomas

Microscopically, the carcinomas consist of well-formed tubules with low-grade nuclei. Lymph node metastases are rare, and prognosis is excellent.

13.4.3.3   Spreading

Direct Spreading

The cancer cells spread directly along the mammary ducts and can involve the corresponding mammary acini or spread along the periductal interstitial space to the periphery and into adipose tissue. As the cancer tissue expands, it can even invade the major chest muscle and the chest wall [9–12].

Lymphatic Metastasis

The tumor tissue metastasizes to the ipsilateral axillary lymph nodes first, then to the subclavian lymph nodes, and retrogradely to the supraclavicular lymph nodes in advanced stages. Breast cancer located in the upper inner quadrant of the breast often metastasizes to the para-internal mammary artery lymph nodes and further to the mediastinal lymph nodes. Occasionally, it can metastasize to the contralateral axillary lymph node. A small number of cases can metastasize to the contralateral axillary lymph nodes through the superficial lymphatic vessels of the chest wall or the deep fascial lymphatic vessels [9–12].

Hematogenous Metastasis

Advanced breast cancer can metastasize to tissues or organs such as the lungs, bones, liver, adrenal gland, and brain through blood vessels.

13.4.3.4 Prognosis

The prognosis of breast carcinoma is related to several factors as follows [9–12]:

1. The size of the primary carcinoma. Invasive carcinomas smaller than 1 cm have an excellent prognosis in the absence of lymph node metastases and may not require systemic therapy.
2. The grade of the carcinoma. The most common grading system for breast cancer evaluates tubule formation, nuclear grade, and mitotic rate to divide carcinomas into three groups. Well-differentiated carcinomas have a significantly better prognosis as compared with poorly differentiated carcinomas. Moderately differentiated carcinomas initially have a better prognosis, but survival at 20 years approaches that of poorly differentiated carcinomas.
3. The histologic type of carcinoma. All specialized types of breast carcinoma (tubular, medullary, cribriform, adenoid cystic, and mucinous) have a somewhat better prognosis than carcinomas of no special type ("ductal carcinomas").
4. The presence or absence of estrogen or progesterone receptors. The presence of hormone receptors confers a slightly better prognosis.
5. The proliferative rate of the cancer. Proliferation can be measured by mitotic counts, flow cytometry, or immunohistochemical markers for cell cycle proteins. Mitotic counts are included as part of the grading system. The optimal method for evaluating proliferation has not been determined. High proliferative rates are associated with a poorer prognosis.
6. Aneuploidy. Carcinomas with an abnormal DNA content (aneuploidy) have a slightly worse prognosis compared with carcinomas with a DNA content similar to normal cells.
7. Amplification of HER2/NEU. Amplification of this membrane-bound protein is almost always caused by amplification of the gene. Therefore, overexpression can be determined by immunohistochemistry (which detects the protein in tissue sections) or by fluorescence in situ hybridization (which detects the number of gene copies). Overexpression is associated with a poorer prognosis.

Recently, breast cancer has been classified into four molecular subtypes: luminal A (ER+/PR+, HER2-), luminal B (ER+/PR+, HER2+), HER2 overexpression (ER-/PR-, HER2+), and basal-like (ER-/PR-, HER2-, CK5/CK6+ or EGFR+). Research shows that ER+, PR+, HER2- breast cancer is well-differentiated and sensitive to hormone therap, and has a good prognosis. ER- and PR-, HER2+ breast cancers are generally poorly differentiated and insensitive to hormonal therapy but sensitive to chemotherapy and have a relatively poor prognosis. Breast cancer with negative expression of all three markers is called "triple-negative" breast cancer. Breast cancer with negative expression of all three markers and positive expression of both CK5/CK6 and EGFR is called basal-like breast cancer, which has poor differentiation, high proliferation activity, early metastasis, and poor prognosis. Therefore, the molecular markers and subtypes of breast cancer are of great significance for guiding clinical treatment and predicting prognosis [9–12].

13.4.4 Gynecomastia

Gynecomastia refers to hypertrophy of the breast due to co-hyperplasia of the glands and stroma of the breast and hyperestrogens due to functional testicular tumors, cirrhosis, or medications that may result in breast development in males. Male mammary gland development can occur unilaterally or bilaterally. Button-like nodular enlargement can be found under the areola, as large as the female adolescent mammary gland. Microscopically, the ductal epithelium is papillary with regular columnar or cubic cell morphology and few lobules. This lesion is easily detected on clinical examination but must be distinguished from rare male breast cancer [9–12].

References

1. Zhou Y. Diseases of reproductive system and breast. In: Li Y, editor. Pathology. People's Health Publishing House; 2013. p. 288–91.
2. Zhang Y. Diseases of the reproductive system and breast. In: Wang L, editor. Pathology. Higher Education Press; 2012. p. 198–201.
3. Kumar V, Abbas AK, Aster JC, et al. Robbins basic pathology. 9th ed; 2011. p. 689–94.
4. Kumar V, Abbas AK, Aster JC, et al. Pathologic basis of disease. 9th ed; 2014. p. 1007–20.
5. Zhou Y. Diseases of reproductive system and breast. In: Li Y, editor. Pathology. People's Health Publishing House; 2013. p. 291–4.
6. Zhang Y. Diseases of the reproductive system and breast. In: Wang L, editor. Pathology. Higher Education Press; 2012. p. 201–3.
7. Kumar V, Abbas AK, Aster JC, et al. Robbins basic pathology. 9th ed; 2011. p. 701–3.
8. Kumar V, Abbas AK, Aster JC, et al. Pathologic basis of disease. 9th ed; 2014. p. 1039–42.
9. Bu H, Li Y. Pathology. 9th ed. People's Medicalpublishing House; 2018.
10. Kumar V, Abbas AK, Aster JC. Robbins basic pathology. 9th Ed. Philadelphia, PA: Saunders Elsevier, 2013.
11. Kumar V, Abbas AK, Fauston N, et al. Robbins basic pathology. 10th Ed Philadelphia, PA: Saunders Elsevier, 2018.
12. Lakhani SR, Ellis IO, Schnitt SJ, et al. WHO classification of tumours of the breast. 4th ed. Lyon: International Agency for Research on Corcer; 2012.

Chapter 14
Endocrine System Diseases

Jing Zheng and Haiyan Niu

Contents

Objectives
1. To describe the morphological and clinical features of pituitary adenoma.

J. Zheng (✉)
School of Basic Medical Sciences and Life Sciences, Hainan Medical University,
Haikou, China

H. Niu
Department of Pathology, The First Affiliated Hospital of Hainan Medical University,
Haikou, China

© Zhengzhou University Press 2024 427
K. Chen et al. (eds.), *Textbook of Pathologic Anatomy*,
https://doi.org/10.1007/978-981-99-8445-9_14

2. To describe the clinical features of acromegaly and giantism.
3. To list the causes of hypopituitarism.
4. To list the causes of adrenal cortical hyperfunction.
5. To describe the etiology, pathogenesis, and clinical features of Addison's disease.
6. To list the tumors of adrenal medulla and cortex.
7. To describe the clinical features and diagnosis of pheochromocytoma.
8. To describe the etiology, clinical features and morphology of hyperthyroidism.
9. To describe the etiology, pathogenesis, and clinical features of goiter.
10. To list the etiology and clinical features of hypothyroidism.
11. To list the types, with pathogenesis, morphology and clinical features of thyroiditis.
12. To classify the pathogenesis, morphology, and clinical features of thyroid tumors.
13. To list the etiology, types and the pathologic changes of diabetes mellitus.

Key Concepts
1. **Hyperpituitarism** is the oversecretion of one or more pituitary hormones due to a pituitary or hypothalamic disorder.
2. **Hyperthyroidism** is a clinical and biochemical condition of hypermetabolism caused by excessive secretion of thyroid hormones. Graves' disease is a hyperthyroidism accompanied by diffuse goiter and ophthalmopathy.
3. **Hypothyroidism** is a state of hypometabolism caused by a chronic insufficiency of thyroid hormones. It has two typical clinical presentations: cretinism and myxedema.
4. **Goiter** is enlargement of thyroid due to compensatory hyperplasia in response to thyroid hormone deficiency.
5. **Carcinoma of the thyroid gland** has four major morphologic types with distinctly different clinical behavior.
6. **Hyperadrenal syndromes** include: Cushing syndrome, characterized by excessive cortisol; hyperaldosteronism; and adrenogenital or virilizing syndromes, caused by excessive androgens.
7. **Addison disease** is a disorder caused by progressive destruction of the adrenal cortex.

Introduction
The endocrine system contains a highly integrated and widely distributed group of tissues, including endocrinal glands and endocrinal cells. Anatomically, the endocrine system consists of six distinct organs: pituitary, adrenals, thyroid, parathyroids, gonads, and pancreatic islets [1]. Dispersed endocrine cells produce polypeptide hormones. These cells have the special biochemical properties, Amine Precursor Uptake and Decarboxylation properties. Therefore, they are also known as APUD cells [1].

Combined with the nerve system, the endocrine system orchestrates a state of metabolic equilibrium or homeostasis of the body. The endocrine system diseases share some common features:

1. Cell shape change is concordant with function change.
2. Local lesion causes general impacts (metabolic imbalance).
3. The categories of diseases include inflammation; tumor; and hyperplasia, hypertrophy and atrophy. Inflammation and tumor can be seen in any system. Hyperplasia, hypertrophy, and atrophy, as the pure morphology in a disease, are the features of endocrine system diseases. Hyperplasia and hypertrophy happen for either over workload or over hormone stimulation. Endocrine organs often receive the hormone stimulation and it easily induces the hyperplasia and hypertrophy.
4. Sometimes, it is difficult to judge the hyperplasia and adenoma.
5. Cellular atypia can be seen in both benign and malignant tumor. The cellular atypia is not the criteria to differentiate both.

14.1 Pituitary Gland

The pituitary gland is a small bean-shaped organ. It is located at the bottom of the brain, within the sella turcica, adjacent to the optic chiasm and cavernous sinus. The pituitary is composed of two distinct components: the anterior lobe (adenohypophysis) and the posterior lobe (neurohypophysis). Pituitary is an important endocrine organ which secretes different hormones to conduct other subordinate endocrine organ's function [2].

Most endocrine cells are in the anterior lobe. The adenohypophysis (anterior pituitary) releases six hormones that are, in turn, under the control of various stimulatory and inhibitory hypothalamic releasing factors: TSH, thyroid-stimulating hormone (thyrotropin); PRL, prolactin; ACTH, adrenocorticotrophic hormone (corticotropin); GH, growth hormone (somatotropin); FSH, follicle-stimulating hormone; LH, luteinizing hormone.

The hormones in the posterior pituitary, including antidiuretic hormone (ADH) and oxytocin,sd are synthesized within the hypothalamus and then transmitted down the nerve axons in the pituitary stalk to the post lobe [3].

14.1.1 Hyperpituitarism and Pituitary Adenoma

Hyperpituitarism is caused by excessive secretion of hormones. It is most common in pituitary adenoma. Other less common causes include anterior pituitary hyperplasia and cancer, secretion of hormones by some extrapituitary tumors, and certain hypothalamic diseases.

14.1.1.1 Pituitary Giantism and Acromegaly (Over Production of GH)

Overproduction of GH in children and adolescents, whose epiphyses have not yet fused, causes excessive growth in the length of bones, and the subject becomes too tall. This condition is called pituitary giantism. Some associated coarsening of the facial features usually occurs in response to the effect of GH on the structure of the facial bones [3].

In adults, excessive GH causes acromegaly. Because the epiphyses have fused, there can be no growth in height, but the GH produces thickening and coarsening of bones and generalized enlargement of viscera. Affected individuals have coarse facial features, large prominent jaws, and large spade-like hands, but they are no taller than normal.

14.1.1.2 Over Production of Prolactin

In a nonpregnant woman, excessive secretion of prolactin may cause spontaneous secretion of milk from the breasts (galactorrhea) and cessation of menstrual periods (amenorrhea) [3]. Galactorrhea results from the effect of the hormone on breast tissue. Amenorrhea occurs because high level of prolactin also inhibits secretion of pituitary gonadotropins FSH and LH, which in turn leads to cessation of ovulation and menstrual cycles.

14.1.1.3 Pituitary Adenoma

Pituitary adenoma is a benign tumor in the anterior lobe of the pituitary gland, which is usually associated with excessive pituitary hormone secretion and the corresponding hyperfunction symptoms. Clinical diagnosis of pituitary adenomas accounts for approximately 10% of intracranial tumors. Many of them were accidentally discovered during routine autopsies.

Morphology

Gross View

Pituitary adenoma can be large adenoma (diameter >1 cm) or microadenoma (diameter <1 cm) [2]. Clinically, they can be functional or asymptomatic. Compared with adenomas related to endocrine abnormalities, asymptomatic adenomas and hormone negative adenomas may attract clinical attention at a later stage. They are more likely to be large adenomas. A common pituitary adenoma is a well-defined soft mass that may be localized in the sellar region in smaller tumors. Larger lesions usually extend upward through the sellar diaphragm to the suprasellar region, where they often compress the optic chiasm and adjacent structures. Hemorrhagic and/or necrotic lesions are common in larger adenomas.

Microscopically View

In the normal anterior pituitary gland, there are at least three different types of staining cells cross distribution. The staining characteristics of these three normal cells are determined by the type and quantity of cellular endocrine products, but pituitary adenomas are usually composed of relatively uniform cells. The cells are polygonal with or without mild nuclear atypia and arrayed in sheets, cords, or papillae with abundant capillaries. Supporting connective tissue, or reticulin, is sparse. This cellular monomorphism and the absence of a significant reticulin network are key morphological changes of pituitary adenomas.

Different cell proliferation will cause different clinical symptoms, but the functional status of the adenoma cannot be reliably predicted from its histologic appearance because of same staining cells secreting different hormones. The functional status can be judged only depending on the immnochemical staining, using six antibodies against six hormones.

From Table 14.1, we can see that there three types of hormone cells show acidophilic, and three show basophilic. Immunohistochemistry (IHC) reveals that prolactin cell adenoma is the most common type, followed by null cell adenoma. Null cell is not chromophobe cell. It is negative against any antibodies of six hormones. Then adrenocorticotropic hormone cell adenoma and gonadotroph cell adenoma. Growth hormone cell adenoma and mixed growth/prolactin cell adenoma are not so popular, but they cause significant symptoms [1].

Clinical Symptoms

Systemic symptoms: Relative endocrinic dysfunction will be different in different cell type's adenoma. Increased prolactin will cause amenorrhea, galactorrhea, low libido and infertility. The corticotroph cell adenoma will cause the hypercortisolism. The gonadotroph cell adenoma will cause the low libido. The excessive

Table 14.1 Pituitary adenoma detected by IHC

Cell origin of adenoma	Frequency (%)	Cell type
Prolactin cell	20–30	Acidophilic
Growth hormone cell	5	Acidophilic
Mixed growth/prolactin cell	5	Acidophilic
Adrenocorticotropic hormone cell	10–15	Basophilic
Gonadotroph cell	10–15	Basophilic
Null cell	20	
Thyroid-stimulating hormone cell	1	Basophilic
Other hormonal adenomas	15	

growth hormone will cause different symptoms in children and teenage and adult. It causes gigantism in children and teenage and acromegaly in adults. No doubt the thyroid-stimulating hormone cell adenoma causes the hyperthyroidism. Sometimes, both silent and hormone-negative pituitary adenomas may cause hypopituitarism as they encroach on and destroy adjacent anterior pituitary parenchyma.

Local symptoms: As the adenomas expand, they frequently erode the sella turcica and anterior clinoid processes. It also can induce bitemporal hemianopsia because of optic chiasma being compressed.

14.1.2 Hypopituitarism

It means hypofunction of anterior pituitary. It may occur with loss or absence of over 75% of the anterior pituitary parenchyma. It is usually caused by nonsecretory pituitary adenoma, ischemic necrosis of the anterior pituitary gland, or removal of the pituitary gland through surgery or radiotherapy. In this condition, the anterior lobe failed to secret enough hormones. Then, the functions of the thyroid gland, adrenal gland, and gonads are impaired because trophic hormone stimulation is lost. *Sheehan syndrome*, also known as postpartum anterior pituitary necrosis, is the most common ischemic necrosis of the anterior pituitary gland in clinical practice. In pregnancy period, the anterior pituitary usually becomes enlarged because of prolactin cell hypertrophy and hyperplasia, but the blood supply for enlarged anterior pituitary does not increase. That makes the enlarged anterior pituitary very vulnerable to ischemia. When heavy hemorrhage or shock occurs, necrosis of anterior pituitary will happen in turn.

14.1.3 Hypothalamus and Posterior Pituitary Disorders

Hypothalamic neurons produce two peptides: Vasopressin (ADH) and Oxytocin. They are stored at the axonal end of the neurohypophysis and released into the circulation under appropriate stimulation. Oxytocin stimulates the contraction of smooth muscle in uterus and around mammary duct during pregnancy. Abnormal synthesis and release of oxytocin were not associated with significant clinical abnormalities. Clinically important posterior pituitary syndrome involves the production of ADH. Diabetes insipidus is a rare disease characterized by the inability of the posterior pituitary to secrete ADH. Due to the lack of ADH, affected individuals are unable to absorb water from the renal collecting duct and excrete a large amount of extremely dilute urine. The concentration of serum sodium and osmotic pressure increased due to the excessive loss of water from the kidney. Large amounts of water must be consumed to compensate for the excessive water loss in the urine and to prevent dehydration.

14.2 Thyroid Gland

The thyroid gland consists of two lateral lobes connected by a narrow isthmus. It is located in the overlying of the upper part of the trachea and is regulated by pituitary thyroid-stimulating hormone (TSH). The four parathyroid glands are located on its posterior surface.

Histologically, the thyroid gland is composed of multiple minute spherical vesicles called thyroid follicles. Each follicle consists of a central mass of eosinophilic protein material called colloid, surrounding by a layer of cuboid epithelial cells called follicle cells. Parafollicular cells exist between follicles, they have clear cytoplasm, so are called "C" cells [4].

Under the influence of TSH, the follicular cells synthesize two hormones called triiodothyronine (T3) and thyroxin (T4), which regulate the body's metabolic process and are also required for the normal development of the nervous system. The term thyroid hormone is a general term referring to both of the T3 and T4.

Figure 14.1 shows the cuboid cells of thyroid follicle under high power of microscope. The pink mass in follicle is colloid which is the secretion of follicle cell—thyroglobulin.

If many TSH molecules bind to the follicle cell, the cell will synthesize more thyroxine. It will hypertrophy to high columnar shape to match overload work. On the other hand, if TSH reduces, the follilcle cells will withdraw to low cuboid and work few. The morphology of follicle cells is helpful for us to judge the function of thyroid (Fig. 14.2).

Fig. 14.1 Thyroid follicle

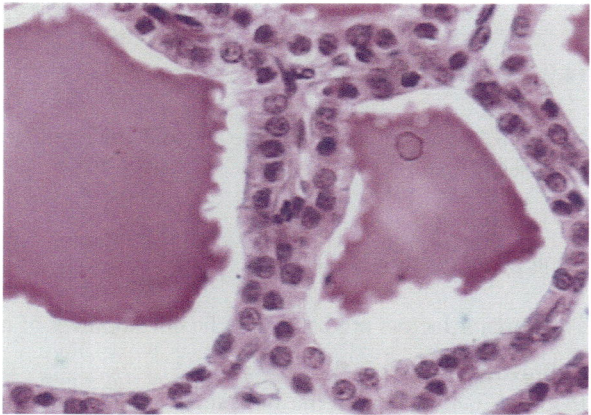

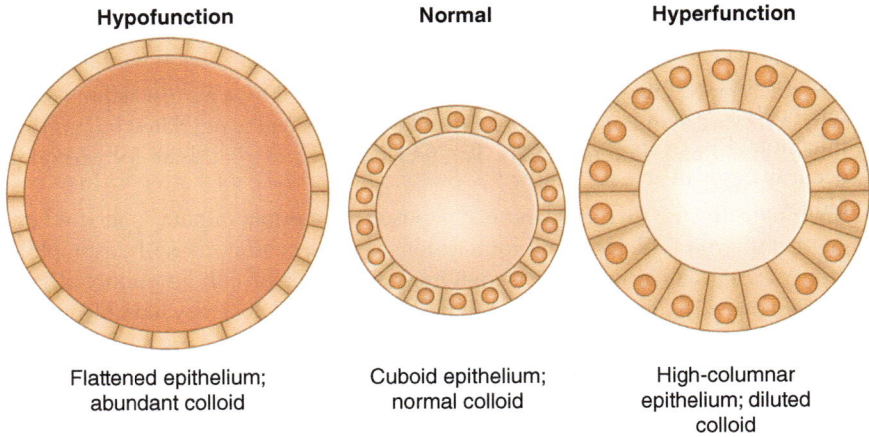

Hypofunction   **Normal**   **Hyperfunction**

Flattened epithelium;   Cuboid epithelium;   High-columnar
abundant colloid   normal colloid   epithelium; diluted
                                                  colloid

Fig. 14.2 Morphology and function of thyroid

14.2.1 Goiter

An enlargement of the thyroid gland is called goiter. The gland maybe uniformly enlarged, called a diffuse goiter, or multiple nodules of proliferating thyroid tissue may form a nodular goiter. A goiter that does not secrete excess thyroid hormone is called a nontoxic goiter. On the other hand, an enlarged gland that produces an excessive amount of hormone and causes symptoms of hyperthyroidism is called a toxic goiter.

14.2.1.1 Nontoxic Goiter

Nontoxic goiter refers to simple thyroid enlarged without thyroidism in majority cases.

Etiology and Pathogenesis

Three major factors predispose to the development of nontoxic goiter:

Iodine deficiency
If iodine is deficient in the diet, not enough iodine will be available to produce adequate thyroid hormone for the needs of the individual. The low level of thyroid hormone leads to a compensatory increase in TSH produced by pituitary. Hypertrophy and hyperplasia and goitrous enlargement will occur in response to TSH stimulation. Also, it is an attempt to extract the meager amount of iodine from the blood more efficiently to make enough hormones.

Deficiency of enzymes required for synthesis of the thyroid hormone or ingestion of substances that interfere with the function of these enzymes
The enzyme-deficient gland is unable to produce sufficient hormone without enlarging. It is the cause of the familial or heritage goiter. The ingestion of substances that interfere with thyroid hormone synthesis at some level, such as calcium and some "natural" food (cabbage, cauliflower and turnips) has been documented to be goitrogenic. Over calcium induces intestine decrease the absorption of iodine and cyanide in some vegetables interfere with the iodine aggregation toward thyroid.

Increased hormone requirements
In some individuals, the thyroid gland may be able to produce adequate hormone under normal circumstances but may be unable to increase its output in response to increased requirement without enlarging, such as in puberty, during pregnancy and under conditions of stress.

Excessive intake of iodine
Excessive iodine occupies most of the functional groups of peroxidase. Then tyrosine oxidation will be interfered and organic process of iodine will be blocked. As a result, inadequate thyroid hormone leads to a compensatory enlargement of thyroid gland.

Morphology

Compensatory increase in TSH produced by pituitary induces recurrent and persistent hyperplasia and hypertrophy of thyroid gland. The process can be divided into three stages [5].

Stage of hyperplasia (diffuse hyperplastic goiter): Under microscope, follicles hypertrophy and hyperplasia with few colloids can be observed. The follicles are lined by crowded columnar cells and the small blood vessels between follicles are hyperemic. In gross, the thyroid gland is diffusely and symmetrically enlarged with smooth surface and soft texture.

Stage of stored colloid (diffuse colloid goiter): Long-term iodine deficiency can cause large colloid storage. However, the change is not uniform throughout the gland. Some follicles are distended with mass colloid, whereas others remain small. The follicular epithelium is flattened and cubical, and colloid is abundant during periods of involution (Figs. 14.3 and 14.4). In gross, the cut surface of thyroid is usually brown, somewhat glassy, and translucent.

Stage of nodules (multinodular goiter): With time, recurrent hyperplasia and involution combine to produce an irregular enlargement of the thyroid, known as multinodular goiter. In gross, there are irregular nodules on the section, containing different amounts of brown gel like colloid (Fig. 14.5). Regressive change occurs frequently, particularly in older lesions, such as hemorrhage, fibrosis, and cystic dilatation. The microscopic appearance includes some colloid-rich follicles

Fig. 14.3 Thyroid follicle in the stage of stored colloid

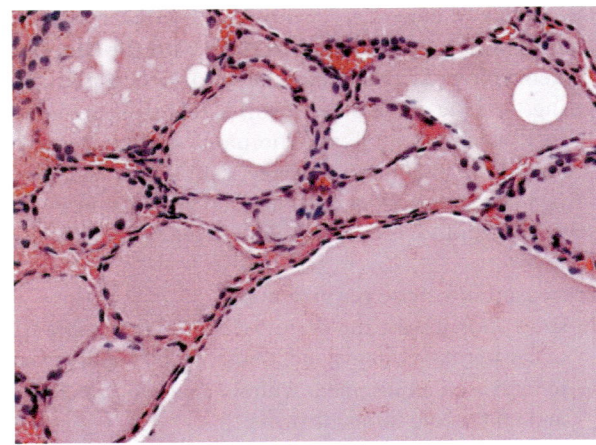

Fig. 14.4 Flattened follicular epithelium and abundant colloid

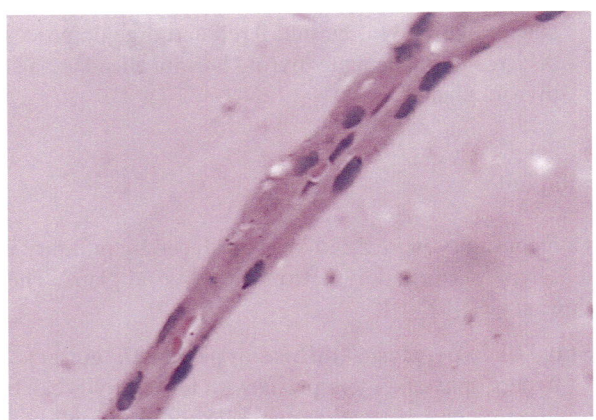

Fig. 14.5 Multinodular goiter in gross

Fig. 14.6 Multinodular goiter in microscopic appearance

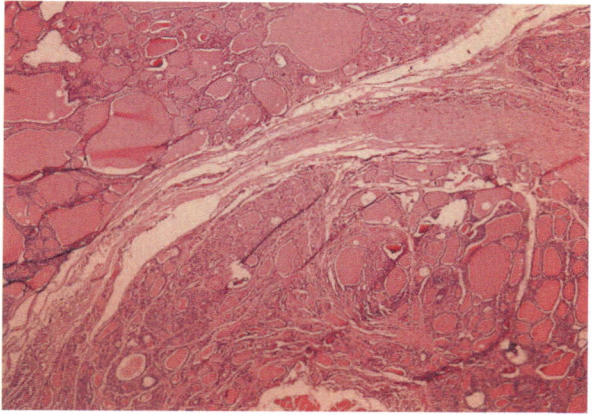

lined by inactive epithelium, hypertrophy and hyperplasia of some follicles, and fibers hyperplasia and separating the follicles to form nodules (Fig. 14.6).

14.2.1.2 Toxic Goiter

Toxic goiter refers to a group of goiters with thyrotoxicosis. Thyrotoxicosis is a hypermetabolic disorder caused by excessive levels of thyroid hormone regardless of cause. The syndrome of thyrotoxicosis is manifested by nervousness, palpitation, rapid pulse, fatigability, muscular weakness, weight loss with good appetite, diarrhea, heat intolerance, warm skin, excessive perspiration, emotional liability, menstrual changes, a fine tremor of hand, eye change, and variable enlargement of thyroid gland. In patients with ophthalmopathy, orbital tissue edema occurs due to the presence of hydrophilic glycosaminoglycan. In addition, there is lymphocyte infiltration, mainly T cells. The orbital muscles are initially edematous, but fibrosis may occur in the later stages of the disease. The incidence rate of this disease is the highest between the ages of 20 and 40, and women are more vulnerable than men.

Etiology and Pathogenesis

Thyrotoxicosis can be caused by different disorders, which include Graves disease, hyperfunctional thyroid adenoma, goiter with hyperthyroidism, some types of thyroiditis, increased TSH from pituitary adenoma, or thyrotropin-releasing hormone (TRH) from thalamus. Graves disease is the most common cause of toxic goiter, which is an autoimmune disease induced by autoantibodies to the TSH receptor. The autoantibodies to the TSH receptor are detectable in almost all patients with Graves disease. Several autoantibodies mimic the action of TSH, resulting in the stimulation of thyroid epithelial cell activity. What propels B cells to make autoantibodies is not clear. CD4+ helper T cells are highly suspected to be starter, many of which are found within the thyroid. Genetic factors are also important in the causation of Graves disease. The genetic susceptibility to Graves' disease is related to the existence of certain HLA Haplotype, especially HLA-B8 and -DR3, as well as allelic variants (polymorphisms) in the genes encoding the inhibitory T-cell receptor CTLA-4 and tyrosine phosphatase PTPN22 [1].

Morphology

In the typical case of Graves disease, the thyroid gland is diffusely enlarged because of the presence of diffuse hypertrophy and hyperplasia of thyroid follicular epithelial cells. The gland is usually smooth and soft, and its capsule is intact (Fig. 14.7). Microscopically, in untreated cases, the follicular epithelial cells are tall, columnar, and more crowded than normal. This crowding usually leads to the formation of small papillae that protrude into the follicular cavity (Fig. 14.8). Compared to papillary thyroid carcinoma, this type of papillae lacks a fibrous vascular core. The colloid is typically pale with prominent peripheral scalloping indicating high turnover. Stroma may show lymphoid infiltrate (Fig. 14.9), sometimes with germinal center formation. Stroma fibrosis occurs later. Radiotherapy can induce atypical changes in the follicular epithelium.

The changes in extrathyroidal tissue include systemic lymphoid hyperplasia. In patients with ophthalmopathy, orbital tissue edema occurs due to the presence of hydrophilic Glycosaminoglycan. In addition, there is lymphocyte infiltration, mainly T cells. The orbital muscles are initially edematous, but fibrosis may occur in the later stages of the disease.

14.2.2 Hypothyroidism

Hypothyroidism is a low metabolism syndrome caused by sero-thyroxin level declined by either structure or functional reason. Hypothyroidism can be divided into two categories: primary and secondary, depending on whether hypothyroidism is caused by inherent thyroid abnormalities or by diseases of the hypothalamus or

Fig. 14.7 Diffusely enlarged thyroid

Fig. 14.8 Small papillae and peripheral scalloping

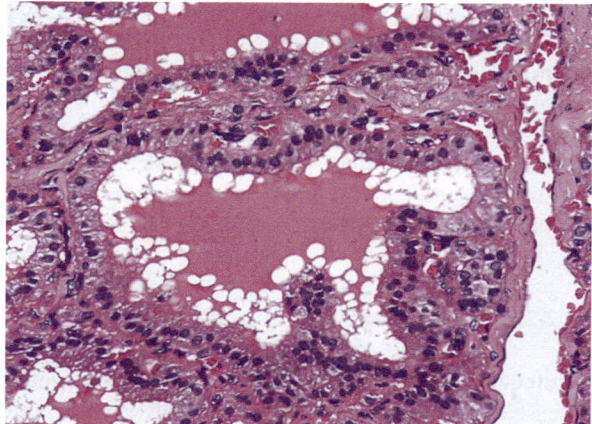

Fig. 14.9 Lymphoid infiltration

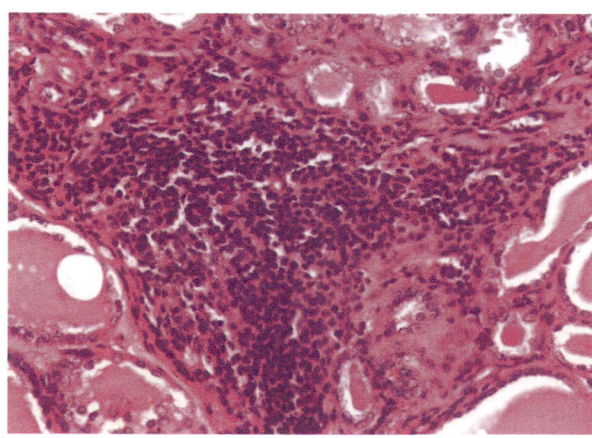

pituitary gland. The common causes of primary hypothyroidism are thyroid surgery, radiotherapy, some drugs, some invasive diseases or thyroiditis that destroy most of the glands. Secondary hypothyroidism is caused by TSH deficiency resulting from any of the causes of hypopituitarism, including a pituitary tumor, postpartum pituitary necrosis, or trauma. Sometimes hypothyroidism also can be caused by TRH deficiency because of hypothalamus disorders.

14.2.2.1   Myxedeyma

Hypothyroidism in older children or adults is sometimes called myxedema. It is manifested by a general slowing of the body's metabolic processes. Frequently, there are localized accumulations of mucinous material in the skin, from which the disease received its name. The manifestations of myxedema include widespread apathy and mental retardation, which may resemble depression in the early stages of the disease. Patients with myxedema are listless, intolerant to cold, and often obese. The edema rich in mucopolysaccharide accumulates in the skin, subcutaneous tissue, and many internal organs, resulting in widening and thickening of facial features, enlarged tongue, and deeper voice. Decreased intestinal motility leads to constipation. Pericardial effusion is common. In the later stage, heart enlargement may lead to secondary heart failure [2].

14.2.2.2 Cretinism

It refers to hypothyroidism that develops during infancy or early childhood. This disease is common in regions around the world where dietary iodine deficiency is prevalent. Iodine deficiency leads to insufficient synthesis and secretion of thyroxine. The clinical features of cretinism include impaired development of the skeleton and central nervous system. The severity of mental disorders in patients with cretinism seems to be directly affected by the time when thyroid hormones deficiency occurs in the uterus [2].

14.2.3 Thyroiditis

Thyroiditis is a group of diseases that occur in the thyroid gland and are characterized by inflammation. These diseases include acute diseases that cause severe thyroid pain and diseases with relatively less inflammatory changes and mainly manifested as thyroid dysfunction. This section focuses on the more common and clinically significant types of thyroiditis: (1) Subacute Thyroiditis, (2) Hashimoto Thyroiditis (or Chronic Lymphocytic Thyroiditis), and (3) Riedel Thyroiditis.

14.2.3.1 Subacute Thyroiditis

Subacute thyroiditis, also known as granulomatous thyroiditis or De Quervain's thyroiditis, is much less common than Hashimoto disease. It clinically presents with sore throat, marked tenderness in thyroid area, fever, and malaise. De Quervain's thyroiditis most commonly affects middle-aged women. Subacute Thyroiditis is considered to be caused by viral infection or post viral inflammatory process. Most patients have a history of upper respiratory tract infection before the onset of thyroiditis. Compared with autoimmune thyroid disease, immune response is not self-sustaining. Therefore, the process of subacute thyroiditis is often self-limited.

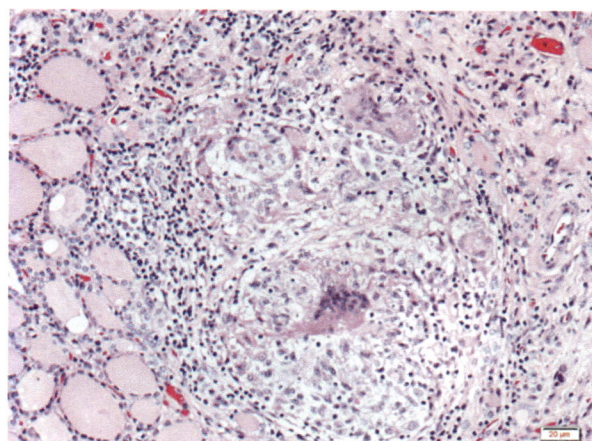

Fig. 14.10 Foreign-body giant cell granulomas

Morphology

The texture of the thyroid gland tissue becomes slightly hard. The thyroid capsule is intact and can be enlarged unilaterally or bilaterally. Histologically, thyroid follicles rupture and colloid extravasation lead to foreign body giant cell granuloma and chronic inflammatory cell infiltration (Fig. 14.10). Granuloma are indistinct and usually noncaseating and surrounded by follicles. Some giant cell can contain ingested colloid materials. Disease healing is achieved through the inflammatory process and fibrosis repair.

14.2.3.2 Chronic Lymphocytic (Hashimoto) Thyroiditis

Hashimoto Thyroiditis is the most common cause of hypothyroidism in areas with adequate iodine levels worldwide. It is an autoimmune reaction mediated inflammatory disease, also known as autoimmune thyroiditis. The peak age of disease occurrence is between 45 and 65 years old. Women are more common than men, with a dominant ratio of 10:1 to 20:1 [2]. Although it is mainly a disease of elderly women, it may also occur in children and is the main cause of non-epidemic goiter in children. The thyroid gland is usually a diffuse and symmetrical enlargement. In some cases, focal enlargement may also be seen, forming a nodular appearance. The capsule is intact and there is a good boundary between the gland and adjacent structures. The cutting surface is light gray and brown. The texture of thyroid gland is hard and somewhat fragile. Microscopic examination reveals marked lymphocytic infiltration of the thyroid parenchyma with well-developed germinal center formation (Fig. 14.11). The follicles are small and atrophic and show marked oncocytic change of follicular epithelium cells distinguished by the presence of abundant eosinophilic, granular cytoplasm, and enlarged, hyperchromatic nuclei, termed Hürthle, or oxyphil cells. Ultrastructurally, the Hürthle cells are characterized by a

Fig. 14.11 Marked
lymphocytic infiltration
and atrophic follicles

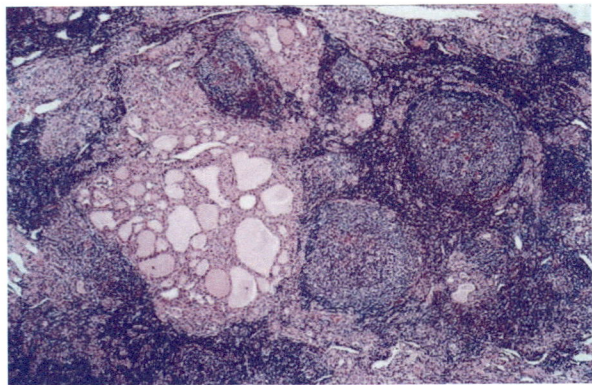

large number of prominent mitochondria. Interstitial connective tissue is increased
and usually abundant. However, unlike in Reidel's thyroiditis, the fibrosis does not
extend beyond the envelope of the thyroid gland.

14.2.3.3   Riedel's Thyroiditis

It is also called fibrous thyroiditis, chronic woody thyroiditis, or invasive thyroiditis.
Riedel thyroiditis is a rare disorder of unknown etiology and more common in older
age women. Its characteristic is extensive thyroid fibrosis, and even the fibrosis
process can involve adjacent neck structures, leading to adhesion between the thyroid and surrounding tissues. Therefore, the thyroid mass formed after fibrosis
needs to be differentiated from the real thyroid neoplasm clinically. In addition, this
type of fibrosis may be related to idiopathic fibrosis in other parts of the body, such
as the retroperitoneum.

Grossly, only part of thyroid gland is involved by a hard, stone-like fibrotic process. The cut surface show solid, fibrotic tissue that often extend beyond the thyroid
capsule into perithyroid tissue. In histology, thyroid parenchyma is replaced by
extensive fibrosis, frequent hyalinization, admixed with focal chronic inflammation
cells. Important feature is presence of chronic inflammation of venous walls within
areas of fibrosis and no giant cell present.

14.2.4   Neoplasm of Thyroid Gland

The thyroid gland produces various tumors, and tumors originating from thyroid
follicles are the most common tumors in the thyroid gland, ranging from localized
benign adenomas to highly invasive anaplastic carcinomas. Fortunately, most solitary thyroid nodules have been proved to be benign lesions, either follicular adenoma or localized non-neoplastic diseases (such as nodular hyperplasia, simple cyst

or thyroiditis lesions). In the following sections, we will describe the main thyroid neoplasms, including various types of adenomas and carcinomas.

14.2.4.1 Adenoma

Thyroid adenoma is a benign tumor derived from follicular epithelium. Like all thyroid neoplasm, follicular adenoma is usually isolated. The vast majority of adenomas are nonfunctional. Therefore, in morphology and clinical practice, they may be difficult to distinguish from thyroid proliferative nodular areas and less common follicular carcinoma. Although the vast majority of follicular adenomas are nonfunctional, a small part of them produces excessive thyroid hormones ("toxic adenoma") and cause clinically significant symptoms of high thyroxine.

Morphology

The typical thyroid adenoma is a solitary round to oval tumor with a complete fibrous capsule, which compress the adjacent non-neoplastic thyroid. On the cut surface, it shows a firm, homogeneous, grey-white or brown mass (Fig. 14.12). These characteristics are important for distinguishing multinodular goiter. Nodular goiters usually contain multiple nodules on its section (although the patient may be clinically manifested as isolated dominant nodules), and lacks a good capsule. Most nodular goiters do not compress adjacent thyroid parenchyma (Table 14.2).

Under a microscope, the constituent cells are arranged in uniform follicles (Fig. 14.13). The growth pattern of follicles in adenoma is usually very different from that of surrounding non neoplastic glands, which is another distinctive feature of nodular goiter. For the latter, the abnormal nodules and uninvolved thyroid parenchyma show similar growth patterns. Tumor cells are uniform and have clear cell

Fig. 14.12 Thyroid adenoma with a complete fibrous capsule

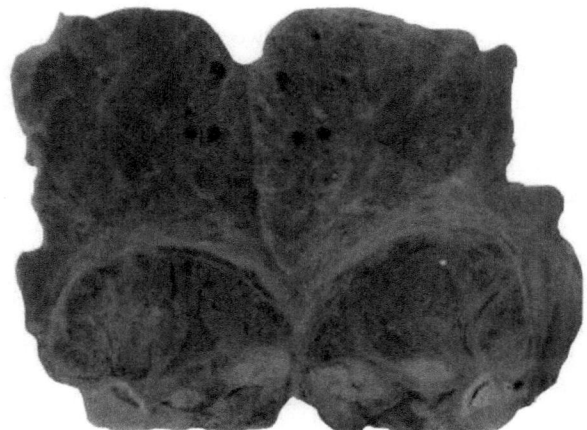

Table 14.2 Discrimination of thyroid adenoma and nodular goiter

Parameters	Thyroid adenoma	Nodular goiter
Nodular	Single	Multiple
Capsule	Intact	Not intact
Glands in capsule	Concordant	Not concordant
Glands in and out of capsule	Different	Same
Compressed gland near the outside of capsule	Yes	No

Fig. 14.13 Uniform follicles in thyroid adenoma on the left side

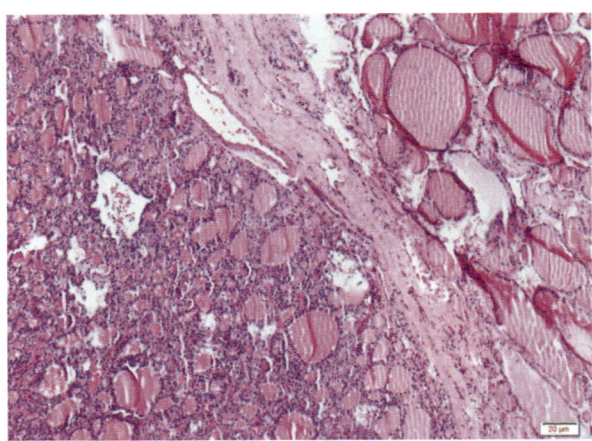

boundaries. Similar to endocrine tumors at other anatomical sites, even benign follicular adenomas may sometimes exhibit focal nuclear pleomorphism, atypia, and prominent nucleoli (endocrine atypia); this itself is not the basis for diagnosing malignant tumors. One of the main criteria for distinguishing between benign and malignant endocrine tumors is the biological behavior of the tumor. The hallmark of all follicular adenomas is the presence of a complete and well-formed capsule around the tumor. Therefore, careful evaluation of the integrity of the capsule is crucial for distinguishing between follicular adenoma and follicular carcinoma, which manifest as capsule and/or vascular invasion (see below). Thyroid adenomas can have different patterns or subtypes: **simple adenomas** exhibit mature thyroid follicles and normal amounts of colloid. **Colloidal adenomas** are similar to simple adenomas, except that the follicles are larger and contain more colloids. The characteristic of **embryonic adenomas** is a trabecular pattern, where poorly developed follicles contain little or no colloid. **Fetal adenomas** are similar to embryonic adenoma but often characterized by the formation of small follicles containing a small amount of colloid. **Hürthle cell adenomas**, where tumor cells obtain bright eosinophilic granular cytoplasm (changes in eosinophilic or Hürthle cells). The above histological subtype classification is based on morphology and has no clinical significance.

Clinical Features

Most thyroid adenomas present as painless nodules, which are usually found during routine physical examinations. Large lumps may cause local compression symptoms, such as difficulty swallowing caused by compression of the esophagus. The Assistive technology for preoperative evaluation of suspected painless nodules include ultrasound and fine needle aspiration biopsy. However, due to the need to assess the integrity of the capsule, only careful histological examination after complete removal of the thyroid mass can make a clear diagnosis of thyroid adenoma. Thyroid adenoma has a good prognosis and almost no recurrence or metastasis.

14.2.4.2 Carcinoma

The main subtypes of thyroid cancer are as follows: papillary carcinoma, follicular carcinoma, medullary carcinoma, and anaplastic carcinoma. Papillary carcinoma, follicular carcinoma and most of anaplastic carcinoma are derived from follicular epithelial cells. Thyroid Medullary carcinoma is derived from C cells adjacent to thyroid follicles. The most common type is papillary thyroid cancer. Due to the unique clinical and biological characteristics of each subtype of thyroid carcinoma, these subtypes will be described separately below.

Papillary Carcinoma

Thyroid papillary carcinoma (PTC) is the most common type, accounting for 85% of the incidence rate of thyroid cancer. It can occur at any age and is more common in children or young women (before the age of 40). Some patients have undergone neck X-ray treatment during childhood. The tumor grows slowly and can be localized within the thyroid gland for several years. The tumor cells can diffuse from the primary part to other parts of the gland and cervical lymph nodes through lymphatic vessels. The incidence rate of thyroid papillary carcinoma is increasing year by year. The prognosis of papillary thyroid cancer varies with its subtypes.

Morphology

PTC may manifest as solitary or multifocal lesions, with clear boundaries and even enveloped by a capsule. The lesion may include areas of fibrosis and calcification, sometimes manifested as cystic lesions. Tumors often present as gray white, hard, and granular sections. Papillary cancer can only be diagnosed after microscopic examination. A papillary architecture, complex branching true papillae (contain fibro-vascular axis), usually accompanied with formation of psammoma body, exists in many cases (Fig. 14.14). Some papillary cancers do not form typical papillary structures. The diagnosis of papillary carcinoma is more important based on nuclear characteristics. The nucleus of papillary carcinoma cells is large, oval, or

Fig. 14.14 Complex branching true papillae and psammoma body

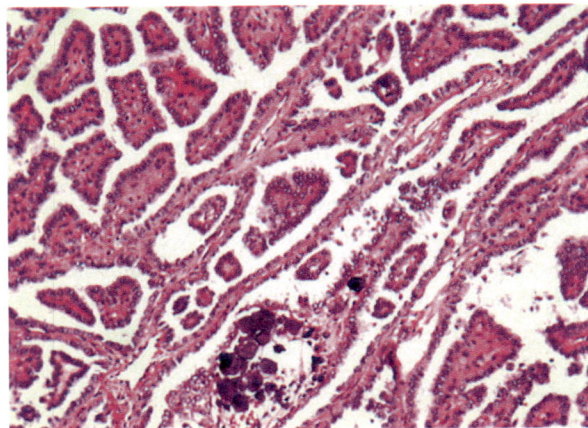

Fig. 14.15 Ground-glass nuclei

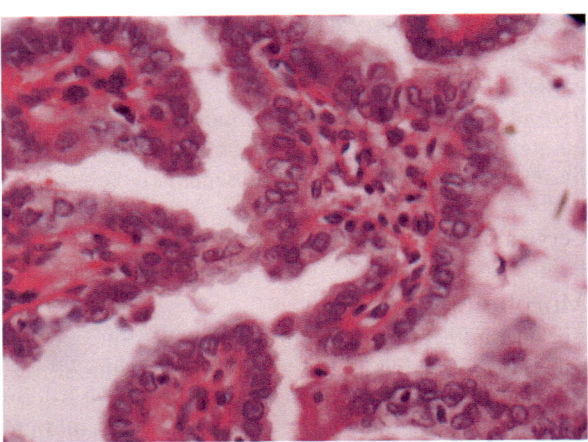

irregular and contains very fine and dispersed Chromatin, which gives them an optically clear appearance (Fig. 14.15), giving them the name of "Ground glass", with crowding/overlapping, nuclear groove or nuclear pseudoinclusions. PTC subtypes have received more attention in recent years, with the most common being classic and follicular subtypes. The classic type accounts for about 50% of PTC. Other common subtypes of PTC include diffuse sclerosis, high cell subtypes, and columnar cell subtypes, which are generally considered to have poorer prognosis compared to classical and follicular subtypes.

Clinical Features

The most common manifestation of papillary carcinoma is a painless mass in the neck. The lump is either located within the thyroid gland or formed due to cervical lymph node metastasis. However, regional lymph node metastasis usually does not have an adverse impact on long-term prognosis. In a few patients, hematogenous metastasis occurs at the time of diagnosis. In general, the prognosis is less favorable among elderly patients and in patients with invasion of extra-thyroidal tissues or distant metastases.

Follicular Carcinoma

Follicular carcinoma refers to a malignant epithelial tumor with follicular cell differentiation, lacking the characteristics of papillary thyroid carcinoma. It is the second most common type of thyroid carcinoma. Compared to papillary carcinoma, follicular carcinoma usually occurs in older patients.

Morphology

Follicular carcinoma usually manifests as well-defined nodules, similar to follicular adenoma, which may not be distinguishable from follicular adenoma on gross examination. Moreover, under a microscope, most follicular cancers are composed of fairly uniform cells, forming small follicles that are reminiscent of normal thyroid gland. Therefore, tissue structural and cellular atypia are not the most important for the diagnosis of follicular carcinoma, and careful evaluation of the presence of malignant biological behavior is crucial. Extensive histological sampling of the tumor boundary or capsule is required to exclude capsule (Fig. 14.16) and/or vascular invasion (Fig. 14.17). The widespread invasion of adjacent thyroid parenchyma or nerves makes the diagnosis of cancer evident in some cases (Fig. 14.18).

Fig. 14.16 Capsular invasion

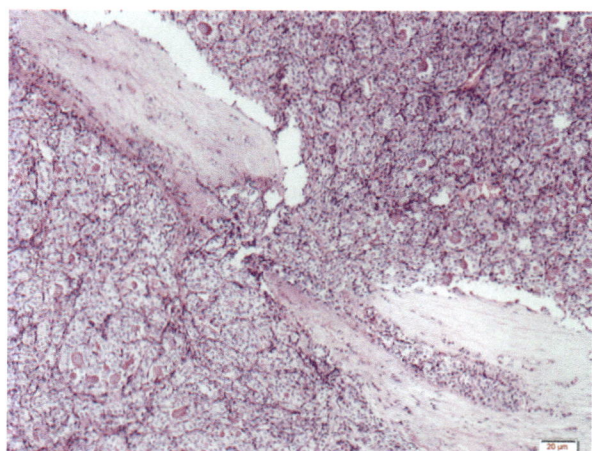

Fig. 14.17 Vascular invasion

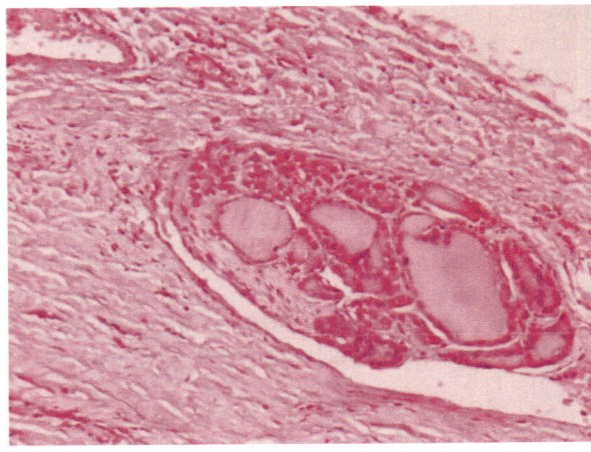

Fig. 14.18 Extensive invasion of nerves

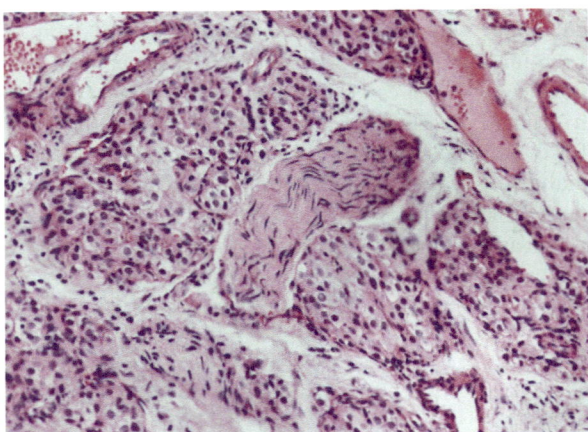

Clinical Features

The most common manifestation of follicular carcinoma is solitary "cold" thyroid nodules (nodules without radioactive development). These tumors often metastasize to the lungs, bones, and liver through the bloodstream. Compared to papillary carcinoma, regional lymph node metastasis is not common.

Medullary Carcinoma

Medullary carcinoma is a malignant neuroendocrine neoplasm composed of cells with C-cell differentiation, characterized by not containing thyroglobulin (TG) (Fig. 14.19) but calcitonin. Medullary carcinoma may be an isolated nodule or a

Fig. 14.19 TG (−) in medullary carcinoma

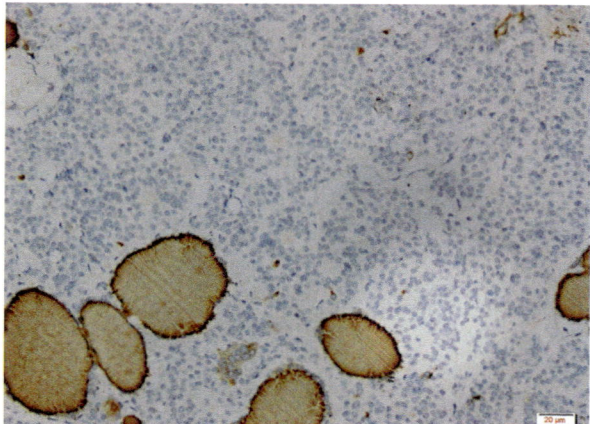

Fig. 14.20 Medullary carcinoma

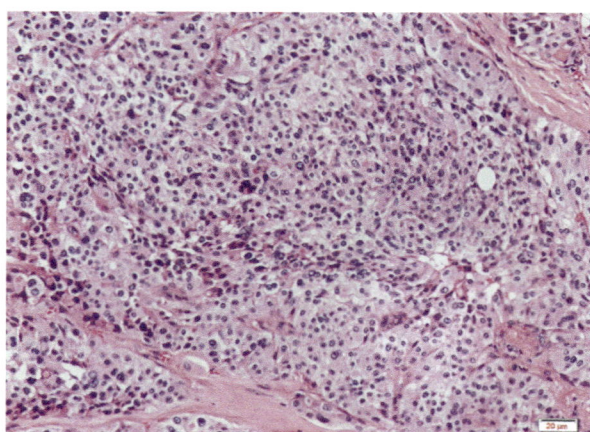

multiple lesion involving both lobes of the thyroid gland. Larger lesions typically include areas of necrosis and bleeding and may extend through the thyroid capsule to infiltrate surrounding tissues. Microscopically, myeloid carcinoma is composed of polygonal, circular, or spindle-shaped cells that can form nests, trabeculae, and even follicles (Fig. 14.20), separated by varying amounts of fibrous vascular stroma. In many cases, the deposition of acellular amyloid protein from calcitonin molecular variants in adjacent stroma is a prominent feature of these tumors. By immunohistochemical method, calcitonin is easily proved in the cytoplasm and matrix amyloid protein of tumor cells (Fig. 14.21).

Fig. 14.21 Calcitonin (+)
in medullary carcinoma

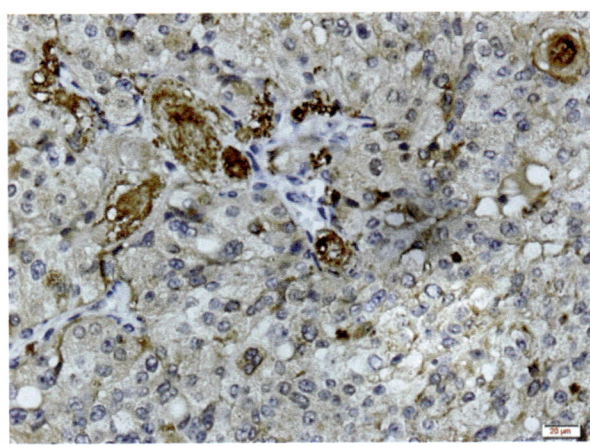

Anaplastic Carcinoma

Thyroid anaplastic carcinoma, also known as undifferentiated thyroid cancer, is one of the most invasive human tumors with a high mortality rate and complete or partial undifferentiation. Patients with anaplastic cancer are older than those with other types of thyroid cancer, with an average age of 65 years. Anaplastic cancer generally manifests as a large mass that rapidly grows and invades adjacent neck structures outside the thyroid capsule. Under the microscope, these tumors are composed of highly anaplastic cells that may exhibit several histological patterns, including (1) large pleomorphic giant cells; (2) spindle shaped cells with a sarcomatous appearance; (3) mixed spindle and giant cell lesions; and (4) small cells, similar to small-cell carcinoma in other parts. True small-cell carcinoma is unlikely to exist in the thyroid, and most "anaplastic small cell" tumors are eventually proven to be medullary carcinoma or malignant lymphoma. Some tumor lesions may have papillary or follicular differentiated areas, indicating that they originate from well-differentiated cancers.

14.3   Adrenal Gland

The adrenal gland is a crucial endocrine organ in the human body. The adrenal glands are located above the kidneys, one on each side, and together they are enveloped by the renal fascia and adipose tissue. The left adrenal gland is semilunar, and the right adrenal gland is triangular. The total weight on both sides of the adrenal gland is about 30 g. From the side view, the gland is divided into two parts: the adrenal cortex and the adrenal medulla. The surrounding part is the cortex, and the inner part is the medulla. They are different in development, structure and function, so they are actually two kinds of endocrine glands [4].

14.3.1 The Adrenal Cortex

The adult adrenal cortex is solid and golden, accounting for 90% of the total weight of the gland. The adrenal cortex is divided into three layers: the outer layer is the zona glomerulosa, which accounts for 15% of the cortex. The cells are arranged in a ball shape and secrete mineralocorticoid. The second layer is the zona fasciculata, which accounts for 75% of the cortex. The cells are arranged in strips and secrete glucocorticoids. The third layer is the zona reticularis, which accounts for about 10% of the cortex and secretes sex hormones. Under the stimulation of adrenocorticotropic hormone (ACTH), the zona reticularis can be widened, and the zona fasciculata can be narrowed accordingly. Adrenal cortex diseases can be divided into diseases related to cortical hyperfunction and diseases characterized by cortical hypofunction.

14.3.1.1 Adrenocortical Hyperfunction (Hyperadrenalism)

Since there are three basic types of corticosteroids secreted by the adrenal cortex (glucocorticoids, mineralocorticoid, and sex hormone), there are three different types of hyperadrenergic clinical syndrome: (1) Cushing's syndrome, which is characterized by excessive cortisol; (2) hyperaldosteronism; and (3) adrenogenital or virilizing syndromes caused by androgen excess. Due to the overlapping functions of certain adrenal steroids, there may be similarities in the clinical characteristics of certain syndromes.

Cushing Syndrome

This disorder is caused by any condition that produces an elevation in glucocorticoid levels. The glucocorticoids excess causes disturbances of carbohydrate, protein, and fat metabolism. The blood glucose rises. Protein synthesis is impaired, and body proteins are broken down, which leads to loss of muscle fibers and muscle weakness. Bones become weaker and more susceptible to fracture as the protein breakdown leads to loss of the connective tissue framework of the bones. The amount and distribution of body fat is altered. Fat tends to accumulate on the trunk, whereas the extremities appear thin and wasted because of muscle atrophy. The skin becomes thin and bruises easily. Stretch marks (striae) often appear in the skin as fat deposits accumulate in the subcutaneous tissues of the trunk. The face appears full and rounded, which is sometimes called a "moon face." Salt and water are retained because of the increased output of mineralocorticoids, leading to an increase in blood volume and a rise in blood pressure.

Four distinct conditions may give rise to this syndrome:

An ACTH-producing tumor of the pituitary, which stimulates the adrenal gland to enlarge and produce excess hormone. The most common cause of a corticosteroid excess is a small ACTH secreting pituitary adenoma, and this condition is called Cushing's disease, accounts for more than half of the cases of spontaneous, endogenous Cushing syndrome.

A corticosteroid-hormone-producing tumor of the adrenal cortex.

Administration of large amounts of corticosteroid hormone to treat diseases that respond to the hormone, as may be required to help suppress the immune response in recipients of organ transplants or patients with autoimmune diseases or to help induce remission in patients with leukemia.

A malignant tumor, such as a small-cell lung carcinoma, produces ACTH or a similar protein that resembles the "real" hormone.

There is no fixed pattern of morphological changes in the adrenal gland, which depends on the etiology of hypercortisolemia. The morphological changes of the adrenal gland can be manifested as one of the following abnormalities: (1) cortical atrophy; (2) diffuse hyperplasia; (3) nodular hyperplasia; and (4) adenomas, rarely carcinomas.

Overproduction of Aldosterone

Aldosterone promotes absorption of salt and water by the kidneys in exchange for potassium which is excreted, and its secretion is regulated primarily by the rennin-angiotensin-aldosterone. In roughly 80% of cases, primary hyperaldosteronism is caused by an **aldosterone-secreting adenoma** in one adrenal gland, a condition referred to as Conn syndrome. The clinical manifestations of primary hyperaldosteronism are those of hypertension and hypokalemia. The aldosterone excess produced by the tumor promotes excessive absorption of sodium and excessive excretion of potassium by the kidneys. Since water is absorbed along with the sodium, the blood volume increases along with the sodium concentration and the blood pressure also rises along with the blood volume. The excessive hormone-induced excretion of potassium lowers blood potassium, which impairs neuromuscular function and leads to muscle weakness. The high aldosterone output exerts a negative feedback effect on rennin production by the kidneys, and plasma rennin falls.

Overproduction of Adrenal Sex Hormones

Adrenal gland dysfunction associated with abnormal production of sex hormone is uncommon. This may result from congenital hyperplasia of the adrenal gland or from an adrenal sex-hormone-producing tumor. Adrenal tumors that elaborate sex hormones are rare. When such a tumor develops, however, either androgen or estrogen may be produced. The clinical features depend on the age of the individual when the tumor becomes manifest and on the sex of affected person. In a child, the tumor produces precocious puberty, and the character of the sexual development depends on the type of hormone elaborated. In adults, an estrogen-producing neoplasm elicits no hormonal symptoms in women but induces feminization in men. An antrogen-secreting tumor masculinizes a woman but causes no hormonal symptoms in a man.

14.3.1.2 Adrenal Insufficiency

Adrenal cortex insufficiency or hypofunction may be caused by primary adrenal disease (primary adrenal insufficiency) or reduced adrenal stimulation due to ACTH deficiency (secondary adrenal insufficiency). The patterns of adrenal cortex insufficiency can be divided into the following categories: (1) primary acute Adrenal cortex insufficiency (adrenal crisis), (2) primary chronic adrenal cortex hypofunction (Addison's disease), and (3) secondary adrenal cortex hypofunction.

Acute Adrenocortical Insufficiency

Various types of stress can increase the normal secretion of cortisol by the adrenal gland, which is about two to seven times higher than usual. Under severe stress, blood cortisol can be higher than 1 mg/L to meet the needs of the body. When there is primary or secondary, acute or chronic adrenal cortex hypofunction, the normal amount of cortisol cannot be produced, and the secretion of cortisol cannot be increased correspondingly during stress, so a series of acute clinical manifestations of adrenocortical hormone deficiency can be produced. Acute adrenal cortex insufficiency is most common in the following clinical conditions:

1. Chronic adrenal cortex hypofunction (Addison's disease)
   Acute impairment of adrenal cortex function caused by infection, trauma, surgery and other stress conditions, or withdrawal of hormone.
2. Long-term massive adrenocortical hormone treatment
   Inhibit the function of the hypothalamic pituitary adrenal axis, and even after discontinuing medication for 1 year, its function remains in a low state, especially with poor responsiveness to stress. Therefore, patients who have received long-term corticosteroid treatment will also suffer from acute adrenal cortex dysfunction if they do not supplement or increase the hormone dose in time when encountering stress.
3. After adrenal surgery
   Adrenalectomy for adrenal cortex hyperplasia or extra adrenal diseases (such as metastatic breast cancer), or after the removal of adrenal adenoma, the remaining adrenal gland often atrophies. If the hormone is not supplemented or the hormone dose is not increased correspondingly under stress, it can also cause acute adrenal cortex dysfunction.
4. Acute adrenal hemorrhage
   The common cause is severe sepsis, mainly caused by Neisseria meningitidis sepsis, which is related to disseminated intravascular coagulation (Waterhouse-

Friedrichsen syndrome). Septicemia caused by other bacteria, epidemic hemorrhagic fever, and so forth can also be accompanied by adrenal hemorrhage.

Chronic Adrenocortical Insufficiency

Chronic Adrenal cortex hypofunction can be divided into two categories: primary and secondary. The primary, also known as Addison's disease, is due to insufficient secretion of adrenocortical hormone caused by autoimmune, tuberculosis, fungal infection or tumor, leukemia and other reasons that destroy most of the bilateral adrenal glands. Secondary cases refer to insufficient secretion of Corticotropine-Releasing Factor (CRF) by the hypothalamus or ACTH by the pituitary gland. In general, clinical manifestations of adrenocortical insufficiency do not appear until at least 90% of the adrenal cortex has been compromised. Chronic adrenal cortex hypofunction is more common in adults. Tuberculosis is more common in men than in women. Idiopathic autoimmune disorders are more common in women than in men. The main clinical manifestations are weakness, pigmentation of skin and mucosa, weight loss, hypotension, anorexia, nausea, vomiting, water electrolyte metabolism disorder and nervous system damage.

In patients with primary adrenal diseases, elevated levels of adrenocorticotropic hormone precursor hormone stimulate Melanocyte, resulting in pigmentation on the surface of skin and mucosa. The face, armpits, nipples, areola, and perineum are particularly common areas of pigmentation. In contrast, hyperpigmentation did not occur in patients with secondary Adrenal cortex insufficiency. The decrease of Mineralocorticoid (Aldosterone) activity in patients with primary adrenal insufficiency will lead to potassium retention and sodium loss, leading to hyperkalemia, hyponatremia, volume failure, and hypotension. On the contrary, secondary adrenal hypofunction is characterized by insufficient secretion of cortisol and androgen, but the synthesis of aldosterone is normal or close to normal.

14.3.1.3 Adrenocortical Neoplasms

Adrenocortical Adenomas

Adrenal cortex adenoma has been described in the preceding discussion of Cushing's syndrome and hyperaldosteronism. Most cortical adenomas do not cause hyperthyroidism or even obvious clinical manifestations, so they are usually discovered by chance during autopsy or abdominal imaging examination for unrelated reasons. In fact, the interesting term "adrenal incidentaloma" has quietly entered medical dictionaries to describe these incidentally discovered tumors. Usually, adenomas are small, with an average diameter of 1–5 cm and a complete capsule [5]. On the section, due to the presence of lipids in tumor cells, adenomas are usually yellow to yellowish brown (Fig. 14.22). Microscopically, adenoma is composed of cells similar to normal adrenal cortex cells. The nucleus is often small and may also exhibit a

Fig. 14.22 Yellow
adrenocortical adenoma

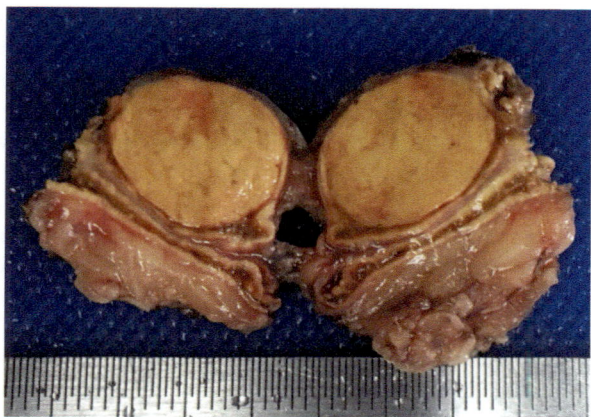

Fig. 14.23 Vacuolated
cytoplasm

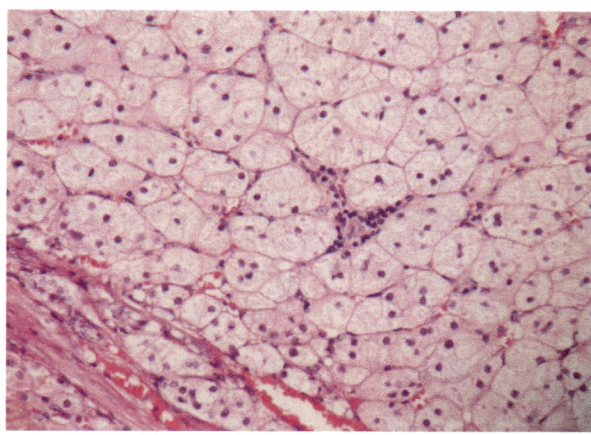

certain degree of pleomorphism ("endocrine atypia"). The tumor cells are arranged
in acinar, nested, or short strip-like structures, with the stroma separated by slender
fibers and blood vessels. The cytoplasm of tumor cells varies from eosinophilic to
vacuolated (Fig. 14.23), depending on their lipid content; mitosis activity is usually
not obvious.

Adrenocortical Carcinomas

Adrenal cortex carcinoma is a rare tumor that may occur at any age, including childhood. The tumor is generally large in size, often more than 100 g and occasionally
more than 1000 g. It shows aggressive growth, unclear boundary, brownish-yellow
or Pleochroism section, soft texture, often bleeding, necrosis, and cystic change.
Microscopically, poorly differentiated tumors exhibit significant cell atypia, often
with multinucleated tumor giant cells and a lot of mitosis. A well-differentiated

adenoma is difficult to distinguish from an adenoma if the tumor is small in size and has a capsule. The differences between the two can be referred to as follows: (1) Cortical cancer is commonly characterized by extensive bleeding and necrosis, while adenomas rarely have necrosis. (2) Destruction of the capsule, invasion of blood vessels, and surrounding organizers are generally cancer. (3) There are many mitotic images, and those with a field of view greater than 2/10 magnification are mostly malignant, while adenomas have few mitotic images. (4) The cancer has extensive and obvious nuclear atypia, multinucleated tumor giant cells, large nucleoli, and inclusion bodies in the nucleus. (5) The volume and weight of tumors have certain reference value, with adenomas often having a diameter of less than 5 cm and a weight of less than 50 g. Adrenal carcinoma has a strong tendency to invade the adrenal vein, venae cava, and lymphatic vessel vessels. Regional and peripheral lymph node metastasis and distant hematogenous spread to the lungs and other organs are common. Bone metastasis is unusual. The median survival period of the patient is approximately 2 years.

14.3.2 Adrenal Medulla

The adrenal medulla is located in the center of the adrenal gland. From the perspective of embryogenesis, medulla and sympathetic nervous system come from the same source, which is equivalent to a sympathetic ganglion. It is innervated by the preganglionic fibers of the great splanchnic nerve (belonging to Sympathetic nervous system), forming the sympathetic nervous system adrenal system. The glandular cells in the adrenal medulla are large and polygonal, arranged in clusters or irregular cable networks around the blood sinuses. The cells contain small particles, and after being treated with chromium salt, some particles react brown with the chromium salt. The cells containing such particles are called chromaffin cell. These cells synthesize and secrete catecholamine, mainly adrenaline. The most important diseases of adrenal medulla are tumors, including neuronal tumors (including neuroblastoma and more mature ganglioma) and tumors composed of chromaffin cell (pheochromocytoma).

14.3.2.1 Pheochromocytoma

Pheochromocytoma is a tumor originating from chromaffin tissue of neuroectoderm, which mainly secretes Catecholamine, such as norepinephrine and epinephrine. The characteristic manifestation of this disease is paroxysmal or persistent hypertension. Some patients may experience severe heart, brain, and kidney damage due to long-term hypertension or may experience a life-threatening crisis due to sudden onset of severe hypertension. However, if diagnosed and treated in a timely and early manner, it is a curable secondary hypertension.

Morphology

Pheochromocytoma ranges from a small lesion confined to the adrenal gland to a massive hemorrhagic mass weighing several kilograms. On the cut section, the smaller pheochromocytoma is yellowish-brown and well-defined, compressing the adjacent adrenal gland. Larger lesions are often hemorrhagic, necrotic, and cystic, often involving the adrenal gland [5]. Microscopically, pheochromocytoma is composed of polygonal to spindle chromaffin cells and their supporting cells, which are divided into pits through rich vascular networks (Fig. 14.24). The cytoplasm of tumor cells is usually basophilic, with a fine-grained appearance (Fig. 14.25). Because there are particles containing catecholamine, the cytoplasm of tumor cells can be colored by various silver staining methods. Chromogranin will always be positive in the cytoplasm of tumor cells by IHC test (Fig. 14.25). Both capsular and vascular invasion may be encountered in benign lesions, and the presence of mitotic figures does not imply malignancy. Therefore, the final diagnosis of malignant pheochromocytomas is entirely based on the presence of metastasis. These may involve regional lymph nodes as well as more distant sites, including liver, lung, and bone.

Clinical Features

The main clinical manifestation of pheochromocytoma is hypertension. Moreover, the clinical manifestation of this type of hypertension is often described as a sudden and sharp increase in blood pressure, which is related to tachycardia, palpitations, headaches, sweating, tremors, and fear. Although elevated blood pressure is the characteristic clinical manifestation of pheochromocytoma, in actual clinical work, less than half of pheochromocytoma patients will have isolated paroxysmal hypertension attacks. In approximately two-thirds of patients, hypertension occurs in the form of chronic persistent elevation, although there are often unstable factors of

Fig. 14.24 Polygonal chromaffin cells

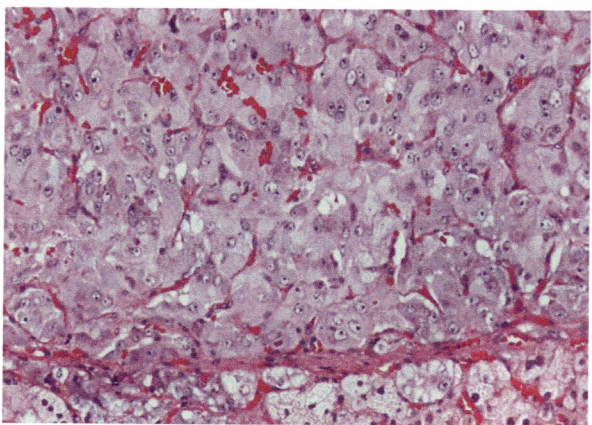

Fig. 14.25
Chromogranin (+)

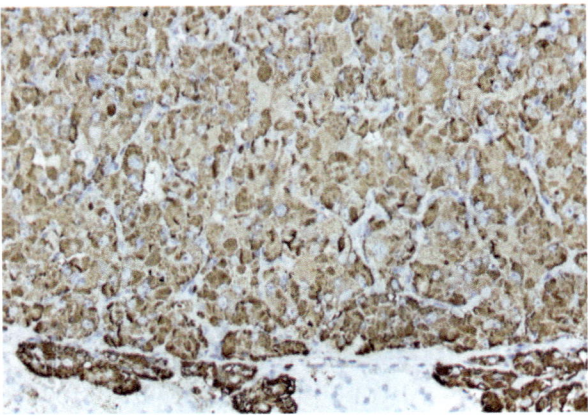

hypertension. Whether persistent or sporadic, hypertension is associated with an increased risk of myocardial ischemia, heart failure, kidney damage, and cerebrovascular accidents. In some cases, pheochromocytoma may secrete other hormones, such as adrenocorticotropic hormone and somatostatin, which may be related to the clinical features related to secretion. The laboratory diagnosis of pheochromocytoma is based on evidence of increased urinary excretion of free catecholamine and its metabolites (such as vanillin and adrenaline). The isolated benign pheochromocytoma was treated by surgical resection after the patients were given adrenergic blockers before and during the operation. Multifocal lesions may require long-term medication treatment for hypertension.

14.4   Endocrine Pancreas

The endocrine part of the pancreas consists of pancreatic islets, the islets of Langerhans. Pancreatic islets are cell clusters composed of endocrine cells, distributed between pancreatic acini. Newborn pancreatic islets account for approximately 10% of the pancreatic volume and weigh approximately 300–450 mg. The adult pancreas, on the other hand, has approximately 1 million pancreatic islets, accounting for approximately 1.5% of the pancreatic volume [4] and weighing approximately 1500 mg. Therefore, based on the proportion of pancreatic islets to pancreatic volume or body weight, the weight of pancreatic islets decreases with age. The relative distribution density of pancreatic islets is highest in the tail of the pancreas, followed by the body, and lowest in the head of the pancreas. The size of pancreatic islets varies, with small ones consisting of only over 10 cells, whereas large ones can contain hundreds of pancreatic islet cells. Pancreatic islet cells are distributed in clusters and cords, with abundant distribution of porous capillaries between cells.

Human pancreatic islets are mainly composed of four types of cells—β, α, δ, and PP cells [4]. The cells can be differentiated morphologically by their staining properties, by the ultrastructural structure of their granules, and by their hormone content.

The β cell produces insulin, which is the most potent anabolic hormone known, with multiple synthetic and growth-promoting effects; the α cell secretes glucagon, inducing hyperglycemia by its glycogenolytic activity in the liver; δ cells contain somatostatin, which suppresses both insulin and glucagon release; and PP cells contain a unique pancreatic polypeptide (vasoactive intestinal peptide, VIP) that exerts several gastrointestinal effects, such as stimulation of secretion of gastric and intestinal enzymes and inhibition of intestinal motility.

14.4.1 Diabetes Mellitus

Diabetes mellitus is not a single disease but a group of metabolic disorders sharing the common underlying feature of hyperglycemia. Hyperglycemia is caused by defects in insulin secretion or impaired biological effects, or both. Long-term hyperglycemia leads to chronic and secondary damage and dysfunction of various tissues, especially the eyes, kidneys, heart, blood vessels, and nerves.

14.4.1.1 Classification

The vast majority of cases of diabetes fall into one of two broad classes, depending on whether the diabetes results primarily from insulin deficiency or from inadequate response to insulin. Type 1 diabetes is characterized by an absolute deficiency of insulin secretion caused by pancreatic β-cell destruction, usually resulting from an autoimmune attack. Type 1 diabetes accounts for approximately 10% of all cases. Type 2 diabetes is caused by a combination of peripheral resistance to insulin action and an inadequate compensatory response of insulin secretion by the pancreatic β cells ("relative insulin deficiency"). Approximately 80–90% of patients have Type 2 diabetes.

Pathogenesis of Type 1 Diabetes Mellitus

Type 1 diabetes is an autoimmune disease in which islet destruction is caused primarily by T lymphocytes reacting against as yet poorly defined β-cell antigens, resulting in a reduction in β-cell mass. Type 1 diabetes most commonly develops in childhood, becomes manifest at puberty, and is progressive with age. Most individuals with Type 1 diabetes depend on exogenous insulin supplementation for survival, and without insulin, they develop serious metabolic complications such as acute ketoacidosis and coma. At present, the etiology is not clear, which may be caused by the combined effect of genetic susceptibility and environmental factors. The possible causes are as follows: (1) Autoimmune system defects can be detected in the blood of type 1 diabetes patients with a variety of autoantibodies, such as glutamate decarboxylase antibody (GAD antibody), islet cell antibody (ICA antibody), and so forth. These abnormal autoantibodies can damage the insulin secreting β cells of human islets, making them unable to secrete insulin normally. (2) Genetic defect is the basis of type 1 diabetes, which is manifested in HLA antigen and other gene loci on the sixth chromosome. (3) Virus infection may be a trigger. Because patients with Type 1 diabetes often have a history of virus infection in a period of time before onset, and Type 1 diabetes often occurs after the prevalence of virus infection [2].

Pathogenesis of Type 2 Diabetes Mellitus

The pathogenesis of Type 2 diabetes is not very clear. At present, it is believed that many lifestyles are important factors leading to Type 2 diabetes, including obesity and overweight (BMI higher than 25), insufficient physical activity, unhealthy diet, excessive stress and urbanized life. Nevertheless, genetic factors are even more important than Type 1 diabetes. Most cases of diabetes have multiple gene changes.

These different genes may increase the risk of Type 2 diabetes. The two metabolic defects of Type 2 diabetes are (1) decreased responsiveness of peripheral tissues to insulin (insulin resistance) and (2) β cellular dysfunction, manifested as insufficient insulin secretion in the face of insulin resistance and hyperglycemia. In most cases, insulin resistance is the primary event and is followed by increasing degrees of β-cell dysfunction.

Insulin resistance is defined as resistance to the effects of insulin on glucose uptake, metabolism, or storage. Insulin resistance is a characteristic feature of most individuals with Type 2 diabetes and is an almost universal finding in diabetic individuals who are obese.

14.4.1.2 Morphology of Diabetes and Its Late Complications

Although the occurrence of diabetes is related to the abnormality of endocrine pancreas, the pancreas may not show significant pathological changes. The important morphological changes of diabetes are related to many late systemic complications. In individuals with diabetes under strict control, onset may be delayed. However, in most patients, morphological changes may be found in arteries (big vessels disease), small vessel basement membrane (microvascular disease), kidneys (diabetes nephropathy), retina (retinopathy), nerves (neuropathy) and other tissues. These changes are seen both in Type 1 and T type 2 diabetes.

Pancreas

Lesions in the pancreas are inconstant and rarely of diagnostic value. Distinctive changes are more commonly associated with Type 1 than with Type 2 diabetes.

Reduction in the number and size of islets is most often seen in Type 1 diabetes, particularly with rapidly advancing disease. Eosinophilic infiltrates may also be found, particularly in diabetic infants who fail to survive the immediate postnatal period.

In Type 2 diabetes, there may be a subtle reduction in islet cell mass, demonstrated only by special morphometric studies. Amyloid replacement of islets in long-standing Type 2 diabetes appears as deposition of pink, amorphous material beginning in and around capillaries and between cells. At advanced stages the islets may be virtually obliterated; fibrosis may also be observed. This change is often seen in long-standing cases of Type 2 diabetes.

Macrovascular Disease

Diabetes exacts a heavy toll on the vascular system. The hallmark of diabetic macrovascular disease is accelerated atherosclerosis affecting the aorta and large and medium-sized arteries. Except for its greater severity and earlier age of onset, atherosclerosis in diabetics is indistinguishable from that in nondiabetics. **Hyaline arteriolosclerosis**, the vascular lesion associated with hypertension, is both more prevalent and more severe in diabetics than in nondiabetics, but it is not specific for diabetes and may be seen in elderly nondiabetics without hypertension.

Microangiopathy

One of the most consistent morphologic features of diabetes is diffuse thickening of basement membranes. The thickening is most evident in the capillaries of the skin, skeletal muscle, retina, renal glomeruli, and renal medulla. However, it may also be seen in such nonvascular structures as renal tubules, the Bowman capsule, peripheral nerves, and placenta.

Diabetic Nephropathy

The kidneys are prime targets of diabetes. Three lesions are encountered: (1) glomerular lesions; (2) renal vascular lesions, principally arteriolosclerosis; and (3) pyelonephritis, including necrotizing papillitis.

The most important glomerular lesions are capillary basement membrane thickening; diffuse mesangial sclerosis, and nodular glomerulosclerosis (Fig. 14.26). The glomerular capillary basement membranes are thickened throughout their entire length. Diffuse mesangial sclerosis consists of a diffuse increase in mesangial

Fig. 14.26 Diffuse
mesangial sclerosis

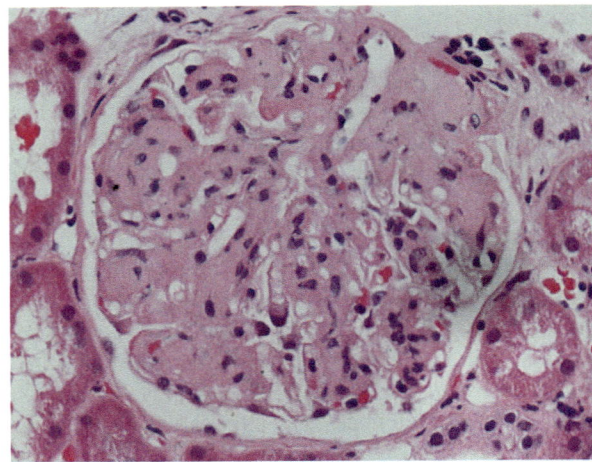

Fig. 14.27 PAS staining
positive

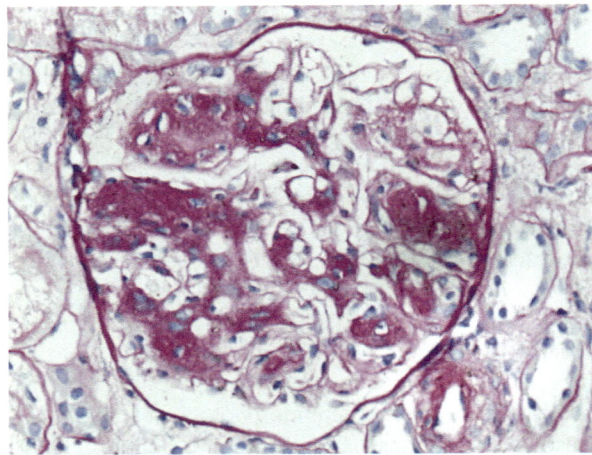

matrix along with mesangial cell proliferation and is always associated with basement membrane thickening. When glomerulosclerosis becomes marked, patients manifest the nephrotic syndrome, characterized by proteinuria, hypoalbuminemia, and edema. **Nodular glomerulosclerosis** describes a glomerular lesion made distinctive by ball-like deposits of a laminated matrix situated in the periphery of the glomerulus. These nodules are PAS positive and usually contain trapped mesangial cells (Fig. 14.27). This distinctive change has been called the Kimmelstiel-Wilson lesion, after the pathologists who described it [2].

Pyelonephritis is an acute or chronic inflammation of the kidneys that usually begins in the interstitial tissue and then spreads to affect the tubules. Both the acute and chronic forms of this disease occur in nondiabetics as well as in diabetics but are more common in diabetics than in the general population, and once affected, diabetics tend to have more severe involvement. One special pattern of acute pyelo-nephritis, necrotizing papillitis (or papillary necrosis), is much more prevalent in diabetics than in nondiabetics.

Ocular Complications of Diabetes

Visual impairment, sometimes even total blindness, is one of the more feared consequences of long-standing diabetes. The ocular involvement may take the form of retinopathy, cataract formation, or glaucoma. Retinopathy, the most common pattern, consists of a constellation of changes that together are considered by many ophthalmologists to be virtually diagnostic of the disease. The lesion in the retina takes two forms: nonproliferative (background) retinopaty and proliferative retinopathy. Nonproliferative retinopathy includes intraretinal or preretinal hemorrhages, retinal exudates, microaneurysms, venous dilations, edema, and, most importantly, thickening of the retinal capillaries (micro-angiopathy). The so-called proliferative retinopathy is a process of neovascularization and fibrosis. This lesion leads to serious consequences, including blindness, especially if it involves the macula. Vitreous hemorrhages can result from rupture of newly formed capillaries; the resultant organization of the hemorrhage can pull the retina off its substratum (retinal detachment).

Diabetic Neuropathy

The most common mode of involvement is symmetrical neuropathy around the lower limbs, which affects motor and sensory functions, especially the latter. Other forms include peripheral neuropathy, which can lead to intestinal and bladder dysfunction, sometimes sexual impotence, and diabetes mononeuropathy, which may be manifested as sudden foot drop, wrist drop, or isolated cranial nerve paralysis. Changes in the nervous system may be caused by microvascular disease, increased permeability of capillaries supplying nerves and direct axonal injury caused by changes in sorbitol metabolism [2].

14.5 Neuroendocrine Tumors of Diffuse Neuroendocrine System (DNES NET)

14.5.1 Diffuse Neuroendocrine System

The neuroendocrine system exists in two phenotypes, either as discreet organoid aggregates (pituitary, adrenal, parathyroid) or as disseminated, nonuniform distribution of cells. This latter group has been assigned the term diffuse neuroendccrine system (DNES) [4]. DNES cells are a group of small cells that are individually dispersed among the whole body are known collectively by several names: Argentaffin and argyrophilic cells—because they stain with silver stains; APUD cells—because some of them can take up the precursors of amines and decarboxylate them; or Enteroendocrine cells—because they secrete hormone-like substances and are located in the epithelium of the enteric (alimentary) canal. Some of these cells are individually designated according to the substance that they produce. Generally, a single type of DNES cell secretes only one hormone, although occasional cell types may secrete two different hormones. Cells of the DNES have been localized not only in the digestive tract but also in the respiratory system and in the endocrine pancreas. Additionally, some of the secretory products synthesized and released by these DNES cells are identical with neurosecretions localized in the CNS. The significance of their diverse location and the substances they produce is only incompletely understood.

14.5.2 Neuroendocrine Tumors, NETs

Neuroendocrine tumors originate from diffuse neuroendocrine cells. Tumors express neuroendocrine markers and exhibit other neuroendocrine characteristics. Neuroendocrine tumor is a group of heterogeneous tumors. From pathological morphology to biological behavior, there are great differences between different tumors. Neuroendocrine neoplasms/tumors (NET) once considered "rare" have been steadily increasing in incidence and prevalence over the past three decades. DNES NETs occur most frequently in the lung and gastro-entero-pancreatic (GEP) regions. NETs currently represent 2% of all cancers and GEP-NETs represent the second most prevalent gastrointestinal neoplasm after colorectal cancer.

The clinical manifestations of this relatively uncommon disease are protean and nonspecific, thereby leading to alternative diagnoses and the average lag between first symptoms and diagnosis of NET of about 7 years. It is therefore understandable that 60–80% have metastases at presentation, with the liver being the most common distant metastatic site. There is another important consequence of metastases, the "carcinoid syndrome" (diarrhea, abdominal pain, sweating, flushing, bronchospasm, tachycardia, and fibrotic heart disease). According to cell sources, tumor is divided into two types: nerve type and epithelial type. Pheochromocytoma and paraganglioma are nerve type. And some NETs originating from gastrointestinal tract or pancreas belong to the epithelial type.

References

1. Mohan H. Textbook of pathology. 7th ed. Jaypee Brothers Medical Publishers (P) Ltd.; 2015. p. 782–820.

2. Kumar V, Abbas AK, Aster JC. Robbins basic pathology. 9th ed. Saunders, an imprint of Elsevier Inc.; 2012. p. 715–63.
3. Crowley LV. Human disease (pathology and pathophysiology correlations). 5th ed. Jones and Bartlett Publishers International; 2001. p. 647–71.
4. Xiaorong C, Chen X. Histology and embryology. 2nd ed. Hefei: China University of Science and Technology Press; 2014. p. 104–14.
5. Yulin L. Pathology. 9th ed. Beijing: People's Health Publishing House; 2018. p. 299–312.

Chapter 15
Diseases of the Nervous System

Baohua Niu and Suxia Wu

Contents

B. Niu (✉) · S. Wu
School of Basic Medical Sciences, Henan University, Zhengzhou, China

© Zhengzhou University Press 2024
K. Chen et al. (eds.), *Textbook of Pathologic Anatomy*,
https://doi.org/10.1007/978-981-99-8445-9_15

Objectives
1. To grasp the lesion characteristics of epidemic cerebrospinal meningitis and epidemic encephalitis B.
2. To be familiar with the etiology, infectious pathways, pathogenesis, and clinicopathological associations of epidemic cerebrospinal meningitis and epidemic encephalitis B.
3. To understand the etiology, pathogenesis, pathologic changes, and clinicopathologic associations of poliomyelitis, Alzheimer's disease, Parkinson's disease, and central nervous system tumors.
4. To understand the etiology, pathologic changes, and clinicopathologic associations with common complications (intracranial hypertension, cerebral edema, and brain tumors) of ischemic encephalopathy, obstructive cerebrovascular disease, intracerebral hemorrhage, hydrocephalus, and neurological disorders.

Key Concepts
1. Red neuron
2. Satellitosis
3. Neuronophagia
4. Microglial nodule
5. Gitter cell
6. Senile plaque
7. Waterhouse-Friderichsen syndrome
8. Neurofibrillary tangles

Introduction
The nervous system is the processing mechanism which communicates information from the outside world to the body. Diseases of the nervous system have notable characteristics [1]:

1. Identical lesions can occur at different sites, yet they have different clinical manifestations and consequences.
2. The most frequently occurring and critical complications of central nervous system diseases are an increase in intracranial pressure, cerebral edema, and hydrocephalus.
3. There are various types of pathogens which can infect the brain, including bacteria, viruses, parasites, rickettsia, helix, and so forth. Bacterial and viral infections are the most common.
4. Tumors of the nervous system can occur in the cells of the coverings, from cells intrinsic to the brain, or other types of cells within the skull. They may also spread from elsewhere in the body.
5. Depending on where in the brain they occur, even low-grade or benign tumors can have a poor clinical outcome.

15.1 The Principal Pathologic Changes of Nervous System Diseases

The nervous system is comprised of neuroglial cells (astrocytes, oligodendrocytes, and ependyma), blood vessels, and microglia. It is estimated that over half of human genes are contained within the nervous system.

15.1.1 The Principal Pathologic Changes of the Neuron and Nerve Fibers

The neuron is the principal functional unit of the central nervous system and one of the most complex cells in any of the bodily systems. Neuronal damage can cause ischemic anoxia, infection, and toxicosis in the brain.

15.1.1.1 The Principal Pathological Changes of the Neuron [1–3]

1. Acute neuronal injury

    Pathologic reactions of neurons have several well-characterized forms. The most common is coagulation necrosis, which presents together with acute hypoxia-ischemia injury, infection and toxicosis. Neuronal necrosis is characterized by a loss of cytoplasmic ribonucleoproteins and denaturation of cytoskeletal proteins, leading to the development of intense cytoplasmic eosinophilia (known as red neurons) in hematoxylin and eosin (H&E)-stained sections. The Nissl body in cytoplasm also disappears in this situation. Coagulation necrosis occurs together with the same nuclear changes seen in other organs, including condensation of nuclear material (pyknosis) and loss of nuclear staining (karyolysis). Finally, ghost cells are the outline or trace of a residual cell.

2. Simple neuronal atrophy

    Simple neuronal atrophy is the process from neuronal degeneration to neuronal death, and is characterized by condensation of nuclear material (pyknosis) and loss of nuclear staining (karyolysis). Unlike red neurons, the Nissl body in cytoplasm does not disappear. Neurodegenerative disorders are the main cause of simple neuronal atrophy, including multisystem atrophy and amyotrophic lateral sclerosis (ALS). Loss of atrophic neurons is invisible in early stages, while glial cell proliferation appears in later stages.

3. Central chromatolysis

    Central chromatolysis is a common reaction to viral infection, hypoxia-ischemia, Vitamin B deficiency, and axonal injury. It is characterized by the Nissl substance being dispersed from the cell's center to its periphery, perikaryon enlargement, peripheral displacement of the nucleus, and nucleolus enlargement.

4. Inclusion body formation

Inclusion bodies in the neuronal cytoplasm or nucleus can be observed in some viral infections and degenerative diseases. The form and size of inclusion bodies and coloring also vary between diseases; for example, in Parkinson's patients, the Lewy corpuscle can occur in the cytoplasm of the substantia nigra neurons, while inclusion bodies can occur simultaneously in the nucleus and cytoplasm of those with cytomegalovirus infection.

5. Neurofibrillary degeneration

Neurofibrillary degeneration, also known as neurofibrillary tangles, is where thickened neurofibrils condense and tangle around the nucleus, which can be observed by silver staining.

15.1.1.2   The Principal Pathologic Changes of the Nerve Fiber [1]

1. Axonal injury and axonal reaction

Axonal reaction occurs when there is serious damage to the central nervous axons or peripheral nerve axons. This process is associated with swelling and disintegration of axon, disintegration of the myelin sheath, and cell proliferation reactions. Degenerative changes in an injured axon occur over time and involve the axon's distal regions.

2. Demyelination

Demyelination is caused by Schwann degeneration or myelin injury. Demyelination refers to separation, swelling, fracture, and disintegration of the myelin lamina. The ability of the myelin sheath to remyelinate is limited in the central nervous system.

15.1.2   The Principal Pathologic Changes of the Neuroglia [1]

Neuroglial cells are composed of astrocytes, oligodendrocytes, and ependymal cells.

15.1.2.1   The Principal Pathologic Changes of Astrocytes

Astrocytes, as major supporting cells in the brain, are subject to the most commonly occurring reactive changes. This includes swelling, reactive astrogliosis, inclusion body formation, and so forth.

1. Cellular swelling

Cellular swelling occurs in acute injury, such as hypoxia, hypoglycemia, and toxic injuries.

2. Reactive astrogliosis

Reactive astrogliosis is a repairing action which can be likened to a fibrous scar which occurs elsewhere in the body. In contrast to fibroblasts, astrocytes do not produce collagen. Thus, the glial scar is primarily comprised of cytoplasmic processes, with little or no extracellular protein.

3. Corpora amylacea

In H&E-stained sections, corpora amylacea are spherical, almost basophilic, concentrically lamellated structures and can found where there are astrocytic foot processes, in particular in the subependymal, subpial, and perivascular zones.

4. Rosenthal fibers

Rosenthal fibers are round, elongated, brightly eosinophilic bodies in astrocytic cytoplasm. Because of a GFAP gene mutation, in some diseases (e.g., Alexander disease), abundant Rosenthal fibers can be found in the periventricular, perivascular, and subpial locations.

15.1.2.2 The Principal Pathologic Changes of Oligodendrocytes

Satellitosis is characterized by five or more oligodendrocytes which surround degenerative neuron cells, which could be associated with neurotrophy.

15.1.2.3 The Principal Pathologic Changes of Microglia

It is now generally accepted that microglia are mesoderm-derived cells; the primary function of which is to act as a fixed macrophage system in many inflammatory conditions. Microglia cells can be activated in the central nervous system (CNS) as a result of injury.

1. Neuronophagia

This refers to microglia cells which aggregate around cell bodies of dying neurons and engulf injured neurons. Blood-derived macrophages are the main phagocytic cells present in inflammatory foci. For example, in Japanese, encephalitis neuronophagia manifests as a reaction to necrotic neurons.

2. Microglial nodule

Microglia cells aggregate in compact clusters to form microglial nodules because of infections in the CNS, in particular viral infections.

3. Gitter cell

Gitter cells are activated microglia cells which can accumulate abundant intracellular lipid to form cells with foamy cytoplasm.

15.1.2.4   The Principal Pathologic Changes of Ependymal Cells

The inside of the ventricular system is lined by ependymal cells. Disruption or loss of ependymal cells can occur with the local proliferation of subependymal astrocytes to produce small irregularities on the ventricular surfaces, known as ependymal granulations. Viral inclusions in ependymal cells can occur as a result of certain viral infections, particularly cytomegalovirus (CMV).

15.2   Common Complications of Central Nervous Diseases [1, 4]

Raised intracranial pressure, cerebral edema, and hydrocephalus are the complications of central nervous diseases which occur most frequently.

15.2.1   Raised Intracranial Pressure and Herniation

Increased intracranial pressure is defined by a CSF pressure above 2 kPa (the normal value is 0.6–1.8 kPa). Hydrocephalus is associated with the majority of cases and is caused by either a tumor, inflammation (e.g., meningocephalitis and abscess), hemorrhage, intracranial hematoma, or cerebrospinal fluid circulation disorder. The increase in intracranial pressure has three stages: compensatory stage, decompensation stage, and vasomotor paralysis. Brain displacement and encephalocoele deformation can occasionally occur as a result of increased intracranial pressure, leading to part of the brain tissue becoming embedded in the mediastinum cerebri, tentorium cerebelli, or foramina magnum; after which, brain herniation occurs.

1. Subfalcine (cingulate) herniation occurs when the midline of the brain moves oppositely because of the occupying lesions of the unilateral cerebral hemisphere (especially occurring in the frontal lobe, parietal lobe, and temporal lobe). The cingulate gyrus of the same side displaces under the free margin of falx cerebri. Then, the herniated cingulate gyrus is compressed which can result in bleeding and necrosis in compressed brain tissue. The branches of the anterior cerebral artery may also be compressed, causing cerebral infarction of the territory supplied by that vessel.
2. Transtentorial (uncinate, mesial temporal) herniation occurs when the medial aspect of the temporal lobe is compressed against the tentorium cerebelli's free margin. This can result in a number of issues; first, the ipsilateral oculomotor nerve is compressed, causing transiently pupillary contraction; then, pupillary

dilation and impairment of ocular movements occurs on the side of the lesion. Second, progression of transtentorial herniation is often accompanied by hemorrhagic lesions in the midbrain and pons, known as secondary brainstem or Duret hemorrhages, which are linear or flame-shaped lesions usually occurring in the midline and paramedian regions due to tearing of penetrating veins and arteries supplying the upper brainstem. Third, when the lateral shift of mesencephalon occurs, the contralateral cerebral peduncle is displaced under the free margin of tentorium cerebelli, causing compression to the contralateral cerebral peduncle; the changes in the peduncle in this setting are known as Kernohan's notch. Finally, the posterior cerebral artery may also be compressed, causing hemorrhagic infarction of the territory supplied by that vessel (for instance ipsilateral occipital lobe calcarine fissure).

3. Tonsillar herniation is the displacement of the cerebellar tonsils and medulla oblongata through the foramen magnum. This herniation pattern is life-threatening as it causes brainstem compression and compromises vital respiratory and cardiac centers in the medulla oblongata.

15.2.2 Cerebral Edema

Cerebral edema, or brain parenchymal edema, is excessive liquid accumulating in brain tissue and is associated with several pathologic processes, for example, hypoxia, trauma, infarct, inflammation, tumor, and toxicosis. Two principal types are recognized:

1. Vasogenic edema. This is the most common type and occurs when the blood-brain barrier is disrupted and fluids escape from the intravascular compartment into the brain's intercellular spaces.
2. Cytotoxic edema is retention of water and sodium in cytoplasm, caused primarily by generalized hypoxic-ischemic or intoxications.

Usually, both vasogenic and cytotoxic edema can be observed. When macroscopically examined, the edematous brain is heavier than normal and the white matter appears to have obvious edema, the gyri are flattened, the intervening sulci are narrowed, and the ventricular cavities are compressed. If the brain edema is severe, brain herniation can occur. When microscopically examined, and when vasogenic edema occurs, brain tissue loosens, large amounts of liquid accumulate in the widened spaces between cells and perivascular. Conversely, when cytotoxic edema occurs, large amounts of liquid can be observed to accumulate in neurons, glial cells, and vascular endothelial cells, causing enlargement of cell volume and hypochromatic cytoplasm.

15.2.3 Hydrocephalus

Hydrocephalus is an increased CSF volume within the ventricular system, accompanied by continuous dilatation of the ventricles. The majority of cases happen as a consequence of obstruction to the free flow of CSF (termed noncommunicating hydrocephalus), increased CSF production, or decreased absorption (termed communicating hydrocephalus). Communicating hydrocephalus occurs in patients with cerebral cysticercosis, tumors, congenital malformation, inflammation, trauma or subarachnoid hemorrhage, and so forth. However, noncommunicating hydrocephalus can also occur in patients with papillary tumor of the choroids plexus or chronic arachnoiditis.

The pathological changes of hydrocephalus are variable depending on the different lesion sites and degree of lesion. For example, during mild hydrocephalus, mild ventricular dilatation and mild ventricular atrophy can be observed. However, high ventricular dilatation and severe compression of brain tissue occurs during severe hydrocephalus.

15.3 Infections of the Central Nervous System

Bacteria, viruses, rickettsia, Leptospira, fungus, and parasites can all cause CNS infections. Additionally, HIV can cause opportunistic infections (toxoplasmosis, cytomegalovirus infection) or central nervous system lymphoma.

Four principal routes exist by which infectious microbes enter the nervous system [1].

1. Hematogenous spread: This is the most common mean of entry; infectious agents enter through arterial circulation; for example, during sepsis, the infective embolus enters through arterial circulation and then reaches the central nervous system.
2. Local extension: Occurs secondary to an open skull fracture, usually mastoiditis, an infected tooth, or tympanitis, nasosinusitis, bone erosion, and propagation of the infection into the CNS.
3. Direct implantation: Usually traumatic, it is rarely iatrogenic (such as Lumbar puncture).
4. Via the peripheral nervous system: Some pathogens, such as certain viruses (e.g., the rabies virus) invade the peripheral nerves to cause the damage to nervous tissue. The herpes simplex virus invades the central nervous system along the olfactory nerve and trigeminal nerve.

15.3.1 Bacterial Disease

The most commonly occurring intracranial bacterial diseases are meningitis and brain abscess.

15.3.1.1 Meningitis [1, 4]

Meningitis includes pachymeningitis and leptomeningitis. As a result of various antibiotics, pachymeningitis as secondary to cranium infection is greatly reduced. Therefore, meningitis is usually known as leptomeningitis, including infection of the pia mater, arachnoidea, and cerebrospinal fluid. Based on examination of CSF inflammatory exudate, meningitis is broadly classified into purulent meningitis (usually bacterial meningitis), lymphocytic meningitis (usually viral meningitis), and chronic meningitis (usually tuberculous, spirochetal, cryptococcal or fungus).

Epidemic cerebrospinal meningitis is acute suppurative inflammation in the cerebral spinal cord membrane, usually occurring sporadically in winter and spring as a result of meningococcal infection. The majority of patients are children and adolescents. Patients usually display systemic signs of infection based on evidence of meningeal irritation and neurologic impairment, including fervescence, headache, vomit, skin petechia, irritability, clouding of consciousness, and neck stiffness. In the most severe cases, toxic shock can occur; if left untreated, pyogenic meningitis can be fatal.

Etiology and Pathogenesis

Meningococcus possess capsule, which can be resistant to leukocyte phagocytosis. Often, bacteria in a partial secretory of the nasopharynx can invade the bodies of others by droplet propagation. However, few people suffer from bacteremia, septicemia, or meningitis.

Pathological Changes

During disease progression, patients experience three stages:

1. Upper respiratory tract infection: After 2–4 days of bacteria propagating in the nasopharyngeal mucosa, patients display upper respiratory tract infection symptoms. Major pathological changes include mucous congestion and edema, a small amount of neutrophil infiltration and increase of secretion. The condition may get worse in some patients.

2. Septicemia: In this stage, blood culture can be positive, and mucocutaneous petechia or ecchymoses can be observed in most patients. High fever, headache, vomit, and increased neutrophils in peripheral blood can occur in patients as a result of endotoxins.

3. Meningitis stage: The characteristic lesion is suppurative inflammation of the cerebral spinal cord membrane (Figs. 15.1 and 15.2). Hyperemia of cerebrospinal meningeal vessels and yellow purulent exudates within the leptomeninges over the surface of the brain are evident. In severely affected areas, exudation is densest in subarachnoid space and covers the sulci and gyri. In less severe areas, tracts of pus can be found along blood vessels on the brain surface. Because of exudation obstructions, cerebrospinal fluid circulation disorder can occur and cause different degrees of ventricular dilatation. On microscopic examination, arachnoid angiectasia is obvious (Fig. 15.3①), and a large number of neutrophils infiltrate in subarachnoid space (Fig. 15.3②). In untreated meningitis, Gram stain shows that there are varying numbers of the causative organism, although they are often indemonstrable in treated cases. Generally, the parenchyma of the brain is not involved.

Clinical Pathological Correlation

1. Meningeal irritation signs

   Signs of meningeal irritation include neck stiffness and positive Kernig's sign. Neck stiffness is a protective neck spasm, occurring as a result of inflammation involving the arachnoidea, pia mater and pia mater spinalis surrounding the spinal nerve root. As a result of inflammation, nerve roots are pressed through the intervertebral foramen; then, the patient feels pain when the neck or back muscles are moving. Occasionally, infantile patients can show signs of opisthotonus.

Fig. 15.1 Epidemic cerebrospinal meningitis. Hyperemia of cerebrospinal meningeal vessels and yellow purulent exudates within the leptomeninges over the surface of the brain are evident. The exudates are densest in subarachnoid space, covering the sulci and gyri

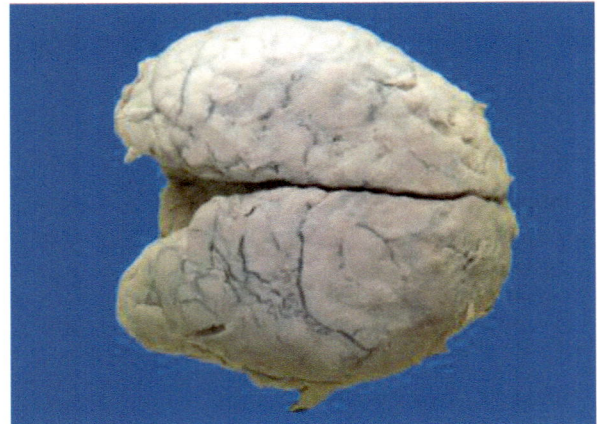

Fig. 15.2 Epidemic
cerebrospinal meningitis.
A thick layer of
suppurative exudate covers
the brain stem and
cerebellum and thickens
the leptomeninges

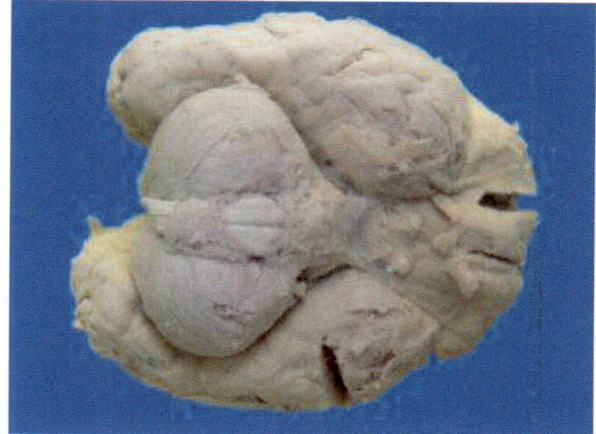

Fig. 15.3 Epidemic
cerebrospinal meningitis.
① Arachnoid angiectasia is
obvious. ② A large number
of neutrophils can be seen
in the subarachnoid

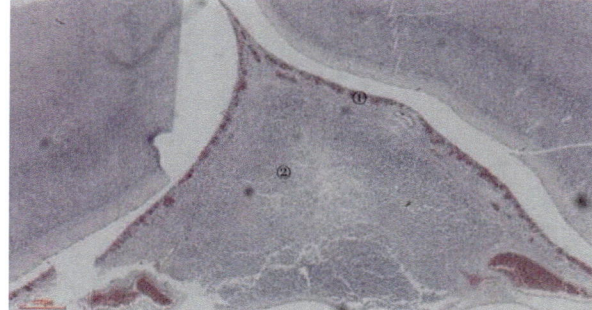

2. Symptoms of increased intracranial pressure

Increased intracranial pressure symptoms include a severe headache, projectile vomiting, papilloedema, and full bregma.

3. Changes of cerebrospinal fluid

CSF is cloudy or purulent with raised neutrophils and protein levels, but glucose content is markedly reduced. Diplococcus meningitidis can be seen on a smear or can be cultured.

Outcome and Complications

Most patients recover if treated in a timely manner and with the extensive use of antibiotics. The current mortality rate has fallen from 70–90% to less than 5%. In very few patients, sequelae are likely, such as hydrocephalus, cerebral nerve damage, and obstructive disease caused by arteritis of the base of the skull. Some patients (most commonly children) will develop fulminant epidemic cerebrospinal meningitis, which is fatal. According to the clinical pathological characteristics, fulminant epidemic cerebrospinal meningitis can be classified into two types:

1. Meningococcal septicemia with severe meningitis

   Septic shock is the major clinical manifestation. Patients present with severe clinical manifestations, such as extensive bleeding of the skin and mucous membrane and ecchymosis, as well as peripheral circulatory failure, known as Waterhouse–Friderichsen syndrome. Recently, Waterhouse–Friderichsen syndrome was considered to result from meningitis-associated septicemia with hemorrhagic infarction of the adrenal glands and cutaneous petechiae. However, it is now considered the result of toxic shock and DIC.

2. Fulminant meningoencephalitis

   In fulminant meningitis, inflammatory cells infiltrate the leptomeningeal vein walls and can extend to brain substance (focal cerebritis). If not treated in a timely manner, it can be fatal.

15.3.1.2   Brain Abscess

Brain abscesses are primarily caused by streptococci and staphylococci. Different infection pathways cause different positions and numbers of abscesses and are caused by hematogenous spread (usually multiple foci, distributed in the various brain sections), direct implantation of organisms, or local extension from adjacent foci (usually single lesion).

The pathological changes of brain abscesses are similar to the extracranial organ abscesses. On macroscopic examination, acute abscesses are discrete lesions with central liquefactive necrosis (Fig. 15.4). However, in chronic abscesses there is exuberant granulation tissue with neovascularization around the necrosis, a surrounding fibrous capsule, and edema. A zone of reactive gliosis with numerous gemistocytic astrocytes exists outside the fibrous capsule.

Fig. 15.4 Brain abscess

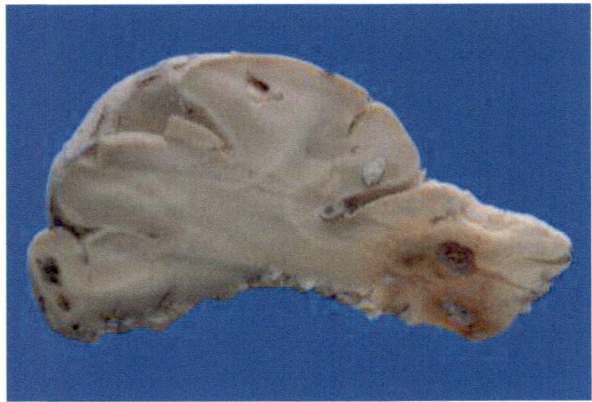

15.3.2   Viral Disease [1, 3]

Viral encephalitis occurs as a result of viral infection of the brain, by viruses such as Herpes virus (DNA viruses, including herpes simplex virus, herpes zoster virus, EB virus and cytomegalovirus, etc.), arbovirus (RNA viruses, including Japanese encephalitis virus, forest encephalitis virus, etc.), enteric virus (small RNA virus, such as poliovirus, Coxsackie virus, ECHO virus, etc.), rabies virus, HIV, and so forth.

Epidemic encephalitis B is an acute infectious disease caused by the Japanese encephalitis virus. It is generally spread by mosquitoes, with pigs serving as a reservoir for the virus. It most commonly occurs in the summer and early autumn. Patients develop general symptoms, such as ardent fever, somnolence, twitch, and coma. Approximately 50–70% of patients are children (in particular those under 10 years old).

15.3.2.1   Etiology and Pathogenesis

Patients with encephalitis B and host (livestock, fowl) act as reservoirs. Culex, aedes, and anopheles act as the carriers of disease. In China, Culex (Culex) tritaeniorhynchus is the main vector of the disease. Immune defense and the blood-brain barrier are important in the occurrence of the disease.

15.3.2.2   Pathological Changes

Epidemic encephalitis B is primarily lesions of brain parenchyma inflammation, occurring most severely in the cerebral cortex, basal ganglia, and thalamus. On gross appearance, leptomeningeal hyperemia and edema, as well as widened gyri and narrowed sulc, can be observed. In severe cases, petechial hemorrhage and semitransparent softening of the brain are apparent. On microscopic examination, characteristic histologic features of epidemic encephalitis B are perivascular and parenchymal mononuclear cell infiltration (lymphocytes, plasma cells, and macrophages) (Fig. 15.5), degeneration and necrosis of nerve cells (satellitosis and neuronophagia) (Fig. 15.6), glial cell reactions (including the formation of microglial nodules), and softening of the brain (Fig. 15.7).

15.3.2.3   Clinical Pathological Correlation

Because of the extensive involvement of nerve cells and inflammatory damage of the cerebral parenchyma, most patients have neurologic symptoms including somnolence or coma. Severe disease is characterized by upper motor neuron damaged symptoms, for example, muscle tone enhancement, tendon hyperreflexia, twitch

Fig. 15.5 Epidemic
encephalitis
B. Perivascular
mononuclear cell
infiltration

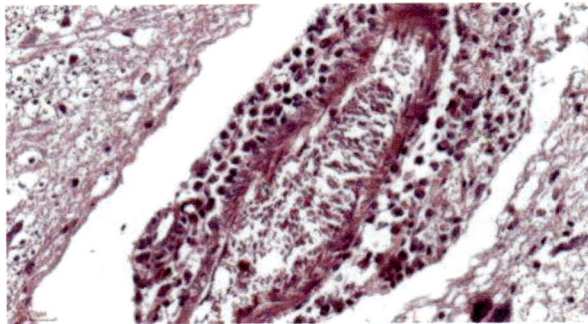

Fig. 15.6 Neuronophagia

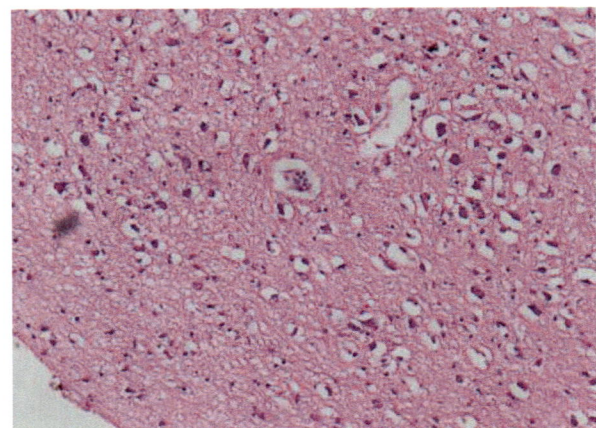

Fig. 15.7 Epidemic
encephalitis B. Softening
of the brain

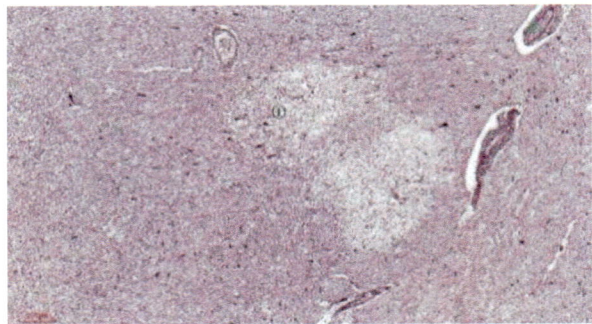

and spasm, and so forth. Some patients experience a headache or vomiting. Severe intracranial hypertension can cause hernia cerebri, which is fatal.

Most patients can recover if treated. However, a few patients have neurologic sequelae or even die.

15.4 Neurodegenerative Diseases

Neurodegenerative diseases are chronic, progressive disorders characterized by the gradual loss of neurons in discrete areas of the CNS, the underlying mechanism of which is unknown. General characteristics include: selective loss of neuron cells, whereby involved neurons may be atrophic or dead. Astrogliosis might also be seen. Therefore, patients usually have specific clinical manifestations. Secondly, special neuropathologic findings can be seen in different degenerative diseases, e.g. Lewy bodies, and neurofibrillary tangles. Accumulation of abnormal proteins, such as Aβ protein and Tau protein in Alzheimer's disease and α-synuclein in Parkinson's disease (PD), can also be found.

Common clinical features of degenerative diseases are [1]:

1. Dementia is the main clinical manifestation Alzheimer's disease and Pick's disease and affects the cerebral cortex.
2. Cinesipathy is the main clinical manifestation of Huntington's disease, PD, Progressive supranuclear palsy, and Multiple system atrophy (MSA), which are involved with the basal ganglia and brainstem.
3. Asynergy is the main clinical manifestation of Friedrich asynergy and ataxia telangiectasia, which are involved the cerebellum and spinal cord.
4. Amyosthenia is the main clinical manifestation of amyotrophic lateral sclerosis and Duchenne-Aran disease, which are involved with motor neurons.

15.4.1 Alzheimer's Disease [1]

Alzheimer's disease is a cortical degenerative disease characterized by progressive dementia, and is the most common type of dementia. Most patients begin to show symptoms after the age of 50. The incidence of the disease increases with age, with prevalence approximately 47% for the 85–89-year-old patients. The disease usually becomes clinically apparent with decaying cognitive factors including memory, intelligence, orientation, discretion, disturbance of emotion, and behavior disorder. Within 5–10 years, the patient often dies of secondary infection and systemic failure.

15.4.1.1 Etiology and Pathogenesis

The cause of Alzheimer's is mostly unknown. Some cases can be caused by genetic factors, education level, metal ion damage, and secondary transmitter change. Five to ten percent of cases are familial. Recent studies indicate that genes on chromosomes 21, 19, 14, and 1 are related to Alzheimer's disease. Familial Alzheimer's disease shows that the accumulation of Aβ protein in the brain plays the major role in Alzheimer's disease, which is derived from the abnormal degradation of amyloid precursor protein.

15.4.1.2 Morphology

On macroscopic examination, cortical atrophy and widening of the cerebral sulci are most prominent in frontal, temporal, and parietal lobes. Ventricular enlargement commonly occurs after severe atrophy.

Alzheimer's disease major microscopic abnormalities are neuritic plaques, neurofibrillary tangles, granulovacuolar degeneration and Hirano bodies [1].

1. Neuritic plaque, also termed senile plaque, is an extracellular structure. Senile plaques are focal, spherical collections of dilated in shape, 20–200 μm in diameter. They are dystrophic neurites around a central amyloid core. In H&E staining, eosinophilic clumps can be surrounded by a clear halo, and irregularly argentaffin granules or filamentous substance can be observed at the periphery. Immunohistochemistry show that the dominant component of the plaque core is Aβ.
2. Neurofibrillary tangles, intracellular abnormalities, are filament bundles in the neuronal cytoplasm, visible as basophilic fibrillary structures with H&E staining. Neurofibrillary tangles are commonly found in cortical neurons, particularly in the entorhinal cortex, as well as pyramidal cells of the hippocampus, the amygdala, the basal forebrain, and the raphe nuclei.
3. Granulovacuolar degeneration presents as a small vacuole in the neuronal cytoplasm; argyrophilic granules can be observed in small vacuoles. It is most commonly found in hippocampal pyramidal cells.
4. Hirano bodies are rodlike eosinophilic bodies with actin as their major component, and are most commonly within hippocampal pyramidal cells.

15.4.2 Parkinson's Disease [1]

PD is a slow progressive disease characterized as a lesion of striatal substantia nigra, typically occurring those aged 50–80 years old.

15.4.2.1 Etiology and Pathogenesis

PD is characterized by a progressive and selective degeneration of dopaminergic neurons of the substantia nigra, the mechanisms of which are not fully understood.

Epidemiologic evidence indicates that environmental exposure can increase the risk of PD. MPTP (1-methyl-4-phenyl-1,2,3,6-tetrahydropyridine) produces moderate to severe parkinsonism in humans because of substantia nigra neuron death. Six genes have so far been found to be linked to PD, and gene PARK-1 is the most important one and is related to α-synuclein.

15.4.2.2 Morphology

Typical macroscopic findings are pallor of the substantia nigra and locus ceruleus. On microscopic examination, the pigmented, catecholaminergic neurons in these regions are reduced. Often, some remaining neurons contain single or multiple, cytoplasmic, eosinophilic, round inclusions known as Lewy bodies, which are composed of fine filaments, densely packed in the core but loose at the rim.

Clinical manifestations include tremor, myotonus, exercise reduction, posture and gait instability, starting and stopping difficulties, masklike face, and so forth. In the later stages, some patients can develop dementia. The relationship between Alzheimer's disease and PD with subsequent dementia development still needs to be clarified. The course of PD is usually over 10 years, with the patient usually dying from secondary infection or injuries resulting from a fall.

15.5 Anoxia and Cerebrovascular Disease

The most prevalent neurologic disorder in regard to both morbidity and mortality is cerebrovascular disease.

15.5.1 Ischemic Encephalopathy

Ischemic encephalopathy is severe brain damage caused by hypotension, cardiac arrest, hemorrhage, hypoglycemia, asphyxia, and so forth.

15.5.1.1 Influence Factor

A number of factors influence prognosis. First, sensitivity to anoxia varies with various brain segments; for example, gray matter is more sensitive than white matter to anoxia. Second, local vascular distribution and vascular status are related to the injury site. Moreover, subsequent brain injury depends on the degree and duration of ischemia and anoxia, and patient survival time.

15.5.1.2 Morphology

Typical pathological changes include central chromatolysis, red neuron, disintegration of myelin and axon, repair of glial scar, and so forth. This can be observed in the brain tissue of moderate hypoxic patients.

15.5.2 Obstructive Cerebrovascular Disease

Cerebral infarction results from a local arrest or reduction of cerebral blood flow. Usually, thrombotic or embolic obstruction is the cause of vascular obstruction. Cerebral infarction is usually anemic infarction. The gray matter in the infarct area is reduced after a few hours of infarction, and 1 week later, necrotic tissue can be liquefied.

15.5.3 Brain Hemorrhage

Brain hemorrhage includes intracerebral hemorrhage, subarachnoid hemorrhage, and mixed hemorrhage.

15.6 Demyelinating Diseases [1]

In the CNS, the axon is tightly wrapped in the myelin sheath, ensuring the rapid propagation of nerve impulses. Demyelinating diseases are characterized by damaged myelin, with axons being relatively well preserved. Myelin regeneration ability in the CNS is limited; secondary axonal injury can cause severe results. The disease's clinical manifestation depends on the capacity of regenerating myelin and the degree of secondary damage to axons that occurs during the course of the disease.

15.6.1 Multiple Sclerosis (MS)

The most common demyelinating disease is multiple sclerosis and usually affects women. The clinical course of the illness evolves as relapsing and remitting episodes of neurologic deficit during variable time intervals. On each of these occasions the relapses may be different, causing different symptoms of the nervous system. MS is now considered as an autoimmune disease, influenced by genetic and environmental factors.

Multiple, well-circumscribed, glassy, gray-tan, and irregularly shaped plaques occur in horns of the ventricle and paraventricular white matter on macroscope examination. The plaques are also located in the optic nerves and chiasm, brain stem, ascending and descending fiber tracts, cerebellum, and spinal cord. On microscope examination, perivascular demyelinating is one of the main changes of the disease. Monocytes and lymphocytes present as perivascular cuffings at the outer edges of the lesion in the early stages. In active plaques, ongoing myelin breakdown with abundant macrophages which contain myelin sheath debris (termed foamy cells) can be observed. In quiescent plaques where the infiltrating inflammatory cells disappear, no myelin is found, and there is a hyperplasia in the number of astrocytes.

MS lesions occur anywhere in the CNS and, therefore, can induce various clinical manifestations; certain patterns of neurologic symptoms and signs are commonly observed.

15.6.2 Acute Disseminated Encephalomyelitis

Acute disseminated encephalomyelitis (ADEM) is a demyelinating disease of the CNS usually presenting as a monophasic disorder associated with multifocal neurologic symptoms and disability. Viral infections or vaccinations are believed to induce ADEM and these include measles virus, rubella virus, chickenpox virus, vaccinia vaccine, and hydrophobia vaccine. Symptoms usually begin 1–3 weeks post-infection. Major symptoms include fever, headache, nausea and vomiting, confusion, and coma. The disease rapidly worsens over hours to days; 80% patients recover well, and 20% have a fatal outcome.

On microscopic examination, there is perivascular demyelination accompanying the inflammatory response, a characteristic of ADEM. Although the lesions rapidly progress, the axons are well preserved.

15.6.3 Acute Necrotic Hemorrhagic Encephalitis

Acute necrotic hemorrhagic encephalitis is a rare, rapid progressive, and fatal disease in young people and children. The majority of lesions occur in the cerebral hemispheres and brain stem and are characterized by lesions similar to cerebral fat embolism-brain swelling with white matter punctate hemorrhage. When microscopically examined, the disease appears as focal necrosis of small vessels with peripheral hemorrhage, perivascular demyelination, and infiltration of neutrophil, lymphocyte, and macrophage, in addition to cerebral edema and meningitis. Unlike ADEM, the necrosis of the acute necrotic hemorrhagic encephalitis is more extensive.

15.7 Tumors of the Nervous System [1–5]

15.7.1 Tumors of the Central Nervous System

CNS tumors include primary (75%) and secondary tumors (25%). The most common primary brain tumors are gliomas (50%), meningiomas (20%), pituitary adenomas (15%), and nerve sheath tumors (8%).

CNS tumors possess the following distinct characteristics:

1. There is currently no precancerous lesion or carcinoma.
2. Benign or malignant tumors can infiltrate the brain and cause serious clinical deficits and also have a poor prognosis.
3. Prognosis is influenced by the anatomical location of the tumor; for instance, benign meningioma can compress medulla and cause cardiac arrest.
4. The lesion can spread along the subarachnoid space; however, there are very few tumors which metastasize outside of the CNS.
5. Different types of intracranial tumors can develop the same clinical manifestations, for example, epilepsia, paralysis defect of field vision, and high intracranial pressure.

15.7.1.1 Gliomas

Gliomas, which derive from glial cells, include astrocytic tumors, oligodendroglial tumors, and ependymal tumors. Astrocytic and oligodendroglial tumor growth occurs by progressive infiltration, invasion, and penetration of the surrounding tissue, however, ependymal tumor is inclined to be a solid tumor.

Astrocytic Tumor

Astrocytic tumor is the commonly occurring type of glioma and accounts for 30% of primary tumors, including various types of astrocytomas with different clinico-pathological characteristics: pilocytic astrocytoma (WHO grade I), subependymal giant cell astrocytoma (WHO grade I), pleomorphic xanthoastrocytoma (WHO grade II), diffuse astrocytoma (WHO grade II), anaplastic astrocytoma (WHO grade III), glioblastoma (WHO grade IV), and Gliaomatosis cerebri.

Genetic alterations have been identified in astrocytoma, for instance, in the inactivation of TP53, RB, P16^{INK4a} and the heterozygosity loss on chromosome 10. Inactivation of TP53 is the commonly occurring gene alteration in astrocytomas. PDGF-A and its receptor can be overexpressed. Transition from lower-grade astrocytoma to glioblastoma is associated with DCC gene disruption.

Macroscopically, tumors are gray, and they expand and distort the brain. Some tumors can seem to be demarcated from the surrounding brain tissue, however, infiltration beyond the outer margins is always present. The tumors vary in size from a few centimeters to enormous lesions, which can replace the entire hemisphere. The cut surface of the tumor is either firm or soft depending on the quantity of glial fiber; on occasion gelatinous, cystic degeneration and hemorrhage may occur.

On microscopic examination, the tumor cells have various sizes and shapes; different astrocytomas exhibit different characteristics of cell density, nucleus, vascular proliferation, and necrosis. Astrocytomas are GFAP-positive.

Diffuse astrocytomas account for 10–15% of astrocytic tumors and are characterized by good differentiation and slow growth. The average age of onset is 30–40 years old and usually occurs in the cerebral hemispheres, frontal lobe, and lobi temporalis. Common signs and symptoms include seizures, headaches, and focal neurologic deficits. Histologically, the tumor cells are well-differentiated, densely cellular, and show occasional nuclear pleomorphism (Fig. 15.8), with mitotically active cells being rare. Several different subtypes of diffuse astrocytomas are recognized, including fibrillary, protoplasmic, gemistovytic, and mixed cells type. The most common subtype is fibrillary astrocytoma. The mean survival time post-operation is 6–8 years. Diffuse astrocytomas tend to progress to more advanced grades, such as glioblastoma.

Fig. 15.8 Astrocytoma

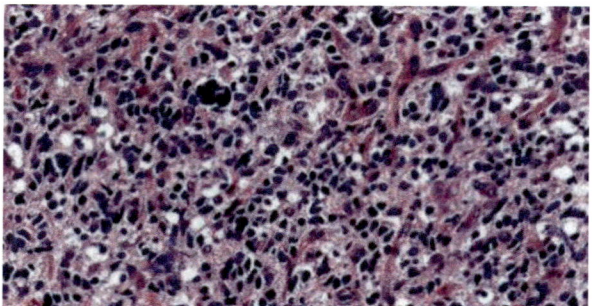

Fig. 15.9 Glioblastoma
(pseudopalisading)

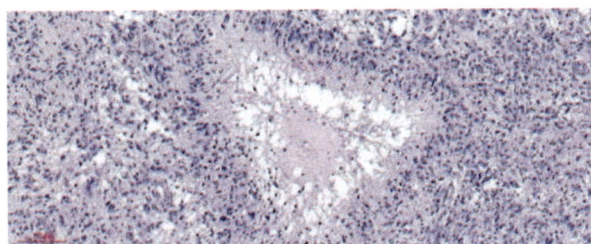

Anaplastic astrocytoma has a poor prognosis. The histologic characteristics are an increase of tumor cell density, evident nuclear atypia, nuclear hyperchromatism, increased mitosis, vascular endothelial cell proliferation, and so forth. Anaplastic astrocytoma tends to progress to glioblastoma.

Glioblastoma is the most aggressive cancer, which starts in the brain; it can be divided into primary and secondary categories and frequently occurs in the lobi temporalis, parietal lobe, frontal lobe, and occipital lobe. Histologically, it appears similar to anaplastic astrocytoma and is accompanied by necrosis and vascular proliferation. However, in contrast to anaplastic astrocytoma, necrosis in glioblastoma occurs in hypercellularity areas with highly malignant cells which aggregate along the edges of the necrotic regions, known as pseudopalisading (Fig. 15.9). Occasionally, vascular proliferation is characterized by tufts of piled-up vascular cells. When vascular cell proliferation is extreme, the tuft forms a ball-like structure, termed the glomeruloid body.

The worse prognosis correlates with rapid growth of the tumor. Mean survival time post-operation is only 12 months.

WHO grade I–II tumors (pilocytic astrocytoma, pleomorphic xanthoastrocytoma, and subependymal giant cell astrocytoma) usually have slow growth, relatively benign behavior, and favorable prognosis. The tumors are usually well-circumscribed with the surrounding brain, typically occurring in children and young adults. On microscopic examination, pilocytic astrocytoma is comprised of bipolar cells with long, thin "hairlike" processes; often present are Rosenthal fibers, eosinophilic granular bodies, and microcysts.

Oligodendroglial Tumor

Oligodendroglial tumors include oligodendroglioma and anaplastic oligodendroglioma. Oligodendroglioma, occurring most frequently in 40–45-year-old adults and derives from oligodendrocytes, is considered WHO grade II. The lesions are primarily located in the cerebral hemispheres, particularly in the frontal lobe. On macroscopic examination, oligodendrogliomas are gray masses, soft, well-circumscribed but invasive growth, often with cysts, focal hemorrhage, and calcification. On microscopic examination, the tumors are comprised of sheets of regular small and round cells with spherical nuclei (like normal oligodendrocytes) surrounded by a clear halo of cytoplasm. The tumor cells array dispersively, with an occasional formation of secondary structures, in particular with perineuronal satellitosis. It usually contains a delicate network of anastomosing capillaries, and can co-occur with different degrees of calcification and psammoma body formation. Histochemical and immunohistochemical staining shows that galactolipid, carbonic anhydrase isozyme C, CD57, MAP, or MBP are all positive. The most common genetic alterations in oligodendrogliomas are loss of heterozygosity for chromosomes 1p and 19q.

Oligodendroglioma is sensitive to chemotherapy. Because the tumor is slow growing, oligodendroglioma patients have an average survival of approximately 10 years post-operation; the most common presenting signs and symptoms are epilepsia or focal paralysis. In contrast to oligodendroglioma, anaplastic oligodendroglioma grows rapidly, and the mean length of survival time after operation is only 3.5 years.

Ependymal Tumor

Ependymal tumors include ependymoma and anaplastic ependymoma.

Ependymoma is considered as WHO grade and arises from the ependyma-lined ventricular system, including the oft-obliterated central canal of the spinal cord. It typically constitutes 2–9% of the neuroepithelial tumors, and most often occurs in children and adolescents. On macroscopic examination, ependymomas are gray masses, well-circumscribed, globular, or lobulated, often with cysts, focal hemorrhage, and calcification (similar to an oligodendroglioma). On microscopic examination, ependymomas are comprised of uniform cells, round to oval nuclei, with abundant cytoplasm. The most characteristic histological change is that tumor cells may form gland-like structures resembling the embryologic ependymal canal, with long, delicate processes extending into a lumen (Fig. 15.10). Perivascular pseudorosettes are more often present, whereby tumor cells are arranged around vessels with an intervening zone which consists of thin ependymal processes directed towards the vessel wall (Fig. 15.11). Anaplastic ependymomas occur with increased cell density, high mitotic rates, areas of necrosis, and less evident ependymal differentiation. Ependymomas grow slowly, and patients can survive for 8–10 years.

Fig. 15.10 Ependymoma

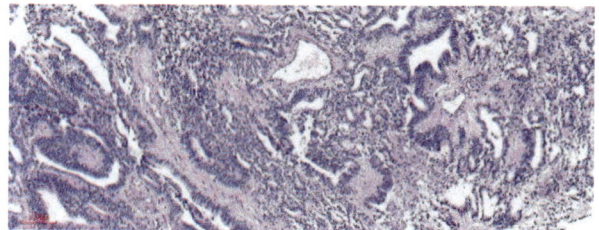

Fig. 15.11 Ependymoma

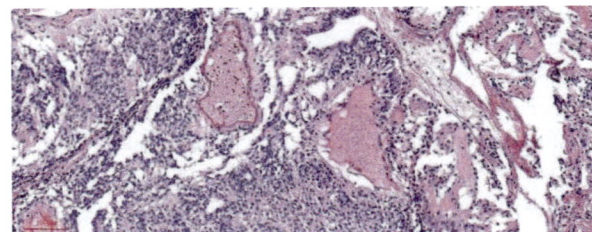

Fig. 15.12
Medulloblastoma

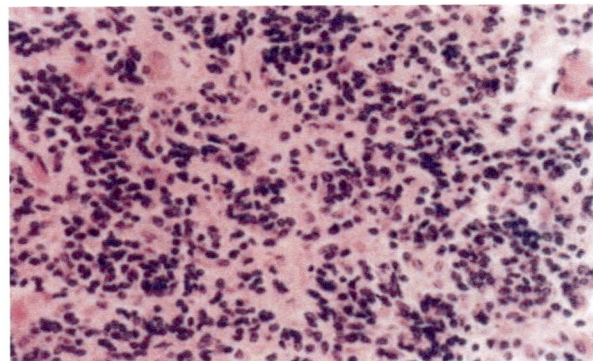

15.7.1.2 Medulloblastoma

The most common embryonal tumor of the CNS is medulloblastoma and occurs predominantly in children. It derives from the embryonal external granular layer cells of cerebellum or subependymal stromal cell. It accounts for 20% of the child tumors and is considered as WHO grade II. Up to 75% of medulloblastoma occurring in children is located in the vermis of the cerebellum and can extend to the fourth ventricle. On macroscopic examination, it usually presents as a soft, grayish mass. On microscopic examination, medulloblastoma is usually extremely cellular; tumor cells are small, round, or ovoid, with little cytoplasm and hyperchromatic nuclei. Occasionally, a pathological mitotic figure can be observed. The typical structure of medulloblastoma is that of the Homer Wright rosettes, characterized as tumor cells arranged around the center of the argyrophilic fiber (Fig. 15.12). Bidirectional differentiation of the tumor cells can be seen using an electron microscope. Moreover, the tumor can express neuronal (e.g., Syn and NSE) and glial (GFAP) phenotypes examined with immunohistochemical assay. The most common genetic alteration, occurring in medulloblastoma, is the isochromosome 17q and chromosome 17 trisomy. The tumor is highly malignant without a good prognosis. However, with operation and regular adjuvant therapy, the 5-year survival rate may be as high as 75%.

15.7.1.3 Neuronal and Mixed Neuronal-Glial Tumors

1. Gangliocytoma and gangliogima

    Most such tumors have slow growth, are well differentiated, and are considered as WHO grade I (gangliocytoma) or WHO grade I–II (gangliogima). However, the glial component of gangliogima can become anaplastic; then, anaplastic ganglioglioma occurs and behaves as grade III lesions. Gangliocytomas are well-circumscribed masses with a small tumor size, slightly hard in quality, grayish red cut surface, and have focal calcification and small cysts in the temporal lobe. On microscopic examination, neoplastic ganglion cells with various nuclei (mononuclear, binucleate, or multinuclear) are irregularly clustered and are randomly orientation of neurites. Additionally, plasmosome and Nissl bodies can be observed in tumor cells, and sometimes medullated fibers and amyelinated nerve fibers can also be mistaken amongst the tumor cells. Immunohistochemical staining of the tumor tissue shows that GFAP is positive of glial cells, and neuronal proteins, neurofilaments, and synaptophysin are positive of ganglion cells.

2. Central neurocytoma

    Central neurocytoma is WHO grade II with an average age of onset of 29 years old. It is usually located in the frontal part of the lateral ventricle, which occasionally extends to the lateral ventricle or third ventricle. On microscopic examination, central neurocytomas are comprised of densely uniform small cells, a round nucleus, and transparent cytoplasm. Pseudorosette can be seen. Synaptophysin and NeuN are positive in immunohistochemical staining.

15.7.1.4 Meningiomas

Meningiomas account for 15–26% of intracranial tumors and are the most common meningeal primary tumors originating from the meningothelial cells of the arachnoid. They are mainly benign tumors occurring in adults, with slow growth and which are easily excised. They seem to be the best prognosis in central nervous tumors and are WHO grade I.

On macroscopic examination, meningiomas are usually attached to the dura with expansive growth and can be located on the cerebrum convex surface. The tumors are usually rounded or lobulated masses in appearance and have a well-defined dural base compressing the underlying brain, although it can be easily separated from it. The mass surface is usually encapsulated with thin, fibrous tissue termed diolame. On the cut surface, the tumor is gray in appearance with a tenacious texture and occasionally gritty-textured; gross evidence of necrosis or extensive hemorrhage is almost not present. On microscopic examination, various histological types are found. These include syncytial, a concentric cell arrangement with a whorled appearance (Fig. 15.13), and the whorls can experience hyaline degeneration and become calcified with psammoma bodies (Fig. 15.13); fibroblastic, with elongated

Fig. 15.13 Meningiomas

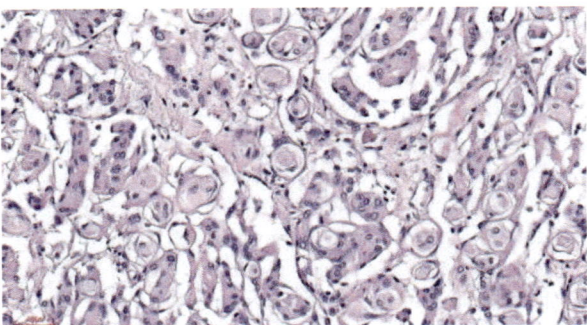

cells and abundant collagen deposition between them; and transitional, which share features of the syncytial and fibroblastic types. There are also rarer patterns. Anaplastic (malignant) meningioma, considered as WHO grade III, is a highly aggressive tumor which the appearance of a high-grade sarcoma. Mitotic rates are often extremely high and necrotic. All meningiomas express vimentin, and most express EMA.

15.7.2 Peripheral Nerve Tumors

In general, peripheral nerve tumors can be ascribed to two categories. One arises from neurilemmal, including schwannoma and neurofibroma; the other is accompanied by different degrees of nerve cell differentiation, such as neuroblastoma and ganglion cell neuroma. Schwannoma and neurofibroma are briefly described as follows.

15.7.2.1 Schwannoma

Schwannoma (neurilemoma), a benign tumor, arises from the neural crest-derived Schwann cell or lemmocyte and is considered WHO grade I. Neurilemoma is the most common tumor occurring in the canalis spinalis, accounting for 25–30% of intraspinal tumors. The most common location of schwannomas within the cranial vault occurs in the vestibular branch of the eighth nerve (termed the acoustic neurinoma) and the cerebellopontine angle. The most common location of schwannomas in the peripheral nerve is the flexor side nerve trunk of extremities.

Schwannomas are round or lobulated, well-circumscribed, encapsulated masses attached to a nerve, which they can be separated from. Tumors form hoar or luidity masses can have areas of cystic and hemorrhagic change. On microscopic examination, tumors display two growth patterns. In an Antoni A growth pattern, cells are spindle shaped and form a palisade or incomplete whirlpool, termed Verocay bodies (Fig. 15.14). In an Antoni B growth pattern, the tumor is less cellularly dense, with

Fig. 15.14 Schwannoma

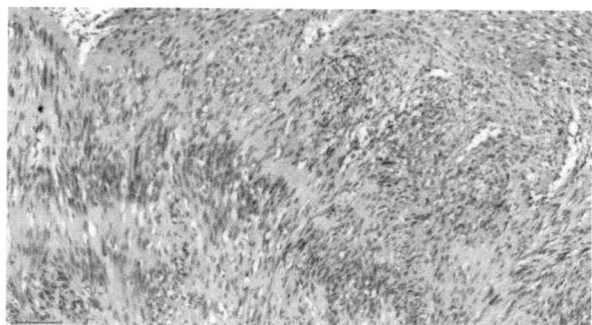

a loose meshwork of cells along with microcysts and myxoid changes. Two patterns of growth usually occur and can be observed in the same tumor, only one of which will express as the main change in different areas. For example, intracranial schwannoma primarily seems to be of the Antoni B growth pattern; however, an Antoni A pattern of growth will appear in neurilemmoma in the spinal canal. The Schwann cell origin of these tumors stems from their S-100 immunoreactivity. Collagen IV and laminin are also usually cell membrane positive.

Clinical features are usually related to the tumor's location and size. Patients who have a small tumor can be asymptomatic; however, if the tumor is larger, paralysis or pain can result from the compression of the nerve. For example, patients with acoustic neuroma often present with hearing loss and tinnitus. Although there are very few tumors which are attached to the brain stem or spinal cord, those which are usually cannot be completely excised, and therefore, there may be a relapse, as most neurilemoma can take radical operation and remain benign tumors.

15.7.2.2 Neurofibroma

Neurofibroma, considered as WHO grade I, can be sporadic or multiple and appears most commonly in the skin or subcutaneous tissue. Neurofibromatosis, or von Recklinghausen disease, is always accompanied by café-au-lait spot and axillary spots.

When macroscopically examined, skin lesions are evident as nodules or polypoids, well-circumscribed, nonencapsulated, diffusely infiltrating into the skin, and subcutaneous. On the cut surface where the tumor is gray and has a vortexed fiber, jelly material can be seen; bleeding and cystic degeneration rarely occur. The stroma of these tumors is highly collagenized and contains myxoid material. The appearance of lesions within peripheral nerves are identically histologic. On microscopic examination, the tumor is comprised of Schwann cells, perineurial-like cells and fibroblasts (interweaving each other), dispersed between nerve fibers. The stroma of these tumors is highly collagenized and contains little myxoid material.

Malignant peripheral nerve sheath tumor (MPNST), which accounts for 5% of soft tissue sarcoma, usually arises from peripheral neurofibroma (especially von

Recklinghausen disease). MPNST is a very aggressive tumor and considered as WHO grade II–IV. When microscopically examined, it is similar to fibroma sarcomatosum, showing more nuclear mitotic images accompanied by vascular proliferation and cell necrosis. Moreover, tumor cell pleomorphism can to be observed.

MPNST occurs most frequently in those aged 20–60 years old. The tumor is generally fast progressing and has a poor prognosis. Five- and ten-year survival rates are 34% and 23%, respectively.

15.7.3   Metastatic Tumors

Central nervous system metastatic tumors represent approximately 20% of intracranial tumors. The most common primary malignant tumor is lung cancer (responsible for approximately 50% of all metastases); the second is breast cancer (15% of all metastases), then melanoma, kidney cancer, gastrointestinal tract malignant tumors, choriocarcinoma, so forth. Metastatic tumors present as mass lesions and can sometimes be the first manifestation of cancer.

Hematogenous metastasis is the most common metastatic route of intracranial metastatic tumors. Cerebrum and endocranium are the sites usually involved in intracranial metastatic disease, whereas spinal metastasis often occurs in the epidural space, pia mater spinalis, and spinal cord. When examined microscopically, metastatic tumors are similar to primary tumors and usually occur together with hemorrhage, necrosis, cystic degeneration, and liquefaction. The boundary between tumor and brain parenchyma is also well defined.

References

1. Yulin L, editor. Pathology. 8th ed. Beijing: People's Medical Publishing House; 2013. p. 326–48.
2. Kumar V, Abbas AK, Aster JC. Robbins basic pathology. 9th ed. Philadelphia: W.B. Saunder; 2012. p. 811–49.
3. Qihui Z, Gengyin Z, editors. Pathology (bilingual textbook). Beijing: Peking University Medical Press; 2014. p. 407–31.
4. Tonghua L, editor. Diagnostic pathology. 4th ed. Beijing: People's Medical Publishing House; 2018. p. 1112–80.
5. Louis DN, Ohagaki H, Wiestler OD, et al. WHO classification of tumours of the central nervous system. 4th ed. Lyon: IARC Press; 2007. p. 10–251.

Chapter 16
Infectious Disease and Deep Mycosis

Miao Wang

Contents

Objectives

1. Master the basic pathological changes of tuberculosis, the composition of tuberculosis nodules, the type of tuberculosis, its morphological characteristics, and outcome.

M. Wang (✉)
Beijing Friendship Hospital affiliated to Capital Medical University, Beijing, China

© Zhengzhou University Press 2024
K. Chen et al. (eds.), *Textbook of Pathologic Anatomy*,
https://doi.org/10.1007/978-981-99-8445-9_16

2. Master the basic pathological changes of typhoid fever; understand the similarities and differences between typhoid fever and intestinal tuberculosis.
3. Master the basic pathological changes of bacillary dysentery and the composition of pseudomembrane.
4. Familiarize basic pathological changes of Leprosy, leptospirosis, hemorrhagic fever with renal syndrome and rabies.
5. Understand the basic pathological changes of sexually transmitted disease and deep mycosis.

Key Concepts
1. Infectious disease and deep mycosis are inflammatory diseases caused by different kinds of pathogenic organisms. They are characterized by the basic pathological changes of inflammation, which are alteration, exudation, and proliferation.
2. Tuberculosis is a chronic infectious disease caused by *Mycobacterium tuberculosis*. Granulomas with caseous necrosis is the main pathological feature. The lung is the most common site of tuberculosis. Pulmonary tuberculosis is divided into primary tuberculosis and secondary tuberculosis.
3. Typhoid fever is an acute systemic infection caused by *Salmonella typhi*. The main pathological change is the extensive proliferation of mononuclear phagocytes. It is most pronounced in the ileum.
4. Bacillary dysentery is an acute pseudomembranous inflammation caused by Shigella. Patients present with headache, fever, blood and pus in the stools, and tenesmus.
5. Leprosy is a chronic inflammatory disease caused by *Mycobacterium leprae*. The disease mainly involves the skin and peripheral nerves.
6. Leptospirosis is caused by Leptospira. The pathological changes of leptospirosis are acute systemic toxic injury and usually involves capillary, which results in distribution of circulation.
7. Rabies is a viral disease that mainly infects the central nervous system. The characteristic lesion is eosinophilic inclusion, called Negri body.
8. Gonorrhea is an acute suppurative inflammation, caused by the gram-negative diplococcus *Neisseria gonorrhoeae*.
9. Condyloma acuminatum is caused by Human papilloma viruses (HPV). In the genital tract, HPV Types 6, 11, 16, and 18 are especially common. Clear vacuolization of the prickle cells (koilocytosis) is a characteristic of HPV infection.
10. Syphilis is a chronic systemic infection, caused by spirochete, *Treponema pallidum*. The main pathologic change pattern is a type of vasculitis involving small arteries, termed obliterative endarteritis. In late syphilis, another pattern of tissue injury is a form of granulomatous necrosis, known as gumma.
11. Acquired syphilis may progress through three distinct phases: primary, secondary, and tertiary Syphilis. The primary lesion is termed as a chancre. The manifestations of secondary syphilis typically include a combination of generalized lymph node enlargement and a variety of mucocutaneous lesions. The charac-

teristic lesion of tertiary syphilis is the gumma. The lesions of tertiary syphilis can involve all of organ system, especially in cardiovascular and nervous system.

12. Mycoses are caused by fungi. They can infect the skin or deeper tissues, including major organ systems. Fungal infections usually lead to necrosis accompanied by acute and chronic mixed inflammation.

13. Candidiasis is caused by genus *Candida*. There are three common presentations: lesions of skin and mucous membranes; deep tissue infections; systemic candidiasis. Histologic sections showed that yeast forms and pseudohyphae were the characteristics of fungus.

14. Cryptococcosis is a systemic fungal infection caused by *Cryptococcus neoformans*. More than 95% of cryptococcal infections involve the meninges and the brain.

15. Aspergillosis is an opportunistic infection caused by Aspergillus. Lung disease is the most characteristic of Aspergillosis.

16. Mucormycosis is an acute suppurative inflammation caused by fungi Mucor, Rhizopus, or *Absidia* species organisms. It is commonly characterized by hyphae growing in and around vessels.

Introduction

Infectious diseases are a group of diseases caused by pathogenic organisms and can be epidemic among certain population. In 1999 WHO (World Health Organization) report, infectious diseases are currently the world's biggest killer of children and young adults. They account for more than 13 million deaths a year, which is half of all deaths in developing countries. In developed countries, infectious disease also plays an important role in morbidity and mortality, secondary to noninfectious diseases, such as atherosclerosis, cancer, and senile dementia [1]. In recent years, due to the application of gene diagnosis technique and effective antibiotic, diagnosis and treatment of infectious diseases have been improved.

Etiology: Most microorganisms are harmless. A few of microorganisms are pathogenic, that is, can cause disease. (1) Viruses themselves lack protein synthetic machinery; they can only replicate within the host cells. However, viruses can induce host cellular damage, death, and malignant transformation. (2) Bacterial endotoxins and exotoxins are two important causes of bacterial injury. Endotoxins are derived from their lipid components, especially from gram-negative bacteria. Endotoxins can damage endothelial cells and disturb the coagulation system as well as cause cell necrosis, fever, and endotoxic shock. Endotoxins cannot be neutralized by antibodies. Exotoxins are proteins secreted by living bacteria. They often have specific enzyme activities and may act on specific tissues or organs. Exotoxins can be neutralized by antibodies. (3) Spirochetes: Leptospirosis and syphilis are typical spirochete-caused infectious disease. (4) Fungi: Mycotoxins (aflatoxins) produced by some fungi can increase the risk of cancer of the liver.

Transmission: There are three necessary factors for transmission of infectious disease: source of transmission, routes of transmission, and susceptible host. Microorganisms are usually passed from humans to humans and from animals to humans by oral, fecal, respiratory, blood-born, or sexual transmission. In hospitals, infections are often spread to patients by air, water, and in blood products.

Pathogenesis: Tissue damage may be due to the action of toxic substances released by reactive inflammatory cells. Generally, infectious agents establish infection and damage tissues in three ways: (1) They directly cause cell death by contact or entering the host cell. (2) They release endotoxins or exotoxins that kill cells. Exotoxins are enzymes that degrade tissue components. Endotoxins damage blood vessels and cause ischemic necrosis. Nontoxic organisms: These do not produce substances that damage tissue. Disease results from tissue reactions to the multiplication of the organism.

Host factors affect the pathogenesis of infectious diseases. Resistance may be "natural" or "acquired" with disease occurring only when barriers to infection are penetrated and protective mechanisms are overcome. Host cellular responses help to overcome the infection by pathogenic bacteria but, at the same, contribute directly to tissue damage. Active or passive immunization also affects resistance. Certain diseases, drugs, radiation, or other iatrogenic factors may reduce resistance. Inherited or acquired susceptibility to disease may be caused by deficiencies in complement components, leading to inadequate inflammatory responses to infection.

Diagnosis of tissue damage: (1) Morphologic appearances are often nonspecific. (2) Pathogenic microorganisms can be identified by culture of the organism, DNA probe (in-situ hybridization), the polymerase chain reaction (PCR), or immunologic methods.

16.1   Tuberculosis

16.1.1   Overview

Tuberculosis (TB) is a chronic, communicable disease caused by *Mycobacterium tuberculosis*. TB typically attacks the lungs but can also affect other parts of the body. When an active TB patient coughs, sneezes, or spreads respiratory fluid through the air, it is transmitted through the air. Most infections are asymptomatic and latent, but about one tenth of latent infections eventually develop into active diseases, and if untreated, more than 50% of infected people will die [1].

TB is estimated to affect 1.7 billion individuals worldwide, with 8 to 10 million new cases and 1.7 million deaths each year. After HIV, TB is the leading infectious cause of death in the world. Infection with HIV makes people susceptible to rapidly progressive TB; over 50 million people are infected with both HIV and *M. tuberculosis*.

Etiology and Pathogenesis

Patients with pulmonary TB who cough up bacilli in the sputum are the source of transmitted infections. Bovine TB typically involves the oropharynx or intestine because the organism is ingested in milk. *Mycobacterium bovis* infects dairy products and causes human infection through contaminated milk. Most infections are acquired by direct person to person transmission of microbial droplets in the air. TB can also occur through blood infection. Hematogenous transmission can spread infection to more distant sites, such as peripheral lymph nodes, the kidneys, the brain, and the bones. All parts of the body can be affected by the disease, though for unknown reasons it rarely affects the heart, skeletal muscles, pancreas, or thyroid.

Components of the *M. tuberculosis* cell wall (such as cold factor, wax D, complement, heat-shock protein, etc.) and host response (immune response and delayed hypersensitivity) contribute to its pathogenicity [2]. The earliest phase of primary TB (<3 weeks) in the no sensitized individual is characterized by bacillary proliferation with the pulmonary alveolar macrophages and airspaces, which result to bacteremia seeding of multiple sites. The development of cell-mediated immunity occurs approximately 3 weeks after exposure. In the draining lymph nodes, processed mycobacterial antigens are presented in a major histocompatibility class II context by dendritic cell macrophages to CD4+ TH1 cells, which are capable of secreting IFN-γ and activate macrophages. Activate macrophages, in turn, release a variety of mediators with important down-stream effects, including (a) secretion of tumor necrosis factor (TNF), which is responsible for recruitment of monocytes, which in turn undergo activation and differentiation into the "epithelioid histiocytes" that characterize the granulomatous response; (b) expression of the inducible nitric oxide synthase (iNOS) gene, which results in elevated nitric oxide levels at the site of infection. Nitric oxide is a powerful oxidizing agent and results in generation of reactive nitrogen intermediates and other free radicals capable of oxidative destruction of several mycobacterial constituents, from cell wall to DNA; (c) generation of reactive oxygen species that can have antibacterial activity. In summary, immunity to a tubercular infection is primarily mediated by CD4+ TH1 cells, which stimulate macrophages to kill bacteria. This immune response, while largely effective, comes at the cost of hypersensitivity and the accompanying tissue destruction. Reactivation of the infection or re-exposure to the bacilli in a previously sensitized host results in rapid mobilization of a defensive reaction but also increases tissue necrosis [3].

The clinical manifestations of the infection of TB are decided by different reactions. Three considerations are involved in the pathogenesis of TB, including quantity of bacteria, virulence of pathogen, and responsiveness of the body sensitivity and immune response. If TB bacteria gain entry to the bloodstream from an area of damaged tissue, they can spread throughout the body and set up many loci of infection, all appearing as tiny, white tubercles in the tissues. This severe form of TB disease, most common in young children and those with HIV, is called miliary TB. People with this disseminated TB have a high fatality rate even with treatment (about 30%). In many people, the infection waxes and wanes. Tissue destruction and necrosis are often balanced by healing and fibrosis. Affected tissue is replaced by scarring and cavities filled with caseous necrotic material. During active disease, some of these cavities are connected to the trachea, which can be coughed out. It contains living bacteria and can spread the infection. Treatment with appropriate antibiotics can kill bacteria and heal wounds. After healing, the affected area will eventually be replaced by scar tissue [4].

Basic Morphology of TB

1. Exudation: Acute inflammatory exudates can occur when hypersensitivity is very pronounced, such as with early infections and high bacterial virulence. Exudates appear as serous or serofibrinous inflammation. The immediate reaction causes edema, hyperemia, and neutrophilic infiltration. However, most of the cells are mononuclear phagocytes and lymphocytes. This kind of inflammation mainly occurs in the lungs, serous membranes, joints, and meninges. *Mycobacterium tuberculosis* can be seen in exudates and macrophages. The exudate can be completely absorbed or become necrotic or proliferative.

2. Proliferation: The most characteristic lesion in TB is the granuloma with central necrosis. The lesion depends on the balance between proliferation of the tubercle bacillus and the intensity of the host immune response (granuloma formation). When proliferation of the microorganism is predominant, that will be many macrophages containing numerous bacilli; the macrophages will be loosely arranged. This is a common presentation of infection by *Mycobacterium avium* intracellularly. When there is little proliferation of the organism and the host immune response is very strong, there are few detectable organisms and numerous well-formed granulomas with central necrosis; this is a common appearance with infection by *M. tuberculosis* hominis or *M. bovis*.

TB granuloma is a characteristic of TB diagnosis (Fig. 16.1). It consists of aggregates of epithelioid cells, Langerhans giant cells, fibroblasts, and lymphocytes. Caseous necrosis frequently is presented in the center, which shows red-stain amorphous granular substance containing tubercle bacilli. Monocytic infiltrates appear initially. They engulf bacteria and, after about 2 weeks, develop delayed hypersensitivity and epithelioid conversion. Tubercles are characterized by formation of granulomas containing compact collections of mononuclear cells called epithelioid cells, which is spindle shaped or polygonal, with abundant cytoplasm, pale eosinophilic, and ambiguous realm. The epithelioid cell

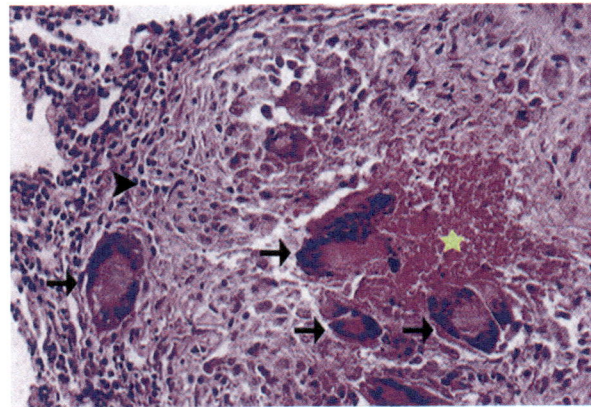

Fig. 16.1 Tuberculosis granuloma. Nodule consists of collection of epitheliod cells with several Langhans' giant cells and lymphocytes. Caseous necrosis is presented in the center

nucleus is round or oval, chromatin understood, even vacuolization, and may have one to two nucleolus. These epithelioid cells arrange in concentric layers. The increase of epithelioid cell activity is conducive to the phagocytosis and killing of *M. tuberculosis*. The majority of epithelioid cell can fuse together and transform into Langerhans' giant cells, which are giant multinucleated giant cells (up to 300 μm in diameter) and rich in cytoplasm. The number of nuclei varies from a dozen to more than a 100 and arrange in garland or horse's hoof form or assembled together. The typical tubercles are composed of lymphocytes with epithelioid cells and Langhans' giant cells in the center. When tubercles coalesce, the center is often necrotic, replaced by eosinophilic, structureless granular material (caseous necrosis), and some lymphocytes. Granulomas with central necrosis are characteristic of TB and have diagnostic value. Fibrous scarring develops around caseous tubercles. With modern therapy, all residual evidence of previous disease may disappear except for a residual scar. Calcification commonly occurs.

3. The single nodule is very small, with a diameter of about 0.1 mm, which is difficult to detect by naked eyes and X-ray. Three or four nodules can be fused into one large nodule. Grossly, the lesions often resemble well-circumscribed nodules. On cut surface, these are gray-white or yellow, homogenous, or greasy. They have necrotic centers with a dry or greasy cheese-like (caseous) appearance. The center could be liquid. Liquefaction may facilitate dissemination of TB in the body.

3. Necrosis: The lesion of necrosis appears when there are numerous organisms or high virulence and the people are in low host-immunity state or strange hypersensitivity reaction. Necrotizing lesions are called caseous necrosis because they are yellowish or milky white, soft and brittle, and generally resemble dried cheese. Microscopically, caseous necrosis showed red strained homogeneous amorphic material or some granular material with cell debris. Caseous necrosis is of equal diagnostic importance in TB. Large areas of caseous necrosis are a sign of TB progression.

These three basic pathological processes are interrelated and also can interchange according to host conditions.

Transformation of Basic Lesions

1. Healing or repair

   (a) Absorption and dissipation are main healing ways. The effusion is absorbed through lymphatics vessel. Small necrosis lesions can be absorbed and dissipated by energetic therapy. Chest X-ray shows that the lesions are dense inhomogeneous cloud shadow with unclear edge. Clinically, the patients are called on the mend and take a turn for cure.

   (b) Fibrosis, fibrous encapsulation and calcification of lesions are main repair ways. Tubercle with small necrosis can undergo fibrosis and heal by organization. Large necrotic focus can be encapsulated by peripheral fibrosis tissue and the caseous center condensed to a hard calcified nodule. Chest X-ray shows the fibrotic lesion is stripe or star-like scar with clear edge, and the density of calcified lesions is higher in the center. Clinically, it is called calcification induration stage.

2. Exacerbation

   (a) Infiltration. When TB gets worse, the lesion appears exudation and necrosis around focus. X-ray examinations show lesion shadow and unclear margin. Clinically, it is called infiltration and extension stage.
   (b) Dissolution and dissemination. Liquefied necrosis excludes mainly through natural tracts, such as branchi, urinary tract, and so on. Cavity necrosis spreads by branchi, lymphatics, or blood stream and results in some new lesions in other sites. X-ray shows uneven shadow density. The bright region and new lesions can be observed. Clinically, it is called dissolution and spread stage.

16.1.2   Pulmonary TB

The lung is more susceptible to TB than other organs. The pattern of host response depends on whether the infection represents a primary first exposure to the organism or secondary reaction in already sensitized host. Pulmonary TB is classified to primary pulmonary TB and secondary pulmonary TB [1].

16.1.2.1   Primary Pulmonary TB

Primary TB is the form of disease that develops in a child who has not been previously exposed to tuberculous bacillus or immunized by Bacillus Calmette-Guérin (BGC) vaccination. Therefore, it is also called the childhood type TB, but primary lesions can also occur in the elderly and immunocompromised people, who may lose sensitivity to tubercle bacillus and may develop primary TB more than once.

Primary TB is characterized by formation of a Primary complex (also called a Gohn complex) (Fig. 16.2). Usually, the affected part of the lung is subpleural. In the middle of the lung, the lymph nodes extend to the hilar lymph nodes. The lesion consists of three components: primary lesion (Ghon focus), tuberculous lymphangitis, and tuberculous lymphadenitis (in the hilar lymph nodes). Primary lesion is usually a 1–1.5 cm area of round, gray-yellow inflammatory consolidation, which locates in lower segment of the upper lobe or the upper segment of the lower lobe, close to the pleura. The center of Ghon focus usually undergoes caseous necrosis. Tubercle bacilli can drain through lymphatic vessel to the regional hilar lymph nodes, which result in tuberculous lymphangitis and tuberculous lymphadenitis. Chest X-ray shows primary complex lesions are dumb bell-like shadow, in which the lymphangitis is in strip-like shape.

Fig. 16.2 Primary complex. The lesion consists of three components: primary lesion (Ghon focus), tuberculous lymphangitis, and tuberculous lymphadenitis (in the hilar lymph nodes)

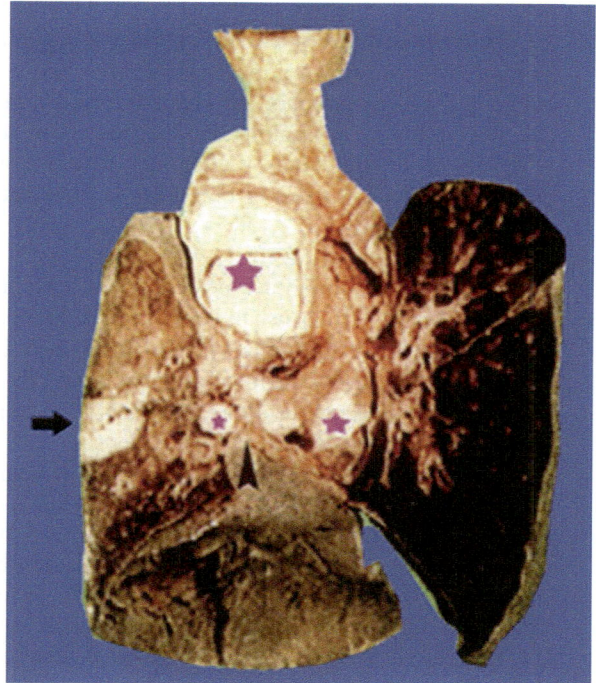

Primary pulmonary TB is usually asymptomatic or manifested as a mild flu-like illness. In the first few weeks, it is also lymphatic and hematogenous dissemination to other parts of the body. In 98% of cases, immunity stops disease progression and healing occurs. The lesions heal by fibrosis and may calcify; despite seeding of other organs, no lesions develop. In 2% of cases, rapidly progressive pulmonary disease causing extensive caseous consolidation of the lung usually occurs only in malnourished or immunodeficient children [1].

16.1.2.2 Secondary Pulmonary TB

This form of TB occurs in people who have immunity to TB before the infection occurs. It usually occurs in an adult as a result either of reinfection or reactivation, with the latter more common. Secondary TB tends to produce more damage to the lungs than primary TB. The subsequent course of the secondary lesions is variable. It is classified to following types according to the pathological features and clinical course.

1. Focal pulmonary TB

    It is the earliest lesion of secondary pulmonary TB and can be derived from the action of the primary or more often secondary period of TB. The most common site is the lung apex, one or more small focus of consolidation. The lesions are one or more small apical foci, usually a 0.5–1.0 cm in maximum diameter, well-circumscribed, and grayish-white to yellow in color. Histologically, small epithelioid cell granulomas are characterized by caseous necrosis and fibrosis. Patients usually are asymptomatic. They may spontaneously heal or treat, leading to fibrous calcified nodules, remaining stationary, or progressing. Transmission through lymphatic vessels or blood can be prevented by allergic reactions and cellular immunity.

2. Infiltrative pulmonary TB

   This is the most common type of secondary TB and usually transformed from focal pulmonary TB. It usually occurs by infection through the airways after inhaling sputum from another person with open TB. The lesions are usually located in upper part of the lungs (subclavicular or apical region). Active cellular immunity causes acute reactions with marked caseous necrosis, which cause the lesion enlargement. On chest X-ray, there are soft "woolly" infiltrates, which show the cloud like shadow with indistinct boundary under the clavicle. When the disease develops progressively, the caseation lesions enlargement can result in irregular acute cavities. Pneumothorax erosion makes the central cheese like material empty, resulting in spontaneous pneumothorax or tuberculous empyema. The erosion of the bronchus will make the cheese like center disappear and form a rough and irregular cavity lined with cheese like material, which is difficult to separate by fibrous tissue. Blood vessel corrosion causes hemoptysis. Clinically, patients can have toxic symptoms, with chronic cough and hemoptysis. With adequate treatment, the process may be arrested, although healing by fibrosis often distorts the pulmonary architecture. If healing does not occur, there may be progressive enlargement of the lesion with liquefaction necrosis, cavitation; and bronchogenic spread. The lesions may penetrate the pleura, cause pneumothorax or empyema, or transform into chronic fibrous caseous TB with cavity.

3. Chronic fibro-cavitative pulmonary TB

   It is often result from infiltrative pulmonary TB and is a most common form of chronic pulmonary TB in an adult. This is a more severe reaction to infection than infiltrative pulmonary TB. It may affect one, more, or all lobes of both lungs. The upper lobes of lung contain multiple thick-walled chronic cavities of varying size (Fig. 16.3). The thickness of their walls can be more than 1 cm. Histologically, the wall of cavity is lined by a yellow-green caseous material, tuberculous granulation tissue, and fibrous tissue successively. There may be coexisting bronchial disseminated tuberculous lesions and diffuse fibrosis in the pulmonary tissues. Old and new lesions often coexist. The upper lung has more lesions and older lesions than the lower lung. In the late period, the lung becomes small, and indurated with pleural extensive adhesion, and the function of the lung may be severely damaged.

   Chronic fibro-cavitative pulmonary TB is called open TB because the sputum is often infective. Infection is transmitted through infected sputum and can lead to TB of the larynx and intestines. Patients may have hemoptysis if the necrosis

Fig. 16.3 Chronic
fibro-cavitative pulmonary
tuberculosis. The upper
lobe of the lung contains
multiple thick-walled
chronic cavities of varying
size. The thickness of their
walls can be more
than 1 cm

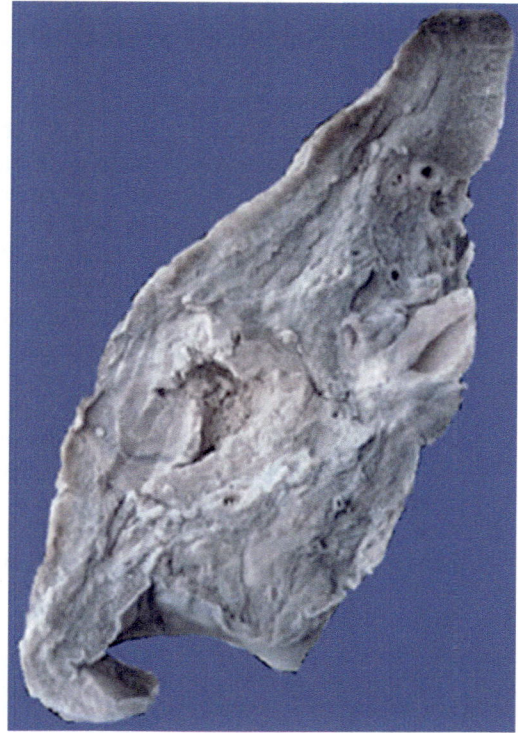

involves blood vessels. The habitus of the patient with long time proceeding of fibrous cavernous pulmonary TB is specific and is named habitus phthisicus. Pulmonary congestion and acrocyanosis are evident. In the final stage, patients develop cor pulmonale due to pulmonary hypertension.

Through the combined application of a variety of anti-TB drugs, the patient's resistance is enhanced and the smaller cavity is contracted. The bigger cavity wall of granulation tissue is gradually replaced by fibrous scar tissue and covered by bronchial epithelium. Thus, although the cavities still exist, they contain no microphyte and are called openly healing.

4. Caseous pneumonia

It represents rapid serious progression of TB. This usually occurs in debilitated immunodeficient or highly sensitized patients. Dissemination of large numbers of organisms in the focus via the bronchial tree can spread rapidly throughout large areas of lung parenchyma and producing a diffuse bronchopneumonia or lobar exudative consolidation ("galloping consumption"). Grossly, one lobe or an entire lung may be affected (Fig. 16.4). It is characterized by extensive caseous necrosis. According to the size of the lesion, caseous pneumonia is classified to lobular caseous pneumonia and lobar caseous pneumonia. Microscopically, there is marked cellular infiltration and necrosis. A large number of tubercle bacilli are usually present. The alveoli are filled with an exudate that contains monocytes and lymphocytes. Patients are in dangerous.

5. Tuberculoma

   The coalescence of caseous necrosis can form a large solid spherical mass (usually 2–5 cm in diameter) (Fig. 16.5). They may be single or multiple and commonly in the upper lobe. Lesions are usually well defined. On chest X-ray, it is easily mistaken for a tumor. This pattern may be caused by infiltrative-pneumonic and focal TB. The prevalence of tuberculoma among all forms of pulmonary TB is 6–10%. In the treatment, the body's resistance increases, causing the lesions become limited, concentrated, and no longer progress. However, this process does not fully heal and accurately delineates the tight formation residuals. Therefore, tuberculoma represents a quiescent disease.

6. TB pleuritis

   It can be divided into moist tuberculous pleuritis (exudative tuberculous pleuritis) and dry tuberculous pleuritis (proliferative tuberculous pleuritis).

   Moist tuberculous pleuritis is an inflammation characterized by exudation. In exudative pleurisy, the basic changes of protein ratio are different. The examination of pleural effusion is very important to determine the nature and etiology of pleural effusion. Serous tubercular effusion is usually transparent, yellowish

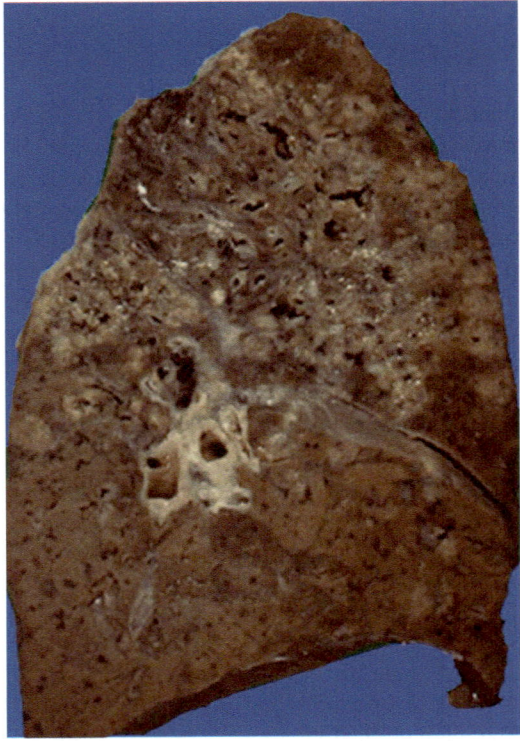

Fig. 16.4 Caseous pneumonia. One lobe of the lung is affected

Fig. 16.5 Tuberculoma. The coalescence of caseous necrosis form a large solid and spherical mass (usually 2–5 cm in diameter)

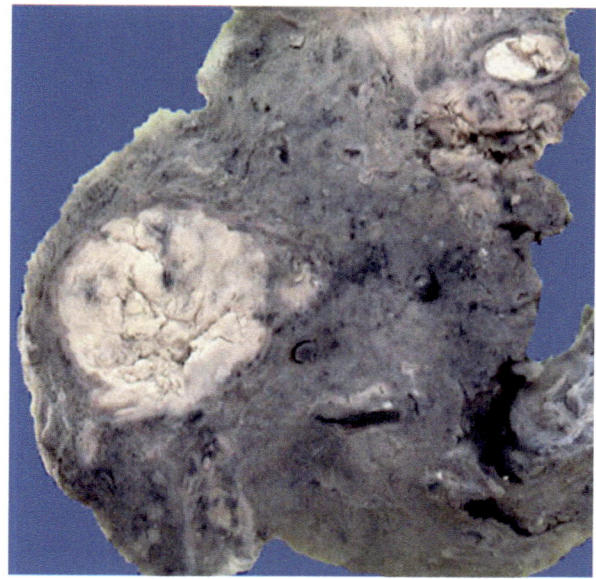

color, with densities from 1015 up to 1025, and contents of protein from 3% to 6%. At the acute stage of exudation, the pleural effusion was dominated by lymphocytes (50–60%), with a small number of eosinophils, erythrocytes, and mesoderm cells. With adequate therapy, effusion may be absorbed. If effusion contains much of the cellulose, it is difficult to absorb and may cause thicken and adhesion of pleura because of organization.

Dry tuberculous pleurisy is a kind of inflammation which is dominated by proliferative lesions. These lesions are usually located in the apex of the lung with granuloma and do not involve the formation of pleural effusion. The basic clinical displays are pain in the chest, dry cough, infringement of general condition, and subfebrile temperature. The location of the pain depends on the site of the injury. When breathing deeply, coughing, and pressing intercostals, the pain is aggravated. During the physical examination of the patient, the injured part of the chest was delayed when breathing, and there was a faint dull sound during percussion. Dry pleurisy can heal at the end of treatment. Sometimes, it gets relapsing character.

16.1.2.3 Miliary TB

Miliary pulmonary disease occurs when organisms drain through lymphatics into the lymphatic ducts, which empty into the venous return to the right side of the heart and thence into the pulmonary arteries. Individual lesions are either microscopic or small, visible (2 mm) foci of yellow-white consolidation scattered through the lung parenchyma (the word miliary is derived from the resemblance of these foci to millet seeds). Miliary lesions may expand and coalesce to yield almost total consolidation of large regions or even whole lobes of the lung. With progressive pulmonary TB, the pleural cavity is invariably involved and serous pleural effusions, tuberculous empyema, or obliterative fibrous pleuritis may develop.

Acute Systemic Miliary TB

It results from the hematogenous dissemination of pulmonary lesions, where large numbers of mycobacteria usually enter one of the pulmonary veins, pass through the left heart, enter the arterial systemic circulation, and spread to the general various organs. The gross appearance is multiple, scattered, about 1–2 mm in diameter, fairly uniform in size, round, and well-defined gray-white nodules (Fig. 16.6). Microscopically, the nodules were mainly proliferative, often without giant cells but with central necrosis. Occasionally, the lesions are exudative, with necrosis in the center. Clinically, the patients present with high fever, night sweats, loss of appetite, hepatomegaly, and splenomegaly. With adequate treatment, the prognosis is favorable. Some patients will die from tuberculous meningitis.

Fig. 16.6 Military tuberculosis of lung. The lesions are multiple, scattered, approximate 1–2 mm in diameter size, fairly uniform size, round, gray-white, and well-demarcated nodules in the lung

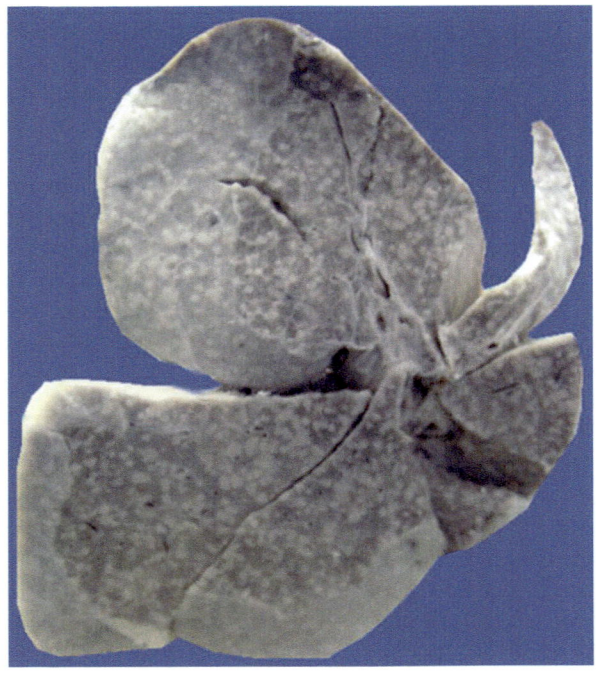

Chronic Systemic Miliary TB

Acute systemic miliary pulmonary TB was not cured after 3 weeks or when *M. tuberculosis* invaded the blood for a long time. In gross appearance, the lesions are uneven in the organ. Microscopically, nodules are often accompanied by exudation, hyperplasia, and necrosis. The patients were mainly adults with a long course.

Acute Pulmonary Military TB

When the caseous necrosis of hilar lymph node is involved directly or through thoracic duct, it enters the systemic vein and spreads to the lung through the right heart and pulmonary artery. The pulmonary lesions in this condition are part of acute systemic miliary TB. In gross appearance, both lungs are in congestion with dark-red in color on the cut surface and numerous gray or gray-yellow miliary nodules in size rising on surface of lung. Individual lesions are either microscopic or small.

Chronic Pulmonary Miliary TB

It often occurs in adults when tubercle bacillus in lesions of extrapulmonary-TB drains into blood periodically. The course of disease is long, and the lesions are new or old lesion and small or big together. The main pathological change is proliferation.

16.1.3 Extrapulmonary TB

It is usually caused by blood borne transmission of pulmonary infection. It can also occur after ingestion of *M. tuberculosis*, such as in milk. Any organ can be infected, for example, brain, liver, gastrointestinal tract, spleen, kidney, adrenal glands, bones, and genitourinary organs.

16.1.3.1 Intestinal TB

Primary intestinal TB usually occurs in children who drink contaminated *M. tuberculosis* milk. It may develop to intestinal primary complex: (1) a primary intestinal lesion, (2) intestinal tuberculous lymphangitis, and (3) mesenteric tuberculous lymphadenitis. Secondary intestinal TB usually occurs in adults (21–40 years old) and most often occurs in the ileocecal region. There are two morphological forms of the disease: the ulcerative type and the hyperplastic type.

Ulcerative Intestinal TB

The organisms are trapped in mucosal lymphoid aggregates of the small and large bowel, which then undergo inflammatory enlargement with ulceration of the overlying mucosa, particularly in the ileum (85%). Grossly, irregular ring- or band-shaped ulcers can be seen in the mucous membrane with some caseation on the bottom, and these ulcers are perpendicular to the long axis of the ileum. Microscopically, caseous necrosis occurs on the serous membrane surface, and granulation tissue is below the necrosis. The most common complication of ulcerative intestinal TB is intestinal stenosis caused by scar contraction and adhesion with adjacent tissues. Perforation and hemorrhage occur from the erosion into peritoneal and vessels in the ulcer base.

Proliferative Intestinal TB

The type disease only occurs in about 15% of patients. It is characterized by formation of tuberculous granulation tissue and fibrous tissue proliferation in the intestine, which leads to the intestinal wall thicken. Usually, the clinical symptoms are intestinal cavity stenosis and obstruction. It is also characterized by formation of multiple polyps and the diagnosis should be differentiated from carcinoma.

16.1.3.2 Tuberculous Meningitis

It is caused by bloodstream spread of *M. tuberculosis*. In children, tuberculous meningitis is usually the result of bloodstream spread of primary pulmonary TB. In adults, most cases are result from TB of lung, bone, joint, urinary, and genital system. Some cases are caused by rupture of tuberculoma in brain.

The base of the brain is usually the most obviously affected region. The brain surface is covered with small tubercles. Yellow gelatinous exudate accumulates in the subarachnoid space of pons, interpedunculare cistern, optic chiasm, and sylvian fissure. There is only a moderate increase in cellularity of the pleiocytosis, which is made up of mononuclear cells or a mixture of polymorphonuclear and mononuclear cells. It may also result in a well-circumscribed intraparenchymal mass (tuberculoma), which may be associated with meningitis. Meninges adhesion may obstruct the flow of cerebrospinal fluid and lead to hydrocephalus. When the inflammation is in the brain stem subarachnoid area, cranial nerve roots may be affected. The symptoms will mimic those of space-occupying lesions. Fever and headache are the cardinal features. Confusion is advanced symptom but coma bears a poor prognosis will be poor prognosis.

16.1.3.3   TB of Genitourinary System

Renal TB is usually a blood-borne infection from TB of the pulmonary. The tubercle lesions often occur at the cortico-medullary junction or in renal pyramids and papillae. With the enlargement of tubercles, the caseous necrosis coalesces into necrotic irregular cavities and results in nonfunctioning kidneys. On the cut surface, kidney shows several cavities in different sizes with coarse wall, yellowish-white caseous necrosis attaches inside (Fig. 16.7). Microscopically, tuberculous lesion is a collection of reticuloendothelial cells, phagocytosis, and giant cell formation, which is surrounded by infiltration and thin collagenous fibers. Necrotic material is often discharged through the ureter. Calcification is uncommon. The entire kidney may become infected. Severe cases involve the entire bladder and deep layers of muscle are eventually replaced by fibrosis tissue. Due to loss of the anti-reflux mechanism, the ureteral orifice fibrosis can lead to stricture formation with hydronephrosis or scarification with vesicoureteral reflux.

The male and female genital organs are usually infected by hematogenous dissemination. In the male, infection of the seminal vesicles and prostate can also occur from the urinary tract. Common patterns of infection include tuberculous salpingitis (infection of the fallopian tubes), tuberculous endometritis, tuberculous prostatitis, and tuberculous epididymitis. Epididymis TB can result in male sterility. Females are more frequently affected. TB of the uterus, fallopian tubes and ovaries are most common. TB of fallopian tubes is one of the causes of female sterility.

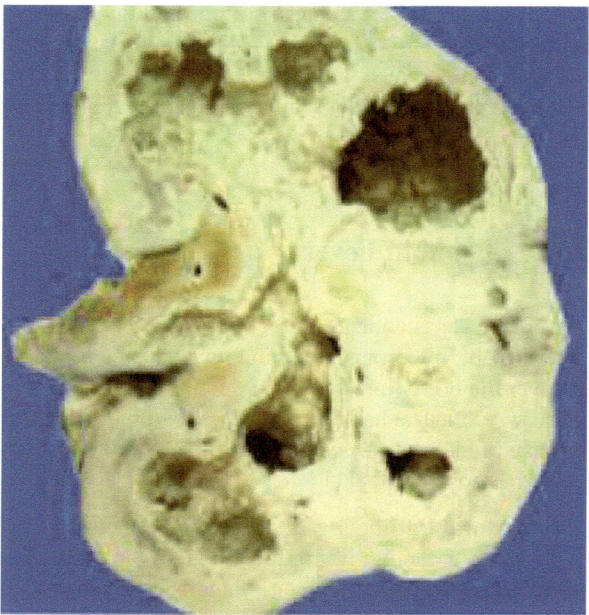

Fig. 16.7 Tuberculosis of kidney. Several cavities in different size with coarse wall and yellowish-white caseous necrosis inside of the kidney

16.1.3.4   TB of Bones and Joints

TB of bones and joints often occur in children and teenagers by hematogenous dissemination.

Bone infection complicates an estimated 1–3% of cases of pulmonary TB. Long bones and vertebrae are most common sites. The lesions are often solitary but also can be multicentric, particularly in patients with underlying immunodeficiency. The lesion usually erodes the peripheral soft tissue and results in the formation of "abscess," which is the liquefaction of necrotic material formed by the side of the bone. Because the "abscess" is not red, hot, and painful, it is named "cold abscess." When abscess erodes into the skin, a sinus tract forms. TB of the vertebral bodies is an important form of osteomyelitis. Infection at this site causes vertebral deformity and collapse, with secondary neurologic deficits.

TB involves joints that matters most spine, hips, knees, feet, elbows and occasionally other joints as well. The patients are young adults or children over 6 years. Spinal TB has a tendency to spread along fascial planes (psoas abscess). In a synovial joint, the disease starts from the synovium and grows slowly over the cartilage; it then extends through the cartilage into the underlying bone, which decalcifies. Sometimes, cold abscesses and sinuses are formed.

16.1.3.5   Tuberculous Lymphadenitis

It is common in children and young adults. The cervical lymph node is the most frequent form. Structure of lymph node is destroyed and replaced by tubercles in different size. The characteristic morphological element is the tuberculous granuloma. There is caseous necrosis at the center of the tubercle surrounded by epithelioid cell, Langerhans giant cell and lymphocytes. Lymphadenitis usually affects the localized group of nodes. Enlarged and adherent lymph nodes can form a big lump (Fig. 16.8).

16.2   Typhoid Fever

Typhoid fever is an acute infectious disease caused by *Salmonella typhi*. It is characterized by hyperplasia of mononuclear-macrophages in the reticuloendothelium all over the body (intestinal and the mesenteric lymph nodes, liver, spleen and bone marrow, etc.). The hallmark of clinic feature is persistent, high fever, splenomegaly, relative bradycardia, rose-colored spots on the trunk, and reduction of blood neutrophils and eosinophils [1].

Etiology and Pathogenesis
S. typhi is a facultative intracellular organism and infects only humans. It has Somatic antigen ("O" antigen), flagellar antigen ("H" antigen), and surface antigen

Fig. 16.8 Tuberculosis of lymph node. Structure of lymph node is destroyed and replaced by tubercles in different size. Enlarged and adherent lymph nodes can form a big lump

("V1" antigen). Wildal's test is the traditional serologic test and for the diagnosis of typhoid fever. It also can release endotoxin, which is the main etiological factor.

The source of infection are patients and healthful carriers. Fecal-oral transmission is the main route of infection. Flies, raw shellfish, and patients' hands are especially important vectors. The infection results from contamination of food and water with feces from a symptomatic case or a symptomatic carrier of typhoid. The ingested bacillus invades the small intestinal mucosa, where it is taken up by macrophages and transported to regional lymph nodes and multiplies in the intestinal lymphoid tissue during the 10-day incubation period. Then they enter the bloodstream through the thoracic duct and are carried to the liver, bone marrow, spleen, and kidneys. Ingestion by phagocytic cells causes hyperplasia of the mononuclear phagocytic system. *S. typhi* reenters the intestinal lumen by way of biliary excretion. The most prominent change is in Peyer's patches and solitary lymph follicles of the distal ileum. Healing usually begins about at the end of the third week and is complete by the fifth week in uncomplicated cases. The main mechanisms of the disease are proliferation of bacteria, coupled with an antigen-antibody hypersensitivity reaction.

Intestinal Morphology

The most obvious pathological changes of intestinal lymphoid tissue were Peyer's patches in the distal ileum and solitary lymphoid follicles in cecum. The whole course is divided into four stages, and each course lasts about 1 week.

1. The stage of medullary swelling

In the first week of infection, there is swelling of the solitary lymphoid follicles and Peyer's patches. Peyer's patches in the terminal ileum become sharply delineated, with plateau-like elevations up to 8 cm in diameter and enlargement of draining mesenteric lymph nodes. The surface of the lymphoid tissue is convoluted, like the surface of the brain; thus, it is also called medullary swelling (Fig. 16.9). Microscopically, Peyer's patch shows congestion and marked proliferation of macrophages. These cells have abundant pale cytoplasm with round or oval eccentric nuclei. Macrophages have increased phagocytic activity and may contain ingested red blood cells, lymphocytes, and cellular debris. These so-called "typhoid cells" have diagnostic significance. Intermingled with the phagocytes are lymphocytes and plasma cells. These cells aggregate to form typhoid nodules or granulomas. Clinically, patients often suffered from blood diseases and rise of typhoid cells.

2. The stage of necrosis

Fig. 16.9 Typhoid fever of the intestine. (The stage of medullary swelling) The surface of the lymphoid tissue is convoluted, like the surface of the brain

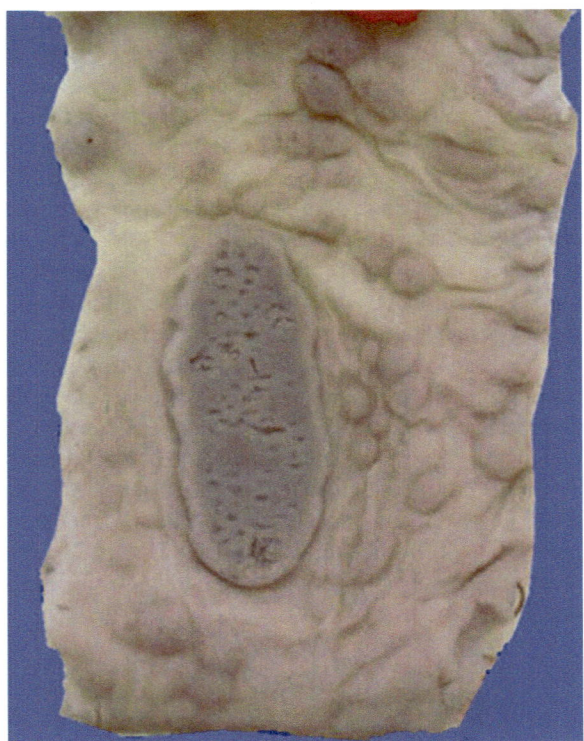

The stage of necrosis occurs in the second week. Lymphoid tissue undergoes necrosis, forming grayish-white or grayish-green areas. Necrosis can extend deeply to the muscular propria and even the serosal coat. The development of focal necrosis can lead to softening and rupture of the lymphoid patches. The clinical manifestations were persistent high fever, bradycardia, hepatosplenomegaly, rose-colored spots in the skin, leukocytopenia, and diarrhea. The more serious complication of extensive necrosis is small vowel perforation, usually accompanied by fatal systemic peritonitis.

3. The stage of ulceration

The stage of ulceration occurs in the third week. Ulcers occur as longitudinal ulcers overlying the Peyer's patches in the ileum. The mucosal ulcers tend to take the shape and extent of the underlying lymphoid patches (round or oval). Necrotic mucosa sloughs off, leaving round or oval ulcers. They have slightly elevated margins. Ulcers may penetrate to the serosa and cause intestinal perforation and peritonitis. Erosion of blood vessels at the ulcer base can cause severe massive hemorrhage. Perforation and hemorrhage are both serious complications of typhoid fever. Typhoid ulcers can be differentiated from tuberculous ulcers. Tuberculous infection spreads along submucosal lymphatics to form circumferential ulcers. Ulcer healing may result in stricture. Typhoid ulcers, on the other hand, lie parallel to the long axis of the intestine, and hence healing will not cause strictures.

4. The stage of healing (the fourth week)

The fourth week is the stage of healing. Ulcers heal through granulation tissue formation and reepithelization of ulcers. Healing of these ulcerated lesions leaves a smooth scar, which never shows any narrowing tendency in formation. Clinically, the patient's temperature shows a staircase-like descend to recover normal. However, in the antibiotic era, it is difficult to find the typical four stages (swollen, necrotic, ulcerative, and healing stages).

Changes in Other Organs

The lesions are characterized by an acute proliferative inflammation of reticuloendothelial and lymphoid tissues, for example, bone marrow, spleen, and liver. Sections of the enlarged spleen in typhoid fever show congestion and dilatation of splenic sinuses. The enlarged liver shows sinusoidal dilatation and degeneration and necrosis of liver cells. Typical granulomas can also be found in the spleen, liver, born marrow, and lymph nodes. Gallbladder colonization, which may be associated with gallstones, causes a chronic carrier state.

The clinical features were fever, relative bradycardia, hepatomegaly and splenomegaly, rose spots, and decreased numbers of neutrophils and eosinophils. Due to endotoxemia and bacteremia, typhoid fever is also associated with a wide range of diseases and may occur during the course of the disease. Laryngitis, bronchitis, and bronchopneumonia may occur [4].

16.3   Bacillary Dysentery

It is diarrhea with abdominal cramping and tenesmus, which loose stools contain blood, pus, and mucus. It is an acute pseudomembranous inflammation caused by *Shigella* species.

Etiology and Pathogenesis
Shigella species are gram-negative bacilli. Patients and healthful carriers are sources of infection. The routes of infection are atypical fecal–oral transmission, and flies serve as important vectors in transmission. It is transmitted by food and water contaminated by *Shigella*. Most of *Shigella* is killed by gastric acid, so few can get into the small intestine. The organisms invade directly the intestinal mucosa by producing endotoxin and cause ulceration of intestinal mucosa. Endotoxin enters blood stream and results in systematic toxemia. Exotoxin, which is produced by *Shigella dysenteriae*, is the main factor that causes large-volume of watery diarrhea [2].

Pathology and Clinical Features
Sigmoid colon and rectum are the most commonly affected areas. In severe conditions, the entire colon and the terminal ileum can be affected. There are three different types of bacillary dysentery: acute bacillary dysentery, chronic bacillary dysentery, and toxic bacillary dysentery.

Acute Bacillary Dysentery
A typical course of disease includes acute catarrhal inflammation followed by unique pseudomembranous inflammation and ulceration, and eventually healing.

Initial acute catarrh inflammation is an acute inflammation with diffuse hyperemia, edema, punctuate hemorrhage, infiltrating of inflammatory cells (neutrophils), excess mucus secretion, and formation superficial erosion in the mucosa. A fibrinosuppurative exudates first patchily, then diffusely covers the mucosa and produces a dirty gray to yellow pseudomembrane. Grossly, on the mucosal surface of the affected colon initially shows patchy distribution of the pseudomembrane, which then diffusely covers the mucosa, resulting in dirty membranous matter (Fig. 16.10). Microscopically, mucosal gland in areas of mild inflammation is preserved. In severely affected areas, there are superficial necrosis, absent glands, submucosal edema, hemorrhage, and neutrophilic infiltration. The necrotic mucosal epithelium is covered by an acute suppurative and fibrinous exudate, forming a pseudomembrane. The pseudomembrane is composed of fibrins and numerous neutrophils, admixed with red cells, cell debris, and bacteria. When the pseudomembrane sloughs, it leaves shallow ulcers of varying sizes. One week later, the inflammatory reaction in intestinal mucosa increased; the mucosa became soft and brittle; and irregular superficial ulcer appeared (map-like ulcer). With the development of the disease, the ulcer margins transformed into active granulation tissue. When the disease remits, this granulation tissue fills the defect, and the ulcers heal by regeneration of the mucosal epithelium, sometimes forming polyps.

Clinically, patients have severe tenesmus and diarrhea with blood and pus or blood and mucus. Because inflammation stimulates the nerve terminal of rectum

Fig. 16.10 Bacillary dysentery. On the surface of the mucosa of affected colon, pseudomembrane patchily distributes and produces a dirty membranous matter

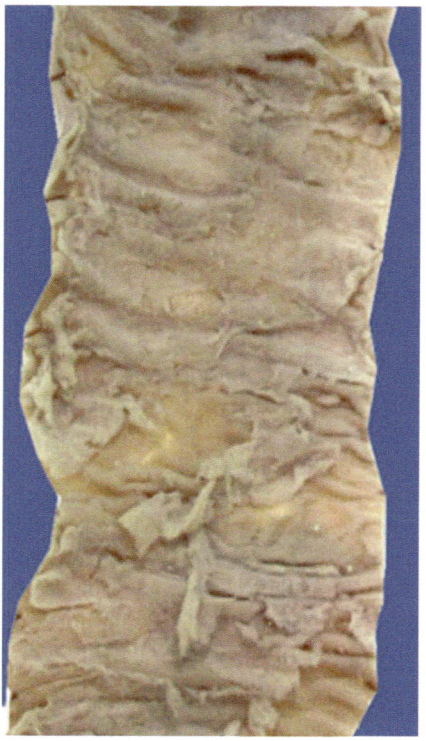

and anal sphincter, patients have tenesmus and the frequency of acute defecation increases. Patients defecate mucoid loose stool and mucoid bloody perulent stool, occasionally accompanied by lamellar pseudomembrane. It is usually a self-limiting disease of 3 days to a week, rarely lasting for a month. Sometimes the disease can transform to chronic bacillary dysentery.

Chronic Bacillary Dysentery

Course of bacillary dysentery has been exceeded more than 2 months. The chronic cases are characterized by repeated injury and repair, ulceration, and healing and can lead to new and old lesions coexistence, chronic ulcer formation, polypoid mucosal irregularity with fibrous scarring, and subsequent stenosis of the bowel.

Clinically, patients have abdominal pain, chronic diarrhea, and intestinal obstruction. Some patients do not have any symptoms or signs, but their stool culture is persistent positive. This type of patients will become chronic carrier and infection sources.

Toxic Bacillary Dysentery

It occurs most commonly in children at the age of 2–7 years. The onset of disease is sudden. Patients have severe systemic toxic symptom and mild intestinal symptom.

Intestinal lesion is catarrh enteritis or follicular enteritis. There are severe general toxic symptoms (high fever, convulsion). Several hours later, toxic shock, coma or brain edema, and respiratory failure quickly appeared.

16.4 Leprosy

It is a chronic inflammatory disease of low infectivity caused by *Mycobacterium leprae*. The disease mainly involves the skin and peripheral nerves. The disease spreads extensively in the world, especially in warm and wet areas [1].

Etiology and Pathogenesis
M. leprae and *Mycobacterium lepromatosis* are the causative agents of leprosy. *M. lepromatosis* is a relatively newly identified mycobacterium isolated from a fatal case of diffuse lepromatous leprosy in 2008. *M. leprae*, an intracellular, acid-fast bacterium, is aerobic and rod-shaped and surrounded by the waxy cell membrane coating, which are characteristics of *Mycobacterium* species. Although, at present, the causative organisms have been impossible to culture in vitro, it has been possible to grow them in animals. Naturally occurring infection also has been reported in nonhuman primates including the African chimpanzee, sooty mangabey, and cynomolgus macaque, as well as in armadillos.

Human-to-human transmission is the primary source of infection. Leprosy was previously thought to be transmitted through intact skin by direct contact. However, current evidence suggests that the bacillus is inhaled into the respiratory tract through prolonged contact with nasal secretions or open ulcers of infected people. Bacilli are taken up by macrophages and disseminated widely throughout the body. Disease results from proliferation of the microorganisms and the host response to them. The disease has a latent period of ~2–5 years. The organisms grow best at lower temperatures and are found mainly in the peripheral nerves and skin. In the skin, leprosy causes atrophy of cutaneous appendages resulting in loss of sweating and loss of eyebrows or hair. In the peripheral nerves, nerve trunks become thick. Patients lose sensation. Limbs also become disabled.

Morphology and Clinicopathologic
The disease is termed a chronic granulomatous disease, and it produces inflammatory nodules (granulomas) in the skin and nerves over time. Leprosy can be classified as three types: tuberculoid leprosy, lepromatous leprosy, and borderline leprosy.

Tuberculoid leprosy accounts for 70% of cases of leprosy disease. The form occurs in patients who have strong immune responses to infection. The characteristic lesion is the result of formation of tuberculoid granulomas. Grossly, the lesions are symmetric and localized to one or more areas of skin, especially the legs and feet. They are annular, flat, and lack sensation. The cutaneous nerves are often palpably enlarged. Microscopically, the granulomatous foci consist of epithelioid cells, Langhans' giant cells, and numerous lymphocytes at the periphery. They closely resemble the lesions of TB. Caseous necrosis is rare but occurs in lesions of nerve stem, which usually affects appendages of the skin and results in hypesthesia and absent sweating. The type of leprosy is less infectious and progresses slowly. The results of treatment and prognosis are better than in other forms. The lepromin skin test is positive. *M. leprae* can be detected by acid-fast stains, but there are fewer organisms than in Lepromatous leprosy.

Lepromatous leprosy accounts for 20% of cases of leprosy disease. The disease occurs when patients fail to develop an immune response. As a result, bacilli proliferate readily and cause more extensive disease than in the tuberculoid form. Lepromatous leprosy presents with granuloma composed of foam cells and few lymph cells. Foam cells originate from macrophage cells, many of which contain lepra bacilli, called lepra cells. Lesions are surrounding small vessels and appendages of the skin and can progress to fuse. There is no cell infiltration layer between epidermis and infiltrative focus, which is one of the features of lepromatous leprosy. Lepromatous leprosy involves the nose, lymph nodes, liver, spleen, and testes. A distinctive deformity of the face is known as leonine face. Intranasal lesions can lead to collapse of nasal bridge or perforation of nasal septum. The lepromin test is negative because patients have no immune response to the organism.

Borderline leprosy is also called as indeterminate leprosy and of intermediate severity. These lesions have features that resemble both tuberculoid and lepromatous forms of the disease. Skin lesions resemble tuberculoid leprosy but are more numerous and irregular; large patches may affect the whole limb, and peripheral nerve involvement with weakness and loss of sensation is common. This type is unstable and may become more like lepromatous leprosy or may undergo a reversal reaction, becoming more like the tuberculoid leprosy form. These lesions may heal or may progress to either tuberculoid or lepromatous leprosy.

16.5 Leptospirosis

It is caused by infection with bacteria of the genus *Leptospira* and affects humans as well as other animals. It is among the world's most common diseases transmitted to people from animals. It distributes worldwide, especially in tropical and subtropical areas. There is a direct correlation between the amount of rainfall and the incidence of leptospirosis, making it seasonal in temperate climates and year-round in tropical climates [1].

Etiology and Pathogenesis
Leptospirosis is transmitted by the urine of an infected animal and is contagious as long as it is still moist. Rats, mice, and moles are important primary hosts. Humans become infected through contact with water, food, or soil containing urine from these infected animals. This may happen by swallowing contaminated food or water or through skin contact. The disease is not known to be spread from person to person, and cases of bacterial dissemination in convalescence are extremely rare in humans. Leptospirosis is common among water-sport enthusiasts in specific areas

because prolonged immersion in water can promote the entry of the bacteria. Surfers and whitewater paddlers are at especially high risk in areas that have been shown to contain the bacteria and can contract the disease by swallowing contaminated water, splashing contaminated water into their eyes or nose, or exposing open wounds to infected water. The etiology of leptospirosis remains unclear.

Morphology and Clinicopathologic
The genus *Leptospira* is thought to comprise many species, which have specific surface antigen and collective inner antigen. Different species of *Leptospira* have different virulence to human and involve different organs. Leptospiral infection in humans causes a range of symptoms, and some infected persons may have no symptoms at all. In clinic, the develop process of leptospirosis is divided into three stages: early, middle, and late. This classification has great significance in guiding the clinical practice, especially the early diagnosis and treatment.

1. Early stage (septicmic stage)
   The symptoms in humans appear after a 1–3 day incubation period. Symptoms of leptospirosis include high fever, severe headache, chills, muscle aches, and vomiting and may include jaundice, red eyes, abdominal pain, diarrhea, and rash. Initial presentation may resemble pneumonia. The first phase resolves, and the patient is briefly asymptomatic until the second phase begins.
2. Middle stage (septicmic with tissue damage stage)
   It occurs after 4–10 days of incubation period, and inner organ damage is seen in this stage. This is characterized by meningitis, liver damage (causing jaundice), and renal failure. Cardiovascular problems are also possible and severe cases commonly die in this stage.
3. Late stage (convalescent stage)
   Most patients recover from illness after 2–3 weeks onset. High fever symptoms gradually subsided. Myocarditis, pericarditis, meningitis, and uveitis are also possible sequelae. Some patients have ocular and neurologic sequelae due to specific immunological reaction.

Pathology
The pathological change of leptospirosis is an acute systemic toxicity, which often involves capillary circulation. It is characterized by degeneration, necrosis, and severe dysfunction of parenchymal organs.

1. Liver: The lesions can be mild and severe. Mild lesions only show slight interstitial edema and congestion of blood vessels and scattered focal necrosis under microscope. In severe cases, jaundice and hepatomegaly may occur. Microscopically, liver necrosis, hepatocyte swelling, steatosis, vacuolation, and infiltration of portal vein into neutrophils were observed. Partial necrosis of liver cells and mitotic, serious liver cells arrange in discrete disorder, which lead to biliary excretion disorder. Necrosis of hepatocytes results in the disorder of coagulation factor synthesis and extensive hemorrhage in skin and mucosa.
2. Kidney: The lesion is mainly interstitial nephritis. The appearance of the renal cortex kidney may have bleeding or hemorrhage spot under the renal capsule.

Microscopically, the main lesions are renal tubular epithelial cloudy swelling and necrosis especially in the distal convoluted tubule and Heinz loop. Renal tubular lumen can be filled with blood cells or transparent tube type, which causes renal tubular dilatation proximal renal tubules. In general, the glomerular lesions are not serious.

3. Lungs: The main lesion is characterized by pulmonary hemorrhage, which is a frequent cause of death in anicteric cases. Hemorrhage can be observed on the surface of lung. Tracheal and bronchial mucosa as well as a large number of interstitial and intra-alveolar tissue also showed obvious edema and hemorrhage.
4. Heart: Hemorrhage can be observed in pericardial membrane. Myocardial cells showed hyalinized degeneration and focal necrosis. Interstitial edema also can be seen. Patients may have tachycardia, cardiac dysrhythmia, and some other symptoms of myocarditis.
5. Striated muscle: Bleeding points can be seen in intercostal muscles, psoas major muscle, gastrocnemius muscle, abdominal muscle, diaphragm, upper wall, and thigh muscles. Microscopically, striated structure of muscle fibers disappears. Particles or hyaline degeneration appear in the muscle fiber cell cytoplasm. Hemorrhage, edema, and inflammatory cell infiltration may occur in stroma.
6. Nervous system: Brain parenchyma and meninges show edema, congestion, hemorrhage, and inflammatory cell infiltration. Nerve cell degeneration can be seen in some cases. Cerebral arteritis may occur in some cases in convalescent stage, especially in children. Cerebral infarction and cerebral atrophy can be seen in subdural or subarachnoid. The main pathological changes are multiple arteritis in pavimentum cerebri and brain parenchyma damage, which result in hemiplegi and aphasia.

16.6   Rabies

Rabies is a viral disease that causes acute encephalitis in warm-blooded animals. The disease is zoonotic, meaning it can be transmitted from one species to another, such as from dogs to humans, usually through the bite of infected animals. For humans, rabies is almost always fatal without post exposure prevention before severe symptoms occur. Rabies virus infects the central nervous system and eventually leads to brain disease and death [1].

Etiology and Pathogenesis
Rabies virus contains five kinds of protein, the glycoprotein (G), a nuclear protein (N), polymerase (L), phosphoprotein (NS), and membrane protein (M). Glycoprotein can bind with acetylcholine binding, which leads to the affinity for neural tissue feature of rabies virus. Rabies virus invades from skin or mucosal damage site. Once it enters a muscle or nerve cell, the virus replicates. Then the virus invades peripheral nerve and spreads to the central nervous system along the peripheral nerve axoplasm. Virus mainly invades neurons of the brainstem and cerebellum.

The rabies virus also can diffuse from the central nervous system to peripheral nerve system, especially invasion in the salivary glands of tongue taste buds, olfactory epithelium. Damage to the vagus nerve nucleus, nucleus of hypoglossal nerve, and other related nerve nuclei, causes respiratory muscle and swallowing muscle spasm, resulting in clinical symptoms such as hydrophobia, dyspnea, and dysphagia; sympathetic nerve stimulation causes increased saliva secretion and sweating; damage of vagus ganglion, sympathetic ganglion, and cardiac ganglia, which causes cardiovascular system function disorder and even sudden death.

Morphology and Clinicopathologic
The lesions are acute disseminated encephalomyelitis, especially in hippocampus of dorsal root ganglion and spinal cord on the bite site. Usually, there is no meningeal lesions. The brain parenchyma showed hyperemia, edema, and bleeding. Nonspecific degeneration and inflammatory changes, such as nerve cell vacuolization, decomposition, hyaline degeneration, and chromatin perivascular infiltration of mononuclear cells, can be seen under low microscopy.

The characteristic lesion is eosinophilic inclusion, called Negri body. Negri bodies are round or oval, dyed cherry red, uniform diameter of about 3 ~ 10 nm and most common in the hippocampus and the Meura Kenno organization the nerve cells in the cerebral cortex. They can also be found in the cone cell layer, spinal cord nerve cells, posterior horn, retinal ganglion cells layer, and ganglion sympathetic ganglion. Under electron microscope, Negri bodies are the virus colony and contain baculovirus particles.

Salivary acinar cells degenerated and monocytes infiltrated. Acute degeneration of pancreatic acini and epithelial cells, parietal cells of the gastric mucosa, adrenal medulla cells, and renal tubular epithelial cells were observed.

Early-stage symptoms of rabies are malaise, headache, and fever, progressing to acute pain, violent movements, uncontrolled excitement, depression, and hydrophobia. Finally, the patient may experience periods of mania and lethargy, eventually leading to coma. The main cause of death is respiratory insufficiency. Because the mortality rate is almost 100%, positive and effective treatment of wounds and preventive vaccination are very important.

16.7   Sexually Transmitted Disease

Diseases transmitted during sexual contact are sexually transmitted diseases (STD), but no STD is acquired solely via coitus. It also can be transmitted by the parents to the fetus or newborn way. The STDs are largely diseases of life-style, and their incidence is higher among patients with multiple sexual partners. As a consequence, the coexistence of many STDs is prevalent in groups with high levels of sexual activity [1].

16.7.1 Gonorrhea

Gonorrhea is a sexually transmitted bacterial infection, caused by the gram-negative, diplococcus *Neisseria gonorrhoeae*. *N. gonorrhoeae* (syn. gonococcus) causes acute suppurative inflammation of the urethra and periurethral glands, which may also extend to the prostate and epididymis. About 90% of males develop a purulent urethral discharge with pain on passing urine as a result of infection, in contrast to females in whom about 70% of gonococcal infections are asymptomatic. Gonorrhoea is a common infection, mainly occurring in 20–24-year-old young adults and has a high infectivity.

Etiology and Pathogenesis

N. gonorrhoeae occurs only in humans. It is morphologically identical to *Neisseria meningitidis*, an important cause of bacterial meningitis. *N. gonorrhoeae* is a facultative intracellular pathogen that binds to and invades epithelial cells through the pili in the envelope of gonococcus and secreted IgA proteo-lystic enzyme. The polysaccharides of capsular contribute to virulence through inhibiting phagocytosis in the absence condition of antigonococcal antibody. In addition, the pathogen releases peptidoglycans and endotoxin, which induce host cell secretion of TNF-α that may cause damages of infectious epithelial cells. Patients and asymptomatic carriers are source of infection. Usually, it is transmitted by the sexual contacts in the adults or by the indirect contact in the children. In addition, gonococcal infection may be transmitted to the fetus during delivery through the birth canal; producing neonatal ophthalmitis. Transmission by fomites such as toilet seats or towels is not likely since *N. gonorrhoeae* cannot survive long outside the human body.

Morphology and Clinicopathologic

The primary acute infection affects the urethra and the periurethral glands in both sexes and the endocervix in the female. Because stratified squamous epithelium is resistant to invasion by *N. gonorrhoeae*, lesions do not usually occur on the external genitalia or in the vagina. Bacteria infect columnar and transitional cells of the mucous membranes and cause an acute suppurative infection, sometimes with abscess formation, then followed by granulation tissue formation and fibrosis. *N. gonorrhoeae* is usually seen easily within the cytoplasm of polymorphonuclear leukocytes at the sites of infection.

In males, acute gonorrhea usually causes severe urethritis with urinary frequency, urgency, and pain on urination. There is a marked purulent exudate. In female patients, acute symptoms are often less severe but chronic infection of the endometrium, fallopian tubes, ovaries, and peritoneum may cause pelvic pain and vaginal discharge. Congenital infection can cause purulent infection of the eyes and result in blindness. Neisseria gonorrhoeae may spread to the whole body by hematogenous routs, which result in a serious systemic infection. Although the disease often has the typical signs and symptoms, clinical manifestations are variant. The clinical manifestations of disseminated gonococcal infection (DGI) include simple skin damage, arthritis, tenosynovitis, or septic arthritis disease spectrum. A few severe DGI can appear endocarditis or meningitis symptoms.

16.7.2   Condyloma Acuminatum

Condyloma acuminatum, also known as genital warts, is caused by Human papilloma viruses (HPV). The incidence rate is increasing. Now it may be the most common disease in patients attending departments of genito-urinary medicine. HPV are a large group of more than 70 double-stranded DNA viruses that can infect squamous epithelium in the skin, mucocutaneous surfaces, and respiratory tract. In the genital tract, HPV Types 6, 11, 16, and 18 are especially common. They are spread by sexual intercourse.

Etiology and Pathogenesis
HPV is a kind of DNA virus. Human is the only host for papilloma viruses. Using molecular hybridization, more than 120 subtypes were identified. About 40 species are involved in reproductive tract infection, and different types of HPV can cause the different clinical manifestations. The incubation period was from 1 to 8 months, average 3 months. Whether there are main clinical manifestations after HPV infection depends on the type of genital epithelium, the site and state of infection, especially the cellular immune status. For patients with immune function defect, especially when the cell immune function defects, such as long-term use of corticosteroids or immunosuppressants, it is very easy to get infected with this disease.

Morphology and Clinicopathologic
In the genital organs, human papilloma viruses cause genital warts (condyloma accuminata), epithelial dysplasia, and squamous carcinoma. The primary lesion is small and soft light red papules as big as needle cap or a grain of rice. Over time, its number and volume increase gradually, eventually changing into papillomatous with cauliflower appearance, rough surface, and soft texture. The characteristic lesion is hyperplastic, fleshy wart, or condyloma acuminatum. This wart occurs most commonly on the glans penis and inner lining of the prepuce or in the terminal urethra. Less often, lesions develop on the shaft of the penis, the perianal region, or the scrotum. Histologically, the epidermis shows papillomatous hyperplasia. Many of the epidermal cells show cytoplasmic vaculation (koilocytes), a feature indicating a viral etiology. Vacuolated cells are the characteristic of the lesion and usually located in the upper of prickle cells. The cell body was big, round hyperchromatic nuclei, perinuclear vacuolization, and pale staining. Basilar cells proliferate, and parabasal cells enlarge with a foamy nuclear chromatin. There is a chronic inflammatory infiltrate and extension of capillary or lymphatic vessels within the dermis. Viral DNA can be identified by in situ hybridization, PCR, and in situ PCR. Viral proteins may be detected by immunohistochemistry.

Dysplasia is characterized by lack of normal maturation (differentiation) of the squamous epithelium from the basal epithelial layer toward the surface, atypical nuclear, and increased mitosis. Dysplasia is a precancerous lesion. It can occur at any site infected by HPV but is especially important in the vulva, vagina, and cervix. HPV Types 16 and 18 are considered to be important causes of genital epithelial dysplasia.

Carcinoma in situ and invasive squamous carcinoma result from progression of dysplasia. HPV infection is considered to be the main cause of carcinoma in the genital tract, especially carcinoma of the penis, carcinoma of the vulva, vagina, and cervix.

16.7.3 Syphilis

Syphilis is a chronic venereal infection caused by the spirochete, *Treponema pallidum*. It has remained an endemic infection in all parts of the world, especially in large cities. There is a strong racial disparity, with African Americans affected 30 times more often than whites [1].

Etiology and Pathogenesis

Treponema pallidum is a fastidious spirochete whose only natural hosts are humans. *T. pallidum* must be stained with silver or viewed by darkfield microscopy to be detected. The usual source of infection is an active cutaneous or mucosal lesion in a sexual partner in the early (primary or secondary) stages of syphilis. The organism is transmitted from such lesions during sexual intercourse across minute breaks in the skin or mucous membranes of the uninfected partner. In cases of congenital syphilis, *T. pallidum* is transmitted across the placenta from mother to fetus, particularly during the early stages of maternal infection. Once introduced into the body, the organisms are rapidly disseminated to distant sites by lymphatics and the bloodstream. Systemic dissemination of organisms continues and the host mounts an immune response. Specific antibody of syphilis is produced in the sixth week after injection and has serological diagnostic value. Syphilis is classified into two types: congenital syphilis and acquired syphilis.

Pathology

Syphilis may affect nearly any organ or tissue in the body. The basic pathologic changes of syphilis are the same, irrespective of the site. The lesions of syphilis vary with the stage of disease. There are two basic microscopic lesions of syphilis.

1. Proliferative arteritis and associated plasma cell-rich inflammatory infiltration. The lesion can occur in all the stage of syphilis. This lesion is characterized by concentric proliferation of endothelial and fibroblastic cells of small vessels, with a surrounding mononuclear infiltrate, known as periarteritis. Local ischemia caused by the vascular changes undoubtedly accounts for some of the local cell death and fibrosis seen in syphilis. Perivascular plasma cell infiltration is of diagnostic significance. Spirochetes are readily demonstrable in histologic

sections of early lesions with the use of standard silver stains (e.g., Warthin-Starry stains).

2. Gumma In late syphilis (tertiary syphilis), another pattern of tissue injury is a form of granulomatous necrosis. Because of its rubbery texture, such a lesion is known as a gumma. It is an irregular, firm mass of necrotic tissue surrounded by resilient connective tissue, which is similar to tubercle. Histologically, the gumma is characterized by central coagulative necrosis and peripheral epithelioid cells, admixed with giant cells. The central zone of structured necrosis can usually be distinguished. Granulation tissue was seriously infiltrated by plasma cells and lymphocytes. This focus is enclosed by a fibroblastic wall, in which the small vessels may show endarteritis and periarteritis. Healing of a gumma is by absorption of the necrotic center and by progressive fibrous proliferation, eventually forming scar.

16.7.3.1 Congenital Syphilis

T. pallidum may be transmitted across the placenta from an infected mother to the fetus at any time during pregnancy. Without treatment, 40% of infected infants die in utero, usually after the fourth month. In cases of congenital syphilis, the placenta is enlarged, pale, and edematous. Microscopically, hyperplastic arteritis involving fetal blood vessels, mononuclear inflammatory response (villi TIS), and chorionic immaturity are shown. According to the clinical manifestations, congenital syphilis can be divided into two types: early (prior to age 2 years) and late (after age 2 years) congenital syphilis.

1. Early congenital syphilis for stillborn infants: The most common manifestations are hepatomegaly, bone abnormalities, pancreatic fibrosis, and pneumonitis. Infantile syphilis is defined as congenital syphilis at birth or within a few months after birth. Affected infants present with chronic rhinitis (snuffles) and mucocutaneous lesions, similar to those seen in secondary syphilis in adults. Visceral and skeletal changes similar to those of stillbirth may also exist.

2. Late congenital syphilis: It refers to cases of untreated congenital syphilis of more than 2 years' duration. Classic manifestations include the Hutchinson triad: notched central incisors, interstitial keratitis with blindness, and deafness from eighth cranial nerve injury. Other changes include a saber shin deformity caused by chronic inflammation of the periosteum of the tibia, deformed molar teeth ("mulberry" molars), chronic meningitis, chorioretinitis, and gummas of the nasal bone and cartilage with a resultant "saddle-nose" deformity.

16.7.3.2 Acquired Syphilis

Acquired syphilis may progress through three distinct phases: primary, secondary, and tertiary syphilis. The lesions of syphilis vary with the stage of disease.

1. Primary syphilis: It is characterized by the presence of a chancre at the site of initial inoculation. The primary chancre in males is usually on the penis. In females, multiple chancres may be present, usually in the vagina or on the uterine cervix. The chancre begins as a small, firm papule, and develops into a hard, painless, indurated ulcer with regular, well-demarcated margins and a "clean," moist base. This occurs within 3 months at the site of inoculation. Regional lymph nodes are often slightly enlarged and firm but painless. Histologically, there were granulation tissue, proliferation of small vascular endothelial cells, and infiltration of a large number of plasma cells and lymphocytes. The ulcer can be healed by fibrosis, producing a small scar. Spirochetes may be demonstrable by a silver impregnation stain. Even without therapy, the primary chancre will subside within a few weeks, forming a subtle scar.
2. Secondary syphilis: The lesions of secondary syphilis are very variable in location and result from systemic dissemination of *T. pallidum* with infection of multiple organs. Important sites of these lesions include: (a) Skin. Lesions are usually symmetrical and may be maculopapular, scaly, or pustular. Involvement of the palms of the hands and soles of the feet is common. In moist skin areas, such as the anogenital region, inner thighs, and axillae, broad-based, elevated lesions termed condylomata lata may occur. Pustular destructive lesions occur on the soles of the feet or palms of the hands. (b) Mucous membranes. Superficial mucosal lesions resembling condylomata lata can occur anywhere, but they are particularly common in the oral cavity, pharynx, and external genitalia. These lesions are highly infectious because they contain numerous microorganisms. Mucous patches, termed as grey white membranes with a dull-red margin, are found in the mouth, pharynx and larynx. The ulcer is also called as Snail-track ulcers. Histology showed typical hyperplastic arteritis with inflammatory infiltration of lymphoplasmacytes. (c) Lymphadenitis. Lymph node enlargement is most common in the neck and inguinal areas. Lymph node biopsy showed nonspecific hyperplasia of germinal center accompanied by plasmacytosis. Granuloma or neutrophils were uncommon. There are numerous spirochetes. The blood vessels may show typical syphilitic vasculitis. (d) Other organs. Less common manifestations of secondary syphilis include hepatitis, renal disease, eye disease (iritis), and gastrointestinal abnormalities. The skin and mucous membrane lesions of secondary syphilis subsided in a few weeks, and the patients entered the early incubation period of the disease, lasting for about 1 year. In the early incubation period, the lesions may recur and spread.
3. Tertiary syphilis: It is often detected many years after the secondary stage. The characteristic lesion is the gumma. The lesions of tertiary syphilis can involve almost any organ system. This phase of syphilis is divided into three major categories: cardiovascular syphilis, neurosyphilis, and so-called benign tertiary syphilis.

   (a) Cardiovascular syphilis: It accounts for more than 80% of cases of tertiary disease, and it is much more common in men than in women. The lesion is endarteritis of the vasa vasorum of the proximal aorta. Occlusion of the vasa vasorum results in scarring of the media of the proximal aortic wall, with consequent loss of elasticity. The aortic disease is characterized by slow progressive dilation of the aortic root and arch, resulting in aortic insufficiency and proximal aortic aneurysm. These are characteristically located in the ascending aorta which becomes markedly dilated. The aortic valves are usually involved, causing incompetence (leaking) of the aortic valve. In some cases, stenosis of the coronary orifice is caused by subintimal scarring and secondary myocardial ischemia.

(b) Neurosyphilis: It accounts for only about 10% of cases of tertiary syphilis. Variants of neurosyphilis include chronic meningovascular disease, tabes dorsalis, and a generalized brain parenchymal disease termed general paresis. Lesions can involve the meninges, spinal cord, or cerebral cortex. An increased frequency of neurosyphilis has been noted in patients with concomitant HIV infection.

(c) The so-called benign tertiary syphilis: It is a relatively uncommon form of tertiary syphilis and characterized by the development of gummas in various sites. These lesions are probably related to the occurrence of delayed hypersensitivity. Due to the development of effective antibiotics such as penicillin, gingiva, which was once very common, has become very rare. According to reports, most of these diseases occur in AIDS patients.

16.8   Deep Mycosis

Diseases caused by fungi are called mycoses. Mycoses are classified as superficial, cutaneous, subcutaneous, or systemic (deep) infections depending on the type and degree of tissue involvement and the host response to the pathogen. Deep mycosis usually affects deep layer of skin and internal organs, causing more destructive disease.

Etiology and Pathogenesis
Fungi are weak pathogens and generally do not produce toxins. Fungi cause tissue damage primarily by a hypersensitivity reaction of the host to the parasitic proteins. The fungi are usually saprophytes that become opportunistic pathogens only under special circumstances: (1) Large, long-term, or multiple use of broad-spectrum and efficient antibiotics, such treatment can cause disordered bacterial equilibrium. Latrogenic factors, such as major surgery, can also contribute. (2) Use of hormone or immunosuppressant; (3) Elderly patients; (4) Severe basic diseases, such as leukemia, lung cancer and other malignant tumors, chronic nephritis, uremia, renal transplantation, chronic obstructive pulmonary disease, pemphigus, cerebral hemorrhage, diabetes, and AIDS; (5) Organ transplantation; (6) Organ intubation and catheter, catheter interventional therapy; (7) Radiotherapy, chemotherapy. Deep fungal infections usually heal or remain latent in normal hosts; in

immunocompromised hosts; however, they can spread systemically and invade tissues, destroying vital organs.

Pathology

Mycoses usually cause necrosis accompanied by a mixed acute and chronic inflammatory reaction.

1. Acute suppurative inflammation with neutrophils and macrophages. Some fungal infections form pseudomembranes. Others cause endocarditis with lesions on the heart valves.
2. Chronic suppurative inflammation with micro-abscesses, infiltrates of lymphocytes and plasma cells, and granulation tissue with fibroblasts.
3. Granulomatous inflammation: Some granuloma lesions are similar to tuberculous granuloma, but massive neutrophil cells aggregate in the granuloma center.
4. Gangrenous inflammation: Some necrotic lesions have different sizes and accompany with conspicuous hemorrhage.
5. Fungous septicemia: Septicemia occurs when fungus infection enters the bloodstream. Untreated septicemia can quickly progress to sepsis, which is a serious complication of an infection characterized by inflammation throughout the body.

16.8.1 Candidiasis

Candidiasis is a common fungal infection caused by *Candida albicans* which has over 20 species of yeasts. *C. albicans* can be found in normal human skin, oral, vaginal, and gastrointestinal candidiasis. Candidiasis is an autogenous infection and can occur in all parts of the body. There are three common presentations of candidiasis:

1. Skin and mucosal superficial candidiasis are common. Warm, moist environment is conducive to the growth of fungi, so skin candidiasis occurs in the armpits, groin, finger (toe), and around the anus. Vaginal candidiasis may occur in patients with diabetes and pregnant women because blood glucose and diabetic and urine increased. In the esophagus, candidiasis is almost always associated with one of the risk factors identified above. Skin and mucosa albicans infection often form white flaky, irregular pseudomembrane in the skin and mucosal surfaces. Candidiasis of the mucous membrane of the mouth is known as thrush.
2. Deep candidiasis is commonly secondary to chronic wasting disease, severe malnutrition and malignant tumor. Infection often occurs in the digestive tract, respiratory tract, heart (endocardium), kidney, brain, spleen, or other major visceral organs. There are multiple microscopic abscesses composed of fungal organisms (yeasts and hyphae), necrotic debris, with neutrophilic infiltrates. Uncommonly, a granulomatous reaction can be seen.
3. Systemic candidiasis results from dissemination of the organism in the blood stream. It can be life-threatening or fatal. In internal organ, candidiasis is

characterized by conspicuous tissue necrosis, micro-abscess, and subsequent granuloma.

Diagnosis: Candidiasis is diagnosed by identifying the fungus on microscopic examination or culture. In tissue sections, blastospore and pseudohyphae of Monilia is diagnosis feature of candidiasis. These spores and hyphae are blue in the HE staining. By using gram or silver staining, spores and hyphae can be seen more clearly.

16.8.2 Cryptococcosis

Cryptococcosis is a systemic fungal infection, caused by *Cryptococcus neoformans*. It usually infects the central nervous system, lung, skin, and bone. Cryptococcus neoformans can not only be isolated from soil, pigeon droppings, and fruit, but also can be isolated from healthy human skin, mucosa, and feces. Environmental pathogens mainly transmit through respiratory tract into the body to cause disease. Majority of cryptococcosis is secondary, especially in patients with Hodgkin's disease and leukemia.

Cryptococcosis is associated with chronic inflammation with a chronic granulomatous reaction. The lesions vary with the stage of disease. In the early stage, the lesion is jelly like because pathogens produce a lot of capsular substance, which can inhibit granulocyte tendency and phagocytosis. Thus, inflammation is mild in lesions. Neutrophil are rare and only a small number of lymphocytes and histiocytes infiltrate. In the late stage, granuloma formed. Fibrous tissue proliferates, and there is a great deal of macrophage, foreign body giant cells and lymphocytes infiltration. The majority of macrophages and foreign body giant cell cytoplasm may have cryptococcal. Fungi invade vessels leading to hematogenous spread and thrombosis.

Over 95% of cryptococcal infections involve the meninges and the brain. Meninges become thick in the base of the brain and jelly-like substance fill the subarachnoid space. In late period, there are more cellular exudation and granuloma form. Meninges are adhesion with brain tissue, which can affect cerebrospinal fluid circulation. Along the perivascular space, the infection also can invade brain tissue (mainly in the gray matter and basal ganglia) and form many small cavities. Microscopically the cystic cavity is the enlargement of perivascular space. Cavities are filled with cryptococcal and glue-like substance excretion. Sometimes vascular inflammation and thrombosis can occur in small vessels, which lead to the surrounding brain tissue ischemia and softening. Glial cells may have mild hyperplasia. Later, granulomatous lesions can occur in the meninges, brain, and spinal cord. The onset of cryptococcal meningitis is slow. Clinically, its symptoms are similar to tuberculous meningitis, and it is easy to be misdiagnosed. Brain parenchymal lesions can confuse together. Cerebrospinal fluid examination for pathogens may confirm the diagnosis.

Diagnosis: The organism has a prominent capsule that can be seen on micro-scopic examination. In HE staining tissue sections, cryptococcal is pale red and is not easy to be seen. The PAS or silver stains can demonstrate the yeasts but fail to stain the polysaccharide capsule. As a result, the organism appears to be surrounded by a halo. This appearance is helpful for diagnosis.

16.8.3 Aspergillosis

Aspergillosis is an opportunistic infection, caused by the fungus *Aspergillus fumi-gates*. This is a common environmental fungus that grows in the soil and is present in the air. Aspergillus spores enter the body through the respiratory tract. In most cases, Aspergillus is a conditioned pathogen, and pathogenic is only based on lower body resistance. Aspergillus can cause disease in many parts of the body, such as skin, ears, eyes, nasal cavity, heart, brain, kidney, respiratory tract, and gastrointes-tinal tract. Lung disease is the most common. Aspergillosis occurs in three relatively distinct forms:

Allergic bronchopulmonary aspergillosis occurs when fungal spores are inhaled and initiate an immune response in the airways. Bronchi and bronchioles contain fungal organisms and show chronic inflammation of the walls of the airways. Allergic bronchopulmonary aspergillosis is similar to clinical manifestations of bronchial asthma. The pathogenesis may be related to allergic reaction induced by aspergillus antigen, which can cause the bronchial spasm and secretion of sticky sputum. Serum IgE and precipitation of antibody against aspergillus antigen can be detected.

Aspergilloma (fungus ball) occurs in people with pre-existing lung disease. The fungal infection creates a cavity 1–7 cm in diameter, filled with a dense mass of fungal hyphae and surrounded by fibrous tissue. The wall of the cavity is composed of collagenous connective tissue, infiltrated by lymphocytes and plasma cells.

Invasive aspergillosis occurs most often in the lungs of patients with impaired neutrophil function. In this situation, aspergillus tends to invade vessels, causing thrombosis, pulmonary infarction, and hematogenous spread of the infection. The lesions are associated with suppurative inflammation, micro-abscesses, and granu-lomas. Hyphae can be observed in the small abscesses and necrotic focus. Aspergillus often causes thrombosis, vascular obstruction, ischemia and necrosis of the tissue. The chronic lesions are fibrous tissue proliferation, including a large number of lymphocytes and mononuclear cell infiltration, and miscellaneous most macro-phages and foreign body giant cells. A large number of mycelium can be observed in the lesions. The infection may spread to other organs such as the brain, kidney, or heart. Aspergillosis of the heart valve usually produces vegetations.

Diagnosis: Aspergillus showed blue purple by HE staining. With the PAS stain-ing, aspergillus was clearly visible.

16.8.4 Mucormycosis

Mucormycosis is a fungal infection caused by fungi *Mucor*, *Rhizopus*, or *Absidia* species organisms. They cause serious sporadic opportunistic infections in patients with underlying diseases, such as diabetic acidosis and acute leukemia. Mucormycosis also occurs in patients treated with corticosteroids or cytotoxic drugs.

Mucormycosis is an acute suppurative inflammation, which spreads rapidly and causes disseminated diseases. The fungi vascular invasion causes thrombosis and infarction. This disease is often characterized by hyphae growing in and around vessels. Vascular invasion allows hematogenous spread of the fungus. Mucormycosis frequently involves the sinuses, brain, or lungs as the areas of infection. Cerebral mucormycosis is the most common type of the disease and can cause death in the short term.

Mucorales initially invade nasal mucosa and then diffuse around, involving the eyeball, soft tissue, blood vessels and nerves. Mucorales invasion causes eyelid infarction and local skin gangrene. Then, it can spread into the cranial cavity and cause orbital, internal carotid artery cavernous sinus thrombosis, fungal meningitis, and cerebral necrosis. The development is rapid and the early diagnosis is very important.

Diagnosis: In the lesions, there are generally no spores, but hyphae in HE staining by hematoxylin stain can be easily observed. PAS or silver stains show non septate wide fungal hyphae with marked right-angle branching.

References

1. Yi L, Zhu L, Hong B. Pathology. 9th ed. People's Health Publishing House; 2018.
2. Nash AA, Mims CA, Stephen J. Mims' pathogenesis of infectious disease. 5th ed. Academic; 2015.
3. Kumar V. Robbins and Cotran: pathologic basis of disease. 7th ed. Oxford: Elsevier; 2014.
4. Liebermeister K. Pathology and treatment of the infectious diseases. Nabu Press; 2010.

Chapter 17
Parasitosis

Zhengshun Xu and Hongwei Ai

Contents

Parasitosis refers to a disease resulting from parasitic infestation. Three necessary conditions for the prevalence of parasitosis are source of infection (person or animal infected by parasite), route of transmission (living environment that is suitable for

Z. Xu (✉)
Medical College, Henan University of Science and Technology, Zhengzhou, China

H. Ai
Institute of Forensic Medicine, Henan University of Science and Technology, Zhengzhou, China

© Zhengzhou University Press 2024
K. Chen et al. (eds.), *Textbook of Pathologic Anatomy*,
https://doi.org/10.1007/978-981-99-8445-9_17

parasite), route of infestation, and vulnerable group (individuals who are lack of immunity against parasitic infestation or individuals with low immunity). Therefore, the spread of parasitosis from humans to humans, animal to animal, or humans to animal will be not only affected by biological factors, but also natural factors and social factors. The prevalence of parasitosis is also featured by territoriality, distinct seasonality, parasitic zoonosis, and other natural focus.

Parasitosis can be divided into acute and chronic type, and chronic type is frequently seen. Some hosts infected by parasite but showed no symptoms are called recessive infection or parasite carriers; in some cases, parasites will lodge in tissues and organs other than the common sites, and this phenomenon is called ectopic parasitism. When human body is infected by parasite, different symptoms may present according to the difference of parasite virulence and host's resistance. The main influence and damage of parasite on the host are as follows: (1) mechanical injury: parasite will lead to local damage, oppression or obstruction during the course of lodging in the host's body, migration, growth and reproduction, and removal from the body; (2) toxic effect: the metabolite and secretion of parasite and the decomposition product of dead parasite will produce toxic effect on the host; (3) immunologic injury: the antigenicity possessed by the secretion, excrement of parasite and the decomposition product of parasite could induce the immune response of host, presenting protective immunity or leading to immunopathological changes; (4) nutrition taking: parasite absorbs the nutrition from the host. If the nutrient substance is absorbed, the host will be lack of nutrient, and the immunity will weaken.

Parasitosis is widespread around the world, especially in areas with poor economy and living condition. It is commonly seen in the developing countries in tropical and subtropical areas. Some parasitic diseases are very rampant at present, severely harming people's health. This chapter will only cover amoebiasis, schistosomiasis, *Clonorchiasis sinensis*, pulmonary type paragonimiasis (paragonimiasis), and echinococcosis (hydatid disease).

17.1 Amoebiasis

Amoebiasis is an infection caused by any of amoebas of entamoeba histolytica. The amoeba is mainly lodging in the colon, and from blood circulation it can also reach to the liver, brain, lung and skin where it can cause amoebic ulcer or amoebic abscesses. Also, amoebiasis may become a systemic disease after affecting various tissues and organs. Amoebiasis is usually transmitted by the fecal-oral route, and the source of infection is the cyst form of the parasite found in feces. It is often epidemic in regions of the world, especially in tropical and subtropical areas, with an infection rate of 0.37–30%.

17.1.1 Intestinal Amoebiasis

17.1.1.1 Etiology and Pathogenesis

Intestinal amoebiasis, also known as amoebic dysentery, is caused by entamoeba histolytica's infection in colon. Symptoms may include abdominal pain, diarrhea, or tenesmus. The life cycle of entamoeba histolytica can be divided into cystic stage and trophozoite stage. In the infection stage, there are four mature cysts, whereas the pathogenic stage belongs to trophozoite stage. Cyst is usually seen in the feces of patient with amoebiasis, and taking food and water contaminated by cysts is the main route of human infection. When the cyst enters the digestive tract and reaches to the ileocecal junction, excystation will occur under the digestion of alkaline intestinal juice and trophozoite will come into being, with a size of about 10–60 μm. Trophozoite obtains nutrition at the upper end of colon and starts proliferation and gradually becomes the early stage of cystica. Trophozoite can swallow red blood cell and tissue fragments, leading to ulcer after invading and damaging tissue of intestinal wall.

17.1.1.2 Pathological Change and Clinical Manifestation

Cecum and ascending colon are the main parts of pathological change, followed by sigmoid colon and rectum. Basic pathological changes include alterative inflammation based on histolysis and liquefaction. It is characterized by flask-shaped ulcer and can be divided into acute stage and chronic stage.

Pathological Change of Acute Stage

Gross finding: Multiple bullate necroses or ulcers, which are grayish with needle shape surrounded by congestion and hemorrhage belt, are noted on the surface of intestinal mucosa in the earlier stage. The necrosis focus enlarges, presenting round-button shape. Trophozoite obtains nutrition from the necrotic tissue and red blood cells of human body, breeding in the mucous layer of intestine and entering into the submucous layer after destructing the mucosal muscular layer. Tissues in the submucous layer are loose, and thus, it is good for amoeba to spread around. Flask-shaped ulcer with sneak margin will take shape when the necrosis tissue is liquefied and falls off. The mucous membranes of the ulcer are normal or manifest mild catarrhal inflammation. If the focus continues to expand, large ulcer with sneak margin will come into being. Some severe ulcers may affect intestinal wall, or even the serosal layer, leading to intestinal perforation and triggering peritonitis.

Under the microscope, pathological changes are mainly characterized by necrosis, dissolution, and liquefaction of the tissues. The inflammatory reaction is mild in the surrounding area of focus, and only congestion, hemorrhage, and infiltration of few lymphocytes, plasmacytes, and macrophagocytes are noted. Amoeba trophozoite can be observed at the junction of ulcer margin and normal tissue, also in the venule of intestinal wall. Trophozoite on the tissue slice is usually round in shape and larger than the macrophagocyte in terms of volume. There is a spherical vesicular nucleus in the trophozoite with a diameter of 4–7 µm. The cytoplasm usually contains glycogen vacuole or swallows red blood cell, lymphocyte, and fragment of tissues.

Pathological Change of Chronic Stage

Due to the coexistence of new and old pathological change, necrosis, ulcer, granulation tissue, proliferation and scar occur repeatedly and alternatively, resulting in the formation of hyperplastic polyp in mucous. In the end, the intestinal mucosa may loss its normal shape completely. The intestinal wall will probably be hardened because of the proliferation of fibrous tissue, which may even give rise to lumen stenosis. Sometimes, amoeboma will come into being if there is too much proliferation of granulation tissue. Such symptom is frequently seen in cecum.

   The complications of amebic dysentery are intestinal perforation, intestinal hemorrhage, lumen stenosis, appendicitis, amoeba anal fistula, and so forth. It will also give rise to pathological change in liver, lung, brain, and organs other than the intestine. Intestinal hemorrhage is common as the small vessels of the intestinal wall are destructed by the pathological change.

17.1.2 Extraintestinal Amoebiasis

Amoebic liver abscess is the most important and common complication of intestinal amoebiasis, which usually occurs in amebic dysentery. The amoeba trophozoite under the intestinal mucosa or muscular layer invades the small vein of intestinal wall and reaches to the liver via portal vein. It is rarely seen that it will infect the liver directly. Amoebic liver abscess is usually in single state and is generally located in the right lobe of the liver (80%).

Gross findings: Abscesses differ in size, with the larger one as big as a fetal head. The abscess content is in brown jam shape, composed of liquefactive necrosis substance and obsolete blood. The connective tissue, blood vessel, and bile duct of the portal area of the necrosis that are not completely liquefied with flocculent shape attached on the abscess wall.

Microscopic findings: Liquefactive necrosis, the light red structureless substance, is inside the intracavity. Incompletely liquefactive necrosis tissues, such as inflammatory cell infiltration, are noted in the abscess wall. Trophozoite is revealed at the junction of necrosis tissue and normal tissue, and granulation tissue and fibrous tissue are accumulated at the surrounding area of chronic abscess.

17.2 Schistosomiasis

Schistosomiasis is a parasitosis caused by the infection of cercaria via human skin or mucous membrane when human body is in contact with the infected water of schistosome cercaria. The main pathological change is the formation of granuloma triggered by the ovum. This section will briefly introduce diseases caused by *Schistosoma japonicum*.

17.2.1 Cause of Disease and Route of Infection

The life cycle of *S. japonicum* can be divided into ovum, miracidium, sporocyst, cercaria, schistosomulum, and adult. Adult takes human body or other mammal such as dog, cat, pig, and cattle as a host, while oncomelania as an intermediate host. The development and reproduction stage of miracidium and cercaria are completed in oncomelania. Three conditions are necessary for schistosomiasis transmission, which include ovum's feces in water, breeding of oncomelania, and human body's contact with infected water.

Adult is dioecious and lodges in the portal vein and the mesenteric vein system. Female produces eggs in the vein of mesenterium, where some ova flow to the liver via the blood flow; some ova enter the intestinal cavity via intestinal wall and are excreted out of the body along with the patient or diseased animal's feces; ova deposited in the local tissues without being excreted will be gradually dead and calcified. When the excreted ova enter the water, miracidium will be hatched. Miracidium meets oncomelania, the medium host, in water and invades the soft tissue of breech block. After the stage of mother sporocyst-daughter sporocyst, miracidium will develop into cercaria and leave oncomelania again to the water. The cercariae of *Schistosoma japonicum* are mainly distributed on the water surface. When human or animal is in contact with the infected water, cercaria, based on the action of tissue plasminogen activator excreted by the cephalic gland as well as the mechanical motion of its muscle contraction, develops into schistosomulum after drilling in the skin or mucous membrane, and removing the tail. Schistosomulum enters the venule or lymph vessel, reaching to the lung along with blood flow via the right heart and spreading to the whole body via blood circulation. Schistosomulum that has entered into the mesenteric vein will be able to develop as adult, and the rest of them are dead on the way. It mainly takes about 3 weeks for cercaria to develop into adult. Female and male adults produce eggs after mating, which will develop into mature ovum containing miracidium after about 11 days. When the ovum matures in the intestinal wall, it will destruct the intestinal mucosa and enter the intestinal cavity, and will be discharged out of the body along with the feces. This is the complete life cycle of *S. japonicum*. The lifespan of ovum in tissue is about 21 days, whereas the adult in human body is about 4.5 years on average.

17.2.2 Basic Pathological Change and Pathogenesis

During the process of *S. japonicum* infestation, cercaria, schistosomulum, adult and ovum will do harm to the host, among which the ovum's damage is the greatest. The antigen released by schistosome in different stage induces the host's immune response, and this is the main reason and mechanism leading to the damage.

1. **Cercaria**: Cercarial dermatitis will occur within 3 days when cercaria drills in the skin which presents as a small papule with itching in the invading area. It will fade away automatically. The microscopic findings show dermis congestion, edema and hemorrhage, with infiltration of neutrophils and eosinophils in the early stage and cellular infiltration in the later stage.
2. **Schistosomulum**: Schistosomulum may cause vasculitis and perivascular inflammation when moving in the body. The damage to lung tissue is the most obvious, presenting lung tissue congestion, edema, punctuate hemorrhage, and leukocyte infiltration, but lesion is generally mild and short.
3. **Adult**: Adult's mouth and ventral sucker adsorb on the vascular wall, leading to the damage of vascular wall at the lodging position, as well as endophlebitis and periphlebitis; its metabolite, secretion–excretions, and other antigen stimulate the host to produce corresponding antibody, forming antigen–antibody compound and inducing type III allergic reaction. Mononuclear phagocyte proliferation is noted in the liver and spleen. The dark brown schistosoma pigment, which is always swallowed, is a heme-like pigment formed after hemoglobinolysis under the action of globinase in the adult body. Similar pigment is also seen in the adult's intestinal tract. The surrounding tissues of dead adult will become necrotic. With a large number of eosinophil granulocytes in infiltration, eosinophilic abscess will come into being.
4. **Ovum**: Ovum deposition causes main lesion. Ovum mainly deposits in sigmoid colon, rectum, and liver, and it is also seen in the terminal ileum, appendix, ascending colon, lung, and brain. A deposited ovum can be divided into immature egg and mature egg. Mature ovum contains mature miracidium, which secretes soluble ovum antigen and thus leads to the formation of characteristic ovum nodule (schistosomal granuloma). According to the development of the pathological change, this nodule can be divided into acute ovum nodule and chronic ovum nodule.

   (a) **Acute ovum nodule**: Acute ovum nodule is a kind of acute necrosis and exudative focus caused by mature ovum. Gross findings: Small grayish nodule with a size of chestnut grain to green bean. Microscopic findings: One to two mature ova are noted in the central area of nodule. Sometimes, radial and acidic clava is attached on the ovum surface. It is the antigen–antibody composite formed by the corresponding antibody produced when the soluble egg antigens (SEA) released by ovum miracidium stimulates the B-lymphocyte system. The surrounding area is a combination of structureless granular necrosis substance and a large number of eosinophil infiltrations. It is in abscess shape and thus is known as acidophilia abscess. Rhombus or multifaceted and refractive protein crystal, as well as charcot-leyden crystal, can be seen inside, as they are composed of acidophil granules of eosinophils. Later, the ovum generates granulation tissue layer at the surrounding area, where lymphocyte, macrophagocyte and eosinophils are in infiltration. Eosinophils account for the main proportion. As the course of disease progresses, the granulation tissue layer gradually develops toward the center of ovum nodule, and epithelioid cell layer in radiate array will come into being at the surrounding area of the nodule. Eosinophils decline significantly, constituting acute ovum nodule of later stage. This is the transition stage developing toward ovum nodule.

(b) **Chronic ovum nodule**: After about 10 days, miracidium will be dead in the egg, and the antigen substance it secretes will disappear. The necrosis substance inside the focus will gradually be eliminated by macrophagocyte, and the ovum will be disintegrated and ruptured. The macrophagocyte inside the focus will become epithelioid cell and few foreign body giant cells. Lymphocytes infiltration and granulation tissue proliferation are noted in the surrounding area of focus. It looks like tuberculoid granulation tumor, thus known as pseudotubercle or chronic ovum. In the end, the nodule will be in fibrosis hyaline change. The central fragment of egg shell and calcified dead egg will be retained for a long time.

17.2.3 Pathological Change of Main Organs

1. **Colon**: As adult usually lodges in the inferior mesenteric vein and superior vein of hemorrhoid, the symptom is most obvious in rectum, sigmoid colon, and descending colon. Gross findings: Intestinal mucosa with congestion and edema as well as flat but bullate focus, which is yellow in granular shape with a diameter of 0.5–1 cm, is noted. Necrosis falls off from the central focus, forming superficial ulcer with uneven size and irregular edges. Ovum may fall into the intestinal cavity and appear in the feces.

   In chronic phase, due to the repeated depositing of ovum, ulcer, and intestinal fibrosis will occur again and again in the intestinal mucosa, leading to the thickening and hardening of intestinal wall, stenosis of intestinal cavity, or even intestinal obstruction. Due to the hyperplasia of connective tissues in intestinal wall, it is hard to secrete ovum into the intestinal cavity. This is the reason why ovum is not easily observed in the feces of patients with advanced disease. In addition, in some cases, the atrophy of intestinal mucosa is noted with disappeared plica. Some of them are in polypus hyperplasia.

2. **Liver**: In acute stage, the liver is slightly enlarged, and multiple small grayish white or grayish yellow nodules, in a size of chestnut grain or green bean, are noted on the surface and section. Microscopic findings: Many acute ovum nodules are revealed near the portal area. Liver cells will be atrophied because of

pressure, and the possibility of degeneration and small focus necrosis also exists. The symptoms include congestion of hepatic sinusoid, Kupffer cell proliferation, and swallowing of schistosoma pigment.

In chronic stage, chronic ovum nodule and fibrosis are noted in the liver. In cases of long-term serious infection, there are a large number of fibrous tissue proliferations near the portal area. The liver will gradually be harder and smaller, turning into schistosoma-type liver cirrhosis because of severe fibrosis. The liver surface is not smooth, and is divided into several upheaval areas of different sizes by shallow grooves. In serious cases, thick and large nodule will come into being. In terms of section, the hyperplastic connective tissues are distributed like branches of trees toward the portal vein. Thus, it is known as pipe stem cirrhosis. Microscopic finding shows a large number of chronic ovum nodules in the portal area, accompanied with lots of proliferation of fibrous tissue. The hepatic lobule is not seriously destructed, and the pseudo lobule is not obvious, different from portal cirrhosis.

3. **Spleen**: Gross findings: The spleen is strong but pliable in texture with thickened envelope. The section is dark red with brown siderotic nodule on the surface. It is mainly composed of old hemorrhage focus accompanied with irony and calcareous deposit as well as proliferation of fibrous tissue. Infarction tissues are occasionally found. Microscopic findings: Atrophy of splenic corpuscle reduces and schistosoma pigment deposit is seen in the mononuclear macrophage. Ovum nodule is occasionally found in the spleen.

17.3   Clonorchiasis sinensis

C. sinensis, or clonorchiasis, is a parasitosis caused by the adult of clonorchis sinensis, which inhabits human intrahepatic ducts. This disease is common in Asia.

17.3.1   Cause of Disease and Route of Infection

The development process of C. sinensis includes adult, ovum, miracidium, sporocyst, redial, cercarial, encysted metacercaria, metacercaria, and so forth. The adult usually lodges in the intrahepatic ducts of human, dog, cat, and pig. When adult produces eggs, the ovum will be discharged via the intestinal tract with bile. Ovum containing miracidium could be swallowed by freshwater snail, the first intermediate host, and from its digestive tract, miracidium will be hatched. Miracidium develops into sporocyst in the snail body, and generates many rediae and cercariae via vegetative proliferation. Mature cercaria leaves the snail to the water, invading freshwater fish or freshwater shrimp, the second intermediate host, and develops into encysted metacercaria in the second intermediate host's muscle. When human or animal ingests fish or shrimp containing encysted metacercaria without being cooked, encysted metacercaria breaks out its cyst after the action of digestive juice in the gastrointestinal tract. The cercaria reversely flows to the intrahepatic bile duct along the bile and lodges in and develops into adult. It takes about 1 month from the intake of encysted metacercaria to the appearance of ovum in feces.

17.3.2 Pathological Change and Complications

C. sinensis mainly lodges in the intrahepatic bile ducts. In less severe cases, there are few worms. Ovum is founded in the feces. Adult lodges in the intrahepatic duct and mucosa epithelium exfoliates from the bile duct. The appearance of liver and bile duct is normal. In severely infected cases, the pathological change may cause cholangitis, cholecystitis, calculus of bile duct, cirrhosis, and atypical hyperplasia of biliary epithelia, or even cholangiocellular carcinoma.

1. **Liver**: Intrahepatic bile duct dilation is the most significant lesion. In cases with severe infection, the liver is slightly swollen with increased weight in the gross findings. Bile duct branch dilated because of the mechanical blockage of adult is seen below the envelope. Such pathological change is common in the left lobe as the bile duct of the left lobe is flat and will be easily invaded by the schistosomulum. The intrahepatic large and middle bile duct with varying degrees of dilation and tube wall thickening are noted on the section. The lumen is full of bile and countless adults. According to the microscopic findings, intrahepatic bile duct dilation is noted, and the epithelial cell and submucosal glands show different degree of hyperplasia. In severe cases, papillary, adenomatous, or atypical hyperplasia will occur, which may lead to canceration. Lymphocyte, plasmocyte, and eosinophilia infiltration are found in the tube wall. In chronic disease, significant fibrous connective tissue is noted.

   Adult lodges in the intrahepatic bile duct, leading to cholestasis in the bile duct and causing secondary infection easily. The dead polypide, ovum and deciduous bile epithelia can also become the core of gallstone, which is good for the gallstone formation.

17.4 Pulmonary-Type Paragonimiasis

Paragonimiasis is a disease caused by the schistosomulum spreading in the tissue and lodging of adult worm. The worm mainly lodges in the lung, with causing pulmonary type paragonimiasis, also known as paragonimiasis. The feature of pathological change is formation of sinus and multilocular small cysts in organs and tissues. This disease is prevalent around the world, especially rampant in summer and autumn.

17.4.1 Pathogeny and Route of Transmission of Infection

The chapter will introduce how to cause pulmonary-type paragonimiasis by *Paragonimus westermani*. The symptoms of pulmonary-type paragonimiasis are coughing, coughing with rusty sputum, hemoptysis, and so forth.

The adult worm of *Paragonimus* lodges in the lung of human body and other mammals, such as cat, dog, and pig. Ovum is usually coughed out with the sputum or discharged with the feces. Ovum entering into water will hatch into miracidium and then drill in the freshwater nail, which is the first host. The ovum will develop into cercaria from sporocyst, mother redia and daughter redia. The cercaria could invade into the body of freshwater stone crab or crayfish, the second host, and develop into encysted metacercaria when it leaves the nail within 2 days. Encysted metacercaria is the type of lung fluke infection. If people feed on stone crab or crayfish containing encysted metacercaria, these will enter the digestive tract and develop into schistosomulum after the action of digestive juice and excystation. Schistosomulum, which has strong ability in daily living, is able to enter the abdominal cavity via the intestinal wall by action of excreted enzyme. Most schistosomulums spread along the peritoneum, move upward via the surface of liver, spleen, and stomach. The schistosomulums can reach to the thoracic cavity directly after passing through the diaphragm and finally invade the lung and develop into adults, where they produce cysts and ova. A few of schistosomulum retain at the abdominal cavity and continue to grow and develop mostly into cercariae after penetrating the superficial layer of the liver or greater omentum. It takes about 2 months from entering into the body as encysted metacercaria to producing ova as cercariae.

17.4.2 Pathogenesis and Basic Pathological Changes

The mechanical injury schistosomulum's spreads in tissue and adult's lodging on local tissues is the main pathogenic effect of *Paragonimus*. In addition, immunopathological effects caused by antigen substance, such as metabolite of polypide, and the formation of foreign body granuloma induced by ovum have also played a certain role in the pathogenic effect.

1. **Serositis**: Polypide may lead to fibrinous or serous fibrinous peritonitis or pleuritis when it travels or lodges in the body cavity.
2. **Tissue destruction and sinus tract formation**: Polypide spreads in the tissue with forming tortuous sinus tract, and may lead to necrosis and hemorrhage. Eosinophils and lymphocytes infiltration are noted in the sinusoidal wall under the microscope. Ovum can be developed into fibrosis.
3. **Formation of abscess, cyst, and fibrous scar**: Schistosomulum or adult will lead to tissue bynecrosis and hemorrhage and also severe inflammatory reaction when it lodged in the organ. Apart from eosinophils, there are a large number of neutophile granulocytes forming abscess. The exudative inflammatory cells and

necrosis tissues will disintegrate and liquefy. The abscess content has been formation brown viscous liquid, with necrosis tissues, polypide, ovum, and Charcot-Leyden crystal noted under the microscope. At the same time, granulation tissue hyperplasia surrounding the abscess is noted, and fibrous membrane is taking shape. Worm cysts got its name because of the existence of polypide in the cyst. Ovum entering the cyst and tissue which surrounds the cyst could lead to form foreign matter granuloma. Due to the migration characteristic, polypide could leave the original cyst and continue to destroy other tissues nearby, and form new cyst. Cysts linked with sinus tracts are called multilocular cysts.

17.4.3 Pathological Change and Clinical Manifestation of Main Organs

1. **Lung**: Pleural thickens accompanied by extensive adhesion. The diaphragmatic surface is the heaviest. Worm cysts, larvae or adults, scattered or in group, are found in the lungs. Polypide and ovum could be found in the cyst. The worm cyst of lung often invades the bronchial wall, connecting the cyst and forming pulmonary cavity. If there is secondary bacterial infection, pneumothorax, empyema or even hemothorax will be happen. Obvious pulmonary fibrosis is noted in chronic cases.

17.5 Hydatidosis

Echinococcosis, also known as hydatid disease, is a parasitosis caused by the larva (hydatid cyst) of tapeworms.

17.5.1 Pathogeny and Route of Infection

The adult of echinococcus granulosus lodges mainly in the small intestine of definitive host, such as dog, wolf, and other carnivorous animals. The polypide is small, 2–7 mm in length and hermaphrodite. It is composed of scolex and three somites (immature proglottid, mature proglottid and gravid proglottid). Gravid proglottid contains infectious ovum, and the gravid proglottid will contaminate pasture, vegetable, soil, and water when it is mature in the small intestine of definitive host, split away off the polypide, and expelled from the feces. The ovum and gravid proglottid hatch in the stomach or duodenum when they are taken by the intermediate host like goat, cattle, pig, rabbit, camel, or other livestock and human. Oncosphere will break out from the shell, attaching on the mucous membrane and then drilling in the blood vessel of the intestinal wall. It will finally reach to the liver via the blood flow of portal vein. Therefore, liver hydatid disease is commonly seen. Few could reach to the lung from the liver and right heart.

17.5.2 Pathogenesis and Basic Pathological Change

The damages of hydatid cyst on organs can be divided into three parts, but mechanical damage is the key part: (1) for the space-occupying growth of hydatid cyst, oppression and damage to the surrounding tissue, the degree of damage depends on the volume, quantity, time of lodging, and position of hydatid cyst; (2) when the hydatid cyst fractures, the foreign protein contained in the cyst fluid will lead to allergic reaction inside the body or anaphylactic shock to dead; (3) hydatid cyst will take in the nutrition from host during its growth and development process, affecting the body health to some extent.

Hydatid cyst is composed of cyst wall and cyst content. The cyst wall can be divided into inner layer and outer layer. The outer layer is cuticle layer, which is white and semitransparent. For instance, vermicelli, about 1 mm in thickness, is characterized by absorbing nutrient substance and protecting the germinal layer; lamellar structure with paralleled red staining is noted below the microscope. The inner layer is germinal layer, or germ layer, with a thickness of about 20 μm. It is composed of single-layer or multilayer germinal cell with significant multiplication capacity. Inclusions are produced from this layer to the cyst. Cyst inclusion includes cyst fluid (or hydatid cyst liquid), protoscolex, brood capsule, daughter cyst, granddaughter cyst, and so forth. The germinal cell grows toward the internal bud and could form countless papule in the cyst inner wall, where brood capsule will gradually come into being. Brood capsule, a small cyst with only one germinal layer, contains many protoscolexes inside. Brood capsule falls off and becomes daughter cyst. The cyst is able to produce protoscolex, brood capsule, and granddaughter cyst that has the similar structure of daughter cyst. In older hydatid cyst, there could be hundreds of daughter cyst. Sometimes, germinal layer grows buds externally and forms cysts.

17.5.3 Pathological Change of Main Organ and Its Consequence

1. **Liver**: Frequently seen in the right lobe. It is single but sometimes multiple. The cyst is located at the diaphragmatic surface, extruding toward the abdominal cavity. Hepatic hydatid cyst grows slowly and could lead to pressure atrophy, degeneration, or necrosis when it enlarges gradually. Its proliferation of fibrous tissue forms a layer of fibrous external cyst. The intrahepatic small bile duct and blood

vessel will be often shifted because of compression or encapsulated in the cyst wall.

2. **Lung**: Pulmonary hydatid cyst is caused by oncosphere when it reaches to the heart from liver and blood or burrows through the lung from the organs nearby. Cyst is commonly seen in the right lung, developed in the middle lobe, and located in the surrounding area of lung. It is usually single. Affected by loose lung tissue, rich blood circulation, and negative pressure suction, pulmonary hydatid cyst is developing rapidly and is able to oppress the surrounding lung tissue, inducing lung atrophy and fibrosis.

Chapter 18
Pathological Techniques

Yueping Liu

Contents

Y. Liu (✉)
Department of Pathology, Fourth Hospital of Hebei Medical University, Shijiazhuang, China

© Zhengzhou University Press 2024
K. Chen et al. (eds.), *Textbook of Pathologic Anatomy*,
https://doi.org/10.1007/978-981-99-8445-9_18

Objectives

1. The development of pathology and the progress of pathological technology promote each other. Macroscopic observation with naked eye and morphological observation at the level of optical microscope are the most traditional and basic

technologies for pathological research and learning. With the development of molecular biology research and the establishment of molecular pathology related technologies, some new and advanced technologies have been applied to pathological research and pathological diagnosis of diseases. In order to facilitate readers' learning and reference, this chapter briefly introduces various pathological techniques and methods.

Key Concepts
1. Pathological
2. Techniques
3. Immunohistochemistry
4. In situ hybridization
5. Microscopic
6. Comparative genomic
7. Biochip
8. Bioinformatics

Introduction
The pathology technology is formed and developed on the basis of pathology, which embodies the development and application of the cross penetration and the new technology of the frontier. The contents of pathology technology include general pathological anatomy and fixed production techniques for general specimens, which are the important basis of pathology, basic knowledge, principles, and techniques for preparation of cell, and tissue specimens.

18.1 General Tissue and Cell for Pathological Techniques

18.1.1 General Specimen Preparation

18.1.1.1 Requirement of Material Delivery Organization

The biopsy specimen should be immediately placed in 10% formaldehyde fixation after surgical incision or biopsy.

18.1.1.2 Materials

Based on the detailed examination of the tissue specimens, the location and number of the materials shall be determined according to the needs of the diagnosis. For multicut material, the same tissue "ice pair" and the remaining tissue "ice remnants" must be further compared with routine paraffin sections, with particular emphasis on the " ice pair." The biopsy tissue should be well-preserved because it is always small and usually needs further examination, such as HE staining, special staining, immunohistochemical staining, and electron microscopy analysis.

18.1.2 Tissue Fixation

18.1.2.1 Fixed Precautions

The volume of the fixed liquid should be at least four to five times that of the tissue blocks. The tissue thickness of less than 3 mm is more appropriate. The fixed time of the organization is usually 3 ~ 24 h. The fixed temperature is mostly room temperature. The fixed container should be relatively large so that the fixed liquid can be well infiltrated into the blocks.

18.1.3 Tissue Dehydration Optical Clearing and Waxdip

The tissue contains a lot of water after fixed. It must be dehydrated before immersion and embedding because water and paraffin wax cannot be mixed. Ethanol is most commonly used as a tissue dehydrator which has strong dehydration capacity. To make paraffin can be immersed into the tissue, after dehydration, the tissue must go through a solvent medium, xylene, which is can mixed with ethanol and dissolve paraffin (Table 18.1).

18.1.4 Tissue Paraffin-Embedded

Conventional paraffin embedding method: The tissue is immersed into the melting paraffin wax with warm tweezers. The buried surface must be flat, and the fragmentized tissue should be assembled. The biopsy tissue should be well-preserved because it is always small and usually needs further examination, such as HE staining, special staining, immunohistochemical staining, and electron microscopy analysis. Generally, 56 °C paraffin may be used for embedding. When the embedding is

Table 18.1 Tissue dehydration clearing and wax dipping time

Processing step	Processing time (min)
70% Ethanol	30 ~ 60
80% Ethanol	120 ~ 300
95% Ethanol I and II	120 ~ 300
Ethanol I and II	30
Ethanol III	60 ~ 120
Xylene I, II	30
Xylene III	60
56 ~ 58 °C paraffin wax I	30
56 ~ 58 °C paraffin wax II	60 ~ 120
56 ~ 58 °C paraffin wax III	120 ~ 180

ending, the wax block can be moved into the refrigerator to accelerate the solidification. The contents of the gastroscope specimen were vertical embedding. At last, the tissue is embedded into a wax block.

18.1.5 Paraffin Section

18.1.5.1 General Paraffin Section Process

The process of cutting the tissue block into thin slices (generally 4 ~ 6 μm) is called paraffin section. The tissue blocks are firstly cooled in the refrigerator and then mounted on a slicer fixture. The specimen should be repaired until all the tissue is exposed, and the tissue section should be flattened before the next step.

18.1.5.2 Frozen Section Method

Frozen sections are mostly used in fresh tissue and cold storage tissue blocks. A small amount of OCT or carboxymethyl cellulose is used to cover the tissue. In the temperature of 25 °C below zero, the tissue becomes a frozen block which can be sliced.

18.1.6 Tissue and Cell Sections Staining

18.1.6.1 Basic Principle

The nucleus is stained blue by hematoxylin. The staining of nuclei by hematoxylin is due to binding of the dye-metal complex and DNA. Most of the cytoplasm is eosinophilic which is stained into red by eosin. The terminology of basophilic or eosinophilic is based on the affinity of cellular components for the dyes. Other colors, for example, yellow and brown, can be present in the sample; they are caused by intrinsic pigments, for example, melanin. Some structures do not stain well. Basal laminae need to be handled by periodic acid-Schiff (PAS) stain or silver stain (Table 18.2).

The frozen section was fixed for 1 min with 95% ethanol 95 and 5 mL glacial acetic acid mixture before the program.

18.1.6.2 Application for H&E Staining

HE staining method can be used to observe the morphology and structure of normal and diseased tissues.

Table 18.2 Hematoxylin and eosin (H&E) staining protocol and time

Processing step	Processing time	
	LEICA AUTO STAINER XL	Frozen section
Xylene I	10 min	
Xylene II	10 min	
Ethanol I	1 min	
Ethanol II	1 min	
95% Ethanol I	1 min	
95% Ethanol I	1 min	
90% Ethanol	1 min	
80% Ethanol	1 min	
Water	1 min	1 min
Hematoxylin	1 ~ 5 min	1 ~ 2 min
Water	1 min	20 s
1% Hydrochloric acid alcohol	30 s	20 s
Water	5 min	30 s
Aqueous solution of ammonia	1 min	30 s
Water	5 min	20 s
Eosin	30 s ~ 5 min	20 s ~ 2 min
Water	5 min	10 s
85% Ethanol	20 s	20 s
95% Ethanol I	30 s	1 min
95% Ethanol II	1 min	1 min
Ethanol I	2 min	1 min
Ethanol II	2 min	1 min
Xylene I	2 min	1 min
Xylene II	2 min	2 min
Xylene III	2 min	2 min
Sealing	Yes	Yes

18.1.7 Connective Tissue Staining

18.1.7.1 Masson Trichromatic Staining

The collagen fiber is blue or green, muscle fiber cytoplasm and red cell red, and the nucleus is blue. Connective tissue staining is used to determine the degree of pathological changes and repair of various tissues on pathomorphology. The origin for tumor of spindle cell soft tissue is fibrous, muscular, or neurogenic.

18.1.8 Collagen Fiber Staining

Collagen fibers were bright red, whereas muscle fiber cytoplasm and red cell yellow. The pathological diagnosis of collagen fiber is mainly used in the differential diagnosis of muscle fibers. The spindle cell tumor of the soft tissue can be fibrous or myogenic. In H&E dyeing, it is sometimes difficult to distinguish. The fibrotic tumor was red with the Van Gieson (VG) method, and the myogenic tumor was yellow.

18.1.9 Rhabdomyolysis Tissue Staining

Rhabdomyolysis fiber, cellulose, nucleus, nucleolus, and neuroglia fiber are blue, whereas collagen fiber, reticular fiber, cartilage matrix brown red, and elastic fiber purple. If the tumor is striated muscle differentiation, blue stripes can be found in the tumor cytoplasm.

18.1.10 Carbohydrate Staining

18.1.10.1 Periodic Acid-Schiff Stain

The positive PAS is red and the nucleus is blue. (1) In diabetes, a large number of glycogen appears in the liver cells. (2) Differentiate malignant lymphoma of bone and Ewing's sarcoma of bone. Glycogen granules were found in the cytoplasm of the bone's Ewing's sarcoma cells and PAS was positive; the malignant lymphoma of bone was not glycogen and PAS was negative. (3) Congenital glycogen accumulation disease can be seen a large amount of glycogen deposit in liver, the trial, myocardium or skeletal muscle, and so forth.

18.1.10.2 Mowry Alcin Blue Iodate Scheff Staining (AB-PAS)

The neutral mucus is red, the acid mucus is green and blue, the compound is purple and red, the nucleus is gray-blue. Intestinal type gastric cancer cells secrete acid mucus, which is blue. Gastric phenotype gastric cancer cells secrete acid mucus, which is red.

18.2 Techniques of Histochemistry and Immunohistochemistry

Immunohistochemistry and immunocytochemistry examination is derived from tissue and cytochemical methods. Basic principle is to use the antigen, and the antibody contact can form "antigen–antibody complex" chemical reaction to detect tissue or intracellular antigen (or antibody) technology. Immunohistochemical and immunocytochemical tests have played an important role in improving the level of pathological diagnosis of tumor. At present, the immunohistochemistry autostainer is widely used (Fig. 18.1).

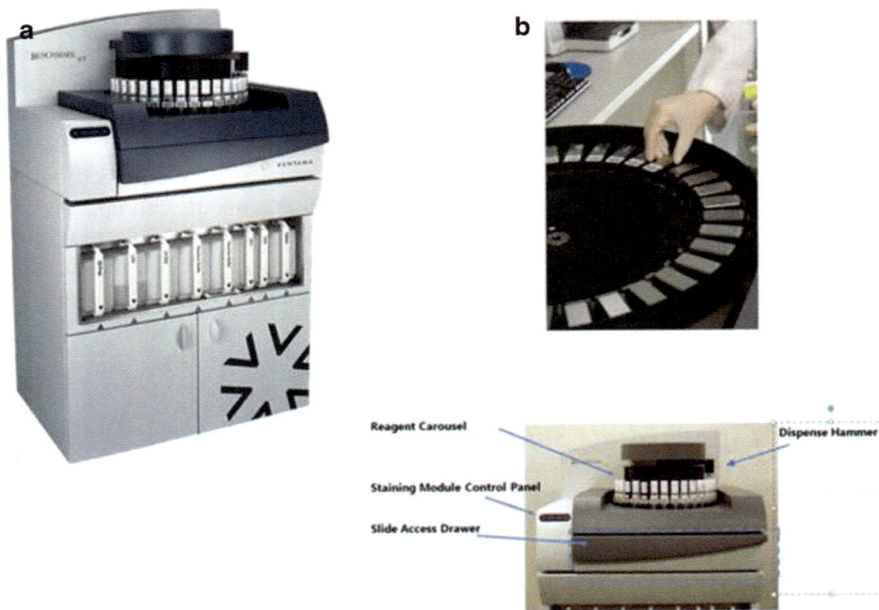

Fig. 18.1 Immunohistochemistry autostainer

18.2.1   Immunohistochemical Staining Procedure

1. Extract antigen (immune-ogen).
2. Monoclonal antibodies were prepared by hybridoma using immunogenicity (rabbit) to prepare antiserum (polyclonal antibody) or immune animal (mouse).
3. Purified antibodies.
4. Antibody marker tissue section (Fig. 18.2).
5. Dyeing reaction.
6. Observe the results.

18.2.2   Common Marker

18.2.2.1   Enzymes

The enzyme labeled should have the following conditions: the substrate is specific and easy to display; The enzyme reaction product is stable and not easy to diffuse. It is easy to obtain a pure enzyme molecule and stable. The enzyme-labeled antibody did not affect the activity of the two. There should be no endogenous enzyme or substrate in the tissue. The commonly used markers are horseradish peroxidase (HRP), alkaline phosphatase (AKP), glucose oxidase (GOD), and so forth.

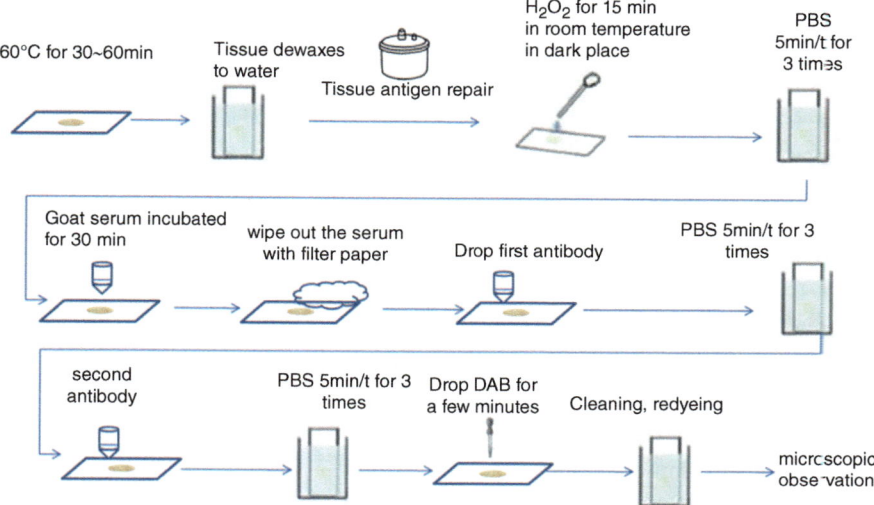

Fig. 18.2 Antibody marker tissue section

18.2.2.2   Fluorescein

It is a substance that produces fluorescence under the excitation of high energy light waves. Commonly used examples include isothiocyanate fluorescein (FITC) and tetramethyl isothiocyanate rhodamine (TRITC).

18.2.2.3   Biotin

The affinity of biotin and lecithin is significantly higher than that of antigen antibody.

18.2.2.4   Metal Markers

Ferritin and colloidal gold are used in immune electron microscopy.

18.2.3   Common Immunohistochemical Staining Method

The commonly used staining methods are direct method, indirect method, peroxidase, anti-peroxidase (PAP) method, and avidin biotin-peroxidase complex (ABC) method (Fig. 18.3).

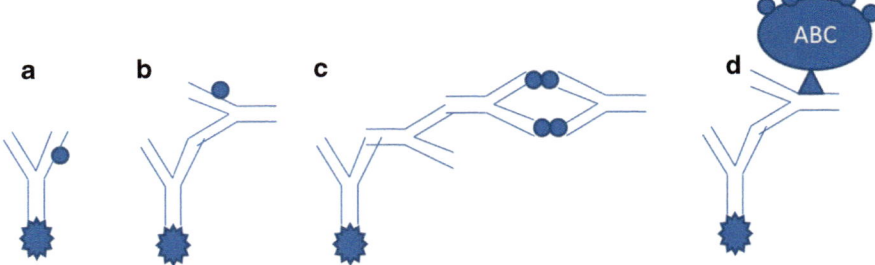

Fig. 18.3 Common immunohistochemical staining method. (**a**) Direct, (**b**) Indirect, (**c**) PAP, (**d**) ABC

18.2.3.1 Direct Method

The fluorescein (immunofluorescence) or enzyme is directly labeled on the first antibody to examine the corresponding antigen. The direct method has the advantages of specificity, but the sensitivity is poor and the antibody is much more.

18.2.3.2 Indirect Method

The second antibody was labeled with a fluorescein or enzyme, a specific antibody, and only racial specificity. The features of this method include the following: (1) It is convenient to pre-bid the two reactance. (2) It is more sensitive than direct, but still poor.

18.2.3.3 Peroxidase Antiperoxidase Complex Method and Bothway PAP

The peroxidase (HRP) immune rabbit/sheep/mouse was first made into rabbits/sheep/mice against HRP. Then, it is combined with HRP to form a stable polygon structure (PAP). The features of this method include: (1) high sensitivity, (2) low background staining (relative), (3) higher sensitivity of the two PAP, but relatively heavy background.

18.2.3.4 Avidin Biotin-Peroxidase Complex Method

The ABC method using avidin-biotin with high affinity, biotin and first form biotin enzyme HRP, again with biotin HRP mixed with avidin according to certain proportion, and form a compound. Also, the biotinylation was first. (1) Replace the avidin with streptomycin, which is SABC method: Strept Avidin Biotin-peroxidase Complex. (2) Link streptomycin and biotin first, which is the LSAB method:

Labeled Streptavidin-Biotin. (3) Streptomycin antibiotic protein linked horseradish peroxidase, which is an S-P method: Streptavidin-Peroxidase conjugated method. If alkaline phosphatase is used to mark the streptomycin, which is the SAP method, then the alkaline phosphatase and horseradish peroxidase are used to mark the streptomycin, which is the DS method. The features of this methods include: (1) high sensitivity (8 ~ 40 times higher than that of PAP); (2) background light, streptavidin is better; (3) method is simple, as time is short. (4) wide range of applications, but also can be used for in situ hybridization and immune electron microscopy.

18.2.4 Attention in the Process of Immunohistochemical Staining

18.2.4.1 Set the Correct Control of Immunohistochemistry

Control principle: compare the first antibody. The same principle should be noticed in the substitution control; positive results have negative control; the negative result has positive control; the staining is clear and the location is accurate. The negative control includes: (1) blank control: first antibody is replaced with PBS. (2) The replacement of the serum control: with the normal serum of the same animal instead of the first antibody. (3) Inhibition of the control: unlabeled antibody combined with the corresponding antigen first. (4) Absorbable control: the purified antigen was used to absorb the antibody. Positive control: a tissue that is known or proved to be positive. Self-control: the use of various tissue components within the tissue section as a control.

18.2.4.2 False-Positive Reactions

Nonspecific reactions: edge phenomena, creases and knife marks, bleeding and necrosis, and so forth. Endogenous peroxidase: erythrocyte, inflammatory cells, degeneration necrotic cells and certain glandular epithelial secretions, and some tissues rich in peroxidase, such as brain, liver, and so forth. The cross-reaction of the antibodies: the antibody itself contains the components of the cross-reaction with the human tissue. The concentration of the reagent is too high or ineffective.

18.2.4.3 False-Negative Reactions

(1) The organization is not fixed or fixed for a long time. (2) Low antibody titer is too low or too long. (3) The antigen in the tissue is blocked by a viscous matrix or secretion. (4) The concentration of DAB or H_2O_2 is improper.

18.2.5 Immunohistochemical Results Determination Method

For the quantitative judgment of positive results, the conventional method is to classify and count according to the color depth or the number of positive cells, grading, and counting statistics with −, +, ++, +++, and so forth. At present, image analysis has been used to measure.

18.2.6 The Application of IHC in Tumor Diagnosis and Differential Diagnosis

The reasons for the application of immunohistochemical staining in the diagnosis of tumor include the following: (1) In conventional biopsy, 5–10% of the most difficult cases are not clearly diagnosed. (2) Tumors with similar morphologic structures are difficult to identify for different tissue sources, such as undifferentiated and malignant lymphoma, and are difficult to be treated. (3) Some metastatic tumors often lack specific histologic features and are unable to determine the primary lesions (such as thyroid cancer and prostate cancer). (4) Through some markers that reflect the proliferation of cells and the malignancy of the tumor, it helps to identify the benign and malignant tumors.

18.3 Electron Microscope Technology

Electron microscope, referred to as electron microscope, is based on principles of electro-optics. Instead of ordinary beam optical lenses, electrophoton beams and optical electron lenses are used for imaging the fine structure of matter magnified under high magnification, high-speed electronics than the wavelength of visible light with short wavelength (the wave-particle duality) and the resolution of the microscope by the restrictions on the use of wavelength. In 1970s, the resolution of transmission electron microscopy was approximately 0.3 nm in size (resolution of approximately 0.1 mm to the naked eye). High-speed electronics operate at wavelengths shorter than visible light (the wave-particle duality). The resolution of a microscope is limited by the wavelength of light used, which is why electron microscopy (about 0.1 nm) is far higher than the resolution of optical microscopy (about 200 nm). Consequently, electron microscopes can reveal atomic lattices of heavy metals and crystals that are too small to be seen by the naked eye (Fig. 18.4).

Fig. 18.4 Electron
microscope. (Source:
Edward C Klatt, Vinay
Kumar MBBS. Review of
Pathology. Fourth Edition.
ISBN 978-4557-5155-6.
Page 193, appendix
Fig. 53, electron
micrograph)

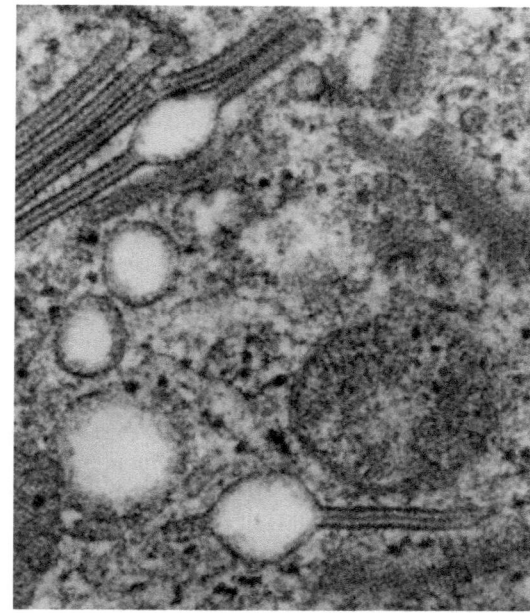

18.3.1 Related Technology of Electron Microscope

18.3.1.1 The Discernable Ability of the Electron Microscope

The resolution of an electron microscope is expressed by the minimum distance between two adjacent points that it can distinguish, which is an important indicator of electron microscopy. The resolution of the microscope can be obtained by 0.6 lambda/sin alpha, where alpha is half of the sharp angle of the electron beam passing through the sample, lambda is the corresponding wavelength.

18.3.1.2 The Composition of the Electron Microscope

The electronic source is a cathode that releases free electrons, and an annular anode accelerates the electrons. The voltage difference between the cathode and the anode must be very high, usually between thousands of volts and three million volts.

A detector is used to collect electronic signals or secondary signals.

18.3.2 The Type of Electron Microscope

Electron microscope can be categorized into transmission form electron microscope, scanning form electron microscope, reflection form electron microscope, and electron beam form electron microscope according to the structure and usage.

Transmission electron microscopy (TEM), which is named after the electron beam penetrates the sample, is amplified with an electronic lens. Its optical path is similar to that of an optical microscope. In such electric mirror, the contrast of this image details is formed through the scattering of electron beams by existing atoms in matter. Thin or low-density areas scatter fewer electrons from the electron beam, resulting in more electrons contributing to the imaging. These areas appear brighter in the image as they pass through the optical column of the objective lens [1].

The digital electron microscope is also known as the video microscope for its physical image that the microscope can see through digital to analog conversion and make it imaged on the screen or computer that the microscope brings. Digital electron microscope (SEM) is a high-tech electronic product which is perfectly integrated with elite optical microscopy technology, advanced photoelectric conversion technology, and liquid crystal screen display technology. Thus, we can improve the work efficiency by observing the microsphere from the traditional eyes to the display on the display. Three-dimensional digital microscope images can produce upright images when observing an object.

18.3.3 The Insufficiency of the Electron Microscope

In the electron microscope, the sample must be observed in the vacuum, so the live sample cannot be observed. The difficulty in analyzing the image is exacerbated by the possibility that the sample will not have the original structure when the sample is processed. Because the projective electron microscope can only observe very thin samples, it is possible that the structure of the surface of the material is different from the internal structure of the material [2]. In addition, the electron beam may destroy the sample by collision and heating [3].

In addition, the price of electronic microscope purchase and maintenance is relatively high.

18.3.4 Comparison of Performance Between Electron Microscope and Optical Microscope

Resolution is an important metric for evaluating the level of an electron microscope. The resolution of an electron microscope is defined by the minimum distance between two adjacent substances that it can recognize. It is related to the incident acute angle and wavelength of the electron beam passing through the substance under observation. The wavelength of visible light is about 300–700 nm, while the wavelength of the electron beam is closely related to the accelerating voltage. Therefore, we can directly observe some atoms and crystals arranged orderly in some heavy metals by electron microscope [4]. Although the resolution of electron microscopes is far better than optical microscopes, it is difficult to see living organisms because electron microscopes require working in a vacuum [5]. Also, the irradiation of electron beam can cause damage to living biological samples. Other issues, such as the brightness of the electron emission and the perfect quality of the electron lens, remain to be further investigated.

The advantages gained:

- Extensible, future oriented platform.
- Improve efficiency and reduce reporting time.
- Continuous supply of captured cells to expert analysis.
- Automatic record storage and data storage.

18.4 Microdissection Technology

18.4.1 What Is Microdissection

Microdissection technology is based on microscope or microscope to cut and separate materials to be selected (tissue, cell group, cell component or chromosome band) through micromanipulation system and collect technology for subsequent research. Microdissection technology is actually a technology of separating and collecting materials for research in the micro field. Therefore, applying this technology is often an important step in many in-depth research works.

18.4.2 Selection and Labeling of Probes

In the study of molecular pathology, two difficult problems are often encountered: First, the selected research materials need to have the same characteristics in a certain aspect, that is, to a certain degree of homogeneity. The vast majority of human tissues are heterogeneous cell clusters, which are made up of many different cells. This homogeneity problem is often encountered in deep research of human tissues, but it is not easy to solve.

18.4.3   The Material of Microdissection

The material of microdissection can be attached to the solid phase in a variety of ways to support a variety of tissue cell component material; its origin is very extensive; paraffin sections and frozen sections, cell preparations, cell smear, cell smears, cell culture, and conventional chromosome preparation can be used.

18.4.4   The Way of Microdissection

According to its development process, it can be divided into four kinds: manual direct microdissection, mechanical assisted microdissection, hydraulic control microdissection, and laser capture microdissection (Laser Capture Microdissection) [6].

18.4.5   Technology Combined with Microdissection

What kind of cells to choose exactly depends on the purpose of this research. However, how to screen the cell that are to be cut and how to analyze the cut cells need to be combined with other methods. It is precisely because it can be combined with a variety of molecular biology, immunology, genetics, and pathology technology to make microdissection technology show vigorous vitality [7]. Based on the requirements of research in microcutting immunohistochemistry, which employs histochemical methods, chemistry, in situ hybridization, in situ end-labeling techniques, PCR, FISH, and special staining methods to identify the need for cutting tissue components, microcutting can be used for extraction of proteins, DNA, and RNA. This allows correlation analysis through Western Blot, Southern Blot, Northern Blot, PCR protein, and nucleic acid.

18.4.6   Influence Factors of Microdissection

Because microdissection is often used in combination with a variety of methods, the factors that affect the final results of microdissection are varied. Therefore, we need to grasp a principle in the process of experiment, that is, all the factors that influence the technology of microdissection should be listed as the influencing factors of the final result of microdissection experiment, which should be considered in the process of experiment. The use of different methods determines the need for different treatment to the microdissection materials. When using microdissection techniques,

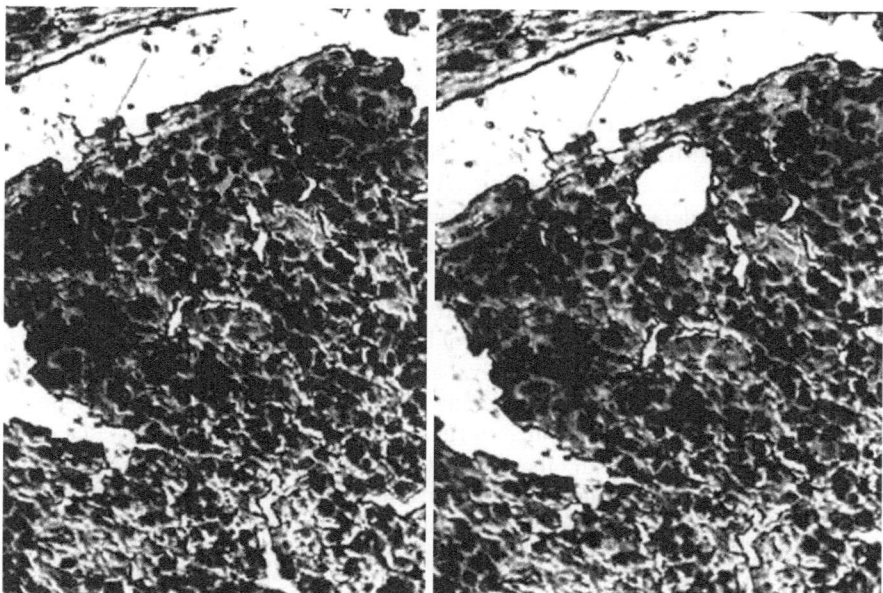

Fig. 18.5 Laser capture microdissection. (Source: Jie Chen, Gandi Li. Pathology. Second Edition. ISBN 978-7-117-13,102-5. Page 515, appendix Fig. 2-6, Laser capture microdissection)

it is important to separate the cells or subcellular components according to the needs the of experimental research, which is influenced by many factors [8] (Fig. 18.5).

18.4.7 The Latest Development of Microdissection

With the help of this revolutionary technology, we can quickly and accurately identify the specific separation of individual or group cells for cellular and molecular biology research in the next step; the precise identification is based on cell morphology, immunohistochemical staining, and histochemical staining methods under the microscope and computer-aided realization. Specific separation is realized by low-energy laser and infrared special cell transfer film. Laser capture microdissection is realized by laser microcell separation system. The system consists of inverted microscope, infrared laser transmitter, vacuum pump, cell transfer film, and computer. Through the microscope in vacuum pump fixed slide, Cells are separated by an infrared laser transfer film attached to the bottom of the centrifugal tube. The cells are picked and transferred into the centrifuge tube and can directly extract the centrifuge tube by dissolution and extraction; the extraction of liquid according to your cell group from the points of discretion and all of the cell separation process, whether the transfer operation or the preservation of samples, without any manual operation, can achieve cell separation without pollution.

18.5 Laser Scanning Confocal Microscopy Technology

Laser scanning confocal microscopy (LSCM) is a high-tech device that combines optical microscopy, laser scanning technology, and computer image-processing technology [9]. It has a high resolution that ordinary optical microscopes cannot achieve, as well as depth recognition capabilities and longitudinal resolution, enabling it to see details in thicker biological samples [10]. Confocal imaging using the illumination point and the detection point conjugate of this feature can effectively inhibit the same focal plane of the nonmeasurement point of stray fluorescence, and nonfocal plane fluorescence from the sample to obtain ordinary optical microscope cannot reach the resolution of both depth identification capability (maximum depth of 200–400 μm typically), and vertical resolution enable you to see details in thicker biological samples. The LSCM enables noninvasive tomography and imaging of samples to visualize and analyze the three-dimensional spatial structure of cells. Combined with other technologies, dynamic observation of live cells, multiple immunofluorescence labeling, or ion-fluorescent labeling can be carried out to study living cells' function and metabolic process.

18.5.1 Main Functions of LSCM

1. Tissue, cell optical section
   This function is also referred to as "cell CT" or "microscopic CT" (Fig. 18.6).
2. Three-dimensional image reconstruction
3. Long-term dynamic observation of living cells
4. Quantitative determination of intracellular acidity and alkalinity
5. Research on intercellular communication, cytoskeleton formation, biofilm structure, and macromolecular assembly by FRAP
   FRAP irradiated a region of cells with high intensity pulse laser to cause the bleaching of fluorescent molecules in the region, and the unbleached fluorescent molecules around the region would diffuse to the irradiated area at a certain rate. The diffusion rate of the fluorescence molecules could be detected directly.
6. Measurement of cell membrane fluidity
7. Photoactivation technology

18.5.2 Requirements and Limitations of LSCM for Samples

Samples for LSCM are preferably samples of cultured cells, such as cell smears or crawlers, or frozen tissue sections. Paraffin-embedded tissue sections are not suitable for this technique. LSCM mainly uses direct or indirect immunofluorescence staining and fluorescence in situ hybridization. The cost of fluorescent labeled probes or antibodies is also high. In selecting this method, we should take full

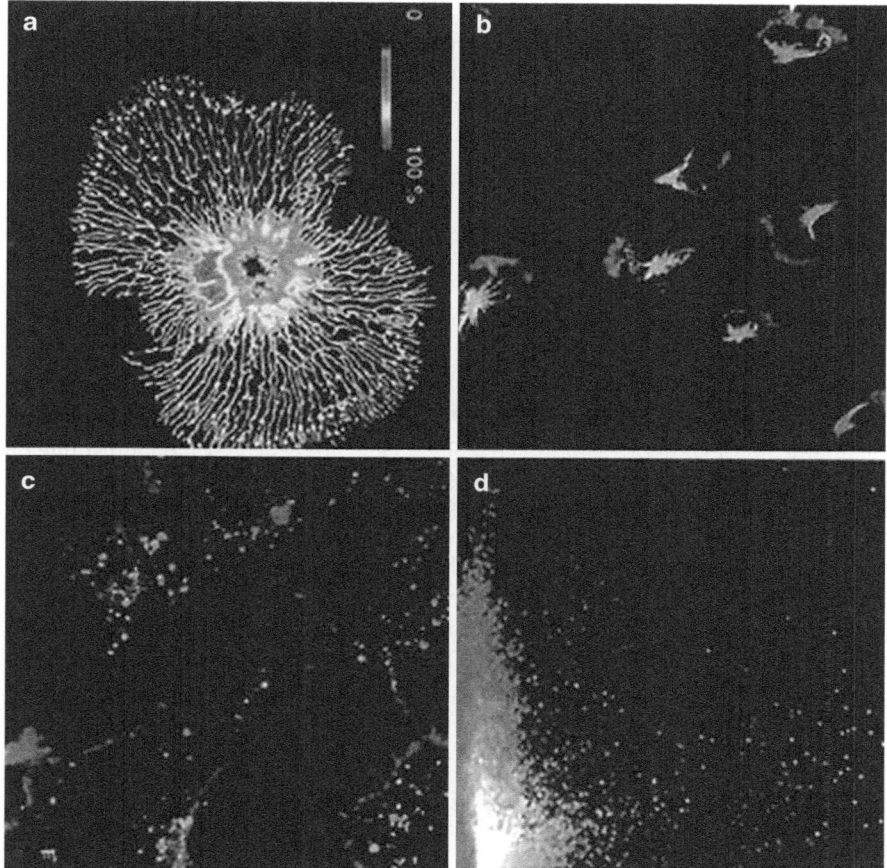

Fig. 18.6 (**a**) Fluorescent-labeled dendritic cells with anti MHC-II molecular antibody staining; (**b**) Actin microfilaments (Rhodamin-labeled, red) and microtubules (FITC-labeled, green) of cytotoxic T lymphocytes; (**c**) Double-labeled staining showed that cathepsin D (FITC-labeled, green) in the cytoplasm of fibroblast and wheat germ agglutinin in cell membrane (Texas Red-labeled, red). (**d**) Double fluorescence staining of fluorescence intensity analysis. (Source: Jie Chen, Gandi Li. Pathology. Second Edition. ISBN 978-7-117-13102-5. Page 516, appendix Fig. 2-7, laser confocal microscopy)

advantage of the advantages of LSCM and avoid using it as an advanced fluorescence microscope.

18.6   In Situ Hybridization (ISH)

In situ hybridization (ISH) is a part of nucleic acid molecular hybridization. It is a technique that combines histochemistry with molecular biology to detect and locate nucleic acids. ISH uses nucleotide fragments of labeled known sequences as probes to detect and locate a specific target DNA or RNA by hybridization on tissue sections, cell smears, or culturing directly. The biochemical basis of ISH is DNA denaturation, renaturation, and base complementary pairing binding. According to the selected probe and the target sequence to be detected, there are DNA–DNA hybridization, DNA–RNA hybridization, and RNA–RNA hybridization.

18.6.1 Selection and Labeling of Probes

The probes used for in situ hybridization include double-stranded cDNA probes, single-stranded cDNA probes, single-stranded cRNA probes, and synthetic oligonucleic acid probes. In general, the length of the probe is 50 ~ 300 bases, and the probe for chromosome in situ hybridization can be 1.2 ~ 1.5 kb. The probe markers have radioisotopes, such as H, 35S, and 33P, which are highly sensitive, but it has a high half-life and radioactive contamination, high cost and time-consuming. Therefore, its use is limited. The nonradioactive probe markers have fluorescein, digoxin, and biotin, although its sensitivity is lower than the radiolabeled probe; because of its stable performance, simple operation, low cost and short time consumed, it is being used more and more widely. The double-stranded cDNA probe can be labeled by gap translation or random primer method, and single strand cDNA probe can be labeled by transcription. The synthetic oligonucleotide probe can be labeled with 51 terminal-labeling methods, that is a tail-labeling method.

18.6.2 Main Procedure of In Situ Hybridization

The experimental materials of in situ hybridization can be routine paraffin-embedded tissue sections, frozen tissue sections, cell smears and culturing cell climbing slices, and so forth. The main procedures include preparation before hybridization, pretreatment, hybridization, cleaning after treatment of hybridization, and detection of hybrids.

18.6.3 Application of In Situ Hybridization

In situ hybridization is mainly used for: (1) The localization of cell-specific mRNA transcription can be used in the research of gene mapping, gene expression, and genome evolution [11]. (2) Detection and localization of viral DNA/RNA in infected tissues, such as EB virus mRNAs, human papillomavirus and cytomegalovirus DNA [12]. (3) The expression and changes of oncogene, tumor suppressor gene, and various functional genes were detected at the transcriptional level. (4) The location of genes on a chromosome. (5) Detection of chromosome changes, such as chromosome quantity abnormality and chromosome translocation and so forth [13]. (6) The study of interphase cytogenetics, such as prenatal diagnosis of genetic diseases and the determination of some genetic carriers of genetic diseases, diagnosis of some tumors and biological dosimetry [14].

18.6.4   Recent Developments and Future Directions in Tissue ISH

In the past few years, ISH has become a powerful clinical and scientific tool for assessing DNA and RNA in the nucleus of intermediate stages of tissue sectioning. Improved hybridization schemes such as the probe availability brought about by the Human Genome Project have shifted the focus from morphological based diagnostic methods to diagnostic methods that combine morphological and molecular features. However, new methods continue to emerge, such as dual color rendering ISH, Fiction, and so on [15].

18.6.4.1   Dual-Color Chromogenic ISH

Double probe FISH technology is a molecular biology technology that has emerged in recent years. It can be used as a clinical diagnosis and scientific research tool. However, at present, this technology has not been fully applied to clinical diagnosis, mainly because it is time-consuming and requires professional and expensive equipment and knowledge. Dual color in situ hybridization (dcCISH) is a technique that uses enzymatic methods to detect hybridization between two isolated probes and targets, and it can be used as a viable alternative to fluorescent dye detection in DNA in situ hybridization. The technique is based on peroxidase or alkaline phosphatase-labeled reporter antibodies using standard immunohistochemical enzyme reaction detection. The advantage of CISH over FISH is that it utilizes a similar working principle to immunocytochemistry amplification. Using a common optical microscope, pathologists can work under bright field conditions while examining tissue morphology and mixed signals. In addition, CISH stained slides can be stored at room temperature, with minimal signal strength loss over time.

As an emerging molecular biology technology, dcCISH technology has been upgraded and improved on the basis of the standard dual probe FISH technology. Figure 18.7 shows the hybridization signal immunohistochemical detection after hybridization. The probe can be labeled with digoxin or biotin, or with fluorescent dyes such as FITC or Texas Red. After hybridization, the probe is labeled with enzymatically conjugated mouse and rabbit antibodies, and enzymatic reactions with appropriate chromophores and substrates result in a strong permanent color (typically red and green) that can be observed using a 40× objective lens.

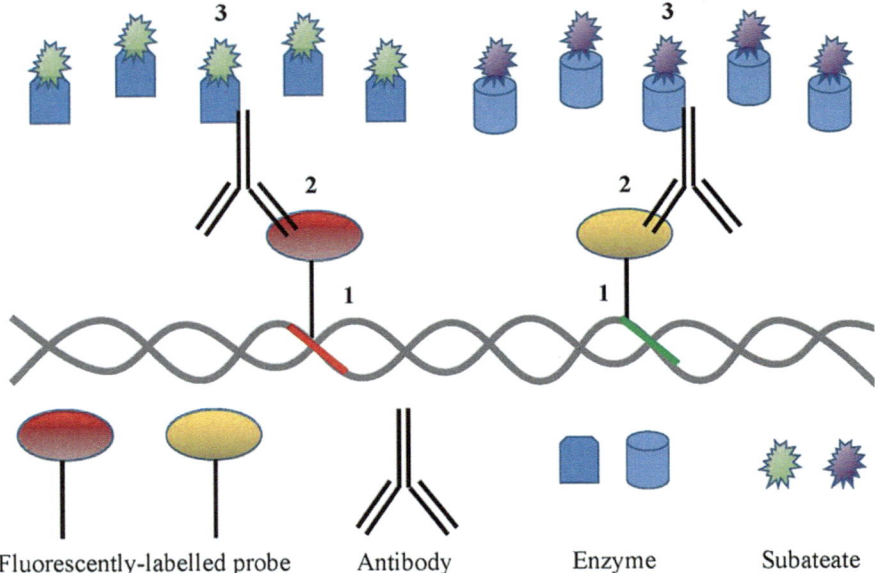

Fluorescently-labelled probe   Antibody   Enzyme   Subateate

Fig. 18.7 dcCISH. (1) Fluorochrome (such as FITC and Texas red)-bound probes are hybridized to the target DNA. (2) A cocktail of AP-conjugated anti-Texas red and HRP-conjugated anti-FITC antibodies are used to detect the hybrids formed. (3) Incubation with an AP-compatible substrate followed by an HRP-compatible substrate provides contrasting colors that can be visualized under bright-field microscope

18.6.4.2   Fiction

Fiction technology was developed in 1992 as a tool for the study of fluorescent immunophenotypes and interphase cytogenetic tumors, which can be used to simultaneously detect immunophenotypic markers and genetic aberrations in cellular preparations. The initial technology was limited by the amount of available fluorescent dyes and the quality of digital imaging and did not receive widespread attention [16]. However, with the improvement of tissue pretreatment methods and the emergence of many new fluorescent dyes, it has been used on FFPE materials for the analysis of lymphoma and the detection of minimal residual disease.

18.6.4.3   The Allen Brain Atlas

Allen Brain Atlas (ABA) is a genome scale ISH project that generates cell level gene expression profiles of the brain and spinal cord of adult C57BL/6J mice [17]. Annotated results for over 2000 genes in this project are provided free of charge. Automated image capture and analysis has provided quantitative expression data that can be directly compared with available microarray datasets. The ABA project uses high-throughput CISH and DIG labeled riboprobes and tyramine amplification to detect mRNA transcripts in tissue slices and localize them to different regions of the central nervous system of mice [18]. This method of genome wide transcription analysis using ISH and its correlation with other genome scale expression platforms will provide valuable data support for the organization and function of normal and abnormal cells and tissues.

18.7 In Situ PCR

In situ PCR technology originated from traditional PCR technology and in situ hybridization technology. In situ PCR technology is the direct amplification of target DNA or RNA fragments without changing the target location and in situ detection of amplification products. In situ PCR is a technique for conducting PCR reactions in tissue cells established by Hasse et al. in 1990. It is an in situ polymerase chain reaction technology that combines PCR technology and in situ hybridization technology. In situ PCR combines the advantages of in situ hybridization with cell localization capabilities and highly specific and sensitive PCR techniques. It can not only distinguish and identify cells with target sequences, but also mark the position of the target sequence within the cell. It has significant practical value in the research of molecular cell level and molecular level and has wide application in many disciplines.

18.7.1 Basic Principle of In Situ PCR

In situ PCR usually uses chemical methods to immobilize specimens to maintain good morphology and structure of tissue cells. Cells are treated with protease for permeability. Digestion of cells using proteases to expose target genes. When PCR is amplified, all kinds of components, such as primers, DNA polymerase, nucleotides, and so forth, can enter into cells or nucleus and can be amplified in situ by RNA or DNA immobilized in cells or nucleus. The amplified products are generally large, or interwoven, and are not easy to pass through the cell membrane or diffuse inside and outside the membrane, and are retained in situ. In this way, the single or low copy specific DNA or RNA sequences in the cell were amplified exponentially in situ. Therefore, amplification products are more easily detected in situ (Fig. 18.8).

18.7.2 Selection of Markers

There are two types of markers: nonradioactive labeling and radioactive labeling. Nonradioactive markers include biotin, peroxidase, digoxin, and so forth. Radioactive markers often use radioisotopes, such as ^{3}H and ^{32}P, and so forth.

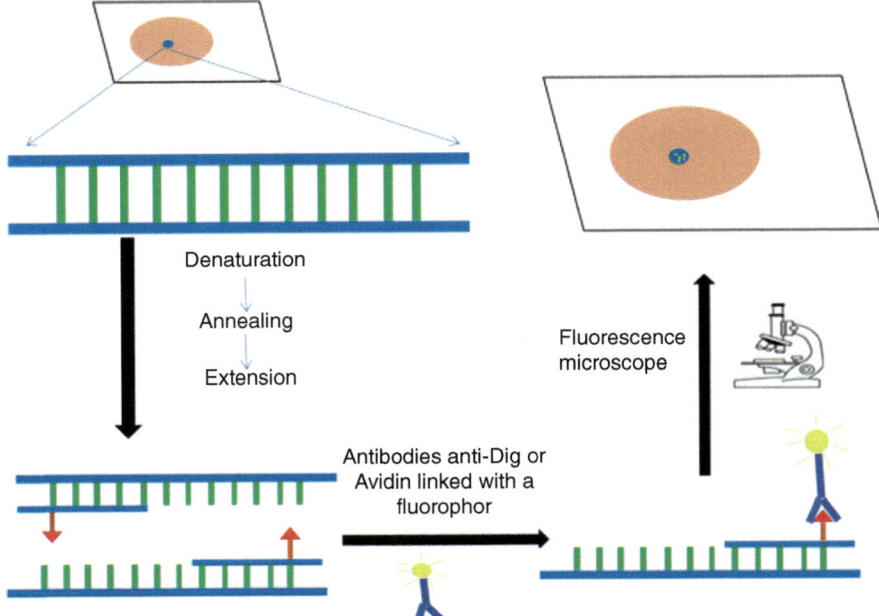

Fig. 18.8 Basic principle of in situ PCR

18.7.3   Types of In Situ PCR

18.7.3.1   Direct In Situ PCR

Direct in situ PCR refers to in situ PCR using labeled single nucleotides or primers in a reaction system.

18.7.3.2   Indirect In situ PCR

Indirect in situ PCR refers to in situ PCR in which the primers and single nucleotides used do not carry any markers, and after the reaction is completed, in situ hybridization is used to detect the specific amplified fragment product.

18.7.3.3   In Situ Reverse Transcription PCR

In situ reverse transcription PCR refers to the process of first performing reverse transcription before performing in situ PCR, using the reverse transcriptional product as a template for amplification.

18.7.3.4   In Situ Loop Isothermal PCR

In situ loop isothermal PCR, which is based on the principle of in situ PCR, has been improved. The penetration treatment conditions are mild, and cells are not easy to break. The specially designed loop primers enable the amplification product to have a large molecular weight, not easy to spread, and the amplification temperature is low and constant. Without using a PCR instrument, the rate of cell loss is greatly reduced.

18.7.4   Application of In situ PCR

Due to its precise localization and high sensitivity, in situ PCR technology has been widely used in research fields such as zoology, botany, virology, microbiology, pathology, oncology, and clinical medicine [19, 20]. In situ PCR is widely used to identify the occurrence and expression of functional genes in cells, including endogenous gene detection, exogenous gene detection, pathogen detection, gene mutation, gene rearrangement, chromosome translocation, identification of gene mutations, and gene expression research.

18.8   Flow Cytometry

Flow cytometry (FCW) is a modern analytical technique for measuring suspended cells or particles in a liquid phase by flow cytometer. Flow cytometry can accurately and rapidly perform multiple parameter and quantitative analysis of individual cells in the flow of cells when the cells remain intact. Moreover, with the help of monoclonal antibody technology and fluorescent dye labeling technology, flow cytometry can not only detect multiple characteristic parameters from a cell, but also can sort a subset of cells with the same characteristics based on a certain parameter for further study.

18.8.1   Main Principles of Flow Cytometry

Flow cytometry has three main components: cell flow chamber and fluid flow drive system, optical system and signal detection system, and data analysis system. The three systems are perpendicular to each other [21]. After stained cells or particles to be tested with specific fluorescent dyes, they are placed into a sample tube and enter a flow chamber. Under the restriction of sheath fluid, cells, or particles are arranged in a single row and ejected from the flow chamber. The laser beam irradiates cells or particles stained with fluorescence, producing scattered light and stimulating fluorescence. These signals are read by the detector and converted into electronic signals, which are output in a standardized format data file. The intensity of the fluorescence signal of forward scattered light (FSC) basically reflects the size of the cell volume. The intensity of the fluorescence signal of laterally scattered light (SSC) represents the intensity of the antigen on the surface of the measured cell membrane or the concentration of substances in its nucleus. Different cell subsets have different cell characteristics. Therefore, based on the two fluorescent signals, cells can be comprehensively analyzed and purified (Fig. 18.9).

18.8.2   *Classification by Flow Cytometry*

18.8.2.1   Conventional Flow Cytometry

Conventional flow cytometry is used for rapid analysis and detection of cell components and particle suspensions. It carefully distinguishes and identifies cells by simultaneously detecting and analyzing multiple signals on the cells in the fluid flow.

18.8.2.2   Acoustic Focusing Cytometers

Acoustic focused flow cytometry uses ultrasound to form a sheath flow of cells for flow cytometry detection. This type of acoustic flow cytometry allows for a larger sample input cell threshold and a lower probability of sample clogging.

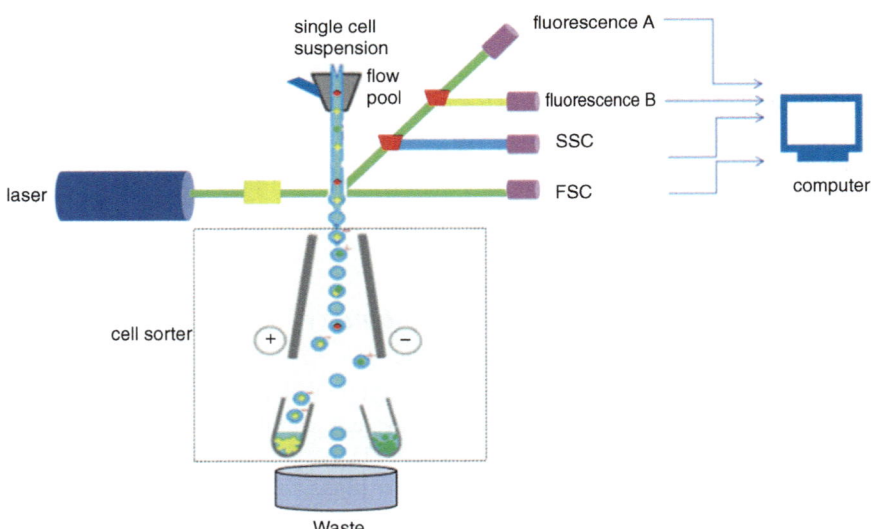

Fig. 18.9 The basic structure of flow cytometry

18.8.2.3   Cell Sorters

Cell sorters is used to quickly separate and obtain target cells. On the basis of analysis and detection, cells are separated through the sorting module.

18.8.2.4   Imaging Cytometers

Imaging cytometer that combines flow cytometry polychromatic detection technology with fluorescence microscope image display technology [22].

18.8.2.5   Mass Cytometers

Mass cytometers is one of flow cytometry technique that utilizes the principle of mass spectrometry to detect multiple parameters of a single cell. It not only inherits the characteristics of high-speed analysis of conventional flow cytometry, but also has the high resolution ability of mass spectrometry detection. This technology not only realizes the simultaneous measurement of dozens of parameters, but also enables the fine clustering of cell populations, and even analyzes cell signal pathways, cell proliferation, cell cycle, and so on.

This technology not only achieves the simultaneous measurement of dozens of parameters but also achieves the fine clustering of cell populations, and may also analysis of signal pathways, cell cycle, cell proliferation, and so on.

18.8.3   Fluorescent Reagent

Commonly used flow fluorescent reagents include organic small molecules, phyco-biliproteins, polymer dyes, tandem dyes, nucleic acid dyes, proliferation dyes, cell reactive dyes, calcium indicator dyes, and so forth.

18.8.4   Data Analysis

Currently, the traditional two-parameter histogram (dot chart) "gated" analysis method is still frequently used. The FCS file format is the data file format for standardized flow cytometry, and all flow cytometry data files have a ".fcs" file extension. However, as the parameters and complexity of flow cytometry experiments increase, new clustering data analysis methods have been used, such as PCA, SPADE, and tSNE. These new data analysis methods enable researchers to extract useful information from high-dimensional data of flow cytometry.

18.8.5   Application of Flow Cytometry

Flow cytometry has been increasingly applied in multiple disciplines [23]. (1) The field of immunology includes immunophenotyping, antigen specific reactions, intracellular cytokine analysis, immunophenotyping, and apoptosis analysis. (2) The field of molecular biology includes fluorescent protein analysis, cell cycle analysis, signal transduction flow cytometry, RNA flow cytometry, cell sorting, and so forth. (3) Other applications include absolute cell counting, quantitative flow cytometry, multiple bead array detection, phagocytosis assay, small particle analysis and sorting, and so forth.

18.9   Image Acquisition and Analysis Technology

Telemedicine, a general definition of this process would be the acquisition, storage and transmission of microscope images from a local site to a remote site for a specific reason. The initial concept was probably driven by researchers in disciplines unincluded pathology or microscopy. The first telepathology system was developed in the United States during 1960s to 1980s. The technology would be unable to deliver a high enough image resolution for diagnostic accuracy comparable to that being achieved by traditional microscopy diagnosis.

However, the rapid growth in Internet, imaging and computing technology has provided a backbone for telepathology infrastructure. There are two main types of telepathology of the advantages/disadvantages for these systems. First, Static Imaging Telepathology, an asynchronous technology that is no simultaneous interaction with the microscope slide. Second, Dynamic Telepathology, which is also known as Real-time Video Imaging. This form can be subdivided into two systems: passive-dynamic and active-dynamic.

18.9.1   Digital (virtual) Microscopy

Automated microscopes represent a role of hybrid technology which bring together various components within the industry to form a device capable of creating virtual slides without being dedicated to that purpose. Due to the microscopes used in these systems, High-specification microscopes and computational techniques (CCD) have the capability to produce very high-quality images.

CCD chip needs the center of the field given by the objective, thus reducing optical aberration to a point considered negligible. When the picture is taken, the slide-moving mechanism puts the slide in the next position for image acquisition while refocusing the slide.

18.9.1.1 Olympus DotSlide

Olympus DotSlide is an automated system (Olympus Vertical BX research microscope) that is scanned by a computer on a glass slide through 1379 × A 1032 pixel camera forms an image. However, many tissues often found in average slide trays are significantly larger than this. The histological specimen can be this size or 25 × 20 mm, traditional cytological specimens can occupy the entire cover glass (50 × 25 mm), a liquid based cytology (LBC) specimen, with a diameter of less than 25 mm.

18.9.1.2 Scanners

(1) Progressive scanning CCD system: Nikon COOLSCOPE II is known as a digital microscope. It has a slide viewing capability and allows for digital image capture. It is not a true virtual slide scanner. (2) Zeiss Mirax system: The system was developed by a team from Semmelweis University in Budapest, Hungary, and sold under the name of Mirax. Mirax DESK is a semiautomatic tool that scans a single slide using an ×20 objective lens. In this series, Mirax SCAN has the largest volume and can be upgraded to a fluorescent module with automatic fluorescent slide scanning capabilities. (3) Aperio ScanScope System: Aperio's ScanScope GL system is an entry-level device commonly used in virtual microscope projects in small laboratories or universities. This machine is a manual single slide scanner. This is suitable for bass environments with an amplification factor of ×400 (via an ×2 amplification converter) (4) Binsong NanoZoomer System: NanoZoomer Digital Pathology (NDP) is a high throughput slide scanning system. The machine currently has two obvious advantages, namely, a true 3D z-stack scan and the option to add fluorescent scanning. NanoZoomer's 3-CCD TDI camera can observe fluorescent tissue samples at low light levels with high resolution. (5) D-Metrix DX-40 imaging system: The DX-40 slide scanner has been developed by D-MetrixInc. (Tucson AZ), which could serve as a digital imaging engine for a very rapid virtual-slide scanner.

18.9.2   The Virtual Slide Format

Virtual microscope technology is constantly developing. However, the image format needs to be optimized. The slide is finally scanned as a virtual slide. This can be considered an image file and are represented and stored in different ways on hardware. IFF, JPEG, or JPEG are the most common general-purpose frameworks. The resulting virtual slides are typically recorded in TIFF format with the manufacturer's suffix. JPEG is the most widely used compression form and has become the standard in the photography industry today. Initially, this format was suitable for the acquisition and effective storage of images captured from traditional optical microscopes.

18.9.3   Image Serving and Viewing

Image server and image viewer are the two core applications delivered by any virtual microscope system. Image service will become the core issue of digital microscopy. Pathologists need to add any new technologies to increase their utilization. Buffering images can be a way to significantly improve transmission speed.

18.9.4   Applications of Virtual Microscopy

Quantitative research is a major issue in standardizing and improving the efficiency of biomedical research. The combination of virtual microscopy and complex image analysis provides an excellent opportunity. Virtual slides allow operators to define accurate protocols and run efficient algorithms across the entire slide or designated areas.

Many factors limit the research of pathological or molecular clinical analysis: (1) Poor quality optimization of slides; (2) The coloration of the reagent is not satisfactory; (3) Typically less than the optimal patient sample size. To alleviate these problems, techniques using tissue microarray (TMA) have been developed.

18.10   Comparative Genomic Hybridization

Proto-oncogenes and tumor suppressor genes play a decisive role in cell proliferation and cell fate (differentiation, aging, and apoptosis). They are influenced by the environment: Overexpression of specific oncogenes can enhance the proliferation of one cell type, but induce apoptosis of another cell type [24].

Microarray technology has considerable data variability due to differences in results caused by DNA extraction, probe labeling, type of platform used, number of analyzed samples, and biometric and statistical analysis methods.

18.10.1   Principles of Array CGH

The principle of array CGH is similar to that of metaphase CGH, which has been widely used for genomic characterization of many solid tumors. Array CGH or matrix CGH was first described in 1997. Array CGH allows high-resolution mapping of amplicon boundaries and smallest regions of overlap and improvement in the localization of candidate oncogenes and tumor suppressor genes, and is only limited by the insert size and density of the mapped sequences used.

18.10.2 aCGH platforms

Currently, there are various aCGH platforms, but it should be emphasized that there is currently no ideal array CGH analysis method. Ordered arrays are manufactured by positioning (using pins) or synthesizing a single probe in an organized pattern on a plane. A random array is constructed by attaching a single probe to beads, then assembling and assembling the beads onto a patterned plane.

18.10.2.1 BAC Arrays

Bacterial artificial chromosomes (BACs), P1 derived artificial chromosomes (PACs), and yeast artificial chromosomes (YACs) have been widely used as large insertion genomic clones for aCGH. The resolution of each BAC array is defined by the number of unique probes it contains. The probe content of the whole genome BAC array ranges from several hundred to 32,000. These platforms provide a strong enough signal for detecting single copy number changes and can effectively assist in extracting DNA from paraffin-embedded (FFPE) tissue.

18.10.2.2 cDNA Array

A preliminary study was conducted on the genome-wide approach to aCGH using cDNA microarrays originally designed for expression profiling. CDNA microarray analysis can only detect known genes and aberrations in EST, as cDNA probes are the only representative of genes expressed on chromosomes. However, cDNA arrays cannot outperform the resolution of currently available alternatives.

18.10.2.3 Oligonucleotide Arrays

There are two main types of oligonucleotide arrays: single nucleotide polymorphism (SNP) arrays and non-SNP arrays, which are composed of single-stranded 25–85 polymeric oligonucleotide elements. Different types of oligonucleotide arrays have different labeling and hybridization schemes and can provide high-resolution measurements of copy numbers.

18.10.2.4 Solexa Sequencing Technology

The Illumina Genome Analyzer (Solexa) uses four proprietary fluorescent labeled modified nucleotides to sequence millions of clusters of genomic DNA on the surface of flowing cells. After fragmentation, the test DNA is connected to an adapter that promotes the binding of single stranded DNA fragments to the surface of flowing cells. This allows the base sequence in a given DNA fragment to be obtained one base at a time. From a molecular genetic perspective, this technology has the potential to resolve several long-standing disputes, such as the mechanism that leads to amplification, and the rearrangement associated with the occurrence of amplified regions.

18.10.2.5   Molecular Inversion Probe arrays

Molecular inversion probes (MIPs) are single oligonucleotides with two flanking inversion recognition sequences that can recognize and hybridize DNA sequences between 41 and 61 bp. Each MIP oligonucleotide has a unique sequence barcode label, which can be annealed to its specific complementary genomic sequence and amplified for detection.

18.10.3   Choosing the Right Platform

The choice of platform is dependent on the types of samples available as this has a direct impact on the quantity, quality and purity (i.e., the proportion of DNA belonging to the cells of interest) of extractable DNA for analysis. Extracted DNA from FFPE is often heavily cross-linked, degraded, fragmented, and heterogeneous and is, therefore, suboptimal for microarray analysis.

18.10.4   Analysis and Validation

The analysis of microarray data always presents statistical errors. Due to the cost and availability of materials, thousands of probe/genomic regions are being studied using relatively small sample sizes.

Regardless of the method used in the analysis of aCGH data, it is crucial to properly plan and validate the amount of data generated that is always and often excessive using in situ or other molecular methods.

18.10.5   Finding the Target

There are two different approaches: top-down and bottom-up research. In addition to detailed aCGH analysis, this process also requires the superposition of array CGH and expression array data (or another relatively high throughput expression profile), protein mass spectrometry, or RNA interference (RNAi) analysis [25–27]. After identifying specific amplification factors and possible driving sequences, the next step is to establish a model to test the genes that cancer cells rely on. Because the biological properties of a large number of oncogenes are closely related to the environment and cell type, it is crucial to identify models that not only contain the amplicons of interest, but also have phenotypes similar to those of the cancer under study.

18.10.6   Clinical Applications

While searching for therapeutic targets, it is widely recognized that biomarkers can be used to measure or predict therapeutic effects [28, 29]. The availability of biomarker complementary diagnostic analysis is both cheap and easy to perform.

18.11   Biochip Technology

Biochip technology is a high-tech emerging in the early 1990s with the Human Genome Project. It refers to the micro-processing and microelectronics technology, solid-phase matrix surface integration of thousands of densely arranged molecular microarray in order to achieve the organization, cells, nucleic acids, proteins, and other biological molecules for efficient, accurate, high-throughput Detection. Commonly used biochips are divided into gene chips, protein chips, tissue chips, liquid-phase chips, and microchip labs (Fig. 18.10).

18.11.1   Gene Chip

Gene chips, also known as DNA chips, DNA microarray; a large number of gene fragments ordered, high-density fixedly arranged on the carrier made of lattice, called gene chip. The principle is that after the labeled DNA to be tested is hybridized with the probe at a specific position on the chip according to the principle of base pairing, the chip is scanned by a laser confocal fluorescence detection system and the like, and the intensity of the hybridization signal is detected to obtain the sample molecule Quantity and sequence information, using computer software for data comparison and analysis, so that the gene sequence and function of large-scale, high-throughput research [30]. Gene chip is the first development in the field of biochips, the most mature and the first to enter the commercial application of technology.

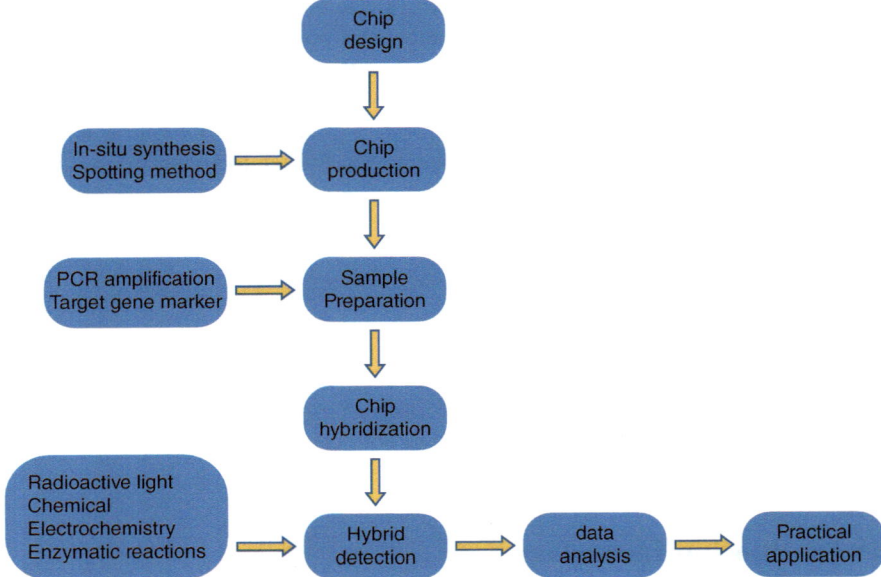

Fig. 18.10 Biochip analysis steps

18.11.1.1 Gene Chip Application

In recent years, gene chip technology has brought many revolutionary innovations in fields such as basic research, clinical diagnosis, drug screening, and guidance of clinical medication and treatment, with the advantages of multiple samples processing in parallel, high-speed analysis, small sample amount, and less pollution.

1. **Gene Function Study**

   (a) Gene expression and regulation

   Gene expression analysis is the most widely used field of gene chip. It analyzes cell gene expression status as a whole, provides a powerful tool for understanding the gene expression related to certain life phenomena, and discusses the gene regulation and the mechanism of gene interaction.

   (b) Detection of gene mutation and polypeptide

   Genetic sequence variation is a major cause of intraspecific and intraspecific differences and is also a genetic basis of disease phenotypes and phenotypic differences under normal conditions. The use of gene chip technology to locate, confirm, and classify these mutations is the basis for the diagnosis of genetic diseases.

   (c) DNA sequencing

   DNA hybridization in the chip sequence and adjacent hybridization sequencing is a new type of efficient and rapid sequencing method. The rationale is that any linear DNA or RNA single strand can be broken down into a series of oligonucleotide fragments, or subsequences, that have a fixed number of bases and overlapped.

(d) Genome and gene research

Recently, some scholars have used this characteristic to study the replication activity of the yeast genome because of the high sensitivity and high accuracy of the gene chip. By comparing the DNA copy number of the yeast genome at 1000 loci in S phase, the origin of replication was found. At the same time, the relationship between replication fork movement in the genome and S-phase replication and transcriptional activity was also studied.

2. **Disease Diagnosis**

(a) Infectious disease diagnosis
(b) Diagnosis of genetic diseases
(c) Tumor diagnosis and classification
(d) Drug resistance test

18.11.2 Protein Chip

18.11.2.1 Principles of Protein Chips

The basic principle of a protein chip is to immobilize a polypeptide, protein, enzyme, antigen or antibody on a solid support to form a molecular lattice by means of mechanical spotting or covalent binding and incubate the protein to be tested with the chip, and then, the fluorescent-labeled protein is reacted with the chip-protein complex.

18.11.2.2 Application of Protein Chips

1. Disease diagnosis and efficacy evaluation
2. Research and development of new drugs
3. Biomolecular interaction studies: (a) antigen–antibody interaction; (b) protein–protein interactions; (c) small molecule–protein interaction; (d) protein–nucleic acid interaction; (e) enzyme–substrate interaction.

18.11.3 Tissue Chips

It can effectively use some precious lesions or puncture specimens to study the expression of specific genes and their corresponding proteins as well as the relationship between the disease.

18.11.4 Liquid Chip

Liquid-phase chip, also known as a suspended array, flow fluorescence technology, is based on xMAP technology, a new biochip technology platform is the rise of the mid-1990s, a new detection technology.

1. Immunological analysis
2. Pathogen detection
3. SNP test
4. Other

18.11.5 Microreactor Lab

The microchip lab has the advantages of highly integrated analysis process, automation, high throughput, fast analysis, less sample required, less cross pollution, low cost, small size, light weight, easy to carry and so on. Its future development trend can even become a personal bioinformatics analysis card or micro-molecular biology laboratory devices, together with the corresponding computer software, as long as the sample drops on the chip, into the computer, the computer will interpret the test result. Its emergence will bring a revolution in such fields as molecular biology, disease diagnosis, curative effect monitoring, development of new drugs, forensic science, and food hygiene supervision.

18.12 Bioinformatics Technology

18.12.1 Introduction to Bioinformatics

18.12.1.1 Introduction

Bioinformatics is a new science that combines biological and informatics methods and uses computers and internet technologies to analyze vast amounts of rapidly accumulated biological data to gain new insights into biological sciences. Bioinformatics analysis includes several steps: biological information access, processing, storage, analysis, and interpretation. The development of bioinformatics is the inevitable result of the development of life science and information science. Its emergence and rapid development are triggering a revolution in the research methods of biological sciences. It will also have a tremendous impact on the development of life sciences and medical examinations.

Since the 1950s, the rapid development of molecular biology has enabled people's understanding of the nature of life to advance to the material basis of life activities—the two major types of biological macromolecules such as nucleic acids and proteins, resulting in a large amount of biological data. A great deal of biological data has been accumulated in the research on the structure, function and interaction of these two kinds of biological macromolecules. Especially with the completion of the Human Genome Project and the development of high-throughput genomic analysis techniques, the output of large-scale and high-throughput biological data has exceeded human's existing analytical capabilities. The use of computer technology and information technology to manage and analyze these massive biological data has become the inevitable trend of development of life science, which led to the biological sciences and information science combined with the cross-cutting edge of the discipline—bioinformatics.

18.12.1.2 Bioinformatics Research Areas

1. Establishment, maintenance and management of various biological databases. It involves the basic knowledge and the classification and application of biological information in the establishment, maintenance and management of a database in information technology.
2. It is an important task of bioinformatics to study efficient statistical tools, analyze algorithms and develop convenient and fast analytical procedures. With the completion of the human genome project, the development of large-scale genome sequencing and high-throughput analysis technology, the longevity of massive biological data has come up but raised new unprecedented requirements for the collection and processing of information. The bottleneck of the speed of computer operation restricts the biological information. The study of science will inevitably require the development of efficient algorithms and procedures to achieve higher analytical capabilities under the existing computing power.
3. Discovering new knowledge from massive amounts of raw biological data. Including sequencing DNA sequences, identifying the coding protein genes, finding regulatory sequences for gene expression, predicting the function of new genes, locating new genes, predicting the structure and function domains of proteins, the expression profile based on accumulated data and knowledge and biochemical metabolic pathways.

18.12.1.3 The Main Task of Bioinformatics

1. Genome-related information collection, storage, management, and provision
2. Discovery and identification of new genes
3. Noncoding region information structure analysis
4. Biological evolution
5. Comparative study of the completed genome
6. Genome information analysis methods

7. Large-scale gene function expression profiling
8. Protein end sequences, molecular space prediction, simulation and molecular design
9. Drug design

18.12.2 Bioinformatics Database

Researchers from around the world bring together biological data from various experiments into several international or national bioinformatics centers, which are important data institutions for bioinformatics research. This section mainly introduces a group of more important international bioinformatics centers.

References

1. Wade CA, McLean MJ, Vinci RP, et al. Aberration-corrected scanning transmission electron microscope (STEM) through-focus imaging for three-dimensional atomic analysis of bismuth segregation on copper [001]/33° twist bicrystal grain boundaries. Microsc Microanal. 2016;22(3):679–89.
2. Xu W, Dycus JH, Sang X, et al. A numerical model for multiple detector energy dispersive X-ray spectroscopy in the transmission electron microscope. Ultramicroscopy. 2016;164:51–61.
3. Schorb M, Gaechter L, Avinoam O, et al. New hardware and workflows for semi-automated correlative cryo-fluorescence and cryo-electron microscopy/tomography. J Struct Biol. 2017;197(2):83–93.
4. Mattarozzi M, Manfredi E, Lorenzi A, et al. Comparison of environmental scanning electron microscopy in low vacuum or wet mode for the investigation of cell biomaterial interactions. Acta Biomed. 2016;87(1):16–21.
5. Rodríguez-Mena R, Türe U. The medial and lateral lemnisci: anatomically adjoined but functionally distinct fiber tracts. World Neurosurg. 2017;99(1):241–50.
6. de Almeida MR, Strömvik MV. Laser capture microdissection: avoiding bias in analysis by selecting just what matters. Methods Mol Biol. 2016;1405:109–19.
7. Castro NP, Merchant AS, Saylor KL, et al. Adaptation of laser microdissection technique for the study of a spontaneous metastatic mammary carcinoma mouse model by NanoString technologies. PLoS One. 2016;11(4):e0153270.
8. Kang L, George P, Price DK, et al. Mapping genomic scaffolds to chromosomes using laser capture microdissection in application to Hawaiian picture-winged drosophila. Cytogenet Genome Res. 2017;152(4):204–12.
9. Paddock SW. Confocal laser scanning microscopy. Biotechniques. 1999;27(5):992–6.
10. Lamprecht A, Schäfer UF, Lehr C. Characterization of microcapsules by confocal laser scanning microscopy: structure, capsule wall composition and encapsulation rate. Eur J Pharm Biopharm. 2000;49(1):1–9.
11. Lawrence JB, Singer RH. Quantitative analysis of in situ hybridization methods for the detection of actin gene expression. Nucleic Acids Res. 1985;13(5):1777–99.
12. Pringle JH, Ruprai AK, Primrose L, et al. In situ hybridization of immunoglobulin light chain mRNA in paraffin sections using biotinylated or hapten-labelled oligonucleotide probes. J Pathol. 1990;162(3):197–207.

13. Lambros MB, Natrajan R, Reis-Filho JS. Chromogenic and fluorescent in situ hybridization in breast cancer. Hum Pathol. 2007;38(8):1105–22.
14. Laakso M, Tanner M, Isola J. Dual-colour chromogenic in situ hybridization for testing of HER-2 oncogene amplification in archival breast tumors. J Pathol. 2006;210(1):3–9.
15. Martinez-Ramirez A, Cigudosa JC, Maestre L, et al. Simultaneous detection of the immunophenotypic markers and genetic aberrations on routinely processed paraffin sections of lymphoma samples by means of the FICTION technique. Leukemia. 2003;18(2):348–53.
16. Korac P, Jones M, Dominis M, et al. Application of the FICTION technique for the simultaneous detection of immunophenotype and chromosomal abnormalities in routinely fixed, paraffin wax embedded bone marrow trephines. J Clin Pathol. 2005;58(12):1336–8.
17. Lein ES, Hawrylycz MJ, Ao N, et al. Genomewide atlas of gene expression in the adult mouse brain. Nature. 2007;445(7124):168–76.
18. Kallioniemi A, Kallioniemi OP, Sudar D, Rutovitz D, Gray JW, Waldman F, Pinkel D. Comparative genomic hybridization for molecular cytogenetic analysis of solid tumors. Science. 1992;258:818–21.
19. Muro-Cacho CA. In situ PCR. Overview of procedures and applications. Front Biosci. 1997;2:c15–29.
20. Nuovo GJ. Co-labeling using in situ PCR: a review. J Histochem Cytochem. 2001;49(11):1329–39.
21. Manohar SM, Shah P, Nair A. Flow cytometry: principles, applications and recent advances. Bioanalysis. 2021;13(3):181–98.
22. McClelland RD, Culp TN, Marchant DJ. Imaging flow cytometry and confocal immunofluorescence microscopy of virus-host cell interactions. Front Cell Infect Microbiol. 2021;11:749039.
23. McKinnon KM. Flow cytometry: an overview. Curr Protoc Immunol. 2018;120:5.1.1–5.1.11.
24. Tan DS, Lambros MB, Natrajan R, et al. Getting it right: designing microarray (and not 'microawry') comparative genomic hybridization studies for cancer research. Lab Investig. 2007;87:737–54.
25. Ylstra B, van den Ijssel P, Carvalho B, et al. BAC to the future! or oligonucleotides: a perspective for micro array comparative genomic hybridization (array CGH). Nucleic Acids Res. 2006;34:445–50.
26. Johnson NA, Hamoudi RA, Ichimura K, et al. Application of array CGH on archival formalin-fixed paraffin-embedded tissues including small numbers of microdissected cells. Lab Investig. 2006;86:968–78.
27. Little SE, Vuononvirta R, Reis-Filho JS, et al. Array CGH using whole genome amplification of fresh-frozen and formalin-fixed, paraffin-embedded tumor DNA. Genomics. 2006;87:298–306.
28. Reis-Filho JS, Simpson PT, Gale T, et al. The molecular genetics of breast cancer: the contribution of comparative genomic hybridization. Pathol Res Pract. 2005;201:713–25.
29. Reis-Filho JS, Pinheiro C, Lambros MB, et al. EGFR amplification and lack of activating mutations in metaplastic breast carcinomas. J Pathol. 2006;209:445–53.
30. Fan JB, Chee MS, Gunderson KL. Highly parallel genomic assays. Nat Rev Genet. 2006;7:632–44.

GPSR Compliance

The European Union's (EU) General Product Safety Regulation (GPSR) is a set of rules that requires consumer products to be safe and our obligations to ensure this.

If you have any concerns about our products, you can contact us on ProductSafety@springernature.com

In case Publisher is established outside the EU, the EU authorized representative is:

Springer Nature Customer Service Center GmbH
Europaplatz 3
69115 Heidelberg, Germany

Batch number: 10091879

Printed by Printforce, the Netherlands